FRIEDEL-CRAFTS CHEMISTRY

INTERSCIENCE MONOGRAPHS ON ORGANIC CHEMISTRY

EDITOR: George A. Olah

CASE-WESTERN RESERVE UNIVERSITY
CLEVELAND, OHIO

Sulfonation and Related Reactions
BY EVERETT E. GILBERT

Peptide Synthesis
BY MIKLOS BODANSZKY AND MIGUEL A. ONDETTI

1,2-Cycloaddition Reactions: The Formation of Three- and Four-Membered Heterocycles
BY LINDA L. MULLER AND JAN HAMER

Mechanistic Aspects in Aromatic Sulfonation and Desulfonation
BY HANS CERFONTAIN

Friedel-Crafts Chemistry
BY GEORGE A. OLAH

Fluorine in Organic Chemistry
BY R. D. CHAMBERS

FRIEDEL-CRAFTS CHEMISTRY

by

GEORGE A. OLAH

C. F. Mabery Professor of Research
Case Western Reserve University

A WILEY-INTERSCIENCE PUBLICATION

JOHN WILEY & SONS
New York–London–Sydney–Toronto

Library of Congress Cataloging in Publication Data:
Olah, George Andrew, 1927-
Friedel-Crafts chemistry.

" A Wiley-Interscience publication."
Based on Friedel-Crafts and related reactions,
edited by the author.
1. Friedel-Crafts reaction.   I. Title.

QD281.A5.043   547'.21   73-754
ISBN 0-471-65315-2

Printed in the United States of America

10-9 8 7 6 5 4 3 2 1

Preface

In 1963–1965 I edited, and wrote in part, with the contribution of many of the active researchers in the field, a comprehensive treatise entitled *Friedel-Crafts and Related Reactions*. This monograph was published in four volumes (six parts) covering more than 5000 pages (with about 20,000 references). Not only have comprehensive reviews of this kind become too expensive for the individual chemist's library, but inevitably fast progress also makes them soon outdated. Thus I agreed with pleasure to the publishers' suggestion that I prepare a shorter review of the Friedel-Crafts field, which according to a suggestion made by Professor R. M. Roberts (*Chem. Eng. News*, January, 25, 1965, p. 96) can indeed be appropriately called *Friedel-Crafts Chemistry*. The present volume comprises reprints of the five general review chapters I wrote for the original version of the book, with additions to bring them up to date until the middle of 1972. New chapters have been added discussing recent general advances in the field for the period 1965–1972, and the mechanistic aspects of the reactions. It is hoped that this single volume of manageable size and more modest price will make the field of Friedel-Crafts chemistry more easily accessible.

The hospitality of the Swiss Federal Institute of Technology, and my friends in Zürich during a sabbatical leave when part of the writing was done, is greatly appreciated, as is the award of a Guggenheim Fellowship which greatly facilitated completion of this book.

June 1972   GEORGE A. OLAH

The Fieser's cats are by now well-known to most chemists. Maybe for a change some of the dog lovers would like to meet Jim, the Olah's cocker spaniel, whose photograph enlightens this page.

Contents

CHAPTER I

Historical*

I. Introduction

In the literature of organic chemistry there is a time-honored custom of designating many reactions by the name of chemists who discovered or developed them. The two men whose names are combined in one of the best known name reactions of chemistry, Charles Friedel and James Mason Crafts, came from two different corners of the world. Charles Friedel, born in Strasbourg, France, and James Mason Crafts, born of American parents in Boston, Massachusetts, proved through their long years of co-operation and friendship that science does not know nationalities or borders.

Surprisingly, although Friedel and Crafts are best remembered by the reaction which bears their names, their work on the use of aluminum chloride in organic synthesis played a comparatively minor role in their lives. Both rendered pioneering service to the development of the chemical science and teaching in their native countries and left a distinctive mark of their outstanding personalities to future generations.

II. Biographical Sketch of C. Friedel and J. M. Crafts

1. Charles Friedel (1832–1899)

Charles Friedel was born at Strasbourg, France, on March 12, 1832. Despite the affluence of his father, a banker, his upbringing and early life were by no means as leisurely as might be expected from the fashion then prevailing. Friedel's father was interested in the growth of scientific knowledge and possessed an active and in-

---

* Written in cooperation with Dr. R. E. A. Dear.

quiring mind; his mother was the daughter of Georges Duvernoy, Professor at the Collège de France and the Museum of Natural History. These circumstances partly explain Friedel's early interest in the experimental sciences and his later choice to continue his career in science rather than to follow his father in the banking profession.

His primary education was completed at the Protestant High School (Gymnasium), followed by instruction in the School of Science of the University of Strasbourg. Louis Pasteur was at that time a faculty member and his lectures undoubtedly encouraged Friedel to continue his work in chemistry and mineralogy. He graduated from the University with bachelor's degrees in letters in 1849 and in science in 1850. He then worked for a year in his father's counting-house, at the end of which time he was convinced that commerce held no attractions for him. The young man, then only twenty, traveled to Paris to broaden his education. There he took up residence with his grandfather (Duvernoy) at the Museum of Natural History. Studies at the Sorbonne were rewarded with a bachelor's degree in mathematics in 1852, followed by a licentiate in the same subject in 1854. One year later he received his licentiate in physical science from the same institution. Meanwhile Friedel, with characteristic energy and enterprise, had begun his collaboration with Wurtz, also a native of Strasbourg, in the laboratory at the Ecole de Médecine on November 10, 1854. At the same time he refused the post of demonstrator at the Ecole des Mines, under Duvernoy, saying he lacked the time to perform the work satisfactorily. But, after attending a series of lectures given by Senarmont on mineralogy, he accepted an appointment in 1856 as curator of the mineralogical collection at the school. The same year he married Emilie Koechlin, daughter of a Mulhouse manufacturer.

In following his dual career, Charles Friedel spent many years of arduous activity divided between mineralogy at the Ecole des Mines, and chemical research under Wurtz. Under Duvernoy at the Ecole des Mines, Friedel was actively engaged in enriching the mineralogical collection and carrying out research. His interests in that field gradually changed from the crystallographical side of the subject towards optics and then to mathematical details; finally the chemical side was rewarded with his attention and thereafter his life was divided between chemistry and chemical mineralogy. Despite the urgings of Pasteur to request admission to the mineralogical section of the Academy of Science, of which Pasteur himself was a member, Friedel preferred to wait for an opportunity to enter the chemical section of the Academy, maintaining that his principal interest was in this branch of science.   He achieved his ambition to become an "immortal" in 1878, filling the vacancy caused by the death of Regnault.

The early researches of Charles Friedel, carried out under the direction of Wurtz, included studies on ketones, lactic acid, and organo-silicon compounds.   In Wurtz's laboratory there was space for sixteen students of whom one-third were usually foreigners, mainly Russians, Germans, Austrians, and Italians.   The laboratories at the Ecole de Médecine were far from ideal but the reputation of Wurtz, comparable to that of Liebig and Bunsen in Germany, was such that a Fellowship in Paris was considered to be a hallmark of a sound scientific education.   In 1861 James Mason Crafts, a mining engineer from Boston, arrived for a period of study under Wurtz.   Crafts was at this time twenty-two years of age, Friedel's junior by almost seven years.   The two men, Friedel and Crafts, struck up an immediate friendship and, finding they had many ideas in common, decided to undertake their research in partnership. Thus between 1863 and 1865 they made joint studies on various organic compounds of silicon.

After Craft's departure from Paris in 1865, Friedel moved into an apartment in the Ecole des Mines.   Here, too, he maintained a small laboratory, a privilege which fell to him through his post as Curator.   In this laboratory Friedel carried on his mineralogical work as well as chemical researches (including those on the aluminum chloride reaction).   It was possibly owing to his extremely busy career, involving both chemistry and mineralogy, that Friedel did not enter for the doctorate until 1869, but then he offered two theses, one upon ketones and aldehydes and the other concerning pyroelectric properties of crystals.   He was by now the author or co-author of almost fifty published papers on various scientific topics.

His career was briefly interrupted by the Franco–Prussian War of 1870, in which Friedel took an active part.   Soon after the cease-fire Friedel was placed in charge of mineralogy instruction at the Ecole Normale Supérieure and in 1876 was appointed Professor of Mineralogy at the Sorbonne, organizing there for the first time a laboratory for mineralogy.   Meanwhile he was still active in the chemical field and from 1874 was again collaborating with Crafts, who had been forced to give up his position at the Massachusetts Institute of Technology because of failing health.   It was during this second period of conjoint effort that the "Friedel-Crafts Reactions" were discovered and investigated.

The death of Wurtz in 1884 left the chair of Organic Chemistry

vacant.   As the most able of Wurtz's colleagues, Friedel was chosen
to succeed him as Professor of Organic Chemistry and Research Di-
rector at the Sorbonne.   In order to undertake this work Friedel relinquished his post as Professor of Mineralogy and henceforth devoted
the majority of his remaining life to building up the French chemical
industry.   Two things spurred him on in this ambition.   The first
was the rapid growth and technological progress of the German
chemical industry which was seriously overshadowing French efforts;
the second was the apparent indifference of the Government to
French inferiority in the field.   It was Friedel's opinion that the lack
of chemists and technicians was detrimental to the nation and he devoted much time and energy to his new project.   The chemical
laboratories at the Sorbonne were even more unsuitable than those
at the Ecole de Médecine (there was, for example, no running water
in the laboratories).   Realizing that his plans for new and greatly
enlarged buildings had but small prospect of immediate realization,
Friedel energetically pursued the construction of temporary facilities
which were large enough to accommodate fourteen students.   Despite their comparative crudity the rooms were nevertheless convenient and Friedel's name was such that there was keen competition
for fellowships.   In 1895 new, enlarged laboratories were built to
Friedel's own design at the Sorbonne, with places for thirty students.
One year later his efforts to increase the number of technical chemists
bore fruit in the founding of the Institut de Chimie.   Here he organized a three-year course in industrial chemistry to be provided, at
Friedel's instigation, by the Municipal Government.

In 1899 his health declined seriously and he was obliged to rest.
Consequently he journeyed to Montauban during the Easter vaca-
tion to visit his daughter, with his second wife, whom he married in
1873, two years after the death of his first wife.   While there his
illness was aggravated and he succumbed peacefully on April 20,
shortly after his sixty-seventh birthday.

During his lifetime, many distinctions and honors came to Charles
Friedel as a result of his active careers in chemistry and mineralogy.
In 1869, the day after defending his doctoral theses he was made
chevalier of the Legion of Honor, and was appointed an officer of the
order in 1888.   He was officer of Public Instruction and was also
Commander of the Orders of St. James of Portugal and of the Crown
of Roumania.   He was awarded the Davy Medal by the Royal
Society in 1880, and the degree D.C.L. (Doctor civilis juris, honoris
causa) was conferred on him by Oxford University in 1894.   In addition the following societies bestowed honorary membership upon

Friedel: The Chemical Society, 1876; Industrial Society of Mulhouse, 1879; Physical Society of Geneva, 1880; Royal Society of Turin, 1882; Academy dei Lincei, 1883; Academy of Munich, 1883; Royal Academy of Lisbon, 1890; Natural History Society of Maxon, 1890; Society Antonio Alzato Mandio, 1890; Royal Society of Brussels, 1892; Society of Physics and Chemistry of Bucharest, 1892; Manchester Literary and Philosophical Society, 1892; German Chemical Society, 1894; Royal Society of Sweden, 1894; British Association for the Advancement of Science, 1895; Royal Society of St. Petersburg, 1895; Physical Society of Frankfurt, 1898.

In 1857 he was one of the founders of the French Chemical Society and served four presidential terms (1870, 1871, 1880, 1888). He was also a co-founder and president of the Mineralogical Society of France and was once president of the Physical Society of France. Friedel was instrumental in initiating the French Society for the Advancement of Science and presided at its first meeting in 1886. He presided at the International Congress of Chemists held at Geneva in 1892 in order to effect the reform of the nomenclature of organic compounds of the fatty acid series and at the time of his death Friedel was in the midst of arrangements for a similar meeting concerning other classes of organic compounds. He was the chief founder in 1899 of the *Revue Générale de Chimie pure et appliquée*, a journal largely devoted to industrial chemistry. Friedel and Wurtz were joint editors of the *Dictionaire de Chimie pure et appliquée* and after the death of Wurtz in 1884 Friedel became the chief editor of the supplements which, during his remaining life, progressed only as far as the letter G.

At the time of his death Friedel was a vice-president of the French Chemical Society and would have been president for the fifth time in 1900, a year during which an international conference of chemists was scheduled to be held in Paris.

2. James Mason Crafts (1839–1917)

James Mason Crafts was born in the city of Boston on March 8, 1839, the son of R. A. Crafts, a wool merchant and manufacturer of woolen goods; his mother was the daughter of a New Hampshire lawyer and politician.

In those days, as today, Boston was a centre of learning and a locality in which there was a general and far-sighted interest in science. This was a fortuitous circumstance for the young man. Boston was also the home of W. B. Rogers, the founder of Massachusetts Institute of Technology and a personal acquaintance of the

Crafts family.   As a child Crafts was usually thought to be of a dili-
gent and studious nature, a trait which was to remain with him
through his life.   After completing his primary education in various
schools he entered the Boston Latin School to prepare for college and
following a year of private study after the school he entered Harvard
University (Lawrence Scientific School) from where he graduated
with a B.S. degree in 1858, at the age of nineteen.   He spent a year
as a graduate student at the Lawrence School studying mining engi-
neering, since mining was then a popular and rapidly growing industry.
However, Crafts soon decided that he would have to travel to the
established European centres of learning if he was to complete his
education.   Consequently in 1859 he journeyed to Freiburg, in Germany, to become an assistant to Plattner, a noted metallurgist of the
day.   While there he became deeply interested in chemistry and the
next year (1860) he was fortunate enough to secure a position as an
assistant to Bunsen, in Heidelberg, and to meet both Kirchhoff and
Helmholtz.   It was during the tenure of Crafts' assistantship that
Bunsen discovered the element rubidium.   In 1861 Crafts began the
first of two sojourns in Paris.

For the next four years he remained at the Ecole de Médecine
studying chemistry under Wurtz and working in collaboration with
Charles Friedel, mainly on organo-silicon compounds.

It was thus with a very sound basic education that Crafts returned
to the United States in 1865 to begin his career.   After a short spell
as a mining inspector in both Mexico and California he decided to
pursue an academic livelihood.   Cornell University had recently
been organized and in 1867 Crafts was chosen as Professor of
Chemistry and then appointed chairman of the chemistry department.   Here he wrote a textbook on "Qualitative Analysis" to meet
the needs of the students at the university, who in those days
undertook a one-year course in chemistry which included four hours
weekly in the laboratory.   The book was dedicated to Wurtz in grati-
tude for the knowledge he had imparted to Crafts in Paris.   Shortly
after this Crafts was married to Clemence Haggerty, of New York
City.   For the next three years Crafts continued in his post at
Cornell University and in 1870 moved to the Massachusetts Institute
of Technology where he took the chair of General and Analytical
Chemistry.   He was extremely interested in stimulating research as
well as in day-to-day teaching and to these ends he caused numerous
innovations and improvements to be introduced into the laboratories, particularly such labor-saving devices that would be counted
as commonplace in today's laboratory.   In addition to teaching

analytical techniques it was also Professor Crafts' duty to give a course in organic chemistry. In this matter he was of the opinion that only the very latest in textbooks should be used and that these should be supplemented by frequent reference to the chemical literature. Thus, under his guidance, the majority of the students selected a textbook by Kekulé, in German, then the most comprehensive work available on the subject. His lectures were also punctuated by frequent literature references, indicating to the students the source of more complete topical information. While at the Massachusetts Institute of Technology Crafts never lost an opportunity to encourage the undertaking of all aspects of chemical research, but unfortunately he was not the most robust of men and in 1874 ill-health forced him to go abroad. Not unnaturally, he returned to Paris and the laboratories of Wurtz. Here he resumed his collaboration with Friedel in various aspects of chemical research. He anticipated that he would be in France for only two or three years and consequently was retained on the faculty at the Massachusetts Institute of Technology. However, the discovery of the Friedel-Crafts reaction in 1877 resulted in such prolific research that by 1880 it was apparent that Crafts would be abroad for a much longer period and he forthwith relinquished his former post. Crafts remained in Paris for seventeen years. In partnership with Friedel and their students and co-workers he published close to one hundred scientific papers in this time, not only on the aluminum chloride reaction but also in his other fields of interest on such topics as thermometry and halogen densities at high temperatures.

In 1891 Crafts again took up residence in the United States. He resumed his former association with the Massachusetts Institute of Technology and in 1892 took over the teaching of organic chemistry. Subsequently, in 1895, he assumed temporary responsibility for the whole chemistry department, being reluctant to abandon his chemical researches in order to take over the administration on a regular basis. Two years later, however, following the sudden death of the President of the Massachusetts Institute of Technology, Crafts was elected to the presidency—a post which required the greater part of his time.

Craft's chief dismay at this time was the standard of the Massachusetts Institute of Technology compared to that of European institutions and he was continually seeking to improve the teaching and research standards. He correctly predicted the need and started the development at the Institute of a renowned graduate school devoted to research in pure and applied science. His duties as President of the Institute became more and more demanding of his time and in 1900 it was necessary for Crafts to choose between the administrative duties connected with his post and his laboratory research, since it was rapidly becoming impossible for him to carry out both tasks in a manner satisfactory to his exacting standards. He chose the latter and thus after three years in office he resigned the presidency and returned to his experimental work. He continued his bench work until 1911, when a severe attack of neuritis abruptly halted its progress. From that time until his death on June 20, 1917, he spent many hours preparing his previously obtained experimental results for publication. His research during this period was in connection with solution catalysis, thermometry and vapor-pressure determinations of various substances. His papers were published in both the United States and France, a custom reminiscent of his days in Paris.

As might be expected Crafts was the recipient of many honors as the outcome of his meticulous scientific work. While in Paris in 1880 he was awarded the Jecker prize for his contributions to organic chemistry and five years later he was appointed a chevalier of the Legion of Honor.

For his investigations in high-temperature thermometry Crafts was awarded the Rumford medal by the American Academy of Arts and Sciences in 1911. He was a long-standing member of this academy and of the National Academy of Sciences. He was also a member or honorary member of the American Chemical Society; the American Association for the Advancement of Science; the Washington Academy of Sciences; the British Association for the Advancement of Science; the Royal Institution of Great Britain. In 1898 Crafts was awarded an honorary LL.D. degree by Harvard University in recognition not only of his scientific achievements but, in addition, of his efforts as President of the Massachusetts Institute of Technology.

III. Historical Review of the Discovery and Development of the Aluminum Chloride Reaction

Alkylation and acylation reactions similar in character to the Friedel-Crafts synthesis were reported in the chemical literature prior to the first publication on the subject by Charles Friedel and James Mason Crafts (14) in 1877 and it is of interest, historically, to investigate the results of the earlier workers. In 1869 Zincke (34), in a paper concerning the synthesis of aromatic acids, reported the formation of diphenylmethane during the attempted synthesis of β-phenylpropionic acid. The attempted reaction was:

$$\text{C}_6\text{H}_5\text{—CH}_2\text{Cl} + \text{ClCH}_2\text{COOH} \rightarrow \text{C}_6\text{H}_5\text{—CH}_2\text{CH}_2\text{COOH} + \text{Cl}_2$$

carried out in benzene solution in a sealed tube with copper or silver metal present to form a metal chloride with the liberated chlorine. Instead he noted the evolution of much hydrogen chloride and the formation of a compound subsequently identified as diphenylmethane. He also noted the formation of some metal chloride.

$$\text{C}_6\text{H}_6 + \text{C}_6\text{H}_5\text{—CH}_2\text{Cl} \xrightarrow{\text{Ag or Cu}} \text{C}_6\text{H}_5\text{—CH}_2\text{—C}_6\text{H}_5 + \text{HCl}$$

Clearly this was a reaction closely related to the later alkylations of Friedel and Crafts. That Zincke's interest was aroused is evidenced by later work (35) where he investigated the reaction of benzyl chloride with benzene, toluene, and xylene. He noted that whereas with copper the reaction had to be carried out in a sealed tube at 150–160° for four to five hours, the use of zinc dust or reduced iron enabled the reaction to proceed below 100° in a flask fitted with a reflux condenser. He represented a straightforward reaction mechanism thus:

$$\text{C}_6\text{H}_5\text{—CH}_2\text{Cl} + \text{C}_6\text{H}_6 \xrightarrow[\text{Zn}]{\text{Fe}} \text{C}_6\text{H}_5\text{—CH}_2\text{—C}_6\text{H}_5 + \text{HCl}$$

$$\text{C}_6\text{H}_5\text{—CH}_2\text{Cl} + \text{C}_6\text{H}_5\text{CH}_3 \xrightarrow[\text{Zn}]{\text{Fe}} \text{C}_6\text{H}_5\text{—CH}_2\text{—C}_6\text{H}_4\text{CH}_3 + \text{HCl}$$

$$\text{C}_6\text{H}_5\text{—CH}_2\text{Cl} + \text{CH}_3\text{C}_6\text{H}_4\text{CH}_3 \xrightarrow[\text{Zn}]{\text{Fe}} \text{C}_6\text{H}_5\text{—CH}_2\text{—C}_6\text{H}_3(\text{CH}_3)_2 + \text{HCl}$$

but noted that "the reaction is of such a peculiar nature that even now it is possible that a sufficient explanation has not been given." Later papers up to the year 1872 (36,37) re-emphasized the use of zinc or reduced iron in place of copper and contained the results of investigations of the structures of the reaction products.

In 1874 Radzivanovskii (32) also reported the condensation of α-phenylethyl bromide with aromatics such as benzene, toluene, and ethylbenzene in the presence of zinc powder. His work thus extended the previous findings of Zincke and indicated the general nature of the condensation of reactive halides with aromatic hydrocarbons.

The first mention of acylation related to the later Al_2Cl_6 method of Friedel and Crafts appeared in 1873 in a preliminary communication by Grucarevic and Merz (22), only to be closely followed by Zincke (38) who claimed to have carried out similar work at an earlier date. Apparently Zincke had tried to produce dibenzoyl (benzil) from benzoyl chloride, in the presence of copper, silver, or zinc, in a reaction similar to his attempted preparation of β-phenylpropionic acid, thus:

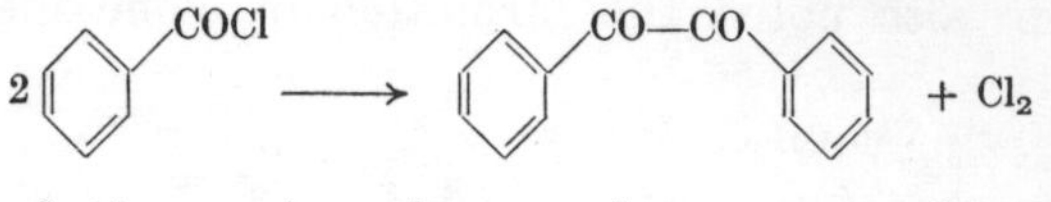

in benzene solution, using the metal to react with the liberated chlorine. Again, however, a strong evolution of hydrogen chloride was noticed and the reaction product was not dibenzoyl, but benzophenone, evidently formed by the reaction:

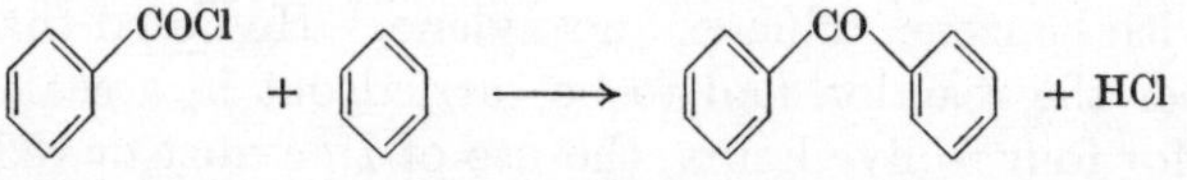

No mention is made of the observation of any metal chloride formation. Similar reactions with high yields were described between benzoyl chloride and toluene, xylene, cymene, and naphthalene. However, the main communication (23) of Grucarevic and Merz in the same year gave a much more detailed account of the acylation reaction. The paper was entitled "Ketones from Aromatic Hydrocarbons and Acyl Chlorides" and confirmed Zincke's earlier work on aromatic hydrocarbons and his preliminary benzoylations of benzene and toluene. In addition the preparation of α- and β-naphthyl phenyl ketone from benzoyl chloride and naphthalene was described:

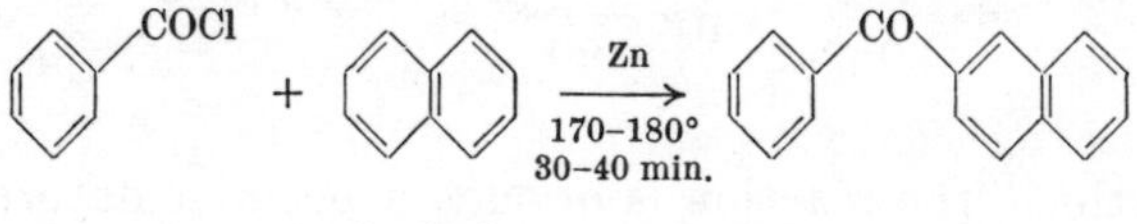

Further preparations included α- and β-dinaphthyl ketones from naphthalene and the corresponding naphthoyl chloride, benzophenone, tolyl phenyl ketone, cymyl phenyl ketone, and a complex diketone from benzoyl chloride and phenol. The latter reaction is represented by the authors thus:

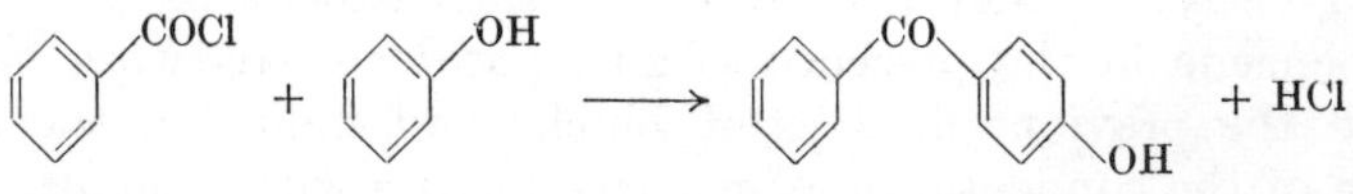

followed by

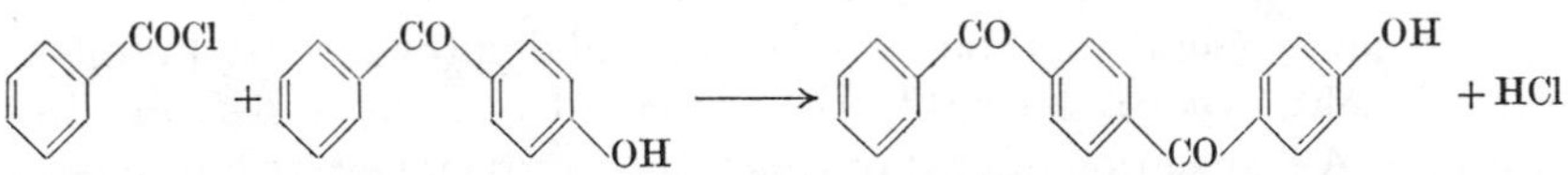

$$C_6H_5COCl + C_6H_5CO\text{-}C_6H_4\text{-}OH \longrightarrow C_6H_5CO\text{-}C_6H_4\text{-}CO\text{-}C_6H_4\text{-}OH + HCl$$

All the above reactions were carried out in the presence of zinc powder, yet no mention is made of the formation of any zinc chloride. No discussion of mechanism is given and the reactions are represented by empirical equations. Yet they are obviously cases of reactions related to the later Friedel-Crafts acylation. In 1876 Doebner and Stackman (13) reported the preparation of salicylaldehyde and o-hydroxybenzophenone from phenol and chloroform and benzotrichloride respectively. Zinc oxide was added before the start of the reaction and it was observed that zinc chloride was present at the conclusion of the experiment.

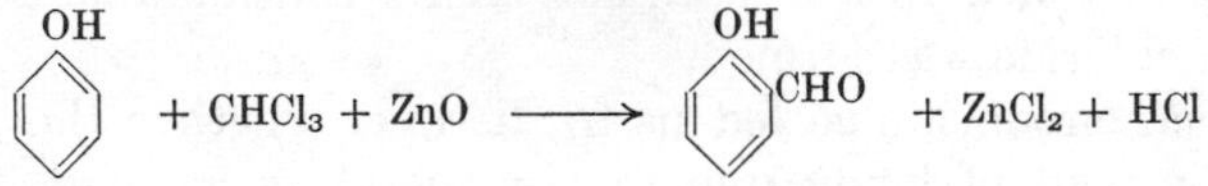

$$C_6H_5OH + CHCl_3 + ZnO \longrightarrow (HO)C_6H_4(CHO) + ZnCl_2 + HCl$$

It should be apparent at this stage that none of the earlier authors was aware of the true nature of the reaction or of its vast potentialities. None of them realized that the metal chloride catalyzed the reaction, although the formation of the chloride had been observed in several instances. It was left to Friedel and Crafts to show that the presence of the metal chloride was essential. Two preliminary oral communications (15,16) were made to the Chemical Society of France in May 1877 of ". . . several experiences relating . . . to the action of aluminum and aluminum chloride on chlorides, particularly amyl chloride; this reaction gave rise to an intense evolution of hydrogen chloride, to some combustible gases from saturated hydrocarbons . . . and to some condensed products" and of ". . . the study of the reaction of aluminum chloride on various chlorides, hydrocarbons, and mixtures of these of varying compositions. For example, aluminum chloride reacts with amyl chloride in the cold, producing hydrogen chloride, hydrocarbons C_nH_{2n+2}, and condensed hydrocarbons less rich in hydrogen. With gasoline one obtains liberation of gas and the formation of condensation products. With a mixture of chloride and hydrocarbon, the formation is established, in good yield, of hydrocarbons from the residues of the hydrocarbon less H and from the chloride less Cl. It is thus that ethylbenzene, amylbenzene, benzophenone, etc., are obtained."

These presentations were followed on June 11, 1877, by a paper

"On a new general method of synthesis of hydrocarbons, ketones, etc." which gave a remarkably clear account of their important discovery. A translation of part of this paper is given here to emphasize the thoroughness of the investigation.

"In a research which we undertook together we were led to study the action of metallic aluminum, in filings or thin leaves, on various organic chlorides. We observed that the reaction, slow at first, and not occurring without the aid of heat, accelerated thereafter becoming almost violent, at which point it was necessary to control it by cooling. The addition of a very small amount of iodine initiated the reaction which was always accompanied by an abundant evolution of hydrogen chloride, which in the case of certain chlorides, such as amyl chloride, was accompanied by gaseous or liquid hydrocarbons, some of which boiled at very high temperatures. Finally the reaction seemed all the more active if a considerable amount of aluminum chloride was formed.

"This last circumstance led us to discover whether the principal reaction, instead of being due to the metal as we had originally supposed, should really be attributed to the metal chloride.

"It was easy to assure ourselves that this was indeed so. When small amounts of anhydrous aluminum chloride were added to amyl chloride, for example, an immediate vigorous evolution of gas was observed, even in the cold. The gas was composed of hydrogen chloride accompanied by gaseous hydrocarbons not absorbed by bromine. In the amyl chloride, in which the first portions of aluminum chloride were dissolved, small droplets were formed, then a dense brown liquid layer, and the reaction seemed to occur mainly at the interface of the two liquids. We had noticed a like action with aluminum, when a similar brown liquid was formed. When the reaction had continued long enough, with the aid of gentle heat to complete it, an amount of hydrogen chloride could be recovered which closely corresponded to the total amount of chlorine originally present in the amyl chloride, and a very varied series of hydrocarbons, from gases to products boiling above the boiling point of mercury was also recovered. The residue was composed of aluminum chloride which sublimed in hexagonal plates or in crystalline crusts when heated for a sufficient length of time in a current of inert gas.

"We have not yet achieved the study of the numerous products of which the first members seem to be of the series of hydrocarbons C_nH_{2n+2} and the higher members seem to be much poorer in hydrogen. At the moment we wish to make these observations: organic chlorides are attacked by aluminum chloride with the loss of hydrogen chloride; most of the products formed, which includes a large proportion of saturated hydrocarbons, are not able to be formed by a simple polymerization of amylene resulting from the subtraction of hydrogen chloride from amyl chloride; it seems preferable that the hydrogen chloride evolved is formed from two molecules of which one furnishes the chlorine and the other the hydrogen, and the two residues then unite.

"This observation led us to attempt the reaction with aluminum chloride under conditions which offered a larger interest of a general synthetic method for furnishing an unlimited number of hydrocarbons and even oxygenated compounds. We thought that on mixing a hydrocarbon with an organic chloride and putting them in contact with aluminum chloride we would be successful in obtaining a reaction having as product the combination of the hydrocarbon radical from the chloride with the hydrocarbon less hydrogen.

"In fact the method worked easily. Having mixed first, amyl chloride with a large excess of benzene, small amounts of aluminum chloride were added. We then saw the occurrence, in the cold, of a regular reaction accompanied by the evolution of hydrogen chloride; two layers were soon formed, the lower being colored brown. As soon as the hydrogen chloride was being evolved only slowly and with the aid of heat, the two layers were separated and treated individually with water. On distillation after drying both gave almost the same products; the only difference being that the clear, upper layer yielded a much larger proportion of benzene and relatively low-boiling hydrocarbons; the lower brown layer, which contained most of the aluminum chloride, gave mainly products boiling at an elevated temperature. It was easy to extract, by a small number of fractionations, a liquid boiling between 185–190°, having the properties of amylbenzene. The latter is formed by a reaction which may be represented empirically by the equation:

$$\text{C}_6\text{H}_6 \;+\; CH_3(CH_2)_4Cl \;\longrightarrow\; \text{C}_6\text{H}_5{-}(CH_2)_4CH_3 \;+\; HCl$$

Immediately we assured ourselves that this method of synthesis is general and is applicable not only to chlorides but also to bromides and iodides.

"In fact, after mixing benzene with ethyl iodide and adding aluminum chloride, we observed the evolution of thick acid fumes containing hydrogen iodide and after treatment with water and

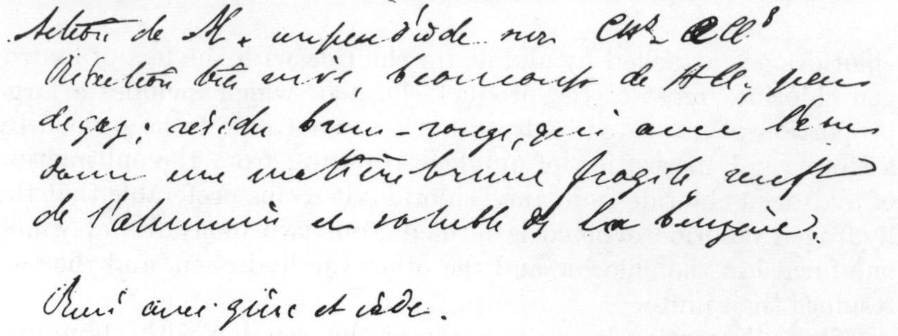

[handwritten notes in Friedel's hand]

Fig. 1.   The first experiment of the Friedel-Crafts type, recorded in Friedel's laboratory notebook.   The page is not dated but three pages later the date April 3 (1877) (in Crafts' handwriting) is recorded (see Fig. 3). (All the illustrations are reproduced in the same chronological order as the original entries in Friedel's records.)

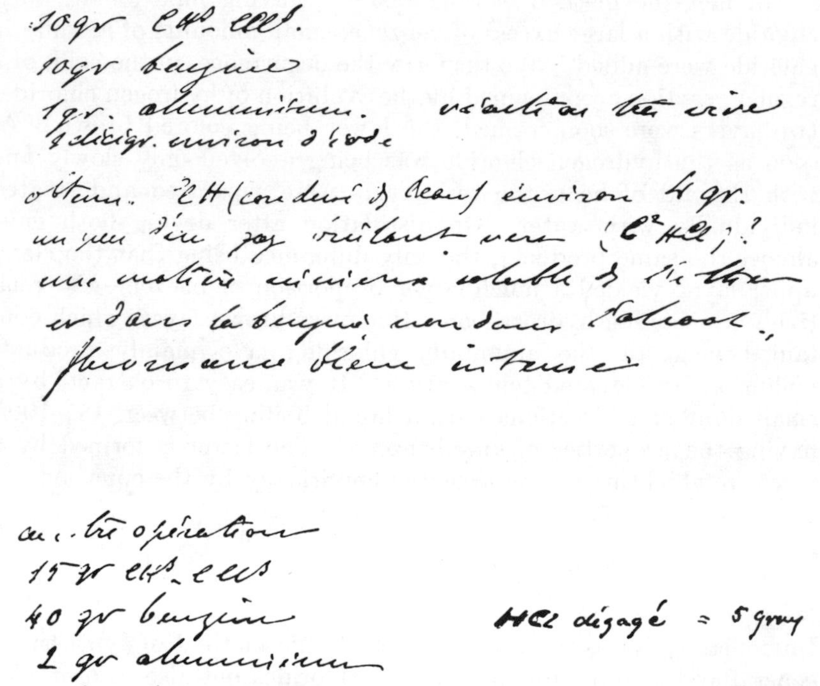

[handwritten notes in Friedel's hand]

Fig. 2.   Condensation of 1,1,1-trichloroethane with benzene in the presence of aluminum and iodine.   The amount of HCl evolved is recorded by Crafts.

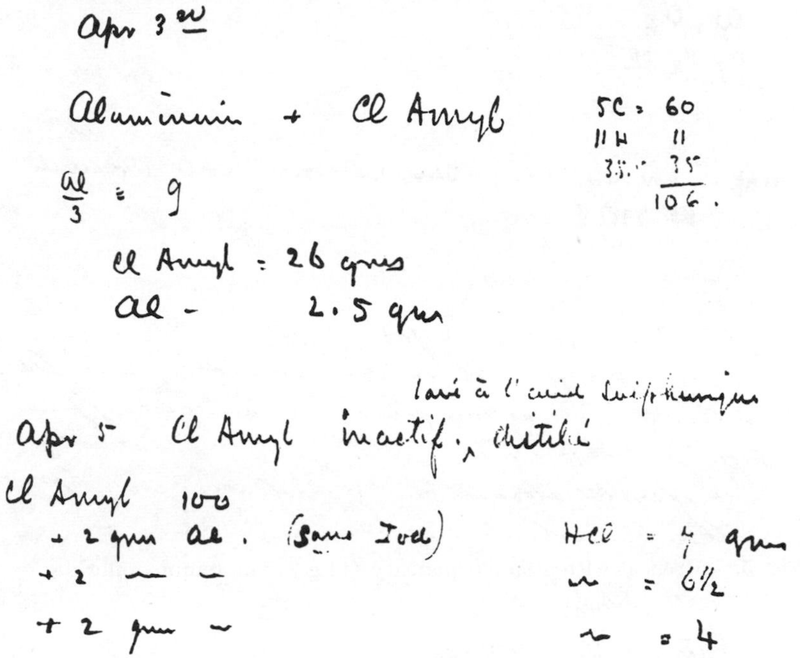

Apr 3ᵈ

Aluminium + Cl Amyl

5C = 60
11 H 11
35. 35
——
106.

$\frac{Al}{3} = 9$

Cl Amyl = 26 gms
Al -   2.5 gr

Apr 5   Cl Amyl inactif, distillé   lavé à l'acide sulphurique

Cl Amyl 100
+ 2 gm Al . (sans Iod)   HCl = 7 gm
+ 2 — —   „ = 6½
+ 2 gm ~   „ = 4

Fig. 3.   The first experiments with amyl chloride.   Both are entries by Crafts.

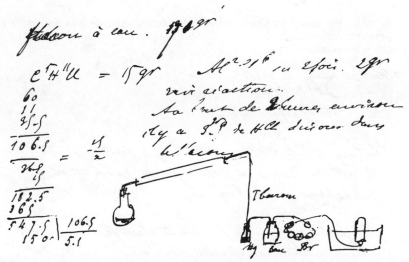

Fig. 4.   The first recorded use of aluminum chloride in such experiments.
The calculation shows that 5.1 g. HCl should be evolved, whereas 3.5 g.
were collected during the experiment.

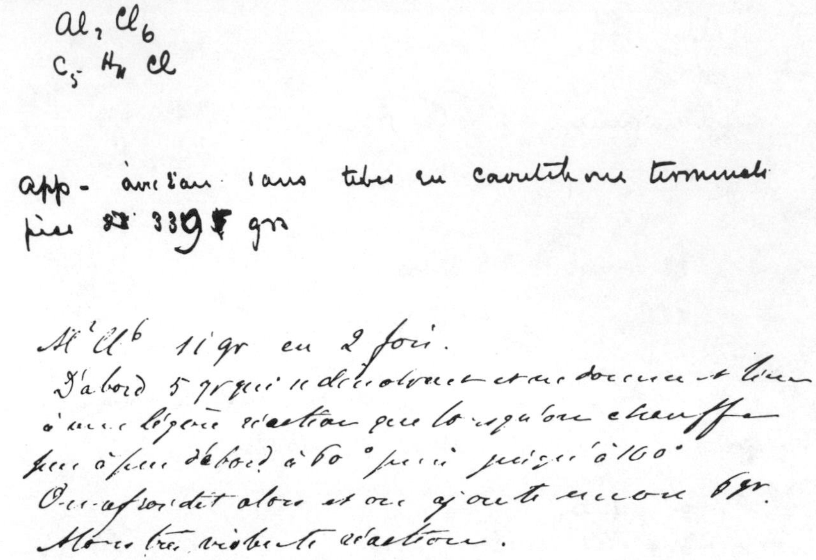

Al_2Cl_6

$C_5H_{11}Cl$

app - ainsian sans tubes en caoutchouc terminale

pue $339\overline{5}$ gr

Al^2Cl^6 11 gr en 2 fois.

D'abord 5 gr qui se dévolvent et se dôrour et bien
à une légère réaction que lorsqu'on chauffe
peu à peu débord à 60° puis jusqu'à 160°
On afroudit alors et on ajoute encore 6 gr.
Alors très violente réaction.

Fig. 5.   A repeat with a larger quantity (11 g.) of aluminum chloride.

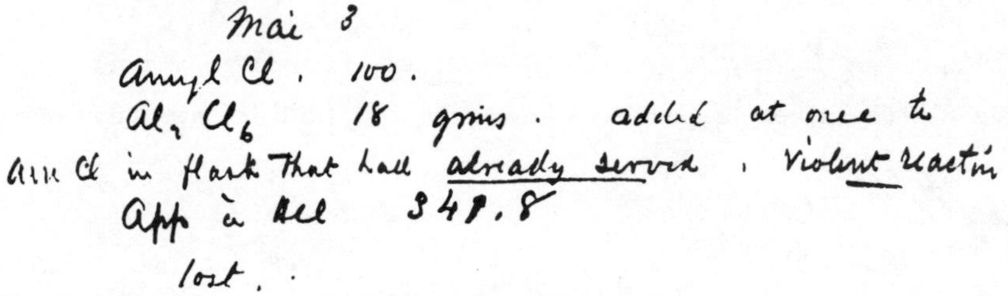

Mai 3

Amyl Cl. 100.

Al_2Cl_6 18 grms.   added at once to

Am Cl in flask that had already served. Violent reaction

app à all 349.8

lost.

Fig. 6.   Another entry by Crafts where aluminum chloride was added to a
previous reaction mixture.   A "violent reaction" was noted followed by
the terse comment "lost"!

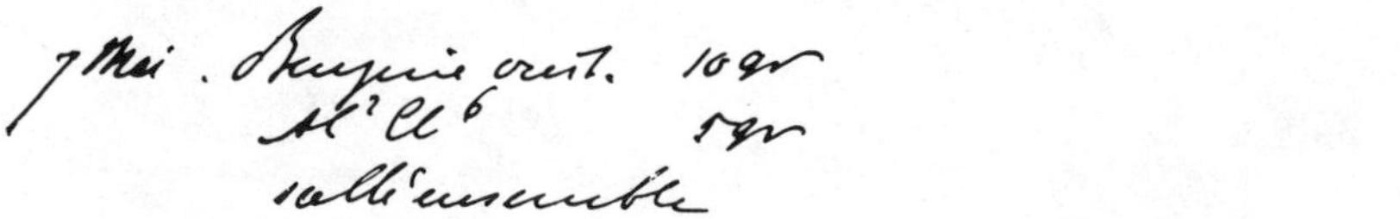

7 Mai. Benzine crist. 10 gr

Al^2Cl^6 5 gr

scellé ensemble

(La benzine et le Al^2Cl^6 semblent réagir lentement à
l'ébullition, mais on obtient peu de produit de t.
plus haut que la benzine)

Un peu de gaz et des produits distillam à des
température croissantes, bien au dessus de 80°.

Fig. 7.   Action of aluminum chloride on benzene in a sealed tube.   Little or
no reaction appears to have occurred.

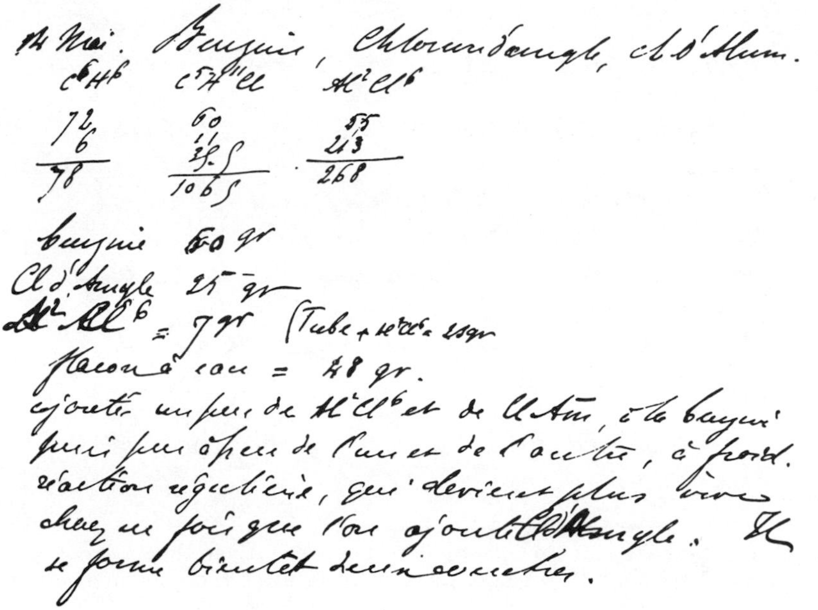

Fig. 8. The classic reaction of benzene with amyl chloride in the presence of aluminum chloride. This has often been quoted as the first Friedel-Crafts reaction when in fact it is recorded 14 pages after the first experiment (Fig. 1).

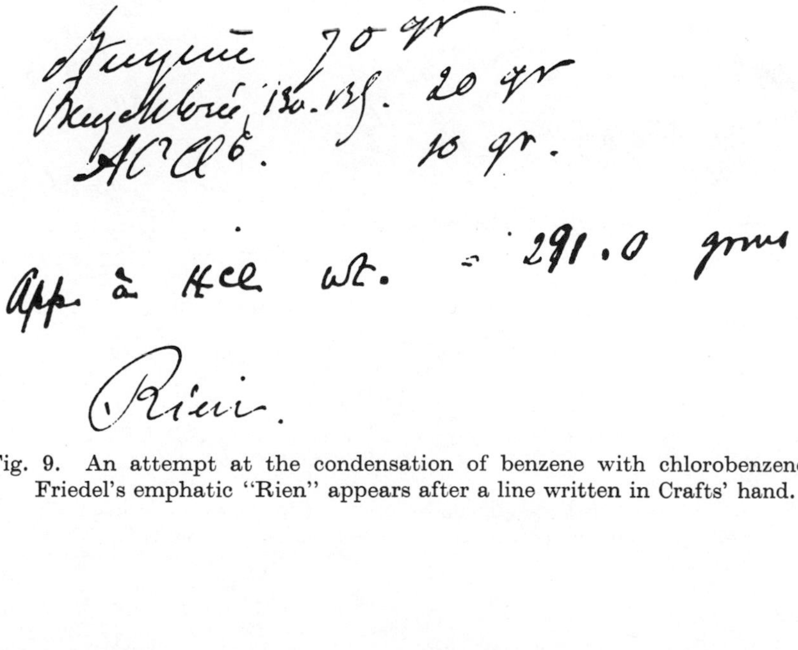

Fig. 9. An attempt at the condensation of benzene with chlorobenzene. Friedel's emphatic "Rien" appears after a line written in Crafts' hand.

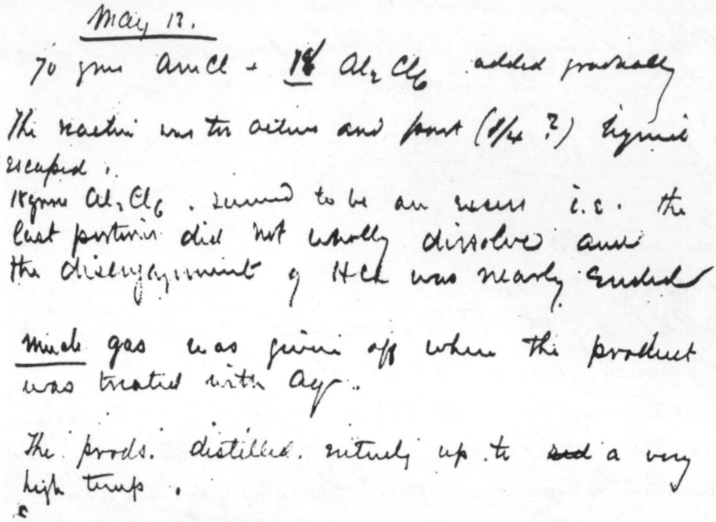

May 13.

70 grms AmCl + 18 Al_2Cl_6 added gradually

The reaction was too active and part (1/4 ?) liquid escaped.

18 grms Al_2Cl_6 seemed to be an excess i.e. the last portion did not wholly dissolve and the disengagement of HCl was nearly ended.

Much gas was given off when the product was treated with aq.

The prods. distilled entirely up to ~~and~~ a very high temp.

Fig. 10. One of the few pages entirely in English. Again the work of Crafts, it is dated May 13 yet appears four pages after the experiment dated May 14 (Fig. 8).

Amylbenzine 185–190°

1gr = Ampoule + 0,7155
0,5155
0,200

0,5gr = tube à eau + 0,8855
0,6875
0,1985

10.97

90gr = app à CO^2 + 4.819
4.1605
0,6585

89.79

$C^6H^5.C^5H^{11}$ = $C^{11}H^{16}$

132
16
148

13200 | 148
89.18

1600 | 148
10.81

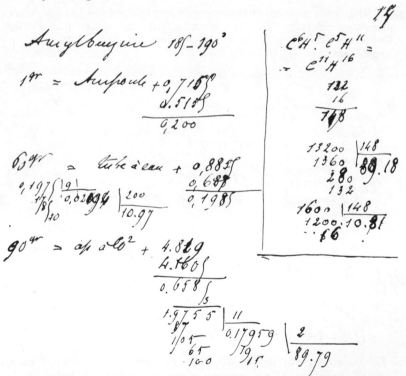

Fig. 11. Analysis of the amylbenzene prepared in the experiment of May 14, 1877. Theoretical percentages are calculated at the right: C, 89.18; H, 10.81. The results obtained were C, 89.79; H, 10.97.

2***

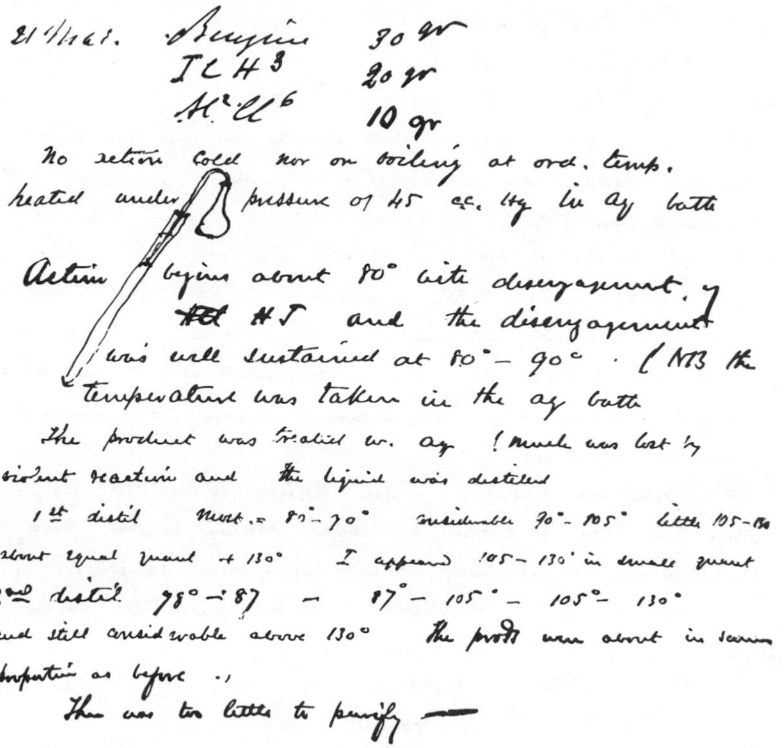

21 Mai. Benzine 30 gr
    ICH^3 20 gr
    Al^2Cl^6 10 gr

No action cold nor on boiling at ord. temp.
heated under pressure of 45 c.c. Hg in a bath

Action begins about 80° with disengagement, y
~~HCl~~ HJ and the disengagement
was well sustained at 80° – 90° . (NB the
temperature was taken in the ay bath

The product was treated w. ay ! much was lost by
violent reaction and the liquid was distilled

1st distil   Most. = 80° – 90°   considerable 90° – 105°   little 105 – 130°
about equal quant + 130°   I append 105 – 130° in small quant

2nd distil 75° – 87   ~   87 – 105°   ~   105 – 130°
and still considerable above 130°   The prod^{ts} were about in same
proportion as before .,

There was too little to purify —

Fig. 12.

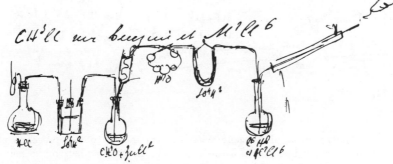

CH^3Cl sur benzine et Al^2Cl^6

Il faut chauffer bien doucement le ballon à
benzine et ajouter il y a un abord d dégag^t
de HCl.
Benzine 240 gr
Al^2Cl^6 24 gr
L'opération dure de 1 h juin à 6 h environ.
Puis on traite par l'eau non fractionné.

Fig. 13.

Br -Methyl   40
Benzine   35.
Al_2Cl_6 12 gms

The Al_2Cl_6 was added at once & after a few moments
reaction became violent . Disengagement of HBr? or H
& distillation CH_3Br cooled with ice and allowed
to set under a pressure about 55 cm Hg . at
ord. temp . The reaction went on for a long time
and was finally stimulated by heating in a bath
up to 60° between 30° - + 60° reaction was It was necessary to continually heat higher to produce the same
apparent reaction

The product was added in small portions to Ag but no
pains were taken to avoid a certain heating so that most of
the remaining CH_3Br evap. & probably some Benzol .

Results after d distil. The two layers were distil
separately for 1st distil the lower gave most higher products
74 - 90 = 1.2) gms   + 95 - 98° 1.2 gms   + 145°. = ?
80 - 82   14.5 " 20.1   98 - 115 = 0.9 "
82 - 84   4.4 "   115 - 125 1. "
84 - 98   2.3 .   125 - 135 1.2 "   (2.4 gms
88 - 95   2.7 "   135 - 138 2.4 "   (+ 185 2ᵈ 15ᵗ d
                             138 - 145 1.7 " + 200° not dist. bt
(95° - 98° (1.2° was destroyed + H_2SO_4 + CrO_3 + gave no Benzoic acid 1.1 gms .
                                                              obt for uppr.
        Total 39. gms —

Fig. 14.

Fig. 12.   The reaction of benzene with methyl iodide in the presence of
    aluminum chloride.   Again a too vigorous reaction has led to extensive
    losses so that after two distillations further purification was impractical.

Fig. 13.   The action of methyl chloride on benzene in the presence of alu-
    minum chloride.   Note the complexity of the apparatus.

Fig. 14.   One of the more completely documented experiments, the reaction
    of benzene with methyl bromide in the presence of aluminum chloride.

Fig. 15.   Action of oxygen on benzene in the presence of aluminum chloride. The reaction products are not identified.

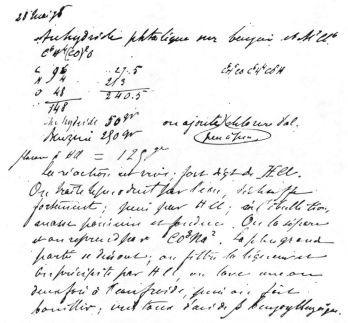

Fig. 16.   Dated (May 28, 1878).   This page describes the reaction of phthalic anhydride with benzene.

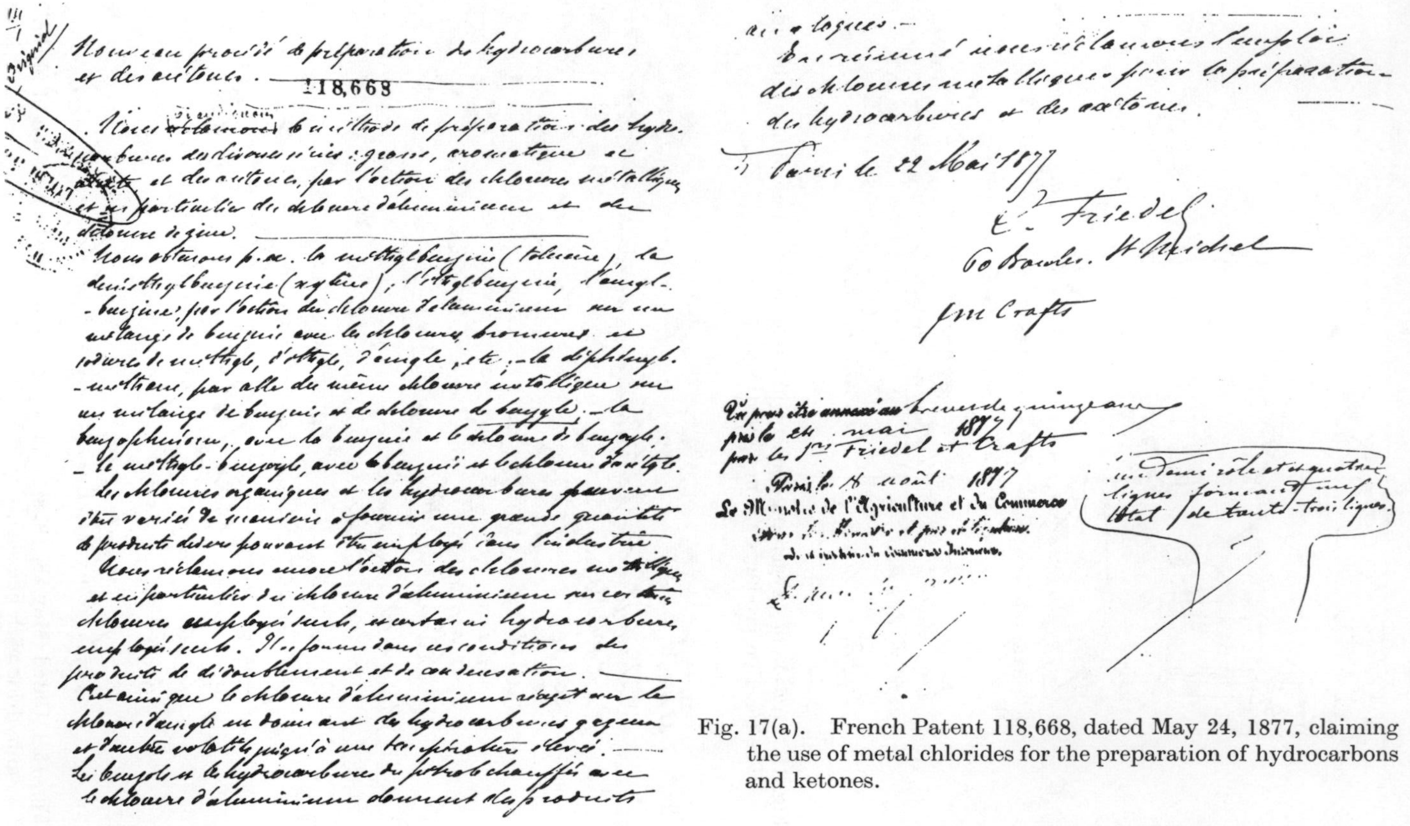

Fig. 17(a). French Patent 118,668, dated May 24, 1877, claiming the use of metal chlorides for the preparation of hydrocarbons and ketones.

BREVET n° 118668, en date du 24 mai 1877.

A MM. FRIEDEL et CRAFTS, pour un procédé de préparation des hydrocarbures et des acétones.

Nous revendiquons la méthode de préparation des hydrocarbures des diverses séries : grasse, aromatique et accite, et des acétones, par l'action des chlorures métalliques, et en particulier du chlorure d'aluminium et du chlorure de zinc.

Nous obtenons, par exemple, la méthylbenzine (toluine), la diméthylbenzine (xyline), l'éthylbenzine, l'amylbenzine, par l'action du chlorure d'aluminium sur un mélange de benzine avec les chlorures, bromures et iodures de méthyle, d'éthyle, d'amyle, etc.; la diphénylméthane, par celle du même chlorure métallique sur un mélange de benzine et de chlorure de benzoyle; la benzophénone avec la benzine et le chlorure de benzoyle, le méthylebenzoyle avec la benzine et le chlorure d'acétyle.

Les chlorures organiques et les hydrocarbures peuvent être variés, de manière à fournir une grande quantité de produits divers pouvant être employés dans l'industrie.

Nous réclamons encore l'action des chlorures métalliques, et en particulier du chlorure d'aluminium, sur certains chlorures employés seuls et certains hydrocarbures employés seuls. Il se forme, dans ces conditions, des produits de dédoublement et de condensation.

C'est ainsi que le chlorure d'aluminium réagit sur le chlorure d'amyle, en donnant des hydrocarbures gazeux et d'autres volatils, jusqu'à une température élevée.

Les benzols et les hydrocarbures du pétrole chauffés avec le chlorure d'aluminium donnent des produits analogues.

En résumé, nous réclamons l'emploi des chlorures métalliques pour la préparation des hydrocarbures et des acétones.

Fig. 17(b).   Text of the patent.

[Fourth Edition.]

A.D.. 1877, 15th DECEMBER. N° 4769.

Treatment of Hydrocarbons.

LETTERS PATENT to Charles Denton Abel, of No. 20, Southampton Buildings, in the County of Middlesex, for the Invention of "IMPROVEMENTS IN THE TREATMENT OF HYDROCARBONS FOR THEIR PURIFICATION AND CONVERSION INTO OTHER PRODUCTS." A communication from abroad by Charles Friedel and James Mason Crafts, both of Paris, in the Republic of France. .

Sealed the 7th May 1878, and dated the 15th December 1877.

PROVISIONAL SPECIFICATION left by the said Charles Denton Abel at the Office of the Commissioners of Patents on the 15th December 1877.

CHARLES DENTON ABEL, of No. 20, Southampton Buildings, in the County of Middlesex. "IMPROVEMENTS IN THE TREATMENT OF HYDROCARBONS FOR THEIR
5 PURIFICATION AND CONVERSION INTO OTHER PRODUCTS." [A communication from abroad by Charles Friedel and James Mason Crafts, both of Paris, in the Republic of France.]

This Invention relates to a method of treating hydrocarbons, whereby they become converted or convertible into other products, as I will now describe.
10 In a vessel or retort is mixed the hydrocarbon with anhydrous or partially hydrated chloride of aluminium, or with oxychloride of aluminium, or with a mixture of either or both these substances with metallic oxides, such as those of zinc, iron, lead, and the like. The hydrocarbon thus operated on at moderate temperatures, varying from 100° to 600° centigrade, becomes transformed into
15 other hydrocarbons or products of oxydation in the course of a few hours. During the process oxygen from the atmosphere or otherwise maybe supplied to the mixture. By this treatment a hydrocarbon, such as petroleum of low grade, is converted into gas, light oils suitable for illumination, and heavy oils containing paraffin; also petroleum, naphtha, and other hydrocarbons containing sulphur, are by the process
20 above described freed from sulphur, and enhanced in value. Naphthaline thus treated is converted into benzol, toluol, and other products. Hydrocarbons such

Fig. 18. British Patent 4769, dated December 15, 1877. "A communication from abroad" covering a remarkably wide field of chemical reactions and refining processes.

as benzol treated as described are caused to combine with cyanogen, forming cyanides which can be converted into acids such as benzoic acid.

In the treatment above described there may be substituted for the aluminium chlorides other metallic chlorides, such as the sesquichloride of iron or the chloride of zinc.   5

The quantity of chloride of aluminium or of the other chlorides mentioned should vary from 5 to 20 per cent. of the hydrocarbons acted upon, according to the extent of the action desired.

SPECIFICATION in pursuance of the conditions of the Letters Patent filed by the said Charles Denton Abel in the Great Seal Patent Office on the 5th 10 June 1878.

CHARLES DENTON ABEL, of No. 20, Southampton Buildings, in the County of Middlesex. "IMPROVEMENTS IN THE TREATMENT OF HYDROCARBONS FOR THEIR PURIFICATION AND CONVERSION INTO OTHER PRODUCTS." A communication from abroad by Charles Friedel and James Mason Crafts, both of Paris, in the Republic 15 of France.

This Invention relates to a method of treating hydrocarbons, whereby they become converted or convertible into other products, as I will now describe.

In a vessel or retort is mixed the hydrocarbon with anhydrous or partially hydrated chloride of aluminium, or with oxychloride of aluminium, or with a 20 mixture of either or both these substances with metallic oxides, such as those of zinc, iron, lead, and the like. The hydrocarbon thus operated on at moderate temperatures, varying from about 100° to 600° centigrade becomes transformed into other hydrocarbons or products of oxydation in the course of a few hours. During the process oxygen from the atmosphere or otherwise may be supplied to the 25 mixture; also organic chlorides may be added with advantage, as they assist the reaction and the formation of definite products. By this treatment a hydrocarbon, such as petroleum of low grade, is converted into gas, light oils suitable for illumination, and heavy oils containing paraffin; also petroleum, naphtha, and other hydrocarbons containing sulphur, are by the process above described freed 30 from sulphur, and enhanced in value. Naphthaline thus treated is converted into benzol, toluol, and other products. Hydrocarbons such as benzol treated as described are caused to combine with cyanogen, forming cyanides which can be converted into acids such as benzoic acid.

In the treatment above described there may be substituted for the aluminium 35 chlorides other metallic chlorides, such as the sesquichloride of iron or the chloride of zinc.

The quantity of the chloride of aluminium or of the other chlorides mentioned should vary from 5 to 20 per cent of the hydrocarbons acted upon according to the extent of the action desired.   40

Having thus described the nature of the said Invention, and in what manner the

same is to be performed as communicated to me by my foreign correspondents, I claim, the process for purifying and converting hydrocarbons by treating them with chlorides and metallic oxides under the influence of heat, substantially as herein described.

5   In witness whereof, I, the said Charles Denton Abel, have hereunto set my hand and seal, this Fifth day of June, in the year of our Lord One thousand eight hundred and seventy-seven.

CHAS. D. ABEL. (L.S.)

distillation, we were able to isolate ethylbenzene, boiling between 133° and 137°. Nevertheless the reaction seemed to proceed less cleanly with the iodide than with the chloride and gave a much larger proportion of high boiling products. These were probably benzene derivatives in which several hydrogen atoms were replaced by ethyl groups, but as yet they have not been completely studied.

"On working with methyl bromide and benzene, under 30 cm. mercury pressure, we observed the same evolution of acid fumes and the production of hydrocarbons boiling at temperatures above benzene and containing toluene or methyl benzene.

"Further, we have obtained toluene and other methylbenzenes in a very easy manner, simply by passing a current of methyl chloride into benzene containing aluminum chloride and heating very gently. Under these conditions there was again an evolution of hydrogen chloride and the methyl radical from methyl chloride attached itself to the benzene. On fractionating the product, after water treatment, large quantities of toluene were obtained, passing over at about 111°, followed by products boiling much higher around boiling points of xylene (137°), mesitylene (163°), durene, or tetra-methylbenzene (190°), and above. It seems that hydrocarbons such as pentamethyl- and hexamethyl-benzene were formed—products heretofore unknown. It seemed clear, at least, that the substitution of the methyl group was not affected by one or more methyl groups already present in the benzene nucleus, otherwise we would have obtained ethylated benzenes or even benzenes containing more complicated aliphatic radicals.

"We have observed further that aluminum bromide and iodide react with organic chlorides and iodides in the same manner as the chloride, forming analogous products.

"In the next communication we shall have the honor of expounding to the Academy several other syntheses of complex hydrocarbons and ketones which we have already carried out, and we shall try to give an interpretation of this reaction, so general yet so unexpected."

Two further papers were presented to the Academy of Science, on June 18, 1877, and July 9, 1877, with the same title as before. In that of June 18, Friedel and Crafts described the preparation of diphenylmethane, triphenylmethane, acetophenone, benzophenone, phthalophenone, and anthraquinone. They concluded that ". . . the examples we have just cited suffice, in our opinion, to show the generality of the new method and at the same time to fix its limitations." On July 9, 1877, five weeks after the first main communication, the use of chlorides other than aluminum was described (18).

It was found that certain chlorides (Zn, Fe, etc.) could replace that of aluminum whereas others (Mg, Co, Cu, Hg, Sb) could not. In this paper the authors showed that the reaction of Zincke was due to the presence of zinc chloride and a reaction mechanism was hypothesized which could apply equally well to the reactions described by Zincke and by Grucarevic and Merz as well as to the Friedel-Crafts synthesis. It was postulated that an intermediate compound was formed between the aluminum chloride and the aromatic hydrocarbon:

$$\text{C}_6\text{H}_6 \; + \; Al_2Cl_6 \; \longrightarrow \; \text{C}_6\text{H}_5\text{-}Al_2Cl_5 \; + \; HCl$$

The complex then reacted with the halogenated hydrocarbon

$$\text{C}_6\text{H}_5\text{-}Al_2Cl_5 \; + \; Cl(CH_2)_4CH_3 \; \longrightarrow \; \text{C}_6\text{H}_5(CH_2)_4CH_3 \; + \; Al_2Cl_6$$

Friedel and Crafts state: "We have not yet any decisive proof to apply in support of the hypothesis we have proposed for the mechanism of the reaction . . . . but to us it seems difficult to find another interpretation. If ours is correct, it will be seen in time that these present reactions, formerly so numerous in science, may be explained by a successive, transitory reaction in which the compound which appears to assist passively is present in small amounts, being regenerated accordingly in such a way as to find it, at the end, in an amount very close to that at the beginning of the reaction."

In the short period of six weeks Friedel and Crafts had published three papers clearly indicating the outline of their alkylation and acylation reactions. The alkylation reaction had been extended to both the aliphatic and aromatic series of hydrocarbons and to alkyl polyhalides. The acylation reaction had been similarly classified. In addition halides other than aluminum chloride were recognized as effective catalysts.

It is pertinent at this point to compare the work of Friedel and Crafts with that of Zincke and the earlier workers. We have seen that somewhat similar reactions with Zn, Cu, and Ag had been carried out previously, but it was only Friedel who recognized the true nature and the tremendous potentialities of the reaction. Friedel himself had inadvertently witnessed such a reaction as early as 1873, as Hanriot (30) has reported in these terms: "About 1873, a foreign student, whose name escapes me, carried out a preparation in Wurtz's laboratory using the Zincke method (action of zinc powder on a mixture of chloride and aromatic hydrocarbon). The reaction soon became violent and Friedel, who happened to be there, helped the young man to separate the liquid from the zinc powder. What caused him and all his assistants great astonishment was to see that the decanted liquid reacted equally violently as when it was on the zinc powder. This peculiar case was discussed at length for some time in the laboratory."

Thus the interest of Friedel in the reaction may be traced back to 1873 and this may have served as the starting point of the work, begun in May 1877 with his collaborator Crafts, on the use of aluminum chloride in organic chemistry.

It is generally believed that the discovery of the Friedel-Crafts synthesis arose from a chance observation on the action of metallic aluminum and iodine on amyl chloride, when the authors were attempting to convert amyl chloride to the iodide.

This belief is based on one of the later publications of Friedel and Crafts themselves (21). However, no record of this particular reaction appears in Friedel's notebooks of 1877 and therefore must have been carried out sometime previously. A study of the original notebooks of Friedel and Crafts shows that the first reaction recorded in 1877 was that between $Al + I_2 + CH_3CCl_3$ and subsequently with the same system and benzene. The page on which this experiment is recorded is undated but three pages later the date April 3, 1877 (in Craft's handwriting), is noted. It is on this latter page that the beginning of the systematic study of the reaction between aluminum and amyl chloride was entered. On the page of April 18 the reaction of aluminum chloride and amyl chloride is entered.

The use of aluminum iodide for the preparation of iodides had previously been reported (24,25,26) by Gustavson.

In this connection it is apposite to discuss a claim regarding the priority of discovery of the reaction. It has been claimed by Rudenko (33) that the reactions are incorrectly named "Friedel-Crafts reactions" and that the importance of the discovery should be attributed entirely to the Russian chemist Gustavson. The article in question gives essentially a summary of Gustavson's scientific work to establish the priority claim.

Gustavson was a student of Mendelcev's, and as such worked in the field of inorganic chemistry on the double decomposition reactions of anhydrides and halogen compounds. Many halides were included in these compounds, but no organic materials, thus this phase of his work can be but remotely connected with the Freidel-Crafts reaction.

In 1874 (24,25,26) Gustavson published a report of a metathesis between carbon tetrachloride and aluminum iodide. This is generally known as the Gustavson reaction. It is claimed by Rudenko that emanating from this study Gustavson showed that many organic chlorides are decomposed by aluminum chloride. However, no supporting references in Gustavson works prior to 1877 can be traced. Following the halogen exchange reaction Gustavson extended his work (27,28) in 1877 to demonstrate the replacement of hydrogen by halogen in aromatic compounds. At the same time by-products resulting from side chain eliminations were observed and in 1878 he published results (29) on the nature of the complex formed between aromatic materials and aluminum halides. Thus up to 1878 no publications concerning Friedel-Crafts alkylation or acylation were submitted by Gustavson although Friedel and Crafts had presented their papers on the subject in the previous year (14).

Gustavson, a well-known Russian chemist of his time, never advanced any priority claim concerning the Friedel-Crafts reaction, which he frequently quoted in his papers. Thus the claims made on behalf of Gustavson some eighty years later without any new data being advanced cannot be taken seriously. It is gratifying to see that attempts to call the reaction the "Gustavson-Friedel-Crafts reaction" (31) were abandoned even in Soviet chemical literature.

It should be pointed out that Friedel and Crafts were well aware of the existence of the Gustavson reaction and had in fact attempted such a reaction between amyl chloride and aluminum iodide. When they observed that the reaction took an entirely different course from that expected they investigated the reaction further and found that aluminum itself or, even better, aluminum chloride would give the same unexpected results. This was an early part of their investigation.

In connection with the priority claim by Rudenko on behalf of Gustavson, it is interesting to note that in the article in question no mention is made of Butlerov who, in 1873 (11) (*i.e.*, four years before Friedel and Crafts' first publication), observed the catalytic effect of boron trifluoride on the polymerization of isobutylene. This work was not pursued at the time or extended to other fields such as alkylation. The discovery of polymerization by metal halide catalyst dates back however to 1839, when Deville polymerized styrene with stannic chloride (12).

Perusal of the notebooks of Friedel and Crafts reveals that a systematic study was made on the action of aluminum chloride on amyl chloride, benzene, a mixture of amyl chloride and benzene, and a mixture of chlorobenzene and benzene. Later the authors, in establishing the versatility of the reaction, studied the condensation of benzene with methyl iodide, methyl chloride, and phthalic anhydride. Their realization of the technical importance of their discovery is evident from the fact that the inventors lost no time in patenting their findings in France (19) and Great Britain (20) and further that during the period 1877 to 1888 Charles Friedel, with the collaboration of James Crafts, published almost fifty papers on the action of aluminum chloride. The patents mentioned above were remarkably far-sighted in their coverage, including the preparation of aliphatic and aromatic hydrocarbons and ketones using metallic chlorides, particularly those of aluminum and zinc, the treatment of crude petroleum in cracking and refining procedures, and the removal of sulfur from petroleum naphtha and other hydrocarbons.

The publications of Friedel and Crafts covered a wide field of applications of their reactions both in the aliphatic and aromatic field and are listed at the end of the chapter. Papers published in the first year after the discovery included descriptions of the reaction of phosgene with benzene to yield benzophenone or benzoyl chloride and benzoic acid; the preparation of phthalophenone and anthraquinone from phthaloyl chloride and benzene; the aluminum chloride catalyzed air oxidation of benzene to phenol; the formation of thiophenol and diphenyl sulfide from sulfur and benzene; the preparation of durene from benzene and methyl chloride. Further papers described the action of carbon dioxide and sulfur dioxide on benzene and the formation of o-acylbenzoic acids from phthalic anhydride and acyl chlorides. After studying the catalytic effect of various metal chlorides on the reaction, when it was noted that zirconium tetrachloride was as effective as aluminum chloride, the authors continued the task of determining the scope of the reaction. This resulted in the preparation of o-toluidine, mono- and di-benzoyldurene, benzophenone, triphenyl carbinol, ditolylethane, and anthracene, to name but a few of the compounds. In addition a few reactions, such as that of phthalic anhydride with aromatic hydrocarbons and chloroform with benzene, were subjected to detailed study and further papers on the decomposing action of aluminum chloride were published. Two comprehensive reviews, in 1884 and 1888, were written and published, which included not only Friedel and Crafts work on the subject but also that of other chemists.

Friedel's last paper before his death concerned the cracking of normal saturated hydrocarbons to lower homologous members.

From this foundation the Friedel-Crafts reactions have spread to cover their wide and complex field of today. As we are approaching the centennial of the discovery of the reactions we can only marvel about the progress made and speculate what the next century may have in store.

Acknowledgments

We acknowledge our indebtedness and express our thanks to M. Edmond Friedel, Directeur de l'Ecole Nationale Supérieure des Mines, Paris, for allowing us access to, and providing a complete microfilm of, the original notebooks of his grandfather, Charles Friedel.

IV. Bibliography of C. Friedel and J. M. Crafts in the Field of the Aluminum Chloride Reactions

1877

1. Sur une nouvelle méthode générale de synthèse d'hydrocarbures, d'acétones, etc. *Compt. rend.*, 84, 1450; **85**, 74.
2. Synthèse de l'acide benzoique et de la benzophénone (with E. Ador). *Compt. rend.*, 85, 673; *Bull. Soc. Chim. France*, **28**, 482.
3. Action de Al_2Cl_6 sur le chlorure d'amyle. *Bull. Soc. Chim. France*, **27**, 482.
4. Action de Al_2Cl_6 sur divers chlorures et carbures. *Bull. Soc. Chim. France*, **27**, 530; **28**, 50.
5. Hexaméthylbenzine. *Bull. Soc. Chim. France*, **28**, 147; **28**, 529.
6. Action de $COCl_2$ sur la benzine en présence de Al_2Cl_6 (with E. Ador). *Bull. Soc. Chim. France*, **28**, 482.
7. Action de l'oxychlorure de carbone sur le toluène en présence de chlorure d'aluminium (by Crafts and Ador). *Compt. rend.*, **85**, 1163; *Bull. Soc. Chim. France*, **30**, 215.
8. French Patent 118,668, 24 May.
9. British Patent 4769, 15 December.
10. Réaction de la benzine en présence de chlorure d'aluminium. *Bull. Soc. Chim. France*, **32**, 3.

1878

11. Nouvelle méthode générale de synthèse organique. *Rev. Scientifique*, (2) **12**, 820.
12. Fixation directe de l'oxygène et du soufre sur la benzine et sur le toluène. *Compt. rend.*, **86**, 884.

13. Fixation directe de l'acide carbonique, de l'acide sulfureux, de l'anhydride phtalique sur la benzine; synthèse de l'acide benzoique, de l'hydrure de sulfophényle et de l'acide benzoylbenzoique. *Compt. rend.*, **86**, 1368.
14. Action du chlorure de cyanogène sur la benzine en présence de Al_2Cl_6. *Bull. Soc. Chim. France*, **29**, 2.
15. Formation d'anthraquinone par l'action du chlorure de phtalyle sur la benzine en présence de Al_2Cl_6. *Bull. Soc. Chim. France*, **29**, 49.
16. Synthèse du phenol par l'oxydation de la benzine en présence de Al_2Cl_6. *Bull. Soc. Chim. France*, **29**, 99.
17. Formation de la benzine et de l'hydrure d'anthracine par l'action du Al_2Cl_6 sur la naphtaline. *Bull. Soc. Chim. France*, **29**, 99.
18. Synthèse du crésylol et de l'acide benzoique. *Bull. Soc. Chim. France*, **29**, 242.
19. Formation du disulfure de diphénylène. *Bull. Soc. Chim. France*, **29**, 338; **29**, 481.
20. Fixation directe de l'oxygène et du soufre sur la benzine et sur le toluène en présence de Al_2Cl_6. *Bull. Soc. Chim. France*, **29**, 434; **31**, 463; *Compt. rend.*, **86**, 884.
21. Formation du durol dans l'action de CH_3Cl sur la benzine en présence de Al_2Cl_6. *Bull. Soc. Chim. France*, **29**, 481.
22. Action de SO_2 sur un mélange de benzine et de Al_2Cl_6. *Bull. Soc. Chim. France*, **29**, 530; **30**, 1.
23. Action des anhydrides acétique et phtalique sur la benzine en présence de Al_2Cl_6. *Bull. Soc. Chim. France*, **30**, 2.
24. Action des chlorures metalliques sur un mélange de chlorure de benzyle et de benzine. *Bull. Soc. Chim. France*, **30**, 146.
25. Ortho toluidine obtenue par l'action de CH_3Cl su run mélange fondu de chlorhydrate d'aniline et de Al_2Cl_6. *Bull. Soc. Chim. France*, **30**, 531.
26. Synthèses à l'aide de chlorure d'aluminium. *Assoc. Franc., Comptes Rendus*, 7e session (Paris). 370.
27. Action of aluminum chloride on petroleum. *Chemische Industrie*, **1**, 411.

1879

28. Fixation du méthyle sur la diméthylaniline sous l'influence de Al_2Cl_6. *Bull. Soc. Chim. France*, **31**, 194.
29. Formation du benzonitrile. *Bull. Soc. Chim. France*, **32**, 547.
30. Sur quelques dérivés du durol (α-tétraméthylbenzine) (with E. Ador). *Compt. rend.*, **88**, 880.

31. Action de l'anhydride phtalique sur la naphtaline en présence de chlorure d'aluminium (by Crafts and Ador). *Compt. rend.*, **88**, 1355; *Bull. Soc. Chim. France*, **34**, 531.

1880

32. Synthèse de l'hexaméthylbenzine et de l'acide mellique. *Compt. rend.*, **91**, 257.

1881

33. Sur les combinaisons de l'anhydride phtalique avec les hydrocarbures de la série de la benzine. *Compt. rend.*, **92**, 883; *Bull. Soc. Chim. France*, **35**, 290; **35**, 503.
34. Action du bromure d'éthylène sur le toluène en présence de chlorure d'aluminium (by Friedel and Balsohn). *Bull. Soc. Chim. France*, **35**, 52.
35. Action de CCl_4 sur C_6H_6 en présence de Al_2Cl_6 (by Friedel and Vincent). *Bull. Soc. Chim. France*, **36**, 1.

1882

36. Sur la préparation du triphénylméthane. *Bull. Soc. Chim. France*, **37**, 6.
37. Action décomposante de Al_2Cl_6 sur les carbures du pétrole. *Bull. Soc. Chim. France*, **37**, 49.
38. Action du chlorure d'aluminium sur un mélange de toluène et de chlorure de benzyle. *Bull. Soc. Chim. France*, **37**, 530.
39. On some decompositions produced by the action of chloride of aluminium. *J. Chem. Soc.*, **41**, 115.

1883

40. Action de Al_2Cl_6 sur la naphtaline. *Bull. Soc. Chim. France*, **39**, 195.
41. Action de Al_2Cl_6 sur la benzine à 200°. *Bull. Soc. Chim. France*, **39**, 306.
42. Action du chloroforme sur la benzine en présence de Al_2Cl_6 (with Vincent). *Bull. Soc. Chim. France*, **40**, 97.
43. Préparation d'aluminium-phényle. *Bull. Soc. Chim. France*, **40**, 162.
44. On the decomposing action that chloride of aluminium exerts on hydrocarbons. Brit. Assoc. Rep. 53rd meeting, Southport. 468.

1884

45. Synthèse du ditolylméthane. *Bull. Soc. Chim. France*, **41**, 273.
46. Action du chlorure de méthylène sur le toluène et sur la benzine. *Bull. Soc. Chim. France*, **41**, 322.
47. Synthèse de diphénylmèthane chloré. *Bull. Soc. Chim. France*, **41**, 370.
48. Sur une nouvelle méthode générale de synthèse des combinaisons aromatiques (Mémoire première). *Ann. Chim. Phys.*, (6) **1**, 449.
49. Action décomposante exercée par le chlorure d'aluminium sur certains hydrocarbures. *Assoc. Franc., Comptes Rendus*, 13e session (Blois), Pt. 2, 137.

1885

50. Action de CH_3Cl sur le toluène en présence de Al_2Cl_6. *Bull. Soc. Chim. France*, **43**, 50.
51. Action de chlorure de benzyle sur CS_2 en présence de Al_2Cl_6. *Bull. Soc. Chim. France*, **43**, 53.
52. Sur l'action décomposante exercée par le chlorure d'aluminium sur certains hydrocarbures. *Compt. rend.*, **100**, 692.
53. Action de l'aluminium sur le chlorure d'aluminium (by Friedel and Roux). *Compt. rend.*, **100**, 1191.
54. Action décomposante d'Al_2Cl_6 sur divers hydrocarbures. *Bull. Soc. Chim. France*, **43**, 196.
55. Action de CH_3Cl sur l'orthodichlorobenzine en présence de Al_2Cl_6. *Bull. Soc. Chim. France*, **43**, 305; **45**, 390.
56. Formation d'anthracènes méthylés. *Bull. Soc. Chim. France*, **44**, 97.
57. Sur l'action du chlorure de méthylène sur la benzine et sur ses homologues. *Assoc. Franc., Comptes Rendus*, 14e session (Grenoble), Pt. 1, 117.
58. Action du chlorure de méthyle sur l'orthodichlorobenzine en présence du chlorure d'aluminium. *Assoc. Franc., Comptes Rendus*, 15e session (Nancy), Pt. 1, 117.
59. Sur les anthracènes méthylés. *Assoc. Franc., Comptes Rendus*, 15e session (Nancy), Pt. 1, 118.

1887

60. Sur l'action du chlorure de méthyle sur la benzine orthodichlorée en présence du chlorure d'aluminium. *Ann. Chim. Phys.*, (6) **10**, 411.

61. Action du chlorure de méthylène en présence du chlorure d'aluminium sur diverses benzines méthylées. *Ann. Chim. Phys.*, (6) **11**, 263.

1888

62. Sur la densité de vapeur du chlorure d'aluminium et sur le poids moléculaire de ce composé. *Bull. Soc. Chim. France*, **50**, 2; *Compt. rend.*, **106**, 1764; **107**, 63.
63. Sur une nouvelle méthode générale de synthèse, des combinaisons aromatiques (Mémoir deuxième). *Ann. Chim. Phys.*, (6) **14**, 433.
64. Décomposition par le chlorure d'aluminium d'un hydrocarbure linéaire saturé (by Friedel and Gorgeu). *Compt. rend.*, **127**, 590; *Bull. Soc. Chim. France*, **19**, 370.

1900

65. Friedel Memorial Lecture (by Crafts). *J. Chem. Soc.*, **77**, Pt. 2, 993.

References

1. Biographical

The personal history of Charles Friedel and James Mason Crafts was based on information obtained from the following articles. The reader interested in minutiae will find a wealth of information therein.

1. Ashdown, A. A., *Ind. Eng. Chem.*, **19**, 1063 (1927).
2. Ashdown, A. A., *J. Chem. Educ.*, **5**, 911 (1928).
3. Behal, A., *Bull. Soc. Chim. France*, (4) **51**, 1493 (1932).
4. Crafts, J. M., *J. Chem. Soc.*, **77**, 993 (1900).
5. Cross, C. R., *Nat. Acad. Sci. (Washington) Biograph. Mem.*, **9** (5th memoir), 159 (1919).
6. Hackspill, L., *Bull. Soc. Chim. France*, (5) **16**, 555 (1949).
7. Hanriot, M. M., *Bull. Soc. Chim. France*, (3) **23**, 1900.
8. Ladenburg, A., *Ber.*, **32**, 3721 (1899).
9. Willemart, A., *J. Chem. Educ.*, **26**, 3 (1949).
10. Willemart, A., *Bull. Soc. Chim. France*, (5) **16**, 559 (1949).

2. Historical

11. Butlerov, A. M., and V. Gorgainov, *J. Russ. Chem. Soc.*, **5**, 302 (1873).
12. Deville, *Ann. Chim. (Paris)*, **75**, 66 (1839).
13. Doebner, O., and W. Stackman, *Ber.*, **9**, 1918 (1876).
14. Friedel, C., and J. M. Crafts, *Compt. rend.*, **84**, 1392 (1877).
15. Friedel, C., and J. M. Crafts, *Bull. Soc. Chim. France*, (2) **27**, 482 (1877).
16. Friedel, C., and J. M. Crafts, *Bull. Soc. Chim. France*, (2) **27**, 530 (1877).

17. Friedel, C., and J. M. Crafts, *Compt. rend.*, **84**, 1450 (1877).
18. Friedel, C., and J. M. Crafts, *Compt. rend.*, **85**, 74 (1877).
19. Friedel, C., and J. M. Crafts, French Patent 118,668, May 22, 1877.
20. Friedel, C., and J. M. Crafts, British Patent 4769, December 15, 1877.
21. Friedel, C., and J. M. Crafts, *Ann. Chim. Phys.*, (VI) **1**, 449 (1884).
22. Grucarevic, S., and V. Merz, *Ber.*, **6**, 60 (1873).
23. Grucarevic, S., and V. Merz, *Ber.*, **6**, 1238 (1873).
24. Gustavson, G. G., *Ann. Chim. Pharm.*, **172**, 174 (1874).
25. Gustavson, G. G., *Ann. Chim. Phys.*, (V) **2**, 397 (1874).
26. Gustavson, G. G., *J. Russ. Phys. Chem.*, **6**, 109 (1874).
27. Gustavson, G. G., *J. Russ. Phys. Chem.*, **9**, 190 (1877).
28. Gustavson, G. G., *J. Russ. Phys. Chem.*, **9**, 213 (1877).
29. Gustavson, G. G., *J. Russ. Phys. Chem.*, **10**, 268, 365, 390 (1878).
30. Hanriot, M. M., *Bull. Soc. Chim. France*, (3) **23**, 1 (1900).
31. Petrova, A. M., *J. Gen. Chem. U.S.S.R.*, 501 (1954).
32. Radzivanovski, C., *Ber.*, **7**, 141 (1874).
33. Rudenko, M. G., *Trudy Inst. Nefti Akad. Nauk. U.S.S.R.*, **6**, 199 (1955).
34. Zincke, T., *Ber.*, **2**, 737 (1869).
35. Zincke, T., *Ann.*, **159**, 367 (1871).
36. Zincke, T., *Ber.*, **4**, 298 (1871).
37. Zincke, T., *Ann.*, **161**, 93 (1872).
38. Zincke, T., *Ber.*, **6**, 137 (1873).

CHAPTER II

Scope and General Aspects

I. Introduction

Charles Friedel and James Mason Crafts in their original work showed that "anhydrous aluminum chloride could be used as a condensing agent in a general synthetic method for furnishing an infinite number of hydrocarbons." In the continuation of their work (see Chapter I dealing with the history of the reaction) they extended their studies of the catalytic effect of aluminum chloride in organic reactions to various fields, in which they discovered:

1. Reaction of organic halides and unsaturated compounds with aromatic and aliphatic hydrocarbons.
2. Reaction of anhydrides of organic acids with aromatic hydrocarbons.
3. Reaction of oxygen, sulfur, sulfur dioxide, carbon dioxide, and phosgene with aromatic hydrocarbons.
4. Cracking of aliphatic and aromatic hydrocarbons.
5. Polymerization of unsaturated hydrocarbons.

Thus in the early stages of the development of the investigation of the catalytic activities of aluminum chloride in organic chemistry, Friedel and Crafts were able to show that it is a most versatile catalyst effecting alkylations, dealkylations, acylations, polymerizations, and an almost unlimited number of other reactions. No other inorganic compound has produced such a variety of unusual and useful reactions in the organic field as anhydrous aluminum chloride. The diversity of reactions is astounding. Aluminum chloride serves as a catalyst in reactions between aromatic and aliphatic compounds, inorganic compounds and organic compounds; in the degradation of aromatic and aliphatic compounds, in substitution reactions and in addition reactions, and in various other

25

reactions, such as polymerizations. Most of these reactions are classified by chemists as reactions of the Friedel-Crafts type and with time it has become a generally accepted concept that any organic reaction brought about by the catalytic action of anhydrous aluminum chloride, or related catalysts, is a Friedel-Crafts reaction.

II. General Considerations

1. Catalysts

Friedel and Crafts found that besides $AlCl_3$ other metal halides, such as ferric chloride, zinc chloride, and sodium aluminum chloride are active catalysts in Friedel-Crafts type reactions. They also found that aluminum bromide and iodide are active as catalysts in the reaction between chlorinated aliphatic compounds and aromatic hydrocarbons. However, the chlorides of magnesium, cobalt, copper, mercury, and antimony were reported to be without effect (194). Later they found (196) that in the rapid condensation between benzyl chloride and benzene the chlorides of zinc, iron, and cobalt behaved similarly to aluminum chloride, but that the action was not so energetic. Zirconium chloride, on the other hand, gave results which were as good as those secured with aluminum chloride.

The ever-increasing number of investigators following the work of Friedel and Crafts found a large number of other Lewis acid type halides (ranging from ferric chloride (440) and boron trifluoride to zirconium halides) to be active catalysts for the reactions. In further development of the work proton acids such as sulfuric, hydrofluoric, pyrophosphoric acid, etc., were found to catalyze similar reactions. Both types of acid catalysts were, from the very beginning, regarded by many workers as substitutes for aluminum chloride.

Syntheses made possible through the catalytic action of metal halide and proton acid catalyst of the Lewis and Brønsted type now include the preparation of practically all classes of compounds.

In the past it was believed that the action of proton acids, such as H_2SO_4 and HF, differed in kind from that of Friedel-Crafts Lewis acids, such as $AlCl_3$ and BF_3. It was observed, however, that the latter type frequently require a co-catalyst, or proton donor. One of the fastest of all hydrocarbon reactions, the polymerization of isobutene, could not be effected in the absence of a co-catalyst (170). It has been concluded therefore that the types of acid catalysts differ only in degree.

The acidity of a typical Friedel-Crafts acid, HF + 7 wt.% BF_3, was recently determined (395). The value of H_0 (Hammett Acidity

Function) was found to be -16.6, compared with about -11 for 100% H_2SO_4 and -10 for anhydrous HF. Thus, this Friedel-Crafts acid, which presumably has an acidity about the same as that of such acids as $AlCl_3 \cdot HCl$ and $AlBr_3 \cdot HBr$, is more acidic than sulfuric acid by a factor of 10^5. This difference can easily account for the greater catalytic activity in many reactions.

In principle, therefore, there is no difference between the Lewis acid catalyzed "typical" Friedel-Crafts reactions and similar reactions carried out under the catalytic effect of Brønsted acids. Owing to the difficulty of maintaining really anhydrous conditions in Friedel-Crafts systems containing catalysts such as $AlCl_3$ or BF_3, it is very doubtful, with the exception of a very few carefully selected recent physico-chemical investigations, that any of the reactions reported taking place under the catalytic effect of anhydrous aluminum chloride (or related Lewis acid metal halide catalysts) were really fulfilling this condition. Since the reported experimental conditions would generally provide copious amounts of co-catalyst (preferentially water in the form of moisture, oxygen, or other impurities in the reagents) these reactions, according to our present-day knowledge, were taking place under the catalytic effect of proton acids, such as $H^+AlCl_3OH^-$, $H^+AlCl_4^-$, etc.

From a theoretical point of view it is interesting to point out that although Friedel-Crafts reactions involving reagents with atoms having unshared electron pairs (O, N, F, Cl, Br, I, S, etc., which are capable of coordinating with the electron-deficient central atom of the Lewis acid catalyst) can be carried out under strictly anhydrous conditions, reactions involving reagents not containing similar atoms (*e.g.*, alkylation or polymerization with olefins) fail under the effect of anhydrous Lewis acid metal halides catalysts. These reactions always need the presence of a co-catalyst, the role of which is to produce a proton acid (or provide a cation other than a proton to initiate the reactions). On the other hand, although proton acids are capable of effecting most of the Lewis acid catalyzed reactions, they are frequently much milder in activity when used alone and need higher reaction temperatures even to achieve comparatively low yields.

The question of catalysts in the Friedel-Crafts reactions will be treated in Chapter III in more detail. However, at this point it is necessary to emphasize that in order to limit the scope of the reactions related to the Friedel-Crafts type to be treated in the present book, the contributing authors agreed to consider only reactions which take place with both Lewis and Brønsted-Lowry

acid catalysts alike and omit those which are specifically carried out
with proton acid catalysts and cannot be achieved with Lewis acid
type acidic halides.   We felt justified in this selection by the fact
that the original work of Friedel and Crafts unquestionably laid
the foundation of the catalytic effect of anhydrous aluminum chloride
and related metal halide catalysts.   Consequently the alkylation
and acylation reactions based on their work, as well as a substantial
number of related reactions, are generally called Friedel-Crafts
reactions.   The realization that these reactions were generally
taking place under conditions where promoters or co-catalysts
always were present (thus the active catalysts were proton acids and
not the Lewis acid halides themselves) extended the field of the
Friedel-Crafts type reactions to the closely related reactions carried
out with proton acid catalysts such as H_2SO_4, HF, etc.   With
our present-day knowledge separation of the two types of reaction
cannot be justified.

2. Definition of Term "Friedel-Crafts Reactions"

Friedel and Crafts could not have foreseen the countless applications, theoretical discussions, and reviews which have followed
the discovery in 1877 of their aluminum chloride reaction.   The
original scope of their reaction has been extended to cover every
conceivable variation of reagent, catalyst, and affiliated reactions.
It is not surprising therefore that these reactions form a large part
of the more general problem of electrophilic reactions.

To define clearly the scope of the Friedel-Crafts reaction is thus
not easy.   C. C. Price gives the following definition in the *Encyclo-
pædia Britannica*: "The Friedel-Crafts reaction is commonly considered as a process of uniting two or more organic molecules through
the formation of carbon to carbon bonds under the influence of
certain strongly acidic metal halide catalysts such as aluminum
chloride, boron trifluoride, ferric chloride, zinc chloride, etc. . . ."

The observation that aluminum chloride is not by any means the
only specific catalyst in the reaction was included in the very first
papers of Friedel and Crafts.   They found that ferric and zinc
chlorides as well as the double salt sodium aluminum chloride could
also be employed but that these were less reactive than aluminum
chloride itself.

To approach the problem of definition it is necessary to come to a
clear understanding that there is not one but a number of reactions
bearing the general name of "Friedel-Crafts" and a large number of
reactions related to this type.   In a general sense we may say that

today we consider Friedel-Crafts type reactions to be any substitution, isomerization, elimination, cracking, polymerization, or addition reactions taking place under the catalytic effect of Lewis acid type acidic halides (with or without co-catalyst) or proton acids. One of the original characteristics of the reaction, namely that hydrogen halide should be evolved in the course of the reaction is by no means a limiting condition any more. It is felt to be appropriate to maintain the name "Friedel-Crafts" for these reactions in honor of the achievement of the original inventors of the aluminum chloride reaction and at the same time to use the term in a more general sense, pointing out that reactions related to the "Friedel-Crafts reactions" are to be included.

It is also unnecessary to limit the scope of the Friedel-Crafts reaction to the formation of carbon–carbon bonds. The formation of carbon–oxygen, carbon–nitrogen, carbon–sulfur, carbon–halogen, carbon–phosphorus, carbon–deuterium, carbon–boron, and many other types of bonds all conform with the general Friedel-Crafts principle.

It is interesting to call attention to the fact that although Friedel and Crafts made their original observation on a reaction involving replacement of hydrogen in an aliphatic compound, i.e., amyl chloride, the main emphasis of the reaction was concerned first of all with aromatic compounds. The preparation of aliphatic compounds involving Friedel-Crafts methods was of minor importance until World War II, when isomerization of paraffins and cycloparaffins and the polymerization of olefins achieved considerable importance. The development of the aliphatic chemistry of Friedel-Crafts reactions stems largely from advances made in the production of relatively pure aliphatic hydrocarbons and their utilization for motor fuels (Ipatieff and co-workers) and the increasing importance of polyolefins. Ethylbenzene, needed for the manufacture of styrene, detergent alkylates, and related products, helped to push aromatic Friedel-Crafts reactions into large-scale production.

3. Composition of Reaction Systems

Under ideal conditions, that is where no complications or side reactions occur, a Friedel-Crafts reaction mixture involves the following components:

(a) The substance to be substituted.
(b) A reagent that supplies the substituent. This may be an olefin, alkyl halide, alcohol, acid halide, or anhydride, etc.

(c) A catalyst, which may be a Lewis acid type acidic halide or a proton acid in the Brønsted-Lowry sense.

(d) A solvent; the function of which is sometimes taken over by excess of the substrate or reagent. Solvents are generally of the non-ionizing type, e.g., CS_2, CCl_4, etc., although solvents with higher dielectric constants are also employed, e.g., nitrobenzene, nitromethane, etc.

(e) The substituted product formed in the reaction (alkylated, acylated product, etc.).

(f) The by-product conjugate acid HX, where X originates from the catalyst.

Of the possible combinations of these constituents, many give rise to complexes that play an important role in governing the results of a given reaction. These Friedel-Crafts complexes will be considered in Chapter IV.

4. Atoms to be Substituted

Contrary to the general belief that in Friedel-Crafts reactions only hydrogen is substituted by an alkyl or acyl group, an increasing number of examples prove that an atom or group other than hydrogen may also be replaced. Besides the well-known disproportionations, de- and transalkylations, there are instances when alkyl groups are directly substituted in Friedel-Crafts reactions.

Hennion and McLeese (260) observed the acylation of p-di-t-butylbenzene with acetyl chloride gives 72% p-t-butylacetophenone. Nightingale (461) also described aromatic acylation of alkylbenzenes giving acetophenone and dialkylacetophenones in addition to the expected p-alkylacetophenone. These results could, however, be explained by disproportionation of the alkylbenzenes with subsequent acylations.

As hexaalkylbenzenes are not isomerized and dealkylation takes place only under vigorous conditions, these are suitable compounds for investigation of Friedel-Crafts substitution of alkyl groups.

The nitration of hexaethylbenzene gives p-dinitrotetraethylbenzene (208,578,577). Hexamethylbenzene gives o-dinitrotetramethylbenzene (578,646).

Pentamethylbenzene loses a methyl group when nitrated in chloroform with fuming nitric acid or with a mixture of nitric and sulfuric acids: 1,2,3,4-tetramethyl-5,6-dinitrobenzene is formed (221). Pentaethylbenzene behaves similarly to give 3,6-dinitrotetraethylbenzene.

Other examples of the displacement of an alkyl group are the formation of nitrotoluene in the nitration of cymene (2,353), the elimination of alkyl groups in the nitration of amyl- and butyl-xylenes (109), and the replacement of an isopropyl group in di- and tetra-isopropylbenzenes (39,457).

Nightingale (460) gives a list of compounds from which Br, CO_2H, CH_3, C_4H_9, other alkyl, and acyl groups may be eliminated in the course of dinitrations if the relevant substituents are *meta* to the first nitro group.

Observations by Smith and his co-workers (577,578,579,580,581) have suggested that the replacement of a methyl or ethyl group is preceded by attack on the group by nitric acid to give an alkyl nitrate. Thus bromodurene (I) on nitration gives an alkylnitrate (II) which is converted to III by the action of concentrated sulfuric acid.

Hexaethylbenzene when brominated with iodine as catalyst gives *p*-dibromotetraethylbenzene (208).

Hexaethylbenzene, chlorinated in carbon tetrachloride solutions in the presence of $FeCl_3$, gives hexachlorobenzene (275).

The acylation of hexaethyl- and hexamethylbenzene with a number of acyl chlorides and anydrides was recently investigated by Hopff (278). The reaction proceeds with elimination of an alkyl group and formation of pentaalkylphenyl ketones and γ-pentaalkyl-benzoyl acids respectively.

Replacement of groups other than hydrogen has not yet been observed in aliphatic Friedel-Crafts reactions, although there is no theoretical reason why such substitution should not occur.

5. Division of Reactions

A. According to Formation of Carbonium-Ion Type Reagents

The Friedel-Crafts type reactions according to Baddeley (26) can be divided generally into interactions of two major types: (1) Interactions in which a carbonium ion or a positively polarized complex, formed from the interaction of reagent molecules or co-catalysts with Lewis or proton acid type catalyst, reacts with an unsaturated (sp^2 or sp^3) carbon atom; (2) Interactions in which a hydride ion is abstracted from a saturated (sp^3) carbon atom to form a carbonium ion (or positively polarized entity) capable of producing reactions of the Friedel-Crafts type.

These two main types of Friedel-Crafts reaction systems predominate certain limitations of the reaction. For example, it is clear why cyclization with unsaturated acyl chlorides (having two reactive groups capable to form carbonium ions) may occur, and why alkanes may form ketones with carbon monoxide (the reaction being initiated by hydride abstraction). On the other hand, attempts at cyclization with saturated acyl halides must be fruitless, unless reaction conditions can be discovered which allow a hydride-ion extraction type reaction to occur with the primarily formed ketone. Reactions of the first type are generally involved in substitution reactions of aromatics and in polymerizations of olefins. These reactions have been fairly well investigated. However, much work remains to be done before the potentialities of the reactions in aliphatic and alicyclic chemistry can be fully appreciated, where hydride abstraction generally plays an important role.

B. According to Generalized Type of Reactions

Concerning types of the Friedel-Crafts reactions, in a generalized sense they can be divided into *alkylations* and *acylations*, including reactions of considerable diversity. This division seems the most logical and will be used in what follows.

III. Alkylation (including Isomerization and Polymerization)

The methods of introducing alkyl groups into aromatic, aliphatic, or cycloaliphatic compounds may vary considerably. The reactants may be of the most varied natures, as may also the substrates undergoing substitution, the catalyst needed to achieve the condensation, as well as the solvents and the conditions of the reaction.

1. Alkylation of Aromatic Compounds

A. Alkylating Agents and Catalysts

In the Friedel-Crafts alkylation of aromatics a hydrogen atom (or other substituent group) of an aromatic nucleus is replaced by an alkyl group through the interaction of an alkylating agent in the presence of a Friedel-Crafts catalyst. The most frequently used alkylating agents are alkyl halides, alkenes, and alcohols, although aldehydes, ketones, and various other reagents have also been used (Table I).

Lewis acid type catalysts for aromatic alkylations include

aluminum chloride, ferric chloride, boron trifluoride, antimony pentachloride, zinc chloride, titanium chloride, etc. Those of the Brønsted-Lowry acid type are: HF, H_2SO_4, H_3PO_4, etc. Acidic oxide catalysts of the silica alumina type and cation-exchange resins are becoming increasingly useful as catalysts.

TABLE I. Most frequently used alklyating agents in aromatic alkylations

Alkyl halides	Ethers
Alkenes	Aldehydes and ketones
Alkynes	Paraffins and cycloparaffins
Alcohols	Mercaptans
Esters (of carboxylic and	Sulfides
inorganic acids)	Thiocyanates

The overall reactions involved using alkyl halides, alcohols, and alkenes as alkylating agents in the presence of aluminum chloride may be written

$$Ar\!-\!H + RX \xrightarrow{\;\;AlCl_3\;\;} Ar\!-\!R + HX$$

$$Ar\!-\!H + R\!-\!OH \xrightarrow{\;\;AlCl_3\;\;} Ar\!-\!R + H_2O$$

$$Ar\!-\!H + R\!-\!CH\!=\!CH_2 \xrightarrow{\;\;AlCl_3\;\;} \underset{\underset{\displaystyle Ar}{|}}{R\!-\!CH\!-\!CH_3}$$

The equation written for the alkylation with an alcohol in the presence of aluminum chloride is obviously an over-simplification, since the alcohol can react with the aluminum chloride (465).

$$R\!-\!OH + AlCl_3 \longrightarrow R\!-\!OH\cdot AlCl_3 \longrightarrow RO\!-\!AlCl_2 + HCl$$
$$\longrightarrow R\!-\!Cl + AlOCl$$

It is not necessary for the alcohol–aluminum chloride reaction to go to completion with formation of the alkyl chloride, as the intermediate Lewis acid complex itself contains a sufficiently polarized alkyl group to enter the reaction. However, the reaction depends very much on the conditions used.

Whereas only catalytic quantities of aluminum chloride are necessary for the reaction of alkyl halides and alkenes in Friedel-Crafts alkylations, considerably larger quantities of the catalyst are necessary when alcohols are used as alkylating agents.

B. Activity of Catalysts

The different Friedel-Crafts catalysts vary in activity in alkylation reactions. Although few direct measurements of catalytic activity are available, the general order among the metal halides in alkylation reaction is reported to be: $AlCl_3 > SbCl_5 > FeCl_3 > TiCl_2 > SnCl_4 > TiCl_4 > TeCl_4 > BiCl_3 > ZnCl_2$ (514,518).

The general order of activity of proton acid catalyst is $HF > H_2SO_4 > H_3PO_4$. It should be pointed out, however, that the relative activities of Friedel-Crafts catalysts depend not only on the acid itself but also on the nature of the alkylating agent employed (base) and a number of other conditions. To mention a few examples: sulfuric acid and phosphoric acid are more effective catalysts in alkylations with alkenes and alcohols than with alkyl halides. Boron tribromide and boron trichloride are more effective catalysts in alkylations with alkyl fluorides than boron trifluoride; however, boron trifluoride is the catalyst of choice in alkylations with olefins. Boron chloride and bromide cannot be applied in alkylations with alcohols owing to the fact that they react to give the corresponding borates. The boron halides generally fail to catalyze alkylations with alkyl chlorides and bromides, but boron fluoride shows some activity if a proton donor co-catalyst is present. The choice of a catalyst for a particular Friedel-Crafts alkylation depends upon the activities of both the substrate to be alkylated and the alkylating agent, as well as the solvent, the reaction temperature, and several other conditions.

C. Reactivity of Aromatics

The ease of Friedel-Crafts alkylations varies with the structure of the aromatic nucleus undergoing the reaction. Electron donor *ortho–para* directing substituents generally facilitate the alkylation of aromatic rings whereas electron-withdrawing *meta*-directing substituents usually inhibit Friedel-Crafts alkylations by deactivation of the nucleus. This deactivation can be upset by the simultaneous presence of powerful *ortho–para* directing substituent groups. Thus although nitrobenzene generally cannot be alkylated by a Friedel-Crafts procedure and acetophenone is alkylated in low yield at a reaction temperature of 180° (23), *ortho*-nitroanisole is isopropylated in 85% yield (105).

The normal activating effect of such groups as OR, OH, NR_2, etc., are often partially nullified by complex formation with the catalyst

$$\text{C}_6\text{H}_5\overset{-}{\text{O}}\text{R} + \text{AlCl}_3 \rightleftharpoons \text{C}_6\text{H}_5\overset{\delta+}{\text{O}}\text{R}\uparrow\overset{\delta-}{\text{AlCl}_3}$$

To the extent that complex formation has occurred the catalyst is unavailable for activation of the alkylating agent and at the same time the activating influence of electron-donating substituents on the aromatic nucleus is replaced by a strong deactivating effect. Complex formation of this type is particularly pronounced with aromatic amines and these substances cannot generally be successfully alkylated by Friedel-Crafts methods.

A characteristic feature of Friedel-Crafts alkylation is a general tendency to form di- and polyalkylated products, besides the desired monoalkylates. This is particularly the case when simple alkyl groups such as methyl and ethyl are introduced. Furthermore the orientation of the alkyl groups introduced is dependent on the reaction conditions, particularly upon the proportion of the catalyst, temperature, etc. Isomerizations frequently accompany the alkylations and therefore in most alkylation reactions it is difficult to predict accurately either the configuration of the entering group or the orientation in the nucleus.

Friedel and Crafts showed (193) that one of the simplest alkyl halides, methyl chloride, reacts with benzene to give not only toluene but a rather complex mixture of polymethylated benzenes, including xylenes and hexamethylbenzene. Jacobsen (318) later found that the yield of methylated products could be improved by introducing the methyl chloride under a slight pressure. Anschütz and Immendorf (10) observed that the methylation may be of a varied nature; mono-, di-, tri-, tetra-, and hexamethylbenzenes are all produced in the same reaction in varying proportions depending on the conditions.

Ethyl halides have been observed to behave in a similar manner to yield mono- or polyethyl compounds.

At one time it was thought that the liquid-phase alkylation of aromatics could only be controlled to produce the monoalkyl derivative by the use of a very large excess of aromatics. However, it has been shown that the difficulty does not arise from any great difference in reactivity of alkylbenzenes and benzene (130,476) (Table II). Instead it is due to the separation (when an excess of aromatic is used as solvent) of a dense catalyst layer in which the

TABLE II. Relative rate of aluminum chloride catalyzed isopropylation and benzylation of benzene and alkylbenzenes with propylene and benzyl chlorode in nitromethane

	Isopropylation (130)	Benzylation (476)
Benzene	1.0	1.0
Toluene	2.10	3.20
Ethylbenzene	1.81	2.45
Isopropylbenzene	1.69	2.22 (n-propylbenzene)
t-Butylbenzene	1.40	2.08 (n-butylbenzene)

mono- and higher alkylbenzenes are selectively soluble. As a result these partially alkylated benzenes are exposed to a high local concentration of both catalyst and alkylating agent and tend to undergo further alkylation. Either the use of solvents (such as nitromethane) or small amounts of a material such as diethyl ether to render the reaction mixture homogeneous or the use of relatively high temperatures to permit a reasonable catalyst concentration in the hydrocarbon phase with concurrent vapor-phase addition of the olefin, avoids the difficulty and permits high yields of the monoalkylbenzene (186,188).

D. Dealkylation, Transalkylation, and the Question of Reversibility

It has long been known that the action of aluminum chloride is not restricted to the introduction of alkyl groups into aromatics. It can also be used to remove alkyl groups from alkylbenzenes (dealkylation) (10,199,319).

Therefore it has often been stated that Friedel-Crafts alkylations are reversible (65,106,319,391,432,459,517,518,607). Hexamethylbenzene heated at 190–200° with aluminum chloride gives methyl chloride and a mixture of hydrocarbons: pentamethylbenzene, durene, isodurene, trimethylbenzenes, xylenes, and small amounts of benzene and toluene. If a stream of dry hydrogen chloride is passed through the mixture the dealkylation is easier and more complete.

When methylbenzenes having 1–4 methyl groups are heated with aluminum chloride, dealkylation and alkylation proceed together so that the product is a mixture of hydrocarbons, some with fewer and some with more alkyl groups than the original alkyl benzene. Thus mesitylene yields xylenes and durenes, as well as smaller amounts of toluene and benzene (559).

Alkyl groups containing two or more carbon atoms are readily transferred from one position to another and from one benzene nucleus to another (intra- and intermolecular isomerization). Polyethylbenzenes react with benzene to give ethylbenzene. Isopropyl and t-butyl groups are still more readily transferred and one alkyl group can displace another (64,102).

Similarly, stirring pure ethylbenzene with aluminum chloride at room temperature for a few minutes yields substantial amounts of benzene and polyethylbenzenes (9).

In one of the most important commercial alkylations, in the preparation of ethylbenzene from benzene and ethylene, polyethylbenzenes are produced as well as ethylbenzene. However, the overall information of polysubstituted products is minimized by recycling the higher ethylation products to ethylate fresh benzene (188).

The methyl group is much more resistant in these rearrangements, so that toluene on prolonged reflux with aluminum chloride gives only a little benzene and xylene (187) and similarly, from xylene and benzene, when heated with aluminum chloride, only little toluene is formed (65). Methyl transfer in liquid-phase reactions has been reported (9,10,255,319,435,459). Ethylation of toluene under vigorous conditions has given hexaethylbenzene. The by-product, not isolated, is possibly an ethylated xylene.

True reversibility implies the dissociation of alkylbenzene into benzene and olefin, $ArCRH$—$CH_2R' \rightleftharpoons ArH + CHR{=}CHR'$ or the regeneration of some other alkylating agent. This does not take place, however, with aluminum chloride or other Friedel-Crafts catalysts at reaction temperatures for liquid-phase alkylation. Although benzene may be distilled off rapidly from ethylbenzene heated under a column with aluminum chloride, no uncondensed gas (ethylene) will appear and the residue is highly ethylated (188). Ethyl groups are shifted but there is no reversal of ethylation. The reason is that the thermodynamic equilibrium is so far on the side of alkylation at moderate temperatures that the partial pressure of ethylene is negligible.

Even at 300° hexaethylbenzene with zinc chloride or 97%

phosphoric acid fails to evolve appreciable amounts of ethylene (186).

At higher temperatures in vapour-phase processes there is some indication of a true equilibrium. Most of these processes operate at 300–500° at which the equilibrium is still favorable for alkylation. Transfer of alkyl groups at 454–538° has been studied by Hansford, Myers, and Sachanen (243) and shows only a slight production of olefins in this temperature range. One patent (403) gives data from which an equilibrium constant may be calculated at 550–600°, the value of which is slightly higher than would be expected from the Bureau of Standards thermal data (8).

The equilibrium between cumene, benzene, and propylene is not much more favorable to decomposition, the calculated partial pressure of propylene at 27° being only 10^{-3} mm. Diisopropyltoluene on heating at its boiling point (225°) with sodium aluminum chloride evolves no appreciable amount of propylene (186). The little gas which was produced proved to be propane (252,309,559). However, Schultze produced propylene and butenes by the decomposition of dialkylbenzenes at 400° (560).

In alkylation of benzene with ethanol

$$C_6H_6 + C_2H_5OH \rightleftharpoons C_6H_5C_2H_5 + H_2O$$

reversal of the reactions is negligible at all temperatures (16).

In the alkylation of benzene with ethyl chloride (539)

$$C_6H_6 + C_2H_5Cl \rightleftharpoons C_6H_5C_2H_5 + HCl$$

the maximum equilibrium partial pressure of ethyl chloride below the boiling point of ethylbenzene would be about 2 mm. Jacobsen (319) claimed to have reversed the corresponding methylation by evolving methyl chloride very slowly from m-xylene or more highly methylated benzenes with aluminum chloride at reflux temperatures, using a stream of dry hydrogen chloride. However, the high yields of higher alkylbenzenes obtained by Simons and Hart (570) with alkyl chlorides and hydrogen chloride under pressure and moderately high temperature proves that the equilibrium is almost entirely on the side of alkylation (216). The same conclusion applies to methylation with dimethyl ether (216) and to alkylation with esters.

At moderate temperatures, although the alkylations might be considered reversible kinetically, they are scarcely reversible thermodynamically.

The simple benzene nuclei are, of course, not the only type to undergo alkylation. Substituted benzenes and condensed aromatic rings (*e.g.*, naphthalene, tetralene, etc.) undergo similar substitutions.

Steric hindrance plays an important role in many alkylation reactions. Condon (130) has shown that propylation of monoalkyl benzenes is affected by steric hindrance (besides different basicity). The larger and more complex the alkyl group the slower the propylation reaction. The decreasing velocity of propylation was found to be toluene > ethylbenzene > cumene > *t*-butylbenzene (Table II).

These differences in rates seem to be a measure of differences in steric hindrance at the *ortho*-positions. Thus if the rates of propylation at the *meta*- and *para*-positions are nearly the same for toluene and cumene, the rate at the *ortho*-position in cumene is about $\frac{1}{6}$ of that of toluene.

The structure of the alkyl group also has an important influence on the number of alkyl groups which can be introduced by direct alkylation. Six methyl, ethyl, and *n*-propyl groups can be attached to the benzene ring by the Friedel-Crafts method but only four isopropyl groups. *t*-Butylation normally yields only *para*-di-*t*-butylbenzene as disubstitution product. 1,3,5-tri-*t*-butylbenzene was obtained by the HF + BF_3 catalyzed disproportionation of *t*-butylbenzene (408).

By indirect syntheses, however, 1,2,4-tri-*t*-butylbenzene (284), 1,2,4,5-tetra-*t*-butylbenzene (268), and 1,2-di-*t*-butylbenzene (267) were prepared recently.

E. Factors Influencing Mono- and Polyalkylations

It has usually been assumed (517,518,607) that the tendency toward polysubstitution during Friedel-Crafts alkylation is due to the greater reactivity of the initially produced alkylbenzene toward further substitution. This explanation, however, has been denied by Francis (186), who presented evidence that alkyl groups on a benzene nucleus have only a small effect on the rate of further alkylation. Francis pointed out that the actual alkylation occurs in a heterogeneous reaction system, specifically in the catalyst layer and that the reason for polysubstitution is the preferential extraction of the early reaction product by this catalyst layer. The tendency toward polysubstitution may be minimized by the use of a mutual solvent for the hydrocarbon and catalyst layers, by high-speed stirring, by operating in the vapor phase or by operating at a temperature sufficiently high that aluminum chloride is soluble in the hydrocarbon layer.

F. Orientation

An additional factor complicating the usefulness of Friedel-Crafts alkylations is the orientation involved in the introduction of more than one alkyl group. It was discovered at an early date that alkylation with aluminum chloride and other catalysts yields a considerable proportion of *meta*-dialkylbenzenes, as well as the expected *ortho*- and *para*-isomers (186,517,518). The relative extent of "normal" and "abnormal" orientation has been found to be a function of the conditions of alkylation. In general, the more vigorous the conditions with respect to the activity of the catalyst or the alkylating agent (reaction time and temperature, relative amount of catalyst, absence of solvent, etc.), the greater is the tendency for the formation of the *meta*-derivatives. Thus alkylations catalyzed by aluminum chloride, the most active catalyst, lead to large proportions of *m*-dialkylbenzenes, particularly with large amounts of catalyst at high temperatures or for long reaction times. Alkylation catalyzed with boron fluoride, sulfuric acid, ferric chloride, zinc chloride, and other catalysts in small concentration and preferably in solvents at low temperatures yield chiefly the normal *ortho–para* dialkylbenzenes (144,575). The question of whether the high *meta*-isomer ratios are due to direct substitution involving a low selectivity reaction (87,88,90,91) or to concurrent or consecutive secondary isomerization of the first formed *ortho*- and *para*-dialkylbenzenes (6,476) is still open for discussion.

2. Alkylating Agents in Aromatic Alkylations

A. Alkyl Halides

Friedel and Crafts discovered the catalytic action of aluminum chloride in organic substitution reactions in 1877 in the case of an alkylation reaction with an alkyl halide (195).

Alkyl fluorides are the most reactive alkyl halides in alkylation reactions. Besides aluminum chloride and related catalysts, boron trifluoride was found to be a very suitable catalyst for alkylations with alkyl fluoride (483). Boron trichloride, boron tribromide, and boron triiodide also catalyze alkylations with alkyl fluorides; however, they are inactive in related alkylations with alkyl chlorides or bromides.

Investigation of the reaction of toluene with cyclohexyl halides showed that BF_3 does not activate the cyclohexylation of toluene with cyclohexyl bromide or chloride, but is a very reactive catalyst for the cyclohexylation of toluene with cyclohexyl fluoride (97).

This sharp difference of catalytic activity of boron trifluoride was connected with the comparatively high stability of the BF_4^- ion formed in complex formation in comparison with the BF_3Cl^- or BF_3Br^- ions. However, the isolation of stable tetrachloroborate and tetrabromoborate complexes make this point questionable (214,261, 375,376,378,486,632,633).

The high reactivity of the C–F bond over C–Cl, C–Br, and C–I bonds in aliphatic compounds enables the preferential reaction of alkyl fluorides to occur in a system containing mixed halides (471).

Owing to their availability and low cost *alkyl chlorides* and *bromides* are most frequently used as alkylating agents. *Alkyl iodides* are less suitable owing to the fact that the alkylation reactions are generally accompanied by side reactions and decomposition.

Alkylation occurs most readily with tertiary halides and benzyl halides, less readily with secondary and still less readily with primary halides. In the series of primary halides methyl halides are the least reactive. It is noteworthy that a similar sequence of activity is also observed in alkylations with alcohols, ethers, and esters.

It is generally necessary to use increasingly vigorous catalysts to introduce the alkyl groups in the above sequence. For example reactive halides like benzyl chloride will react with benzene in the presence of a weak catalyst such as zinc chloride, whereas a more inert halide like methyl chloride requires a considerably more powerful catalyst such as aluminum chloride.

The relative reactivity of the alkyl halides is also governed by the halogen atom. For aluminum chloride catalyzed alkylations with either *n*-butyl or *t*-butyl halides, the order of activity is fluorine > chlorine > bromine > iodine (107).

The same order of reactivity has been found for hydrogen fluoride catalyzed alkylation of benzene with cyclohexyl and secondary octyl halides (569). The failure of the first attempts to use boron fluoride as a catalyst for the Friedel-Crafts alkylation with alkyl halides other than fluorides led to the assumption that in the presence of BF_3 it is not possible to obtain alkylation of aromatics with alkyl halides (649).

However, Hennion (259) showed that if BF_3 is used with a small quantity of water, alcohol, or other polar compounds (co-catalyst) alkylation of aromatic hydrocarbons by alkyl chlorides or bromides proceeds quickly and with a high yield of alkyl substituted aromatics. The reaction was first studied on benzene and toluene. It was shown that benzene and toluene are most easily alkylated by tertiary halides and benzyl chloride. Secondary alkyl halides give a comparatively low yield of alkylated products but this may be increased if sulfuric acid is added to aqueous boron trifluoride. Normal primary alkyl chlorides and bromides at atmospheric pressure hardly alkylate aromatic hydrocarbons in the presence of boron trifluoride.

In alkylation reactions with alkyl halides the catalyst is required only in minute amounts. The same is the case in alkylations with olefins, but with alcohols and related alkylating agents (esters, ethers, etc.) considerably larger quantities of catalysts are necessary, because of the complexing and reaction of the catalyst (aluminum chloride, etc.) with the alcohols (465).

Although boron trifluoride or hydrogen fluoride will catalyze alkylations by alkyl halides, these catalysts are much more effective and useful with olefins and alcohols. Reactions carried out with either of these catalysts are distinguished by the lack of colored and resinous by-products which so often accompany the use of aluminum chloride. Besides aluminum chloride, the catalysts most frequently used are ferric chloride, antimony pentachloride, boron trifluoride, zinc chloride, titanium tetrachloride, HF, H_2SO_4, H_3PO_4, P_2O_5, etc.

The ease of Friedel-Crafts alkylation also varies with the structure of the aromatic nucleus undergoing the reaction. Hydrocarbons, aryl halides, phenols, and aryl ethers are the substances most frequently alkylated. Aromatic heterocycles vary greatly in reactivity. Pyrrole, furan, thiophene, and pyridine react in descending order of reactivity in Friedel-Crafts alkylations. Aminofurans are less reactive than aromatic amines. Attempts to use the Friedel-Crafts reactions with basic nitrogen compounds have met with little success, presumably because the nitrogen atom coordinates with the catalyst. It may be supposed that the failure of quinoline to undergo alkylation is due to the effect on the benzenoid ring by the ammonium group in the complex formed. Pyridine also generally resists alkylation.

Di- and polyhalides find widespread application in Friedel-Crafts alkylations. However, in general the reaction cannot be effected selectively, e.g., to affect only one of the halogen atoms present. Under the catalytic effect of aluminum chloride and related Friedel-Crafts catalysts both (or all) halogens undergo subsequent or simultaneous Friedel-Crafts reactions which results in polycondensed or cyclized products.

The reaction of benzene with methylene chloride in the presence of aluminum chloride affords diphenylmethane (231), through the intermediate formation of benzyl chloride.

$$C_6H_6 + CH_2Cl_2 \xrightarrow[-HCl]{AlCl_3} [C_6H_5CH_2Cl] \xrightarrow[-HCl]{C_6H_6, AlCl_3} C_6H_5CH_2C_6H_5$$

Similarly chloroform reacts with benzene in the presence of aluminum chloride to give triphenylmethane. Triphenylmethane is usually prepared, however, from carbon tetrachloride rather than from chloroform (462).

Treatment of benzene and carbon tetrachloride with aluminum chloride leads to the formation of a complex

$$3C_6H_6 + CCl_4 + AlCl_3 \rightarrow (C_6H_5)_3CCl \cdot AlCl_3 + 3HCl$$

The cautious addition of anhydrous ether to this complex effects reduction and formation of triphenyl methane.

$$(C_6H_5)_3CCl \cdot AlCl_3 + (C_2H_5)_2O \rightarrow (C_6H_5)_3CH + CH_3CHO + C_2H_5Cl + AlCl_3$$

Hydrolysis of the complex leads to the formation of triphenylmethyl chloride in 85% yield.

It is interesting to note, however, (as will be discussed in more detail in Chapter VI), that dihalides can be made to react selectively, affecting only one of the functional halogen atoms, if the two halogens are different and catalysts allowing selective reaction are used. For example, 1-fluoro-2-chloroethane reacts with benzene in the presence of boron tribromide as catalyst to give selectively 2-chloroethylbenzene (471).

$$C_6H_6 + FCH_2CH_2Cl \xrightarrow{BBr_3} C_6H_5CH_2CH_2Cl$$

B. Alkenes

Alkylation of aromatic hydrocarbons with olefins in the presence of aluminum chloride as catalyst was first carried out by Balsohn (37) in 1879.

Olefins are industrially the most frequently used alkylating agents because of their cheapness and ready availability. The olefins most frequently employed are ethylene, propylene, butylene, amylene, etc. Tri-, tetra-, and pentamers of propylene, such as dodecylene, have been frequently used in alkylations in recent years to prepare alkylates (dodecyl benzene) as starting materials for detergents. The olefins employed are easily and inexpensively available directly from petroleum cracking. In general, the olefins of higher molecular weight (e.g., amylene) and with branched chains (e.g., t-butyl derivatives) react more readily than do propylene and ethylene. As olefins tend to polymerize in the presence of acid catalysts they are often employed with an excess of aromatic to suppress polymerization. Most Lewis and Brønsted-Lowry acids are active as catalysts in alkylations with olefins. In the case of Lewis acid catalysts, such as aluminum chloride or boron trifluoride, a co-catalyst is usually needed to achieve alkylation. Under absolute anhydrous conditions no alkylation of aromatics with olefins takes place in the absence of co-catalyst. Catalysts are needed only in small amounts in olefin alkylation. Ethylation of benzene with ethylene has been studied very extensively, particularly in industry, ethylbenzene being used in the manufacture of styrene.

Many methods have been devised for controlling the yield (188). Benzene and other aromatics can be alkylated by olefins in the presence of sulfuric or phosphoric acids as catalyst, as shown first by Ipatieff (300). Thus dodecene, nonene, octene, pentene, butene, and propylene, as well as ethylene have been employed for the alkylation of benzene. Accompanying the alkylation some esterification occurs with the formation of dialkyl esters of sulfuric acid, particularly when weaker sulfuric acid is used. In the presence of these esters the reaction product cannot be distilled satisfactorily, since they decompose and form tarry products at about 120–130°. It was found that the presence of phosphoric acid causes decomposition of such esters, so that distillation can then be effected without tar formation (298).

Aluminum chloride is widely applied in many alkylations with olefins. Benzene, toluene, and xylenes are alkylated by olefins and cyclopropane in the presence of aluminum chloride (51). Usually hydrogen chloride or an alkyl chloride is used as promoter.

It must be emphasized that the action of aluminum chloride is quite varied on the product to be formed, because polymerization, isomerization, and dealkylation can occur concurrently. Although it is not so effective as an initial alkylating agent as sulfuric acid, aluminum chloride is, however, a more effective agent to de- and transalkylate polyalkyl aromatics formed as unwanted by-products in the formation of alkyl aromatics.

Hydrogen fluoride is becoming an equally or more widely used alkylating catalyst than sulfuric acid or aluminum chloride. With hydrogen fluoride considerably less tar formation is generally observed and another advantage is the easy recovery of the volatile catalyst, which cannot be achieved either with sulfuric acid or aluminum chloride. Other mineral acids such as phosphoric and hydrochloric acid are also widely employed. In the class of Lewis acid catalysts besides aluminum chloride, ferric chloride, boron fluoride, and other metal halides are the most frequently used.

With a wide variety of catalysts and alkylating agents to choose from, a combination of catalyst and alkylating agent will depend upon relative cost, as well as upon the product composition and the reaction rate which results from their use.

In dealkylation–realkylation (re-forming) as Ipatieff (297) and other investigators have pointed out (162,459,508) certain catalysts, especially aluminum and ferric chloride, facilitate the removal of an alkyl group from a polyalkylated product. An industrial example is the re-forming of the polyalkylated derivatives separated from the high-boiling fractions in the isopropylations of benzene to cumene.

Benzene homologs, polynuclear hydrocarbons, phenols, and phenol ethers generally react more readily than benzene in alkylation with olefins (447). In the reaction of higher molecular weight olefins with benzene in the presence of aluminum chloride cleavage of the olefin may precede alkylation. Thus benzene reacting with di-isobutylene gives *t*-butylbenzene and di-*t*-butylbenzene, following cleavage of the di-isobutylene (308).

Hydroxyfluoboric acid was found to be a particularly effective catalyst in these so-called depolyalkylations (380). Other catalysts, however, tend to promote intact alkylation.

Phenols and naphthols react readily with olefins in the presence of aluminum chloride. The reaction is quite general and proceeds more readily with olefins with three or more carbon atoms (390).

Aluminum powder itself was found to be a very active catalyst for phenol alkylations (239). It is assumed that aluminum and phenol first form aluminum phenolate which is the active catalyst in the reaction. This type of alkylation generally gives a high yield of the *ortho*-isomer. Alkylation of phenol ethers proceeds well by reaction with olefins in the presence of aluminum chloride. *meta*-Cresol methyl ether, for example, with isobutylene gives *t*-butyl-*meta*-cresol methyl ether (656).

Phenol is alkylated by olefins such as amylenes in the presence of cation-exchange resins (316).

Dowex-50 resin (sulfonated styrene–divinylbenzene copolymer) has been found to be a very effective catalyst in alkylation of phenol, catechol, and other oxybenzenes with olefins such as proplyene, 1-butene, isobutylene (471).

C. Alkynes

Acetylene can be used as the alkylating agent for aromatics in the presence of a number of catalysts such as $AlCl_3$, $GaCl_3$, BF_3, $HgSO_4$.

The reaction of acetylene with benzene and alkylbenzenes in the

presence of sulfuric acid, phosphoric acid, or aluminum chloride as catalyst generally gives 1,1-diphenylethanes (158).

$$2C_6H_6 + CH{\equiv}CH \rightarrow CH_3{-}CH(C_6H_5)_2$$

Usually no more than traces of styrene are found in the reaction of benzene with acetylene, although it was claimed that $AlCl_3$ containing 2–5% $FeCl_3$ improves the yield of styrene (159).

This claim, however, could not be substantiated and no selective catalyst has yet been reported to achieve the reaction

$$C_6H_6 + CH{\equiv}CH \rightarrow C_6H_5{-}CH{=}CH_2$$

without causing either immediate polymerization of styrene or further alkylation to 1,1-diphenylethane,

$$C_6H_5{-}CH{=}CH_2 + C_6H_6 \rightarrow (C_6H_5)_2CH{-}CH_3$$

The usual Friedel-Crafts alkylations with acetylene as alkylating agent give complex mixtures, probably because of the polymerizing action of the more vigorous catalysts on the intermediate styrenes.

D. Alcohols

Alkylations of aromatic hydrocarbons by alcohols can be carried out with primary, secondary, or tertiary alcohols with the same catalysts that are used in the alkylation of aromatic hydrocarbons with alkyl halides or olefins.

Alcohols are more reactive in this respect than alkyl halides. In alkylation with alcohols considerable quantities of catalysts are needed (465).

Aluminum chloride (or related Lewis acid catalysts) first complexes with the alcohol, the complex then may lose hydrogen chloride to give an intermediate compound (alkoxyaluminum dichloride) which by elimination of aluminum oxychloride may yield alkyl halides.

$$C_2H_5OH + AlCl_3 \longrightarrow [C_2H_5OH \cdot AlCl_3] \longrightarrow C_2H_5OAlCl_2 + HCl$$
$$C_2H_5Cl + AlOCl$$

Although it is not necessary that the interaction of the Lewis acid catalyst with the alcohol should continue until the formation of the alkyl halide, an equimolar quantity of catalyst is required to form the primary complex. That, indeed, the necessary step is not alkyl halide formation in the interaction of alcohol with the Lewis acid halides is clearly demonstrated in the case of boron trifluoride,

which is a very effective alkylating catalyst for alcohols and which only forms addition compounds without being able to transform the alcohol into the corresponding alkyl fluoride.

Sulfuric acid is also a widely used catalyst in alkylation with alcohols. In this case the alcohol is first esterified by sulfuric acid and the corresponding alkyl sulfate then reacts with the aromatic compound (233).

Benzene, toluene, and xylenes are alkylated by propyl and butyl alcohol in the presence of 70–80% H_2SO_4 at a temperature of 40–80° (422). To obtain good yields of alkylbenzenes ten volumes of acid are required for one volume of hydrocarbon. Hydrogen fluoride has a high catalytic activity in the reaction, especially in the alkylation of benzene with secondary and tertiary alcohols (568).

Tsukervanik and colleagues (438,620,621) have carried out a thorough investigation of alkylation of alkylbenzenes with alcohols in the presence of a number of catalysts. Aluminum chloride and ferric chloride were the catalysts of preference. A major disadvantage is the necessity of using large amounts of the catalysts in order to obtain good yields of alkylbenzenes. The search for more active catalysts has therefore not been abandoned even at the present time. Boron fluoride has definite advantages as catalyst. McKenna and Sowa (410) studied the catalytic action of boron fluoride in the alkylation of benzene and its homologs by alcohols and showed that an equimolecular mixture of the reagent in the presence of BF_3 gives a mixture of mono-, di-, and polyalkylbenzenes on heating to 60° for six hours. The products of alkylation with alcohols are often different from those obtained with the corresponding alkyl halides since alcohols show less tendency to undergo rearrangements. Thus in alkylations with normal alcohols considerable amounts of n-alkyl derivatives are formed, whereas in the corresponding reactions with n-alkyl halides isomerization is predominant giving iso-, secondary, or tertiary products.

It was reported that dialkylbenzenes formed in the boron trifluoride catalyzed alkylations of alkylbenzenes with alcohols are mainly the *para*-isomers together with small amounts of the *ortho*-derivatives (257). However, this statement could not be verified using modern analytical methods (spectroscopy and gas liquid chromatography).

The ease of alkylation of benzene with alcohols occurs in relation to the property of the latter to be dehydrated. The alkylation is therefore appreciably accelerated by the addition of dehydrating agents such as P_2O_5, H_2SO_4, or $C_6H_5SO_3H$ (613).

Especially in alkylation by primary alcohols water is split off with the utmost difficulty in the absence of dehydrating agents. Topchiev and co-workers (613) studied the alkylation of benzene by propyl and butyl alcohols in the presence of $BF_3 \cdot H_3PO_4$ and showed that this catalyst is very active and enables alkylbenzenes to be obtained in yields of more than 80%. The main products of the reaction are monoalkylbenzenes. Using an excess of benzene the disubstituted products amount to less than 5%. When the relative quantity of alcohol is increased the yield of dialkylbenzene increases too. The most satisfactory ratio of alcohol:aromatic:$BF_3H_3PO_4$ is 0.5:1:0.5.

When benzene is alkylated with optically active d-s-butyl alcohol in the presence of BF_3, l-s-butylbenzene is obtained (519). Whether such a conversion takes place by means of optical inversion or racemization has not yet been established, but it may be observed that if this reaction is carried out in the presence of aluminum chloride dl-s-butylbenzene is formed almost exclusively.

The fact that boron trifluoride does not generally effect alkylations with alkyl chlorides in the absence of co-catalysts enabled Bachman and Hellman (20) to carry out a very interesting haloalkylation of benzene and its homologs with halohydrins. If halohydrins with a hydroxyl group at the secondary carbon and a halide at the primary carbon atom react with aromatics in the presence of boron trifluoride the reaction proceeds exclusively at the expense of the alcohol group with the formation of haloalkylbenzenes, with yields of 30–50%.

$$ArH + CH_3\text{—}\underset{\underset{OH}{|}}{CH}\text{—}CH_2Cl \xrightarrow[\text{(P}_2\text{O}_5,\text{H}_2\text{SO}_4)]{BF_3} Ar\text{—}\underset{\underset{CH_3}{|}}{CH}\text{—}CH_2Cl + H_2O$$

Using aluminum chloride as catalyst in alkylations with alcohols there is generally no rearrangement of the carbon chain during the reaction. Thus with n-propyl alcohol benzene yields n-propylbenzene instead of cumene, which is the chief product when an n-propyl halide is employed as alkylating agent. When one mole of propyl alcohol is allowed to react with an excess of benzene in the presence of 1.3 moles of aluminum chloride for 10 hours at 100°, n-propylbenzene is obtained in 52% yield. Di-n-propylbenzene is also obtained in a 37% yield (312,362,621).

If the alkylating agent and catalyst in a given reaction are very reactive and the aromatic compound is relatively inert, extensive degradation or polymerization of the alkylating agent may occur.

On the other hand if the aromatic compound is very reactive toward the catalyst and the alkylating agent is relatively inert, decomposition of the aromatic compound may take precedence over alkylation. Naphthalene, for example, reacts easily in the presence of aluminum chloride to form binaphthyl. For this reason methylation of naphthalene can be accomplished only with low yield.

Rearrangements occur when sulfuric acid (312,422) or boron fluoride is used as catalyst (410,613)

Under vigorous conditions changes more extensive than isomerization of the alkyl group can occur. Although benzene is alkylated normally in good yield with t-butyl alcohol and aluminum chloride at 30°, the products at 80–95° are toluene, ethylbenzene, and isopropylbenzene (465). Alkylation with 2,4,4- or 2,3,3-trimethyl-2-pentanol can proceed to yield both normal and degraded alkylation products, the extent of degradation increasing with the reaction temperature (282).

E. Esters

Esters have not been as widely applied in alkylations by the Friedel-Crafts reaction as alcohols, chiefly because there is no particular advantage in their use. In fact, with esters of organic acids and aluminum chloride as catalyst, a serious disadvantage is the simultaneous acylation which may occur. When alkyl esters of organic acids are used as alkylating agents the product usually consists of a mixture of alkylated and acylated hydrocarbons (78,374,463).

Hydrogen fluoride is one of the most suitable catalysts for the alkylation of aromatic compounds with esters. A 75% yield of diphenylmethane is obtained by treating benzene with benzyl acetate in the presence of hydrogen fluoride.

$$CH_3COOCH_2C_6H_5 + C_6H_6 \xrightarrow{\text{HF}} C_6H_5CH_2C_6H_5 + CH_3COOH$$

Benzene and its homologs can be alkylated by esters of organic and mineral acids in the presence of boron trifluoride (411). Depending on the ester taken, the reaction proceeds at room temperature or above. n- and s-Butyl esters of formic and acetic acid with benzene give s-butyl benzene. Isobutyl formate gives tertiary butylbenzene. n-Alkyl esters react with more difficulty than esters having groups of branched structure, which agrees with their smaller tendency to dissociate into olefin and acid. The dialkylbenzenes formed in the alkylations were reported to be mainly the p- or o-isomers. Vinyl acetate with benzene forms a mixture of resinous products which is difficult to identify. Acetophenone, *p*-ethylacetophenone, 1,1-diphenylethane and 9,10-dimethylanthracene were identified from the complex products (489).

Alkyl esters of inorganic acids like alkyl sulfates, sulfites, phosphates, alkyl orthosilicates $((RO)_4Si)$, alkyl carbonates (R_2CO_3), alkyl borates, alkyl chloroformates, hypochlorites, chlorosulfonates, and chlorosulfites can be successfully applied in Friedel-Crafts alkylation of aromatics (333).

Alkyl arenesulfonates, particularly alkyl tosylates, were also found to be very active alkylating agents.

F. Ethers

Certain types of ethers can be used to alkylate aromatic compounds. Diethyl ether (209), diisopropyl, and di-*n*-butyl ethers (399), react with benzene and its homologs to give alkylated products.

The condensation of ethyl benzyl ether with benzene and aluminum chloride gives a mixture of ethylbenzene, diphenylmethane, and *m*- and *p*-dibenzylbenzene (337).

When alkyl phenol ethers are heated with aluminum chloride the alkyl group moves to the ring and alkyl-substituted phenols are formed (245).

Ethylene oxide, a cyclic ether, has been condensed with aromatic compounds to yield the corresponding β-arylethyl alcohols. The condensation of ethylene oxide and benzene to β-phenylethyl alcohol, a synthetic ingredient of rose perfume, is catalyzed by aluminum chloride (545).

$$C_6H_6 + \begin{array}{c} CH_2 \\ | \\ CH_2 \end{array}\!\!\!>\!O \xrightarrow{\text{AlCl}_3} C_6H_5CH_2CH_2OH$$

Ethylene, propylene, and butylene oxides also react with aromatic hydrocarbons in the presence of BF_3 to form aromatic alcohols. Benzene gives β-phenylethyl alcohol with ethylene oxide but is partly also converted into dibenzyl (282,289,291).

The possibility of alkylating aromatic hydrocarbons with ethers in the presence of boron fluoride arose from the property of these ethers of forming coordination compounds with BF_3, which upon heating split up into olefins and alcohols (566).

The alkylation of benzene with diethyl, diisopropyl, diisoamyl isopropyl, phenyl, ethyl, *n*-propyl, benzyl, and dibenzyl ethers has been studied (428). In carrying out the reactions it is recommended to use 2 moles benzene, 0.5–1 mole BF_3, and 1 mole ether (470).

As a result of the alkylation mono-, di-, and polyalkylbenzenes are obtained. The dialkylbenzenes reported are mainly *para*-isomers with small quantities of *ortho*-isomers. Isopropyl and benzyl ethers are particularly vigorous alkylating agents. With *n*-propyl benzyl ether diphenylmethane is formed.

The fact that boron trifluoride is inactive in attempted alkylations with ethers in absolute anhydrous media was shown by Burwell and Elkin (98). They studied the alkylation of benzene with *s*-butyl methyl ether in the presence of BF_3 in a ratio of $7:1:1$ and showed that if the reagents and catalyst were well dried and placed into a reaction vessel excluding the entry of moisture from the atmospheres, no reaction would occur during reaction times as long as three weeks. But it is sufficient to add to the reaction mixture a small amount of water to start the immediate process of alkylation, with the exclusive formation of *s*-butylbenzene. Toluene was not observed to be formed even in traces. Sulfuric acid, chlorosulfonic acid, and ethanesulfonic acid are more effective activators (cocatalysts) of boron fluoride in this reaction than water. Thus the coordination compound

$$s-C_4H_9\diagdown \atop CH_3 \diagup O \longrightarrow BF_3$$

is probably very stable at room temperature and in order to split it into olefin and carbonium ion respectively the presence of compounds forming even stronger complexes with BF_3 is necessary.

G. Aldehydes and Ketones

Aldehydes and ketones also serve as alkylating agents in Friedel-Crafts type alkylations. Tertiary or secondary alcohols are probable intermediates but are usually too reactive to be isolated.

$$\text{ArH} + \text{RCHO} \xrightarrow{\text{Friedel-Crafts Catalyst}} [\text{ArCH(OH)R}] \xrightarrow{\text{ArH}} \text{Ar}_2\text{CHR}$$
$$\text{ArH} + \text{R}_2\text{CO} \xrightarrow{\hspace{2cm}} [\text{ArCR}_2\text{OH}] \xrightarrow[-\text{H}_2\text{O}]{} \text{ArCR}_2\text{Ar}$$

Benzaldehyde reacts with benzene to give triphenylmethane (545)

$$C_6H_5CHO + C_6H_6 \xrightarrow{H_2SO_4} [C_6H_5CHOHC_6H_5] \xrightarrow[-H_2O]{C_6H_6} (C_6H_5)_3CH$$

and formaldehyde to give diphenylmethane. This type of reaction is used to prepare the important insecticide 2,2-di(chlorophenyl)-1,1,1-trichlorethane (DDT), from chloral and chlorobenzene (138).

$$CCl_3CHO + 2C_6H_5Cl \xrightarrow[ClSO_3H]{H_2SO_4} (p\text{-}ClC_6H_4)_2CHCCl_3 + H_2O$$

Glyoxals, RCOCHO, possess unusually reactive aldehyde groups and may be condensed with aromatic substances in the presence of aluminum chloride under unusually mild reaction conditions with formation of benzoins in good yields (288).

$$RCOCHO + C_6H_6 \xrightarrow[0°]{AlCl_3} RCOCHC_6H_5$$
$$\overset{|}{OH}$$

When formaldehyde is used as the alkylating agent in the presence of hydrogen chloride the reaction results in introduction of a chloromethyl group (207).

$$ArH + C_2HO + HCl \xrightarrow{ZnCl_2} ArCH_2Cl + H_2O$$

A number of experimental procedures have been used to effect chloromethylation. The most successful is that developed by Blanc (60) which involves treatment of the aromatic substance with formalin or paraformaldehyde and zinc chloride, followed by saturation of the reaction mixture with hydrogen chloride. Other procedures use acetic acid, sulfuric acid, or stannic chloride as catalysts (480). Chloromethyl ether may be used in place of formaldehyde and hydrogen chloride as the chloromethylating agent. Bromomethylations can be effected under similar conditions. Chloro- and bromoethylation as well as chloropropylation and chlorobutylations have also been effected in fairly good yield with the appropriate aldehydes. In addition, in chloromethylation and related chloroalkylation reactions secondary alkylation nearly always takes place, particularly in the presence of stronger catalysts, yielding diarylmethanes. Other diarylalkanes may be similarly formed. The secondary reaction is particularly predominant in fluoromethylation which generally results in a high yield of diphenylmethane.

Acetone condenses with reactive aromatic compounds yielding derivatives of the type $Ar_2C(CH_3)_2$. Thus with phenol it forms 2,2-di-(p-hydroxyphenyl)propane (224).

$$2C_6H_5OH + CH_3COCH_3 \xrightarrow[-H_2O]{H_2SO_4} HOC_6H_4\overset{\overset{\displaystyle CH_3}{|}}{\underset{\underset{\displaystyle CH_3}{|}}{C}}C_6H_4OH$$

H. Alkanes and Cycloalkanes

The unique behavior of aluminum chloride and similar strong acid catalysts in isomerizations, dehydrogenations, and cracking of saturated hydrocarbons makes it possible to use paraffins as alkylating agents. Following the theory of Ipatieff and Grosse (304) the primary reaction is the splitting of the paraffin into an olefin and a lower molecular weight paraffin.

$$C_nH_{2n+2} \rightarrow C_{n-z}H_{2(n-z)+2} + C_zH_{2z}$$

The olefin so formed can then act as an alkylating agent. As an example, benzene is alkylated by octane in the presence of aluminum chloride (302) to give butylbenzenes.

$$C_8H_{18} + C_6H_6 \xrightarrow{\text{AlCl}_3} C_4H_9C_6H_5 + C_4H_{10}$$

Treatment of paraffinic hydrocarbons with benzene in the presence of aluminum chloride leads to the formation of various alkyl benzenes by degradation of the paraffins, a reaction which has been termed "destructive alkylation" (232). Most probably the reaction starts according to the general scheme by primary dehydrogenation (*via* hydride abstraction) of octane to octene and depolymerization to two molecules of butene, which then act as Friedel-Crafts alkylating agents.

The alkylation of benzene with cyclopropane as alkylating agent gives a 65% yield of n-propylbenzene.

$$C_6H_6 + \underset{\underset{\displaystyle CH_2}{}}{\overset{\displaystyle CH_2-CH_2}{}} \xrightarrow{\text{AlCl}_3} C_6H_5CH_2CH_2CH_3$$

Aromatic hydrocarbons are also easily alkylated by isoparaffins in the presence of BF_3 and tertiary alkyl fluorides (342). Thus methylcyclohexane and benzene in the presence of t-butyl fluoride and boron trifluoride give methylcyclohexylbenzene and t-butylbenzene. With higher isoparaffins, aromatic hydrocarbons in the presence of HF and BF_3 undergo destructive alkylation, similar to the destructive alkylation of isobutane (47). For example benzene is alkylated by 2,2,4-trimethylpentane in the presence of HF with an addition of 5% BF_3 at temperature of 400° and increased pressure (90 atm.) with the formation of 65% yield of t-butylbenzene. BF_3 alone does not catalyze this reaction, according to the results of Ipatieff and Grosse (230), whereas in the presence of aluminum chloride and zinc chloride the reaction proceeds readily at 25–70° and atmospheric pressure.

Aromatic hydrocarbons are also alkylated by normal paraffins containing not less than five carbon atoms in the molecule, in the presence of HF and 2–20% weight of BF_3 at temperatures of 180–400° and under pressure (388).

I. Lactones

The reaction of benzene and derivatives with lactones of aliphatic monocarboxylic acids generally results in alkylation of the aromatics yielding aralkanoic acids.

The reaction of benzene and derivatives with γ-lactones usually results in an alkylation to yield γ-arylbutyric acids (430).

butyrolactone γ-phenylbutyric acid

When an excess of aluminum chloride catalyst is used, γ,γ-dimethylbutyrolactone and γ-butyrolactone with benzene yield 4,4-dimethyl-1-tetralone and 1-tetralone respectively (11,617). It was proposed that these reactions take place by opening of the lactone ring with resulting alkylation of benzene followed by cyclization *via* acylation.

However, the alternative possibility that lactones can cleave first to acylate aromatics followed by a cycloalkylation must also be considered (92).

The application of lactones in the Friedel-Crafts reaction has also been extended to the synthesis of 1-indanone by treatment of benzene with β-propiolactone (527).

J. Mercaptans

Mercaptans can be used successfully as alkylating agents of aromatics in a manner similar to alcohols.

$$C_6H_6 + RSH \xrightarrow[BF_3]{AlCl_3} C_6H_5R + H_2S$$

Instead of water H_2S is eliminated as by-product, which owing to its low boiling point escapes from the reaction mixture. Consequently less catalyst is needed than in the corresponding alkylations with alcohols.

The relative order of reactivity of mercaptans in alkylations is tertiary > secondary > primary.

Tertiary mercaptans, such as t-butyl mercaptan, alkylate benzene relatively easily (413).

$$C_6H_6 + t\text{-}C_4H_9SH \xrightarrow[\text{BF}_3]{\text{AlCl}_3} t\text{-}C_4H_9C_6H_5 + H_2S$$

Benzyl mercaptan converts benzene into diphenylmethane (381).

However, primary and secondary alkyl mercaptans alkylate aromatics only with difficulty.

K. Sulfides (Thioethers)

Dialkyl sulfides in the presence of Friedel-Crafts catalysts, such as aluminum chloride, yield alkylated products with aromatics. Both alkyl groups of the sulfide can take part in the reaction.

$$2C_6H_6 + RSR \xrightarrow{\text{AlCl}_3} 2C_6H_5R + H_2S$$

As hydrogen sulfide is the volatile by-product smaller amounts of catalyst are again needed than in the corresponding alkylations with ethers (471). Di-t-butyl and dibenzyl sulfide are the most reactive, while simple di-n-alkyl sulfides react less readily.

Silica–alumina type acid catalysts are also effective in carrying out the reactions (178).

$$(t\text{-}C_4H_9)_2S + 2C_6H_6 \to 2t\text{-}C_4H_9C_6H_5 + H_2S$$

L. Thiocyanates

Alkyl thiocyanates, RSCN, behave like alkyl pseudo halides and are effective alkylating agents in the presence of Friedel-Crafts catalysts.

$$C_6H_6 + RSCN \xrightarrow{\text{AlCl}_3} C_6H_5R + HSCN$$

Normal alkyl thiocyanates in alkylation reactions give an unusually high proportion of n-alkylated products instead of the isomerized side chain obtained with n-alkyl halides (471).

M. Sulfenyl Chlorides and Sultones

Alkyl sulfenyl chlorides react with aromatic hydrocarbons in the presence of aluminum chloride catalyst to give thioethers with good yields (83).

$$ArH + ClSR \xrightarrow[-\text{HCl}]{\text{AlCl}_3} Ar\text{—}S\text{—}R$$

(Ar = benzene, toluene, naphthalene, anthracene, xylenes, mesitylene, durene; R = CH_3, C_2H_5).

Sultones can also be effectively applied in the Friedel-Crafts alkylation reactions with aromatic hydrocarbons, in the presence of aluminum chloride, to give good yields of aralkane sulfonates. In the case of 4-hydroxyl-1-butanesulfonic acid sultone 4-aryl-1-alkanesulfonates are formed without rearrangement of the entering chain. With toluene and anisole mixtures of substituted isomers are formed (616).

$$\text{ArH} + \underset{\underset{\text{O}\rule{2.5cm}{0.4pt}\text{SO}_2}{|}}{\overset{\overset{\text{R}}{|}}{\text{HC}}}\text{—CH}_2\text{—CH}_2\text{—CH}_2 \quad \xrightarrow[\text{2., Na}_2\text{CO}_3]{\text{1., AlCl}_3} \quad \text{Ar—}\underset{\underset{\text{R}}{|}}{\overset{\overset{\text{H}}{|}}{\text{C}}}\text{—CH}_2\text{—CH}_2\text{—CH}_2\text{SO}_3\text{Na}$$

N. Deuterium Halides (Aromatic Hydrogen Exchange)

Aromatic hydrocarbons undergo ring deuteration with deuterium halides in the presence of Friedel-Crafts catalysts (like aluminum chloride, zinc chloride, stannic chloride, boron trifluoride, etc.) (343,348).

$$\text{Ar—H} + \text{DCl} \xrightarrow{\text{AlCl}_3} \text{Ar—D} + \text{HCl}$$

Deuterium iodide, bromide, chloride, and fluoride all can be applied in the exchange reaction.

Strong deuteron acids catalyze the reaction equally well and can act both as the deuterium source and catalysts. D_2SO_4, DF, and CF_3CO_2D are the most frequently used.

That aromatic ring deuteration is indeed a typical electrophilic or for this reason Friedel-Crafts type reaction was first established by Ingold (294,295). In more recent years the availability of tritium extended the scope of aromatic hydrogen exchange to ring tritiation.

As a Friedel-Crafts reaction aromatic hydrogen exchange can be classified as an alkylation type reaction with deuterium halides or other deuteron acids (as compared with alkylation with alkyl halides or esters).

$$\text{ArH} + \text{DA} \rightarrow \text{ArD} + \text{HA}$$

Hydrogen isotopes are only means to prove the general phenomena of protium exchange, which can be demonstrated in acid catalyzed hydrogen exchange of deuterated (tritiated) aromatics with proton acids.

3. Cyclialkylation of Aromatic Compounds

Friedel-Crafts alkylation of aromatics with difunctional compounds such as dihalides, unsaturated halides, diols, unsaturated alcohols, diolefins, etc., generally leads partly to open-chain alkylates, but, if conditions are suitable, mainly to cyclialkylation products. These compounds react at each end of the difunctional open chain to attach a new cycle to the aromatic nucleus. It is most probable that such difunctional compounds first form a monoalkyl derivative with the aromatic compound to yield an aryl-substituted monofunctional compound. The latter then undergoes an intramolecular alkylation forming the new ring.

The term "cyclialkylation" was first used by Bruson and Kroeger (95).

Cyclialkylations can be divided into intermolecular reactions (involving a bifunctional alkylating agent and aromatics) and intramolecular reactions (involving ring closure of monofunctional aralkyl compounds in Friedel-Crafts type reactions).

A. Intermolecular Cyclialkylations

a. Dihalides

Ring formation readily occurs in alkylation of aromatics with di- and polyhalides. Of particular interest is the reaction of di- and trihalomethanes with aromatics in the presence of aluminum chloride. In the reaction of dichloromethane with benzene besides diarylmethanes, anthracene derivatives are also formed (379).

$$CH_2Cl_2 + CH_2(C_6H_5)_2 \xrightarrow{AlCl_3} C_6H_4 \begin{array}{c} CH_2 \\ \diagup \quad \diagdown \\ \diagdown \quad \diagup \\ CH_2 \end{array} C_6H_4 + 2HCl$$

Anthracene is obtained through the oxidation of dihydroanthracene in the course of the reaction. Tri- and tetraphenylmethanes result from the reaction of chloroform and carbon tetrachloride with benzene. Tetraphenylethylene is also obtained with the former. Another example of ring formation using dihalides in Friedel-Crafts alkylations is the reaction of benzene with 2,5-dichlorohexane

Intermolecular cyclialkylations of the Friedel-Crafts type have been widely employed in the synthesis of polycyclic structures.

Cyclialkylation of aromatic compounds with 1,4-dihalides in the presence of aluminum chloride provides a convenient method of synthesizing hydroaromatic ring systems. Thus the reaction of benzene with 1,4-dichlorobutane yields tetralin, octahydroanthracene, octahydrophenanthrene, and dodecahydrotriphenylene in varying amounts depending on the conditions used (524).

Treatment of benzene with o-xylylene bromide in the presence of aluminum chloride yields anthracene by cyclialkylation followed by dehydrogenation (574).

b. Diols

Cyclialkylation of aromatic compounds with 1,4-glycols or 1,3-glycols in the presence of aluminum chloride leads to tetralin or indane derivatives respectively (619).

In general comparatively large proportions of catalyst are required to complete the ring closure and the results are less satisfactory than with the corresponding dihalides. When sulfuric acid or boron trifluoride are used as the catalyst, 1,4-glycols (such as 2,5-dimethyl-2,5-hexanediol) yield indane derivatives from rearrangement accompanying cyclialkylation (95).

c. Diolefins

Dienes can also be used in Friedel-Crafts cyclialkylations. For example treatment of phenol with 2,5-dimethyl-1,4-hexadiene gives 5,5,8,8-tetramethyl-5,6,7,8,-tetrahydro-2-naphthol (95).

B. Intramolecular Cyclialkylation

a. Aralkyl halides

Intramolecular alkylation of haloalkyl benzenes is perhaps the simplest intramolecular counterpart of the classical Friedel-Crafts alkylation reaction. The earlier work in this field has been reviewed by Thomas (607). Baddeley and Williamson (31) have compared the inter- and intramolecular alkylation processes and concluded that generalizations applied to intermolecular alkylation of benzenes do not apply to corresponding intramolecular processes. The latter, being comparatively fast, are not usually accompanied by isomerizations which are common in the intermolecular processes. Primary, secondary, or tertiary phenylalkyl chlorides thus undergo ring closure with isomerization.

Similarly, since a carbonyl group promotes ring closure, phenyl haloalkyl ketones will readily enter into intramolecular alkylation.

On the other hand when the reactivity of the aromatic ring has been decreased, as in the case of deactivating substituents, the ring closures require stronger conditions and give rearranged products.

The cyclization of α-bromoaralkyl ketones to indanone is an interesting variation of ring closures of phenyl haloalkyl ketones. These reactions are considered to take place through an α,β-unsaturated ketone as an intermediate. The fact that the α,β-unsaturated ketones will also yield the same indanones supports this view.

2,5,7-trimethyl-1-indanone

b. Aralkenes

When an olefinic double bond is present in an aromatic side chain Friedel-Crafts alkylation results in ring closure and provides a convenient synthesis of condensed ring systems (49).

as-phenyl-o-biphenylethylene   9-phenyl-9,10-dihydrophenanthrene

The treatment of styrene (or substituted styrenes) with acid catalysts produces varying amounts of a saturated cyclic dimer, which was found to be 1-methyl-3-phenylindane. The reaction is now realized to be a cyclialkylation in which the aromatic hydrocarbon containing a conjugated double bond is protonated in acid medium to form the styryl (methyl phenyl carbonium) ion. This ion can then alkylate an added olefin (second styrene molecule) to produce an intermediate with a carbonium ion on an extended side chain.

Intramolecular alkylation of the latter results in the formation of the indane ring system (528,586).

c. Aralkanes

Alkylbenzenes containing an α-tertiary hydrogen, such as in isopropyl and s-butyl side chains, will rapidly transfer this hydrogen as a hydride ion. Such a hydride-ion abstraction produces an α-carbonium ion which is exactly analogous to that formed in the protonation of styrene compounds. Subsequent alkylation of an olefin followed by ring closure produces indanes by much the same process.

Cyclialkylations initiated by hydride-ion transfer were first reported by Ipatieff and Pines (311). They found that p-cymene, on treatment with a branched olefin such as methylcyclohexene, or trimethylethylene in the presence of an acid catalyst, yielded 1,3,3,6-tetramethyl-1-p-tolylindane.

d. Arylaliphatic alcohols

Friedel-Crafts intramolecular alkylation of aromatic alcohols is a widely used method for the preparation of polycyclic hydroaromatic compounds.

Cyclizations of phenylalkyl alcohols to tetralins and indanes have been extensively investigated (530).

CH$_2$CH(OH)CH$_2$C(CH$_3$)CH$_3$ ⟶ (structure with CH$_3$ CH$_3$)

CH$_2$CH(OH)CH$_2$CH$_2$CH$_3$ ⟶ (structure with CH$_3$)

e. Aralkyl carbonyl compounds

Intramolecular alkylation with β-aralkyl carbonyl compounds yields indene derivatives in high yields and there have been many applications of this reaction (121).

$$\underset{CH_2}{\overset{CH_3}{\underset{}{\overset{|}{\underset{}{C=O}}}}} \quad \xrightarrow{AlCl_3} \quad \overset{CH_3}{\underset{CH_3}{}}$$

The cyclodehydration of γ-phenyl ketones provides a convenient preparation of 1,2-dihydronaphthalene derivatives. The latter are convenient intermediates in the synthesis of compounds related to the sterols and in the synthesis of polycyclic aromatic compounds.

(structure C=O) ⟶ (structure)

The reaction is of wide scope and according to the general equation can lead to fully aromatic rings or fused aromatic heterocyclic ring systems (79).

(structures showing A / CH / O=C—R ⟶ A / CH / C—R / OH ⟶ (−H$_2$O) A / R)

(R = alkyl, aryl, or H; A = —C≡C, —O—, —S—, —NH—, —N=C—, —C=N—.)

Some of the heterocyclic systems readily obtainable by this method are thianaphthanes, benzofurans, quinolines, isoquinolines, and indoles.

Cyclodehydration of δ-aralkyl aldehydes and ketones yields benzo-suberene derivatives.   This reaction is particularly useful for syntheses in the colchicine field.   Polyphosphoric acid is the preferred catalyst for these cyclizations.

f. Phenylethylamides (Bischler-Napieralski and related reactions)

The cyclodehydration of acyl derivatives of β-phenylethylamines to 3,4-dihydroisoquinolines by heating with catalysts such as phosphorus pentoxide, phosphorus oxychloride, zinc chloride, aluminum chloride, and polyphosphoric acid is known as the Bischler-Napieralski reaction (59,638).

The Morgan-Walls application of the Bischler-Napieralski reaction leads to phenanthridines (606).

4. Arylations of Aromatic Compounds

In contrast to Friedel-Crafts alkylations the corresponding arylations of aromatics are by no means as well known or easy to carry out.   Three types of arylating agents are so far known to be suitable for the introduction of aryl groups into aromatic compounds in Friedel-Crafts type reactions:

(A) Dehydrogenating condensation of aromatics (Scholl reaction).
(B) Arylation with aryl halides.
(C) Arylation with diazonium halides.

A. Dehydrogenating Condensation (Scholl Reaction)

Dehydrogenating condensations of aryl nuclei have long been known to occur under the action of strong Friedel-Crafts catalysts. In fact, Friedel and Crafts themselves observed the formation of biphenyl from benzene and aluminum chloride.

The elimination of two ring hydrogens accompanied by the formation of an aryl–aryl bond under the influence of aluminum chloride or related Friedel-Crafts catalysts is generally known as the Scholl reaction.

The dehydrogenating condensations can take place either in an inter- or intramolecular way.

Intermolecular Scholl reactions are numerous and range from the formation of biphenyl from benzene,

the formation of perylene from naphthalene (through binaphthyl),

of 2,2'-dipyridyl from pyridine,

to high molecular weight polycondensate aromatics.

The Scholl type reactions generally need higher temperatures and strong acid catalysts. Consequently there are many limitations concerning functional groups and compounds which do not tolerate these conditions. In some cases very complex reaction mixtures are formed. In addition the structure of the products obtained is frequently uncertain because rearrangements or fissions may occur under the reaction conditions used. There is sometimes a possibility that several different condensations may take place.

The yields in intermolecular Scholl reactions are generally low (owing to possible side reactions and cracking).

The intramolecular Scholl reactions are basically cyclialkylations. As an example, 1,2-diphenylethane when heated with aluminum chloride in carbon disulfide gives phenanthrene

Triphenylmethane heated in benzene solution with aluminum chloride gives 9-phenylfluorene.

9-phenylfluorene

Of particular interest to the dyestuff industry is the cyclization of 1,1′-dianthrimide to the carbazole Indanthrene Yellow.

The activity of the catalysts (aluminum chloride) must be somewhat depressed in this case (as with complexing with nitrobenzene) since the presence of a secondary amino group enhances the reactivity.

In the intramolecular Scholl reactions six-membered rings are formed more easily than five-membered rings.

The Pummerer condensation of aromatics with quinones as well as the Anschütz condensation can be considered to be reactions of the Scholl type (dehydrogenating condensations).

B. Aryl Halides

The aromatic carbon–halogen bonds are generally considered to be too unreactive to enable aryl halides to be used as electrophilic arylating agents in Friedel-Crafts type reactions.

Although Friedel-Crafts hydrocarbon syntheses with aryl halides are not common, several instances dealing with reactions of this type have been reported.

Condensation of p-fluorophenol with benzene in the presence of aluminum chloride to give hydroxybiphenyls indicated that fluorine in the ring is more labile than either bromine or chlorine (no similar reactions with bromobenzene and chlorobenzene were achieved) (637).

Reactions of halogenated naphthalenes and benzene take place fairly smoothly. α-Chloronaphthalene reacts with benzene in the presence of aluminum chloride to give a 40% yield of α-phenyl naphthalene, together with some β-phenyl product. A similar reaction takes place with α-bromonaphthalene, but the yield is much lower. β-Chloro- and β-bromonaphthalene in similar reactions give α-phenylnaphthalene preferentially, the arylation reaction obviously taking place under isomerizing conditions leading to the more stable α-products (115).

Halogenated polynuclear hydrocarbons undergo self-condensation in certain cases under the influence of aluminum chloride. 9-Bromophenanthrene, in benzene solution, treated with aluminum chloride at room temperature yields 2,3,10,11-dibenzoperylene, the condensation occurring with intermediate bromine migration (116).

The reaction resembles the Scholl condensation. Here, however, dehydrohalogenation occurs instead of dehydrogenation.

Mono- or di-halogenated chrysenes condense upon treatment with aluminum chloride to give dinaphthoperylenes (631).

Recently it was demonstrated that under suitable conditions simple halobenzenes may be used as Friedel-Crafts arylating agents (460,488). Their decreasing order of reactivity is F ≫ Cl > Br. Aromatic hydrocarbons are arylated with fluorobenzene in the presence of aluminum halides giving the corresponding biphenyls. The directive effects are in agreement with an electrophilic aromatic substitution.

$$2C_6H_5F \xrightarrow[\text{HF}]{\text{AlBr}_3} (o\text{-},p)\text{---}FC_6H_4 \cdot C_6H_5$$

C. Diazonium Halides

A study of the Friedel-Crafts reaction of benzenediazonium chloride with aromatic hydrocarbons and aluminum chloride was made by Mohlau and Berger as early as 1893 (427) who found that with benzene the following reactions occurred:

(1) $C_6H_5 \cdot N_2Cl + C_6H_6 \rightarrow HCl + C_6H_5 \cdot N{=}N \cdot C_6H_5$

(2) $C_6H_5 \cdot N_2Cl + C_6H_6 \rightarrow HCl + N_2 + C_6H_5 \cdot C_6H_5$

(3) $C_6H_5 \cdot N_2Cl \rightarrow N_2 + C_6H_5Cl.$

In reaction (1) the benzenediazonium chloride is simply substituting the phenyl group and azobenzene is formed. A cleavage of nitrogen also occurs, however, thus giving rise to biphenyl, as in reaction (2). Some of the benzenediazonium chloride undergoing cleavage forms chlorobenzene, as shown in reaction (3). The two main reaction products are biphenyl and chlorobenzene.

Phenylation of toluene, biphenyl, and naphthalene has also been carried out with benzenediazonium chloride and $AlCl_3$. With toluene, a mixture of o- and p-methylbiphenyl was secured. The reaction with biphenyl gives mainly p-terphenyl. With naphthalene a mixture of α- and β-phenylnaphthalene is obtained. In all cases chlorobenzene, from cleavage of the benzenediazonium chloride used, is a by-product.

Phenylation of thiophene, pyridine, and quinoline also takes place according to an electrophilic substitution pattern.

The yields of substituted benzenes in all cases are necessarily low owing to simultaneous formation of the other products already mentioned. In using Mohlau and Berger's procedure for the preparation of methylbiphenyl from phenyldiazonium chloride and toluene, a 30% yield of the pure product was secured by Knowles (352).

Benzenediazonium fluoride reacts with aromatic hydrocarbons in anhydrous hydrogen fluoride media to give the corresponding phenylated products in addition to fluorobenzene.

Boron trifluoride added to the system primarily forms benzenediazonium tetrafluoroborate, which on decomposition in aromatic media gives phenylated products.

In 1927 Balz and Schiemann (38) reported the thermal decomposition of aryldiazonium tetrafluoroborates to give aryl fluorides.

$$ArN_2^+ BF_4^- \rightarrow ArF + N_2 + BF_3$$

The Schiemann reaction has since gained widespread application for the preparation of aromatic fluorine compounds (531).

Nesmeyanov and his co-workers (396,397,398,453) investigated the decomposition reaction of aryldiazonium tetrafluoroborates in nitrobenzene and found that *m*-nitrobiphenyl was formed in the reaction in addition to fluorobenzene. They concluded that an ionic type of arylation reaction takes place.

Decomposition of aryldiazonium tetrafluoroborates and also of aryldiazonium tetrachloro- and tetrabromoborates in aromatic solvents leads to electrophilic ring arylation (485).

5. Isomerization in Aromatic Alkylations

A. Skeletal

A characteristic of the Friedel-Crafts alkylations is the tendency of alkyl groups to rearrange during the alkylation process. Thus the alkylation of benzene with *n*-propyl chloride in the presence of aluminum chloride leads to *n*-propylbenzene and isopropylbenzene (cumene). Similarly the alkylation of benzene with *n*-butyl chloride in the presence of aluminum chloride leads to *n*-butylbenzene and sec-butylbenzene. Isobutyl chloride when reacted with benzene gives only *tert*-butylbenzene. The usual tendency in rearrangement of the alkyl groups during alkylation is in the direction primary→ secondary→tertiary. Under the usual conditions of a Friedel-Crafts alkylation the alkyl group rearrangement is initiated by migration of a hydrogen atom and not by direct carbon skeletal rearrangement. Under conditions favoring the formation of more stable carbonium ions isomerization takes place more readily. However, under conditions where an *n*-alkyl halide or *n*-alcohol is able to form only a polarized covalent complex without ionization taking place, the isomerization may be retarded and a considerable amount of *n*-alkyl-benzene is formed. The alkylation of benzene with cyclopropane as alkylating agent was mentioned previously. With aluminum chloride as catalyst or with sulfuric acid at low temperatures the product is *n*-propylbenzene, whereas with sulfuric acid at 65° the product is mainly cumene.

There are two different types of isomerizations taking place in the course of alkylations under Friedel-Crafts conditions. The first involves the alkylating agent (alkyl halide, alcohol, etc.) itself, which may undergo rearrangement during Friedel-Crafts reaction conditions and generally become converted to the most highly branched isomer. The second type of rearrangement involves the alkylated aromatic products themselves.

a. Rearrangement of alkylating agent

n-Propyl chloride undergoes isomerization and is converted to isopropyl chloride under the influence of aluminum chloride. Similarly n-butyl chloride is isomerized to sec-butyl chloride and later to tert-butyl chloride. This fact may be explained by assuming that cleavage of hydrogen chloride takes place with the formation of an olefin, which then reacts with hydrogen chloride in accordance with Markownikoff's rule to form the isomeric halide. In all these isomerizations temperature is an important factor. Higher temperature generally facilitates rearrangements.

b. Rearrangement of alkyl aromatics

The second type of rearrangement involves the alkylated aromatic products. Thus, 2,2-dimethyl-4-n-butylbenzene isomerizes to 1,3-dimethyl-5-sec-butylbenzene and this is converted to 1,3-dimethyl-5-tert-butylbenzene under the action of aluminum chloride. n-Propylbenzene is readily isomerized to isopropylbenzene, as is n-butylbenzene to sec-butylbenzene under the catalytic effect of aluminum chloride or related Friedel-Crafts catalysts. As might be expected, the tendency toward rearrangement of alkyl groups during Friedel-Crafts alkylations is dependent on the alkylating agent and the reaction conditions (catalyst, solvent, temperature, etc.).

The rearrangements are attributed to abstraction of a hydride ion by the acid, thus leaving a carbonium ion followed by an intra-ionic shift of hydride, phenide, and (where applicable) methide ions, giving the most stable carbonium ion in the system. The most stable carbonium ion is usually one which was stabilized by resonance with the aromatic system. Nenitzescu (443) found evolution of molecular H_2 in some of these reactions, thus indicating that removal of a hydride ion was the first step.

Besides the rearrangement of alkylbenzenes leading to more highly branched products there is also the reverse type, first observed by R. M. Roberts (529). n-Propyl-β-C^{14}-benzene rearranges with wet aluminum chloride to α-C^{14}-propylbenzene and the reaction mixture, besides the n-propylbenzene, also contained 3% of isopropylbenzene. That the equilibrium

$$C_6H_5CH_2CH_2CH_3 \rightleftharpoons (CH_3)_2CH(C_6H_5)$$

really exists was proved by Nenitzescu (450,442), who found that isopropylbenzene can be isomerized to *n*-propylbenzene. Secondary butylbenzene was found to isomerize to isobutylbenzene with a ratio in the equilibrium mixture of 1 : 2. Tertiary amylbenzene isomerizes to 2-methyl-3-phenylbutane. If the 2-methyl-3-phenylbutane so obtained reacts at 80° with aluminum chloride for an additional 24 hours, a considerable amount of neopentylbenzene is also produced. As these isomerizations represent thermodynamically controlled equilibria, it must be suggested that the stability of the isomeric phenylalkanes, in contrast to that obtained in Friedel-Crafts alkylations, is in the decreasing order of primary > secondary > tertiary. This observation can explain why during intramolecular isomerization of di- and polyalkylbenzenes the normal alkyl chain can migrate without branching (29,251,345,407), *i.e.*, they represent the thermodynamically most stable isomers.

B. Positional

The isomerization of dialkylbenzenes generally results in a considerable increase of the *meta*-isomer in accordance with the fact that a thermodynamically controlled equilibrium is approached in which the relative ratios of the *ortho*-, *meta*-, and *para*-isomers can be theoretically predicted upon the basis of their thermodynamic stabilities (604).

Similar thermodynamically controlled equilibra are also obtained in the case of treatment of dihalo- (487) and polyhalobenzenes, as well as haloalkylbenzenes (479) with Friedel-Crafts catalyst systems. Aryl groups migrate in a similar fashion to alkyl and cycloalkyl groups. No reported instances of Friedel-Crafts isomerization of nitro, acyl, or sulfo groups have yet been verified.

a. Intramolecular

In the movement of alkyl groups around a ring or from one ring to another the first step, it is generally agreed, is the addition of a proton at the ring carbon atom holding the alkyl group (25,27) to form a sigma complex. The alkyl group being less firmly held can then move intramolecularly by a 1,2-shift (isomerization) or be detached completely as a carbonium ion capable of moving to another ring (disproportionation). Both reactions usually occur together but with methyl groups intermolecular migration is less important. Xylenes can be isomerized without disproportionation (27,406). *p*-Xylene and *o*-xylene can each be isomerized to *m*-

xylene with no direct interisomerization between the *ortho*- and *para*-isomers (5,89,406). With ethyl and isopropyl groups some intermolecular migration always occurs (4,7,99). With *t*-butyl groups intermolecular migration is dominant (3).

b. Intermolecular

Isomerization of alkyl groups is frequently accompanied by disproportionation, thus the isomerization can follow either an inter- or an intramolecular course. Intermolecular isomerization enables higher alkylated benzene derivatives to be used as alkylating agents to produce the monoalkylate. This is the case when in the ethylene alkylation of benzene to produce monoethylbenzene the higher ethylated products formed are recycled to alkylate additional benzene to ethylbenzene.

Isomerization can be shifted beyond thermodynamic equilibrium by an excess of catalyst (406,409). With minor amounts of BF_3 in HF xylenes are converted to the thermodynamic equilibrium: 18% *ortho*-, 60% *meta*-, 22% *para*-xylene. With excess BF_3 the product is almost 100% *meta*-isomer.

Reaction in the acid layer and exchange with the hydrocarbon layer occurs until the hydrocarbon layer reaches thermodynamic equilibrium, but with excess BF_3 all the xylenes are complexed. The complex with *m*-xylene is much more stable than those with the other xylenes; consequently the *m*-xylene content in the acid phase is much higher. Removal of the catalyst by rapid distillation or quenching with water leaves a product that is close to 100% *m*-xylene.

Similarly in the case of the tri- and tetramethylbenzenes excess HF and BF_3 shifts the equilibrium and gives 100% mesitylene and isodurene respectively, the most basic isomers.

These findings explain the fact that many Friedel-Crafts alkylations form *meta*-oriented products in greater than equilibrium amounts. They have practical importance in that an isomerization reaction can be steered to yield a single product.

Isomerization with strong acid catalysts is usually accomplished under mild conditions. Temperatures less than 100° are high enough for methyl groups; ethyl groups migrate at room temperature and isopropyl and *t*-butyl groups at −80°. Under these conditions the alkyl group generally remains intact. With a weaker catalyst, such as silica–alumina, higher temperatures are required. Under such conditions the alkyl group can divide or grow. Thus xylenes and ethylbenzene are interconverted at 515° (63) and the methyl groups of hexamethylbenzene are converted to butyl groups and removed at 344° (601). At these high temperatures it is thought that there is an interconversion of C_6- and C_5-ring carbonium ions, alternately adding and removing a carbon atom to and from a side chain and to and from the ring (133).

As with isomerization, disproportionation can be found beyond thermodynamic equilibrium by an excess of catalyst (387). With HF plus a little BF_3 ethylbenzene is, at equilibrium, converted to the extent of about 50% into benzene and m-diethylbenzene. Formation of a complex with m-diethylbenzene shifts the equilibrium and slows down any further reaction to triethylbenzene.

c. Effect of isomerizations accompanying alkylations

As isomerization and disproportionation are always possible during alkylations, it is necessary to be aware of these side reactions in all cases where Friedel-Crafts conditions are employed. Generally neither the configuration of the entering group, nor the orientation in the nucleus can be accurately predicted. In fact alkylation appears to be in this respect entirely different from acylations, which generally give reliable orientation with no isomerization phenomena. (For a summary of earlier work see C. C. Price (518).)

The Friedel-Crafts alkylation itself must be considered as a kinetically controlled reaction, in which the incipient carbonium ions primarily produced react partly in their original state and partly in a rearranged form. In contrast the rearrangement of the alkylated products is obviously a thermodynamically controlled process. This rearrangement generally takes place at higher temperatures (80–100°) and needs prolonged reaction times, sufficient to obtain the thermodynamic equilibria. Accordingly, these isomerizations (of both intra- and intermolecular nature) of the alkyl group are influenced only by the stability of the starting materials and end products and are independent of the reaction mechanism involved in the individual reactions.

6. Alkylation with Metal and Organometallic Halides

An analogous reaction to Friedel-Crafts alkylation with alkyl halides is possible with certain metal and organometallic halides, particularly in reactions with boron, phosphorus, arsenic, silicon, and sulfur halides.

Direct Friedel-Crafts haloboronation of aromatics with boron trihalides can be achieved in good yield (96,433,615).

$$Ar\text{---}H + BX_3 \xrightarrow{AlX_3} ArBX_2 + HX \quad [X = Cl,Br]$$

Phosphorus trihalide reacts similarly with aromatic compounds to give the corresponding aryl dihalophosphines.

$$Ar\text{---}H + PCl_3 \rightarrow ArPCl_2 + HCl$$

Halosilanes, such as trimethylbromosilane will not react with benzene in the presence of aluminum bromide to produce a significant amount of trimethylphenylsilane (215).

That halosilanes do not undergo reactions analogous to the Friedel-Crafts alkylation may be due to several factors. The halosilanes do not produce a siliconium ion upon interaction with Lewis acid halides, the weak carbon–silicon bond being cleaved instead. Alternatively the siliconium ion even if it is formed may be ineffective in attack upon a benzene ring. Finally, the silylation reaction, if it occurs, could be reversible.

$$(CH_3)_3SiBr + C_6H_6 \xrightleftharpoons{AlBr_3} C_6H_5Si(CH_3)_3 + HBr$$

Most evidence (541) points to the conclusion that siliconium ions or even incipient siliconium ions are not formed by the interaction of aluminum halides with substituted silanes, at least not below temperatures of 80°. However, Sommer (584) recently provided evidence for a limiting siliconium-ion mechanism in solvolysis of chlorosilanes and for the relative stability of the trialkyl siliconium ions, R_3Si^+ as compared with trialkyl carbonium ions, R_3C^+.

Therefore, it is not possible at the present time to rule out entirely the formation of an incipient siliconium ion or differentiate between the possible causes of the failure to effect silylations, corresponding to alkylations, with halosilanes.

Tetramethylsilane or trimethylethylsilane give no traces of phenyltrimethylsilane when they react with benzene in the presence of aluminum bromide at 80°. Extensive disproportionation occurs under these conditions. Phenyltrimethylsilane is not detected in the aluminum bromide catalyzed disproportionation of trimethylsilane in benzene under conditions where the disproportionation of trimethylsilane has reached equilibrium. However, a trace of hydrogen may be detected in the reaction, possibly suggesting trimethylsilylation, but also explicable on the basis of trace impurities.

$$(CH_3)_3SiH + C_6H_6 \xrightarrow{AlBr_3} (CH_3)_3SiC_6H_5 + H_2$$

Silicochloroform, however, reacts with benzene in the presence of catalyst such as boron chloride at high temperatures and under pressure.

$$C_6H_6 + HSiCl_3 \xrightarrow[\substack{275° \\ \text{pressure}}]{BCl_3} C_6H_5SiCl_3 \quad (73\%)$$

There is little information on the mechanism of the reaction and there seems little doubt that in the vapor phase the reaction may be at least partially free radical in type. However, at lower temperature, as 130° in the presence of $AlCl_3$ or BF_3, particularly under pressure, there is pronounced promotion by Friedel-Crafts catalysts, thus ionic processes also play an important part. The reactions are understood to proceed through hydride abstraction (41a)

$$HSiCl_3 \rightleftharpoons SiCl_3^+ + H^-$$

$$C_6H_6 + SiCl_3^+ \rightarrow C_6H_6SiCl_3^+ \rightarrow C_6H_5SiCl_3 + H_2$$

It should, however, be emphasized that silicon halides do not enter Friedel-Crafts type condensations under the conditions normally employed for alkylations with organic halides.

In attempted silylations of benzene and toluene with trialkylchlorosilanes and anhydrous silver tetrafluoroborate or silver hexafluoroantimonate, instead of the silylation reaction

$$C_6H_6 + (CH_3)_3SiCl + AgBF_4 \not\rightarrow C_6H_5Si(CH_3)_3 + AgCl + HBF_4$$

only halogen exchange was observed (471),

$$(CH_3)_3SiCl + AgBF_4 \rightarrow (CH_3)_3SiF + AgCl + HBF_4$$

with formation of the corresponding trialkylfluorosilane.

Friedel-Crafts type metal halide catalysts also effect disproportionation of alkyl organometallics, mainly R_4Pb, R_4Si, R_2Hg.

Sulfur halides, such as sulfur dichloride, react with aromatic hydrocarbons in the presence of Friedel-Crafts catalysts to give diphenyl sulfides (68,71).

$$2C_6H_6 + SCl_2 \xrightarrow{AlCl_3} C_6H_5\text{—}S\text{—}C_6H_5 + 2HCl$$

In the presence of excess sulfur chloride, cyclic thianthrene is formed.

$$2C_6H_6 + 2SCl_2 \xrightarrow{-4HCl}$$

It is also possible to carry out related Friedel-Crafts reactions directly with aromatic hydrocarbons and elemental sulfur (200). From the reaction products thiophenol, diphenyl sulfide, and thianthrene are isolated.

When diphenyl ether reacts with sulfur in the presence of aluminum chloride, phenoxythiine is formed.

Aluminum chloride also acts as a dehydrogenating catalyst in other reactions of sulfur with aromatic compounds, for example in the preparation of dibenzthiophene from biphenyl and sulfur.

Selenous bromide (Se_2Br_2) reacts with benzene in the presence of aluminum chloride to give diphenylselenide, diphenyldiselenide, triphenylselenium chloride, and other products.

Arsenic halides, like arsenic trichloride, react with aromatic hydrocarbons in the presence of aluminum chloride. Triphenylarsine is the main product, with smaller amounts of phenylarsine dichloride and diphenylarsine chloride (377,643).

$$3C_6H_6 + AsCl_3 \rightarrow (C_6H_5)_3As + 3HCl$$

$$2C_6H_6 + AsCl_3 \rightarrow (C_6H_5)_2AsCl + 2HCl$$

$$C_6H_6 + AsCl_3 \rightarrow C_6H_5AsCl_2 + HCl$$

7. Alkylation of Aliphatic Compounds

Although Friedel and Crafts discovered the aluminum chloride catalyzed alkylation reaction in 1877 on an aliphatic system, it was not until the mid-thirties that much interest arose in aliphatic alkylations. It is to the merit of Ipatieff and co-workers (A. V. Grosse, V. I. Komarewski, H. Pines, and L. Schmerling) to have practically rediscovered the Friedel-Crafts alkylation of aliphatic compounds and developed new methods for the preparation of a wide variety of aliphatic alkylates. It was mainly through their efforts that high-octane gasolines became available during World War II.

A. Alkanes

The alkylation of isoparaffins by olefins was first carried out by Ipatieff (305) using aluminum chloride as catalyst. With the help of aluminum chloride it is possible to alkylate all paraffins of normal and iso-structure, except methane and ethane, by means of olefins. However, the reaction is accompanied by side reactions such as polymerization and destructive alkylation, as a result of which the alkylate is a complex mixture of hydrocarbons with a lower octane number than in the product of alkylation with proton acids as catalysts.

Alkanes, particularly isoalkanes, interact with alkenes in the presence of catalysts such as aluminum halides, boron trifluoride, zirconium chloride, hydrogen fluoride, sulfuric acid, and phosphoric acid, to produce higher alkanes by an effective addition of fragments of the alkane across the unsaturation of the alkene (164,315).

Depending upon the particular reactants, catalytic alkylations are carried out at temperatures of $-30°$ to as high as $100°$, and pressures from 1–50 atm. The pressure is usually kept high enough so that at least a part of the reactants are in the liquid state. Although a number of other catalysts of the general Friedel-Crafts type have been used, principal interest centers around aluminum chloride, sulfuric acid, and hydrogen fluoride. Although small differences may exist, the general result of alkylation is the same with any one of these catalysts. The catalytic alkylation is largely confined to the use of isoalkanes (preferably containing tertiary hydrogen). n-Alkanes react with great difficulty and probably with at least partial prior isomerization to isoalkanes. Commercial interest in the alkylation reaction lies in the fact that the highly branched alkanes produced have high octane numbers. The alkane of greatest commercial interest as an alkylation reactant is isobutane. Isopentane is already a liquid with an octane number of 90 and is suitable for use as a motor fuel without further treatment. The alkylation of an isoalkane (isobutane) with an olefin (such as ethylene) is a rather complex reaction accompanied by various side reactions. Among these are: (1) Isomerization of the reactants or products. (2) Dealkylation of the product: the alkylation reaction may "reverse" itself to give a different alkene and a different alkane from the original reactants. (3) Polyalkylation: the product (an alkane) is capable of reacting with more of the alkene thus undergoing polyalkylation.

$$\text{isobutane} + \xrightarrow{\text{ethylene}} \text{hexanes} \xrightarrow{\text{ethylene}} \text{octanes} \xrightarrow{\text{ethylene}} \text{decanes}$$

(4) Polymerization of the alkene: alkenes are subject to polymerization under the influence of Friedel-Crafts reaction conditions. Alkene polymerization competes most successfully with the normal alkylation reaction when too low reaction temperatures are used. (5) Complex formation by the catalyst: highly unsaturated hydrocarbons are formed as by-products during the alkylation reactions; these form complexes with the aluminum chloride catalysts which separate as a reddish-brown "lower layer" during alkylation reactions. (6) Addition of hydrogen halides to the alkene: as in the case of isomerization or polymerization of alkenes, pure, dry aluminum halides (or related catalysts) do not catalyze the alkylation reaction. The presence of hydrogen halides or some other co-catalyst is necessary. In the presence of hydrogen halides, however, addition to alkenes with formation of haloalkanes could take place. (7) Hydrogen transfer: the side reaction of hydrogen transfer involves the overall reaction of two moles of the alkane with one mole of alkene to produce an alkane corresponding to the original alkene and an alkane of twice the carbon content of the original alkane. Thus,

$$2C_4H_{10} + CH_3CH{=}CH_2 \rightarrow CH_3CH_2CH_3 + C_8H_{18}$$

Many of the side reactions however can be minimized by choosing the appropriate reaction conditions (112,549).

The first industrial aliphatic alkylation process was the sulfuric acid catalyzed alkylation of isoparaffins with olefins developed in 1938 (57). Owing to its simplicity, the cheapness of the catalyst, the wide variety of raw materials, and high quality alkylate, this process is still one widely used in industry. In the presence of 96–97% sulfuric acid isobutane, isopentane, and isohexane are comparatively easily alkylated by olefins, except ethylene, with the formation of an alkylate having a high octane number. Isobutane is alkylated with ethylene in the presence of sulfuric acid and salts of silver and mercury, which act as promoters.

Alkylation with sulfuric acid, as with aluminum chloride, is accompanied by isomerization and cracking (58).

Hydro- and dehydro-polymerizations accompany alkylation of alkanes with olefins (437) and therefore in the alkylate there are isoparaffins not only of the same structure and molecular weight which arise from the initial isoparaffins and olefins but also hydrocarbons of smaller and of greater molecular weight.

During World War II hydrogen fluoride was introduced as an alkylation catalyst. It is characteristic of the alkylation of isoparaffins with olefins in the presence of anhydrous liquid HF that isoparaffins are easily alkylated not only by butylene and amylenes, but also with propylene. In the presence of HF the reaction proceeds at room temperature and unlike the reaction with sulfuric acid is not accompanied by side reactions, even with some increase in reaction temperature. Hydrogen fluoride is used in large quantities in relation to the substrate to be alkylated, but as it can be regenerated the total loss of catalyst in the reaction amounts to about 0.2%.

Boron trifluoride was successfully used in the alkylation of isoparaffins with olefins as early as 1935 (201).

In the presence of boron trifluoride isoparaffins having a tertiary carbon atom in the molecule are readily alkylated by olefins. The reaction can be carried out at temperatures of 200° or higher. A small amount of water as co-catalyst is needed in the reaction.

Boron trifluoride mono- and dihydrate, which in effect are hydroxyfluoboric acids, are highly effective catalysts in the alkylation of isoparaffins (isobutane, isopentane) with olefins (218) (ethylene) or mixtures of olefins (ethylene–propylene, ethylene–butylene) (17,53, 93,94,417,502,589,590,591,592,605). The alkylations are carried out in the liquid phase at temperatures between 30–80° and at increased pressure.

The catalytic activity of boron trifluoride is further increased when it is used together with mineral acids, in particular with sulfuric, fluorosulfonic, hydrofluoric, or phosphoric acids.

Sulfuric acid or hydrogen fluoride are strong enough acids to form secondary carbonium ions from propylene or butenes and thus initiate the alkylation. With ethylene, a stronger acid catalyst, such as $AlCl_3$-HCl is needed to initiate and sustain alkylation.

The higher the acidity of the catalyst the greater the amount of self-alkylation of the isoparaffin (547,548). The more acidic catalyst starts more alkylation sequences and effects more hydrogen transfer from the isobutane. Hydrogen transfer from and self-alkylation of isobutane is also at a maximum when an alkyl halide or alcohol (402) is used instead of olefin. The alkyl derivative in these cases must first be converted to an olefin and then alkylated. The chances, therefore, that hydride-ion abstraction from isobutane will occur are much greater than if an olefin had been introduced directly.

Another reaction which had been recognized as occurring

normally in all alkylations (547,548) is alkyl-shift.   Thus with isobu-tane, ethylene and propylene form 2,3-dimethylbutane and 2,3 or (2,4)-dimethylpentane, respectively.

In certain cases the neopentyl configuration can be maintained in alkylations (552).   For example, when ethylene is alkylated with t-butyl chloride in the presence of $AlCl_3$ at low temperature, the product is 1-chloro-3,3-dimethylbutane.

Other secondary reactions in alkylation have been found to be due to the cracking and disproportionation of the main products. Schneider and Kennedy (341,553,554,555) showed that they must arise from a hydrocarbon containing more than 8 carbon atoms. The main alkylation products therefore do not crack directly but first take on another C_4 unit.

The presence of C_5, C_6, C_7, and higher products, the fact that the main hexane product is always the 2,3-dimethylbutane isomer, and the interisomerization of dimethylhexane and trimethylpentanes can all be accounted for by polymerization, alkyl-group shift, and depolymerization reactions.   These reactions are known to occur in the presence of a moderately strong acid.

Recent improvements in commercial H_2SO_4 alkylation have been aimed at cutting down side reactions by having a large excess of isobutane present.   The isobutane, by hydrogen transfer, stops the reaction at the C_8 state and suppresses formation of the heavy alkyl-ate precursor of cracking.   Since the solubility of isobutane in sul-furic acid—where the reaction takes place—is extremely low, efficient mixing is required.   Designs recently developed include (a) improved emulsification of hydrocarbon and acid in contact with a cooling element, and (b) cascade reactors employing multipoint in-jection of olefins, with refrigeration supplied by vaporization of iso-butane from the mixture.   The internal isoparaffin–olefin ratio is thereby kept high.

The solubility of isobutane in HF (about 4 wt.% at room tempera-ture) is much greater than in sulfuric acid.   Efficient mixing is not as serious a problem.   Modern trends in commerical development have therefore been toward simplification of design.

A new catalyst system developed by Evering and Roebuck (175) comprises $AlCl_3$,HCl dissolved in an equimolar $AlCl_3$–ether complex, in which isobutane is much more soluble than in HF.   Side reactions are almost completely eliminated.   With 2-butene the reaction pro-duct is mainly trimethylpentane, predominantly the 2,3,3-, the 2,3,4-, and the 2,2,4-isomers.   With 1-butene the main products are dimethylhexanes.

The remarkable selectivity of this catalyst is probably due to its large solubility for isobutane. Sulfuric acid, with only a limited solubility for isobutane, gives practically the same product with either 1-butene or 2-butene (495). Hydrogen fluoride, with moderate solubility for isobutane, at lower temperatures (220,389) gives a somewhat higher yield of trimethylpentanes with 2-butene than with 1-butene.

Alkyl halides have also been used as alkylating agents for alkanes (253), but this reaction has not been so widely studied as the alkylation with alkenes. Methyl and ethyl bromides have been shown to react with n- and iso-butane to give substantial yields of pentanes and hexanes respectively, together with higher alkanes.

Cycloalkanes, such as methylcyclopentane, cyclohexane, and methylcyclohexane are alkylated by olefins in the presence of catalysts like aluminum chloride, aluminum bromide, and boron trifluoride in a manner similar to alkanes.

B. Alkenes

a. With alkyl halides

Polyhaloalkanes (chloroform, carbon tetrachloride, etc.) alkylate polyhaloalkenes readily in the presence of Friedel-Crafts catalysts (70). The following examples are typical of this type of alkylations:

$$Cl_2C{=}CCl_2 + CHCl_3 \xrightarrow{\text{AlCl}_3} Cl_3CCCl_2CHCl_2$$

$$ClHC{=}CHCl + CCl_4 \xrightarrow{\text{AlCl}_3} Cl_3CCHClCHCl_2$$

Schmerling has shown that alkenes themselves may be alkylated with simple alkyl halides by the proper choice of catalysts and reaction conditions, which minimize competing reactions, principally polymerization (551). The more active catalysts, such as aluminum chloride, require the use of temperatures from about -30–$0°$, while less active catalysts such as zinc chloride permit the use of temperatures as high as $100°$. The nature of the reactant must of course be considered in any specific case. The following examples which illustrate the alkylation reaction here are relatively free from complicating side reactions.

$$(CH_3)_3CCl + CH_2{=}CH_2 \xrightarrow{\text{AlCl}_3} (CH_3)_3CCH_2CH_2Cl$$

$$(CH_3)_2CHCl + CH_2{=}CHCl \xrightarrow{\text{AlCl}_3} (CH_3)_2CHCH_2CHCl_2$$

Schmerling also studied the reaction of t-butyl chloride with cyclohexene in the presence of BF_3 (and a co-catalyst) and found the reaction products to be 1-chloro-1-*t*-butylcyclohexane, chlorocyclo-hexane, and 1-*t*-butylcyclohexane (550).

Relatively few Friedel-Crafts alkylations of alkenes by means of alkyl halides are as free from complications as the examples cited above. Such competing reactions as polymerizations and isomeri-zations are usually involved. When *n*-alkyl halides are used as alkylating agents, isomerization preceding alkylation, leading to secondary or tertiary halides occurs. For example, the alkylation of vinyl chloride with *n*-propyl chloride takes the following course :

$$CH_3CH_2CH_3Cl \xrightarrow{AlCl_3} CH_3CH_2CH_2^+ \rightleftharpoons CH_3\overset{+}{C}HCH_3$$

$$CH_3\overset{+}{C}HCH_3 + CH_2{=}CHCl \rightarrow (CH_3)_2CHCH_2\overset{+}{C}HCl \rightarrow (CH_3)_2CHCH_2CHCl_2$$

Isomerization of the alkylation product and further reaction of the new alkyl halide with alkene may also complicate the alkyla-tion of alkenes.

Chloromethyl ether reacts with alkenes including ketene (593,594).

$$CH_3OCH_2Cl + CH_2{=}CO \xrightarrow{BF_3} CH_3OCH_2CH_2COCl$$

Styrene reacts with secondary and tertiary alkyl halides in the presence of stannic chloride or zinc chloride catalyst to give the corresponding α-haloaralkane, which can be easily dehydrohalo-genated to α-alkylstyrene.

b. With alcohols

The reaction of alkenes with alcohols to form ethers is catalyzed by aluminum chloride, although sulfuric acid is the best of the catalysts which have been investigated (172).

Primary alcohols have been found to be more suitable for the reaction than the secondary alcohols. Tertiary alcohols are the least reactive. In the case of the olefins those which can be derived theoretically from a tertiary alcohol, for example trimethylethylene or isobutylene, are the most reactive.

The lower alcohols react with olefins containing tertiary carbon atoms in the presence of boron trifluoride at atmospheric pressure and a temperature of below 100° (556,563). The reaction proceeds easily when the alcohols are used as coordination compounds with boron trifluoride. 2-Methyl-2-butene with $CH_3OH \cdot BF_3$ forms methyl t-amyl ether according to the reaction

$$CH_3C{=}CHCH_3 + BF_3 \cdot CH_3OH \rightarrow CH_3CH_2C(CH_3)_2OCH_3 + BF_3.$$
$$\underset{\displaystyle CH_3}{|}$$

Methyl alcohol reacts with isobutylene to give t-butyl methyl ether (588).

Ethyl alcohol reacts with propylene as well as ethylene glycol with the formation of the corresponding ethers (386,628).

Alkylation in the presence of boron fluoride catalyst is also possible with cyclic olefins.

Alcohols react with vinyl methyl ketone in the presence of boron trifluoride etherate forming 4-alkoxy-2-butanol (425).

$$CH_3COCH{=}CH_2 + ROH \rightarrow CH_3COCH_2CH_2OR$$

Vinyl ethers and vinyl esters of carboxylic acids give the corresponding acetals with primary and secondary alcohols in the presence of BF_3 simultaneous splitting off of an acid residue

$$CH_2{=}CHOCOR' + 2BF_3 \cdot ROH \rightarrow CH_3CH(OR)_2 + R'COOH + 2BF_3$$

Methyl alcohol, for example, gives an almost quantitative yield of methyl acetal with vinyl acetate in the presence of $BF_3 \cdot CH_3OH$ (139).

Primary and secondary alcohols also react with isopropenyl acetate in the presence of boron trifluoride etherate (141) to give the corresponding ketals.

$$CH_2{=}C(CH_3)OCCH_3 + 2ROH \xrightarrow{\ BF_3 \cdot O(C_2H_5)_2\ } \underset{\displaystyle CH_3}{\overset{\displaystyle OR}{CH_3\!\!\underset{|}{\overset{|}{C}}OR}} + CH_3COOH$$

Halohydrins give halogenated acetals (140) with vinyl esters of carboxylic acids. As an example, ethylene chlorohydrin with vinyl acetate in the presence of $CH_3OH \cdot BF_3$ yields β,β-dichlorodiethylacetal.

c. With mercaptans

Posner showed as early as 1905 (515) that mercaptans can be added to olefins to form dialkyl sulfides, a reaction which is basically the alkylation of an olefin by a mercaptan. In cases such as trimethylethylene, styrene, and α-methylstyrene good yields were obtained in the presence of sulfuric acid (polymerization of the olefins being the main side reaction).

Boron trifluoride and its compounds are suitable catalysts in the reaction of mercaptans with olefins. Vinyl acetate forms alkyl-mercaptoethylacetate with mercaptans in the presence of boron trifluoride etherate.

d. With carbonyl compounds: Prins reaction

Prins discovered in 1919 that alkenes will react with formaldehyde in the presence of acid catalyst to produce derivatives of 1,3-glycols or unsaturated alcohols. The reaction is initiated either by mineral acids (like sulfuric or phosphoric acid) or Lewis acid catalysts (like zinc chloride, stannic chloride, etc.) (14,521).

The course followed in the Prins reaction can be represented according to Nenitzescu (451) leading, depending on conditions, to diacetates or dioxanes besides unsaturated alcohols.

Olefins also react easily with formaldehyde in the presence of boron trifluoride hydrates as catalyst (241,285,536). Depending on the temperature, time of reaction, and composition of the catalyst, 1,3-diols, 1,3-dioxanes, or dienic hydrocarbons are obtained.

Ethylene itself reacts only under drastic conditions. Olefins of the type $R_2C=CH_2$ and $RCH=CH_2$ give mainly 1,3-dioxanes; $RCH=CHR'$ type olefins afford 1,3-glycols, but usually only in low yields (244).

In the extension of the Prins reaction to ketones, several olefins have been used. Typical of such reactions is that of acetone with cyclohexene.

$$+ CH_3COCH_3 \xrightarrow[\text{2. hydrolysis}]{\text{1. } H_2SO_4 \text{ (in } CH_3CO_2H)}$$

2–(2'hydroxypropyl) cyclohexanol

The reaction of acetals with substituted alkenes, such as vinyl alkyl ethers, proceeds in such a way that the alkoxy groups add to the least hydrogenated carbon of the double bond and the rest of the acetal is directed to the most hydrogenated carbon atom. For example, dimethylacetal and vinyl methyl ether in the presence of boron trifluoride etherate form 1,1,3-trimethoxybutane.

$$CH_2=CHOCH_3 + CH_3CH(OCH_3)_2 \rightarrow CH_3CH(OCH_3)CH_2CH(OCH_3)_2$$

Ketals also react easily with vinyl methyl ether (135). The diketals of acetone and cyclohexanol with vinyl methyl ether, in the presence of boron trifluoride etherate catalyst form 1,3-diethoxy-1-methoxy,1,1-methylbutane and 1,1-(2-ethoxy-2-methoxyethyl)-cyclohexane respectively.

C. Alkynes

a. With alkyl halides

Phenylacetylene reacts with secondary and tertiary alkyl halides in the presence of stannic chloride or zinc chloride catalyst to give α-haloaralkenes.

No similar alkylation of aliphatic acetylenic compounds has been reported.

b. With alcohols

Acetylene and its homologs react with alcohols similarly to olefins in the presence of boron trifluoride and its coordination compounds.

However, since these hydrocarbons have a triple bond, two molecules of alcohol are usually added. All monohydric aliphatic alcohols add to acetylene in such a way that alkyl vinyl ethers are first formed and these, in an acid medium, add a second molecule of alcohol according to the Markownikoff rule to form acetals (263).

The addition of cyclohexanol to acetylene in the presence of BF_3 proceeds with difficulty and is accompanied by resin formation (338).

A particular position among substituted acetylenes is occupied by vinylacetylene, in which the double and triple bonds capable of addition reactions are found in a conjugated position. Carothers and co-workers (322) found that monovinylacetylene adds one molecule of methyl alcohol in the 1,4-position in the presence of sodium methylate and by subsequent isomerization of the products methoxybutyne is formed. Rotenberg and Favorskaya (535) found that in the presence of boron trifluoride the reaction takes a different course. Ethyl alcohol adds to the triple bond and forms 2-ethoxy-1,3-butadiene. Methyl alcohol also first adds at the triple bond but then the 2-methoxy-1,3-butadiene so formed splits off a molecule of methyl alcohol and adds it on again in the 1,2-position, with the formation of 2-methoxy-2,3-butadiene.

The reaction of acetylenic carbinols with aliphatic alcohols was investigated by Hennion (205). He found that alcohols from methyl to n-amyl inclusive add in the presence of boron trifluoride etherate to the triple bond of dialkylethynyl carbinols with the formation of ketals of dialkylacetyl carbinols.

$$(CH_3)_2C(OH)C{\equiv}CH + 2ROH \rightarrow (CH_3)_2C(OH)C(OR)_2CH_3$$

D. Active Methylene Compounds

Esters containing active α-hydrogen atoms are alkylated by alkyl halides, alcohols, and ethers in the presence of boron trifluoride catalyst.

Ethyl acetoacetate with benzyl chloride in the presence of boron trifluoride gives ethyl α-benzylacetoacetate (80). A mixture of ethyl acetoacetate and isopropyl alcohol saturated with boron trifluoride and left to react at room temperature gives ethyl α-propylacetoacetate (248).

Ethers also alkylate alkyl acetoacetates in the presence of boron trifluoride. Thus on saturating a mixture of ethyl acetoacetate and diisopropyl ether with boron trifluoride, ethyl α-isopropylacetoacetate is formed. The same product is obtained when ethyl acetoacetate reacts with isopropyl acetate (81). Similarly acetoacetates are alkylated by *t*-butyl alcohol or ether in the presence of boron trifluoride (247).

Hauser and co-workers have made a detailed study of the influence of time and temperature on the yield of alkylated acetoacetic acid esters. The optimum conditions for the reaction were determined (1).

8. Isomerization of Saturated Hydrocarbons

The isomerization of *n*-alkanes (such as *n*-butane and higher alkanes) has assumed great importance in the last 25 years. The need for isobutane as a charging stock for alkylation processes and the necessity for converting low octane number straight-chain hydrocarbons to their higher octane number branched-chain isomers was the reason for interest in alkane isomerization.

In the isomerization of cycloalkanes five- and six-membered rings are favored in the equilibria. Three-, four-, seven-, and higher-membered rings can be considered absent at equilibrium (137,595). Thus cyclopentane does not form cyclopropanes and methylcyclopentane and cycloheptane isomerize irreversibly to methylcyclohexane (507).

Furthermore, in the five- and six-membered ring equilibria the six-membered rings are favored at lower temperatures.

A. Catalysts

The ability to cause isomerization of saturated hydrocarbons is characteristic of acid catalysts. Catalysts of major importance are:

(a) Lewis acid halides, such as aluminum halides and boron fluoride.

(b) Sulfuric acid and related proton acids.

(c) Acidic oxides.

B. Initiation by Carbonium-Ion Type Intermediates

Evidence indicates the same type of carbonium-ion mechanism in the isomerization process with all these catalysts.

The fundamental knowledge needed for the understanding of isomerization (as well as alkylation and polymerization) was supplied early in the 1930's by the carbonium-ion theory of Whitmore (639) and by the subsequent demonstration of the hydride abstraction from alkenes by carbonium ions by Bartlett (42).

Rigorously purified *n*-butane does not measurably isomerize in the presence of $AlBr_3$–HBr at room temperature (509) but the addition of as little as 3 parts per 10,000 of olefin or of traces of oxygen or alkyl halide is sufficient to start the reaction. Formation of a carbonium-ion initiator is therefore indicated:

$$R\!-\!CH\!=\!CH_2 + HX + AlX_3 \rightleftharpoons R\!-\!\overset{+}{C}H\!-\!CH_3 \, AlX_4^-$$

or

$$RX + AlX_3 \rightleftharpoons R^+ \, AlX_4^-.$$

Isomerization can likewise occur at higher temperatures and with excess co-catalyst (509). Hydrogen has also been found among the initial products (62,443). These results indicate that there is another slower initiation mechanism which may be represented as abstraction of hydride ion by a proton:

$$
\begin{array}{ccc}
\quad R & & \quad R \\
\quad | & & \quad | \\
R\!-\!\overset{\displaystyle |}{\underset{\displaystyle |}{C}}\!-\!H + H^+AlX_4^- & \rightleftharpoons & R\!-\!\overset{\displaystyle |}{\underset{\displaystyle |}{C}}{}^+AlX_4^- + H_2. \\
\quad R & & \quad R
\end{array}
$$

The stronger the acid catalyst generally the faster the isomerization. Evering and Roebuck (174) showed that the rate of isomerization of 3- to 2-methylpentane was nearly proportional to the H_0 (Hammett acidity function) of the sulfuric acid catalyst. It has also been shown that isomerization is much faster in the presence of such strong Friedel-Crafts catalysts as $HF\!-\!BF_3$ or $HCl\!-\!AlCl_3$. This is in accordance with the estimate of their acidity being 10^5 times greater than that of HF or sulfuric acid (405).

C. Side Reactions Accompanying Isomerization

Isomerization is always accompanied to some extent by side reactions such as cracking, disproportionation, and conjunct polymer formation. The amount of secondary reaction depends upon the size of the molecule. Butanes crack only to a slight extent, pentanes somewhat more, and hexanes much more. Heptanes do not isomerize without almost complete breakdown.

Side reactions have been controlled with hydrogen (173,314), with aromatics (173,404), with naphthenes (173,404), and with butanes (173,404,562). These compounds also suppress isomerization but they suppress side reactions more strongly. The main products of cracking are pentanes, butanes (with all catalysts), and propane (with strong Friedel-Crafts acids). Pentanes also disproportionate to butanes and hexanes. Methane and ethane are not formed.

Although there is a general agreement that the inhibitors act as carbonium-ion traps and thus suppress all reaction, there is no general agreement on why they suppress side reactions more than isomerization. A recent theory (405) is based on the premise that, as in alkylation, only those molecules containing at least 9 carbon atoms crack readily (341,553,554,555).

Another side reaction common to all acid-catalyzed reactions involving carbonium ions is the formation of a heavy unsaturated oil, which remains dissolved in the acid layer and gradually deactivates the catalyst. Alkyl carbonium ions, besides abstracting a tertiary hydrogen, can readily abstract hydrogen from a position *alpha* to a double bond to form an allylic-stabilized carbonium ion (562). This fragment becomes soluble in the catalyst layer, reacts further by polymerization and hydrogen transfer, and is converted by this hydrogen-transfer reaction into 85% C_3–C_5 saturated paraffins and 15% polyolefinic oil (405) or "conjunct polymer" as it has been named.

Bloch and co-workers (61,266,328) have reported that conjunct polymers are mainly C_5 ring-containing hydrocarbons with a high degree of alkyl substitutions, and with 2 to 3 double bonds, two of which are conjugated. The conjugated double bonds are believed to be distributed between a ring and a side chain. Miron and Lee (426) have shown that the whole conjunct polymer molecule is made up of rings (mostly C_5, some C_6) joined together in either one or two places, with methyl groups on the periphery. The number of carbon atoms per molecule varies from 10 to 27 and the number of rings varies from 1 to 5.

9. Polymerization

A. Catalysts

The first claim of a polymerization of an olefin in the presence of metal halide catalyst was made by Deville (157) who, in 1839, polymerized styrene with stannic chloride. Butlerov and Goryainov (100,101) in 1873 observed that boron trifluoride polymerizes isobutylene. Balsohn (37) observed in 1879 that aluminum chloride transforms ethylene into a viscous oil. A year later Gustavson (234) observed that aluminum bromide is also able to bring about the reaction.

The next known investigation followed in 1902 when Aschan (15) reported on the aluminum chloride catalyzed polymerization of ethylene. Ipatieff (313) extended the work in 1913.

However, until the 1930's Friedel-Crafts polymerization of olefins did not attract much attention (301). Since then a considerable amount of interest has been focused on the aluminum chloride and boron fluoride catalyzed polymerization of olefins. The degree of polymerization of unsaturated compounds depends on the structure of the monomer, the temperature, pressure, time of contact, solvent used, and the character and amount of the catalyst.

The copolymerization of α-olefins and certain vinyl monomers with various catalysts of the Friedel-Crafts Lewis acid type is also well known (497,511).

Effective catalysts in Friedel-Crafts polymerizations include $AlCl_3$, $AlBr_3$, BF_3, $TiCl_4$, $SnCl_4$, a variety of other Lewis acid type halides, and proton acids. Isobutylene, styrene, α-methylstyrene, butadiene, and vinyl alkyl ethers are representatives of the monomers readily converted to high molecular weight polymers by the catalysts mentioned. Propylene, butenes, and other olefins may also be polymerized but the products tend to be relatively low in molecular weight.

B. The Role of Co-catalysts

The mechanism of the Friedel-Crafts polymerization of α-olefins was originally proposed (113) as the direct interaction of the Lewis acid type catalyst with the corresponding olefin.

The inactivity of a pure Friedel-Crafts catalyst for polymerization was first reported by Ipatieff and Grosse (303) in 1936. They reported their experimental discovery that pure dry aluminum chloride does not react with pure ethylene. However, the requirement of an additional component for Friedel-Crafts catalytic activity was not discussed in the literature until ten years later.

Evans, Dainton, Meadows, Norrish, Pepper, Plesch, Polanyi, Russell, Skinner, and others have shown (146,147,148,168,169,170, 171,468,496,513) that in order to achieve polymerization with a Friedel-Crafts type catalyst the presence of a co-catalyst is always necessary. Accordingly, the catalyst–co-catalyst system interacts as a strong proton acid and the electrophilic polymerization proceeds through a simple protonating mechanism.

C. Types of Polymers

High molecular weight polymers are obtained when isobutene is polymerized at low temperatures (between $-80°$ and $-25°$) in the presence of small amounts of a strong acid such as $AlCl_3$ or BF_3. The reaction is over in a fraction of a second (340,609). At low temperatures the side reactions are suppressed and molecular weights up to several million are obtained.

From a practical point of view three main types of Friedel-Crafts polymerization of olefins are recognized (469): (1) conversion of low molecular weight olefins to gasoline-range of olefins (299); (2) conversion to intermediate molecular weight polymers for use as synthetic lubricant oils; and (3) conversion to high molecular weight polymers having 5000 to 100,000 monomer units.

For dimerization of butenes weak acid catalysts such as phosphoric (163,165,654) or aqueous sulfuric acid were used. The reaction takes place in the acid phase. The dimer formed is much less soluble in the acid phase than are the butenes or propene and therefore is scrubbed out by the hydrocarbon phase.

Some termination also occurs by hydride-ion abstraction leading to a partially saturated polymer and a hydrogen-poor acid-soluble oil that gradually deactivates the catalyst (310).

In the case of polymers of sufficiently high molecular weight for use as lubricant oils, the lower molecular weight olefins are contacted with Friedel-Crafts Lewis acid catalysts at 50–100° (522,599). Initiation is by addition to the olefin either of a proton type or of a carbonium-ion type co-catalyst, furnished by an alkyl halide promoter (171). The reaction probably takes place in the catalyst-sludge layer, which provides a medium of high dielectric constant for solvating carbonium ions. Termination may occur by proton expulsion as with the weaker acids, but only after the polymer has grown much longer; this is because the equilibrium in strong Friedel-Crafts acids favors the carbonium ion. Considerable attention has been focused on cyclopolymerization of olefins under Friedel-Crafts conditions.

Ethylene, propylene, and other simple olefins may be polymerized by promoted aluminum chloride to lubricant oils, containing a variety of aromatic hydrocarbons (600).

Ipatieff and Pines (306) observed that the polymerization of ethylene in the presence of a phosphoric acid catalyst gives liquid polymers containing olefinic and paraffinic compounds and also naphthenic and aromatic hydrocarbons.

This type of polymerization where different types of hydrocarbons are formed has been named "conjunct polymerization" (307).

Naphthenes are formed by cyclization of olefins (produced in the polymerization or present in the starting materials) and they can be subsequently dehydrogenated to aromatics.

Under suitable conditions polymerization and dehydrogenation can be combined and directed to increased formation of aromatics. These processes are called "dehydropolymerizations" (332).

IV. Acylation (including Halogenation and Related Reactions)

1. Aromatic Ketone Synthesis

A. Scope

The Friedel-Crafts ketone synthesis implies the introduction of an acyl group into the aromatic nucleus by the action of an acylating agent, usually an acyl halide, acid anhydride, ester, or the acid itself, in the presence of a Friedel-Crafts catalyst.

$$ArH + RCOX \xrightarrow{AlCl_3} ArCOR + HX$$

$$ArH + (RCO)_2O \xrightarrow{AlCl_3} ArCOR + RCOOH$$

$$ArH + RCOOH \xrightarrow{BF_3} ArCOR + H_2O$$

$$ArH + RCOOR' \xrightarrow{AlCl_3} ArCOR + R'OH$$

Aluminum chloride and boron trifluoride are most frequently used, but other catalysts also find application.

Friedel-Crafts acylation of aromatics is of considerable practical value owing to the importance of aryl ketones and aldehydes. Among industrially important ketones made in this way are aceto-phenone, propiophenone, benzophenone, stearophenone, ω-chloro-acetophenone, caproyl resorcinol, etc. Acylating agents in general are more reactive than alkylating agents in Friedel-Crafts type reactions. The reaction of acyl halides with aromatic compounds in the presence of Friedel-Crafts catalyst proceeds more readily than the corresponding alkylations with alkyl halides. Usually it is difficult to introduce more than one acyl group into an aromatic ring. Further, it is characteristic of the Friedel-Crafts acylations of *ortho–para*-directing substituted aromatics, that *para*-derivatives are formed to almost complete exclusion of the corresponding *ortho*-isomers. Exceptions to this are not unknown, however (492).

B. Sequence of Addition of Reagents

Interaction of the aromatic derivative (ArH), acyl halide (R·COX), and aluminum halide (AlX₃) liberates hydrogen halide and produces a complex of aromatic ketone and aluminum halide from which the ketone is liberated by hydrolysis.

$$ArH + R \cdot COX + AlX_3 \rightarrow HX + [ArCO \cdot AlX_3] \xrightarrow{H_2O} ArCOR$$
$$\overset{|}{R}$$

Gradual addition of one component to a mixture of the other two, usually in a solvent, provides the three most general methods of the Friedel-Crafts ketone synthesis.

a. Friedel-Crafts, Elbs procedure

The most widely used procedure is that of Friedel and Crafts (198) themselves, further developed by Elbs (167), in which the catalyst is added as the last reactant.

b. Bouveault procedure

In general, preliminary mixing of the aromatic compound and aluminum halide, followed by addition of the acyl halide (Bouveault procedure) (77,392) is not preferred, since the presence of hydrogen halide (present from impurities, moisture or formed after the first addition of acyl halide) may cause extensive isomerization or disproportionation. Thus gradual addition of propionyl chloride to a mixture of 3-phenylpentane and aluminum chloride in carbon disulfide gives propiophenone and a dipentylpropiophenone.

c. Perrier procedure

In the Perrier variation (500,501) of the reaction the catalyst (aluminum chloride) and the acyl component (acyl chloride) are allowed to react prior to the addition of the substrate.

The Perrier procedure thus consists in preparing the addition compound of the acyl chloride and the catalyst (generally aluminum chloride) which is then allowed to react with the hydrocarbon.

It is generally carried out in carbon disulfide, methylene chloride, or ethylene chloride as solvent, in which the complex $RCOCl \cdot AlCl_3$ is soluble but aluminum chloride itself is practically insoluble. Excess of aluminum chloride is undesirable since it may effect isomerization.

The advantage lies chiefly in the fact that the addition compound is soluble in carbon disulfide. The acylation product, however, forms an insoluble addition compound with the catalyst, which can be separated easily and converted to the ketone by hydrolysis. Tarry impurities remain behind in the solvent. The synthesis of 1-benzoylnaphthalene by this method is particularly to be recommended since the aluminum chloride addition compound of 2-benzoylnaphthalene, always formed as a by-product, is soluble in carbon disulfide and thus does not appear as contaminant of the product. It should be especially noted that the presence of *meta*-directing groups in the acylating agent ordinarily does not interfere. *meta*-Nitrobenzoyl chloride, for example, has been employed successfully in the synthesis of *meta*-nitrophenyl aryl ketones. Another example is the use of the acyl chlorides of the monoesters of dibasic acids in the preparation of keto esters with long side chains.

Stable acyl fluoride–Lewis acid fluoride complexes were isolated and found to be oxocarbonium (acylium) salts (472,478,561).

$$CH_3COF + BF_3 \rightleftharpoons CH_3CO^+BF_4^-$$

$$CH_3CH_2COF + SbF_5 \rightleftharpoons CH_3CH_2CO^+SbF_6^-$$

$$C_6H_5COF + AsF_5 \rightleftharpoons C_6H_5CO^+AsF_6^-$$

They can be used advantageously in a modification of the Perrier synthesis.

The reaction of acyl chlorides (or bromides) with anhydrous silver salts, such as anhydrous silver perchlorate (Prail and Burton), silver tetrafluoroborate, silver hexafluorophosphate, silver hexafluoroantimonate, and silver hexafluoroarsenate (Olah and coworkers), leads to the preparation of acylium salts which are highly reactive acylating agents in ketone syntheses. As these metathetic reactions are carried out in acid-free media (with the exception of the by-product conjugate acid formed in the substitution reaction with aromatics) they are applicable to systems otherwise sensitive to Lewis or Brønsted-Lowry acid catalysts.

C. Amount and Role of Catalyst

Whereas alkylation with alkyl halides requires only catalytic amounts of catalysts (aluminum halide), acylation affords ketones in yields which are proportional to the amount of aluminum halide added. The reaction is complete when little more than a molecular proportion of catalyst is used.

One molecular equivalent of catalyst apparently combines with the acyl halide, giving a 1:1 addition compound, which then acts as the active acylating agent.

Ulich and Heyne (626) concluded that the catalyst is firmly attached to the resulting ketone and is no longer able to activate more acyl halide. This conclusion is not wholly true since halogen exchange between acyl and aluminum halides continues in the presence of excess of ketone. Apparently presence of ketone does not prevent formation of acyl cation but lowers its efficiency as an acylating agent. This deactivation is probably effected by association: $ArCR\!:\!O + {}^+RC\!:\!O^+ \rightarrow [Ar\cdot CR\!:\!O \cdots CR\!:\!O]^+$. Similarly, ability of solvents or other added substances, *e.g.*, *m*-dinitrobenzene, acyl halides, nitrobenzene, nitromethane, diethyl ether, benzophenone, *p*-methoxybenzophenone, to retard or even to suppress Friedel-Crafts reactions is related to their ability to solvate acyl cations and thus to reduce the electron demand and hence the activity for acylation purposes.

A further consequence of association of acylating agents with basic compounds is an increase in the bulk of the reagent and greater resistance to attach at the more sterically hindered positions of aromatic compounds. Thus acylation of chrysene and phenanthrene in nitrobenzene or in carbon disulfide occurs to a considerable extent in a side ring, while that of naphthalene leads to extensive reaction at the less reactive but sterically less hindered β-position. These two consequences of the presence of basic compounds may be avoided by ensuring that they are fully engaged by aluminum halide (24,110).

Acid anhydrides combine with 2 mols. of aluminum halide to afford one mole of acylating agent: $(R \cdot CO)_2O + 2AlX_3 \rightarrow R \cdot COX \cdot AlX_3 + R \cdot CO \cdot O \cdot AlX_2$. A consequence of the reverse reaction, interaction of acylating agent and acyloxyaluminum chloride, is illustrated by the formation of α-tetralone when phenylbutyric acid and acetyl chloride are brought together in the presence of aluminum chloride.

$$CH_3 \cdot COCl \cdot AlCl_3 + C_6H_5 \cdot (CH_2)_3 \cdot CO \cdot O \cdot AlCl_2 \rightleftharpoons C_6H_5(CH_2)_3 \cdot CO \cdot O \cdot CO \cdot CH_3 \cdot 2AlCl_3$$
$$\rightleftharpoons CH_3 \cdot CO \cdot O \cdot AlCl_2 + C_6H_5(CH_2)_3 \cdot COCl \cdot AlCl_3 \rightarrow \alpha\text{-tetralone}$$

To effect intermolecular acylation of this and similar acids it is necessary to protect the carboxyl group by esterification and to employ 2 mols. of aluminum halide: one to engage the ester and the other to provide the acylating agent.

$$C_6H_5(CH_2)_3 \cdot CO_2C_2H_5 \cdot AlX_3 + R \cdot COX \cdot AlCl_3 \rightarrow p\text{-}R \cdot CO \cdot C_6H_4 \cdot (CH_2)_3 \cdot CO_2C_2H_5$$

The rate and the site of substitution with acylating agents derived from monobasic acid anhydrides are essentially the same as those obtained with the corresponding acyl chloride–aluminum chloride complexes. On the other hand dibasic acid anhydrides, even in the presence of an excess of aluminum halide, afford comparatively bulky acylating agents of low reactivity, *i.e.*, they behave as do acyl halide–aluminum halide complexes in nitromethane or nitrobenzene solution. The cause of this behavior probably originates with intramolecular association of the acyloxyaluminum chloride group and acylating agent.

A deficiency of catalyst will lower the overall yield because of incomplete utilization of acylating agent. Insufficient catalyst may also cause self-condensation of the partially complexed ketone, as the formation of dypnone in the acetylation of benzene (108).

An excess of catalyst, on the other hand, often gives rise to large amounts of tar.

D. Isomerization Accompanying Acylation

With homologs of benzene Friedel-Crafts acylations may be accompanied by migration of the substituent groups during the Friedel-Crafts reaction. For example, *ortho-* and *para-*bromotoluene both give 1-methyl-3-bromo-4-benzophenone in benzoylations at 0°. Nightingale has reported an additional number of substituent group migrations during Friedel-Crafts ketone synthesis (459).

However, no isomerization of the carbon chain of the acyl halides (or anhydrides) used in Friedel-Crafts acylations, takes place. Propionyl, butyl, and caproyl halides yield ketones with alkyl groups of configurations which correspond to the branching of the alkyl group in the acyl halide. For example *n-* or iso-butyryl chlorides will yield the corresponding *n-* and iso-propyl phenyl ketones, respectively, with benzene and aluminum chloride. This absence of side-chain rearrangement has made acylations with metal halide catalyst doubly valuable, since on subsequent reduction the ketones give good yields of hydrocarbons of predicted configuration. Johnson's work (326) on the synthesis of the alkylresorcinols has emphasized this point. Thus direct access is obtained to the straight-chain or variously branched homologs of benzene and its derivatives.

Many *ortho-*substituted alkyl aryl ketones, including alkyl and hydroxy substituents are unstable in the presence of hydrogen halide and excess of aluminum halide, especially at raised temperatures, and undergo a variety of irreversible changes which have been classified by Baddeley (26,29) as follows: (1) The ketone may suffer fission into aromatic hydrocarbon and acyl cation, which recombine to provide an isomeric ketone. For example 2-methyl- and 4-hydroxy-2-methyl-acetophenone afford 4-methyl- and 2-hydroxy-4-methyl-acetophenone respectively. α-Naphthyl ketones behave similarly, thus affording the β-isomers. (2) The 2-alkyl group may migrate intramolecularly to the adjacent *m*-position. 2,5-Dimethyl- and 2,5-diethyl-acetophenone and their 6-hydroxy derivatives rearrange to the corresponding 3,5-isomers. (3) In 2,6-dialkylaryl ketones both of the alkyl groups may migrate to the corresponding *m*-positions. Isomerization of 2,4,6-trimethylacetophenone at 100°

and 125° gives the 2,4,5- and 3,4,5-isomers respectively. Migration of 2- and 6-alkyl groups provides ready access to those 3,5-dialkyl- and 3,4,5-trialkyl-phenyl ketones in which the alkyl groups are methyl, ethyl, or n-propyl. The method is not applicable with secondary or tertiary alkyl groups, as these are replaced by hydrogen or methyl under the conditions required for the isomerization. The 2-alkyl group may migrate to the neighboring *meta*-position even when this is already occupied by another alkyl group. The more mobile group will then migrate to the *para*-position. (4) The 2-alkyl group may be eliminated if it is n-propyl or a larger n-alkyl group. 2,5-Di-n-propylacetophenone gives a mixture of the 3,5-isomer and 3-n-propylacetophenone. Secondary alkyl groups are still more readily displaced. m-Isopropyl- and 3,5-diisopropyl-acetophenone are obtained from the 2,5-isomer and excess of aluminum chloride in ethylene chloride solution at room temperature. (5) Dealkylation of the ketone, as in (4) is followed by alkylation only when this is effected intramolecularly. Thus 9-acetyl-1,2,3,4,5,6,7,8-octahydrophenanthrene affords 7-acetyl-2-methyl-4,5-cyclohexeno-indane in excellent yield.

All these reactions are determined by steric interaction and provide ketones in which the carbonyl group is less hindered by vicinal groups (22).

E. Acylation of Substituted Benzenoid Aromatics

Friedel-Crafts acylations proceed successfully not only with the aromatic hydrocarbons themselves but also with many of their substituted derivatives including halo, alkoxy, and even amino derivatives. Nitro and carbonyl compounds do not usually undergo the reaction. As a general rule electronegative groups if present in the hydrocarbons have an inhibiting effect on the reaction. This is most marked with benzene and monocyclic aromatics, so that acylation of nitrobenzene, benzoic acid, and benzophenone, for example, is impossible. Nitro derivatives of phenol ethers, however, are capable of participating in the reaction (596).

a. Deactivating substituents

Halogens attached to the aromatic ring exert a retarding influence, though fluorobenzene is acylated fairly easily. The intensity of the effect increases with the number of halogen atoms present in the nucleus but is not prohibitive, since, for example, it is possible to obtain 3',4'-dichloro-2-benzoylbenzoic acid in 70% yield by the reaction of *ortho*-dichlorobenzene with phthalic anhydride at 100° (503).

Monochloronaphthalenes undergo normal acylations (254), but 1,4-dichloronaphthalene gives a mixture of keto acids (634). The extent of acylation of halotoluenes with acyl chlorides depends upon the position of the halogen in the ring. *para*-Chlorotoluene fails to react with *ortho*-bromoethylbenzoyl bromide. Dihaloaromatic compounds usually react slowly with acyl chlorides (69), but chloroacetyl chloride reacts easily with *meta*-dichlorobenzene.

Groups, such as nitro, keto, sulfo, and cyano, generally prevent the Friedel-Crafts reactions from taking place. Nitrobenzene and methyl benzoate, for example, will not react. Aldehydes and ketones in general are likewise unreactive. This inhibition also explains the observation that usually only one acyl group can be introduced and the consequent ease of obtaining pure products in relatively high yields.

If an activating group, however, is also present it can balance the deactivating effects and reaction may proceed. *o*-Nitroanisole, for example, can be acylated readily.

Aromatic nitriles, in which the cyano group is joined to a nuclear carbon atom, generally fail to undergo the Friedel-Crafts reactions. Such nitriles are partially converted into triazines under the influence of aluminum chloride and acyl chlorides (366,367).

b. Phenols

The presence of phenolic hydroxyl groups alters the behavior of the Friedel-Crafts ketone synthesis. Many Lewis acid type metal halide catalysts, like aluminum chloride, initially react with phenol with evolution of hydrogen chloride and the formation of phenoxy aluminum dichloride.

$$C_6H_5OH + AlCl_3 \rightarrow C_6H_5OAlCl_2 + HCl$$

One equivalent of the reagent is thus consumed and in order to promote condensation with an acyl halide or anhydride an additional amount of catalyst is needed. It must also be mentioned that in the presence of excess phenol the reaction with aluminum chloride can go further to aluminum phenoxide. This is the case in the alkylation of phenol when only a small amount of catalyst is used. The catalytic activity is then due not to the aluminum chloride itself, but to the aluminum phenoxide. Acylations are sometimes accomplished by this procedure, despite the decreased solubility of the aryloxy metal halide compounds in solvents generally used in acylations. The method has, however, serious limitations.

Friedel-Crafts catalysts which do not react with phenols, but form only complexes, are frequently useful in overcoming difficulties encountered with aluminum chloride and other more reactive halide catalysts. Zinc chloride was found by Nencki (439,441) to be a very useful catalyst in the ketone synthesis of hydroxyaromatic compounds, using acetic acid as acetylating agent.

Boron trifluoride is perhaps an even more effective catalyst in ketone syntheses of phenols. As it forms a stable complex with phenol, this can be used as the catalyst in acylations with acetic anhydride. Fluorescein is obtained in almost quantitative yield when resorcinol reacts with phthalic anhydride in the presence of boron trifluoride (412).

Sulfuric acid is an excellent catalyst in acylations with acid anhydrides, as in the formation of fluorescein.

Other proton acids, such as anhydrous hydrogen fluoride and polyphosphoric acid are also useful catalysts in certain acylations involving phenolic compounds.

Methyl 6-methoxy-2,4-dihydroxybenzoate and acetic anhydride in acetic acid solution saturated with boron trifluoride form methyl 3-acetyl-2,4-dihydroxy-6-methoxybenzoate in high yield (338).

Instead of the free phenols themselves, aryl ethers (preferably the methyl ethers) may be used in the ketone syntheses.

Anisole undergoes Friedel-Crafts acylation normally if the reaction mixture is kept at a temperature low enough to avoid cleavage of the methyl group by the catalyst.

$$\underset{\text{OCH}_3}{\bigcirc} + \text{CH}_3\text{COCl} \xrightarrow[\text{CH}_3\text{NO}_2]{2\text{AlCl}_3} \underset{\text{ĊOCH}_3}{\overset{\text{OCH}_3}{\bigcirc}} + \text{HCl}$$

If the free hydroxyacetophenone is ultimately required the product is treated with more aluminum chloride in warm solution to achieve demethylation. The catalyst initially forms only a labile complex with anisole and so is not restrained from catalytic activity. Other catalysts, including boron trifluoride, are also effective in the acylation of phenol ethers (217,416).

The acylation of reactive aryl ethers may be effected in the presence of haloacetic acids as catalysts. The conversion of anisole to p-methoxyacetophenone, for instance, takes place in a 91% yield when acetic anhydride and anisole are allowed to react in the presence of trifluoroacetic acid (455).

Cleavage of the alkyl group in alkoxyaromatic compounds may take place during a Friedel-Crafts acylation reaction. Dealkylation of alkoxyaromatic compounds have been frequently reported in the literature. However, cleavage or migration of acyl groups attached to an aromatic nucleus under the influence of aluminum chloride (or related catalysts) has been observed only occasionally. 9-Acyl-anthracenes, for example, are converted to 1- and 2-acyl isomers. 1-Acetylacenaphthene changes to 5,6- and 3,8-diacetylacenaph-thenes.

c. Fries reaction

The Fries reaction for the preparation of hydroxyaromatic ketones consists of the acid-catalyzed rearrangement of phenol esters. In 1908 Fries discovered (202) that phenol esters of chloroacetic acid rearranged under the catalytic effect of aluminum chloride to give good yields of *ortho-* and *para*-hydroxy-ω-chloroacetophenones. A similar rearrangement was also observed by Fries in the case of phenol acetates (203).

$$\underset{}{\overset{\text{O·OCCH}_3}{\bigcirc}} \xrightarrow{\text{AlCl}_3} \underset{\text{ĊOCH}_3}{\overset{\text{OH}}{\bigcirc}}$$

The reaction was found in subsequent work to be of quite general use and a variety of catalysts other than aluminum chloride, such as

AlBr$_3$, FeCl$_3$, ZnCl$_2$, SnCl$_4$, TiCl$_4$, CrCl$_3$, BF$_3$, CH$_3$C$_6$H$_4$SO$_3$H, H$_3$PO$_4$, HF, etc., were found to be useful.

Aryl arenesulfonates undergo Fries rearrangements similarly to phenol esters.

d. Amines

Friedel-Crafts ring acylation of primary or secondary aromatic amines generally cannot be effected, but is useful in the case of acyl anilides (371). Acetanilide reacts with chloroacetyl chloride to give p-chloroacetylacetanilide in 83% yield (382). In the acyl anilides the nitrogen atom has lost most of its power to coordinate with the catalyst although coordination with the carbonyl oxygen is possible.

Somewhat better results are obtained in the reaction of aromatic amines or their derivatives with alkanoyl halides than with aroyl halides. Zinc chloride has proved to be the most satisfactory catalyst.

Amino groups can also be protected by attaching a para-toluene-sulfonyl (tosyl) group, which is removed after the Friedel-Crafts reaction has been realized (510). The synthesis of 2-aminobenzophenone is illustrative (573).

Only unsatisfactory results are obtained in attempted Friedel-Crafts ketone syntheses of aromatic N,N-dialkylamines. These are, of course, strong bases and yield complex condensation products.

Dimethylaniline is not ring acylated by acid anhydrides, just as it is generally not alkylated by olefins, using boron trifluoride and related catalysts. This is most probably due to the great tendency to form stable inactive complexes with the catalysts such as the boron trifluoride–dimethylaniline complex (416).

F. Heterocyclic Compounds

As would be expected, owing to the high basicity of the hetero-atoms, ring acylation is successful only with the more reactive aromatic heterocycles. Thiophene, which is difficult to alkylate by the Friedel-Crafts method, can be acylated readily. This difference appears to be due to the fact that acylating agents are more reactive than alkylating agents. 2-Acetylthiophene is formed in 83% yield when thiophene is treated with acetyl chloride in the presence of stannic chloride. Another method for preparing the ketone is to condense thiophene with acetic anhydride in the presence of 85% phosphoric acid. When furan is treated with acetic anhydride in the presence of orthophosphoric acid, 2-acetylfuran is obtained in 62% yield. Orthophosphoric acid is an effective catalyst for substitution of furans and thiophenes in general. By the use of phosphorus pentoxide as a catalyst acylation may be realized with the carboxylic acids themselves. By this method 2-benzoylthiophene, for example, can be prepared in 66% yield (246).

2-Acetylfuran has also been prepared by the interaction of furan and acetic anhydride in the presence of boron trifluoride-etherate catalyst (250).

Pyridine and quinoline in contrast are very resistant to ring acylation by the Friedel-Crafts method or, for that matter, by any other method.

G. Di- and Polyacylation

In the ring acylation of aromatic hydrocarbons it is generally difficult to introduce more than one acyl group into the nucleus. This is due to the fact that the acyl group first introduced is present in the reaction media as the aluminum chloride (or related acid catalyst) complex, which produces a powerful deactivating effect on any subsequent electrophilic acylation. The deactivating effect of the first acyl group is, however, counterbalanced in the case of polyalkylated benzenes which can yield diacyl derivatives (44,191,347, 360). Acetylmesitylene, for example, has been converted to diacetylmesitylene.

The activating influence of longer chain alkyl groups is considerably less than that of short-chain groups. In a compound containing an acetyl group *ortho* with respect to two methyl groups, a second acetyl group may enter the nucleus. *para*-Xylene fails to give a diacetyl derivative. Symmetrical triethylbenzenes and ethylmesitylene are capable of giving diacetyl derivatives. Diacetyl derivatives are obtained in nearly theoretical yield from mesitylene, durene, and isodurene, while pseudocumene gives only a monoacyl derivative.

Where the possibility of formation of polyacylated derivatives exists, the time allowed for the reaction largely determines the number of acyl groups introduced into the nucleus.

Naphthalene undergoes diacylation fairly easily. In acenaphthene and biphenyl both rings likewise may be involved. Acetyl chloride and chloroacetyl chloride, for example, yield 4,4'-diacetylbiphenyl and 4,4'-dichloroacetylbiphenyl respectively (576).

H. Replacement and Deacylation during Ketone Syntheses

Acylation differs from alkylation in being virtually irreversible. However, deacylation is comparatively easy to effect in compounds such as acetylmesitylene. Heating with syrupy phosphoric acid converts acetylmesitylene into acetic acid and mesitylene. In addition when attempts are made to introduce a second acyl group which is different from the first, replacement frequently occurs. Thus propionylation of acetylmesitylene yields not acetylpropionyl-mesitylene but dipropionylmesitylene, the acetyl group being displaced. This type of reaction is explained by assuming that in such cases acylation is reversible.

2. Acylating Agents in Aromatic Ketone Syntheses

The most frequently used acylating agents in Friedel-Crafts ketone syntheses are:

acyl halides	ketenes
acid anhydrides	amides
carboxylic acids	nitriles
esters	

The catalysts used are discussed in Chapter III in detail and include amongst others $AlBr_3$, $AlCl_3$, $BeCl_2$, BF_3, $FeBr_3$, $FeCl_3$, $GaCl_3$, $MoCl_5$, PCl_5, PF_5, $POCl_3$, $SbCl_3$, $SbCl_5$, $SnCl_4$, $TiCl_4$, $ZnBr_2$, $ZnCl_2$, $ZnCl_2$–$POCl_3$, H_2SO_4, H_3PO_4, polyphosphoric acid, $HClO_4$, HF, CF_3COOH, $(CF_3CO)_2O$, P_2O_5, $AgClO_4$, $AgBF_4$, I_2.

A. Acyl Halides

For the preparation of alkyl aryl ketones, acylation of aromatics with acyl chlorides, catalyzed by aluminum chloride, is the most frequently used method. The reaction is, however, very general and all four types of acyl halide (fluorides, chlorides, bromides, iodides) can be used.

Phenylacetyl chloride and other related phenylalkanoyl halides behave as typical acyl halides in Friedel-Crafts ketone syntheses and give the expected ketones.

Benzoyl halides and other aroyl halides also react in the usual way with aromatics giving the corresponding ketones.

Acyl fluorides acylate aromatics particularly well in the presence of boron trifluoride, antimony pentafluoride, phosphorus pentafluoride, and arsenic pentafluoride catalysts, although aluminum chloride and many other Lewis acid halides are equally active (472).

$$ArH + RCOF \xrightarrow{BF_3} ArCOR + HF$$

However, when chloride or bromide catalysts are used, halogen exchange of the acyl halide takes place and it is therefore difficult to decide which halide is the effective acylating agent. Halogen exchange also results in the formation of metal fluorides, which are frequently catalytically inactive.

$$3RCOF + AlCl_3 \rightarrow 3RCOCl + AlF_3$$

Thus the catalysts become deactivated and must be used in excess.

a. Relative reactivity

The reactivity of acyl halides in acylation reaction was reported to decrease in the order: acyl iodide > acyl bromide > acyl chloride > acyl fluoride (107). (This is the opposite of the order of reactivity of halogens in alkylation with alkyl halides.) Different order of reactivity was observed with boron halide catalysts. In this case the relative order of activity found was acyl fluoride > acyl bromide > acyl chloride (471).

The relative reactivity orders were not based on kinetic data but on relative yields obtained with various substrates. Yamase and Gotohave (651) have shown that the reactivity sequence with aluminum chloride may also vary with the nature of the acyl halide employed. Whereas acetyl and mesitoyl halides show the same regular order of activity mentioned, in the case of other acyl halides either the bromide or the chloride may be the more reactive acylating agent. Thus it is obvious that other factors (like steric hindrance) besides the electronegativity of the halogen must play some role in the relative reactivity (see Chapter V).

b. Side reactions and decomposition

Preliminary mixing of acyl and aluminum halides must be avoided when it effects undesirable reactions. These may arise in at least two ways: (a) The acyl chloride may contain a group (e.g., an aryl or olefinic group not attached to carbonyl or another deactivating group) which will react either intra- or intermolecularly with the acylating agent. (b) The acyl halide may decompose with evolution of carbon monoxide.

Decarbonylation occurs readily and is a feature of tertiary and α-phenylacyl halides under Friedel-Crafts conditions. Thus pivalyl chloride effects t-butylation, and diphenylacetyl, α-phenylpropionyl, and α-methyl-α-phenylpropionyl chlorides effect aralkylation.

Decomposition of pivalyl and other tertiary acyl halides by aluminum halide in the presence of aromatic compounds has been extensively investigated by Rothstein and Saville (537). With benzene and less reactive aromatics predominant alkylation takes place, following decarbonylation. The evolution of carbon monoxide is minimized and acylation is dominant when reactive aromatics (like anisole) are employed.

ω-Aryl substituted derivatives, such as β-phenylpropionyl chloride or γ-phenylbutyryl chloride, normally cyclize readily to give 1-indanone and 1-tetralone respectively. In the presence of very reactive substrates, however, good yields of ketones may be obtained by intermolecular acylation (103).

c. Unsaturated acyl halides

The Friedel-Crafts ketone synthesis seems to be generally applicable to unsaturated acyl halides, although some unavoidably give complications on account of their unsaturation. For example, with p-xylene, acrylyl chloride gives a large proportion of p-xylylethyl p-xylyl ketone, with some p-xylyl vinyl ketone (436). Hydrindone derivatives also result in place of unsaturated ketones (356).

$$C_6H_6 + CH_2{=}CHCOCl \longrightarrow$$

Longer chain unsaturated acyl halides yield the expected unsaturated ketones. 2,2-Dimethylacrylyl chloride gives a 40% yield of 2,2-dimethylvinyl phenyl ketone with benzene (151).

$$C_6H_6 + (CH_3)_2C{=}CHCOCl \longrightarrow$$

Cinnamoyl chloride behaves as might be predicted on the basis of the presence of a reactive ethylenic linkage and an acyl halide grouping. With benzene it gives 2,2-diphenylethyl phenyl ketone and 3-phenyl-1-hydrindone (362,571).

However, bromobenzene, diphenyl ether, biphenyl (21), or methoxy compounds (571) principally give the expected *para*-substituted unsaturated ketones. Phenylpropynoyl chloride yields an acetylenic ketone (572).

d. Haloacyl halides

In acylation with haloacyl halides although generally only the halogen attached to the carbonyl group reacts with aromatics (119, 240,344) internal condensation may sometimes take place involving a reactive group attached to the aromatic ring and the halogen of the haloalkyl radical. As an example chloroacetyl chloride upon reaction with *p*-tolyl methyl sulfide gives a ketodihydrothionaphthene in 50% yield.

$$\begin{array}{ccc} & CO & \\ & / \quad \backslash & \\ CH_3C_6H_3 & & CH_2 \\ & \backslash \quad / & \\ & S & \end{array}$$

With phenoxyacetic acid *p*-chloroacetylphenoxyacetic acid results, while with diphenyl ether bis-ω-chloroacetylphenyl ether is obtained (372,610).

Bromoacetyl bromide gives monoacylated derivatives with toluene, xylene, and ethylbenzene, but a diacylated product with mesitylene (120,206,372), a behavior which is also shown by bromopropionyl and bromobutyryl chlorides. With phenetole dichloroacetyl chloride gives dichloroacetyl phenetole. The reaction of trichloroacetyl chloride with benzene takes an unusual course resulting in the formation of triphenyl vinyl alcohol, $(C_6H_5)_2C\!=\!C(C_6H_5)OH$, although the normal condensation product, trichloroacetophenone is obtained when moist aluminum chloride is used in the reaction (55).

e. Dibasic acid halides

Acyl chlorides of dibasic acids react with aromatics under Friedel-Crafts conditions to yield diketones. Phosgene, for example, can be condensed with aromatic compounds to give symmetrical diaryl ketones such as Michler's ketone derived from dimethylaniline.

$$2C_6H_5N(CH_3)_2 + COCl_2 \longrightarrow (CH_3)_2N\langle\!\!\!\bigcirc\!\!\!\rangle\!-CO\!-\!\langle\!\!\!\bigcirc\!\!\!\rangle N(CH_3)_2 + 2HCl$$

Oxalyl chloride might be expected to condense with aromatic compounds to yield benzils, but this type of synthesis is realized only with extremely reactive aromatic nuclei, such as anisole. Usually oxalyl chloride breaks down to carbon monoxide and phosgene and yields benzophenones.

Phthaloyl chloride behaves in an anomalous fashion. Instead of the expected *ortho*-diaroylbenzenes it yields the isomeric diarylphthalides. With benzene, for example, the product is diphenylphthalide.

The reason for this behavior is to be found in the fact that phthaloyl chloride is known to exist in two isomeric forms. The normal form being changed to the pseudo form in the presence of aluminum chloride. Thus Friedel-Crafts reactions involve the pseudo form almost exclusively. Fumaryl chloride, on the other hand, is incapable of existing in a pseudo form and reacts normally. Thus with benzene it gives a high yield of *trans*-dibenzoylethylene.

The acyl chlorides of saturated dibasic acids higher than glutaric acid afford the corresponding diketones in satisfactory yields. Adipyl chloride, for example, gives 1,4-dibenzoylbutane.

Owing to the bifunctional nature of dibasic acid halides, cycliacylations frequently take place in their Friedel-Crafts reactions.

B. Acid Anhydrides

a. Monocarboxylic acid anhydrides

Depending on the amount of catalyst used and the reactivity of the acyl group, one or both of the acyl groups of carboxylic acid anhydrides can be utilized in the synthesis of aromatic ketones.

$$ArH + (RCO)_2O + 2AlCl_3 \rightarrow ArCOR \cdot AlCl_3 + RCOOAlCl_2 + 2HCl$$
$$ArH + (RCO)_2O + 3AlCl_3 \rightarrow 2ArCOR \cdot AlCl_3 + AlOCl + 2HCl$$

In addition to the acyl chlorides the acid anhydrides are the most frequently used acylating agents in Friedel-Crafts ketone syntheses.

Besides aluminum chloride, catalysts such as ferric chloride, zinc chloride, boron trifluoride, and trifluoroacetic acid and its anhydride are frequently used in acylations with acid anhydrides.

When "mixed" acid anhydrides are used as acylating agents in the Friedel-Crafts ketone synthesis generally only one of the acyl radicals will be utilized in the reaction.

Acetic formic anhydride gives only acetophenones with benzene and toluene in the presence of Lewis acid halide catalyst ($AlCl_3, BF_3$).

$$ArH + CH_3COOOCH \xrightarrow[BF_3]{AlCl_3} ArCOCH_3 + \underset{\overbrace{H_2O + CO}}{HCO_2H}$$

In the presence of anhydrous hydrogen fluoride as catalyst, however, aldehydes are also formed, although in small amounts.

$$2ArH + 2CH_3COOOCH \xrightarrow{HF} ArCOCH_3 + ArCHO + CH_3COOH + \underbrace{HCOOH}_{CO + H_2O}$$

Mixed acetic–alkanoic or aroic acid anhydrides yield a mixture of both ketones, but the ketone of higher molecular weight is formed preferentially (644).

Acetic benzoic anhydride, for example, reacts with benzene in the presence of aluminum chloride to give benzophenone almost exclusively.

$$C_6H_6 + C_6H_5COOOCCH_3 \xrightarrow{AlCl_3} C_6H_5-CO-C_6H_5 + CH_3CO_2H$$

With asymmetrical aroic acid anhydrides the ketone of lower molecular weight is obtained in larger amount (655).

b. Dicarboxylic acid anhydrides

Anhydrides of dicarboxylic acids, such as succinic and phthalic anhydrides, react with aromatics in the expected way. With benzene they furnish β-benzoylpropionic acid and *ortho*-benzoylbenzoic acid respectively. Another example of succinoylation is the formation of β-(3-acenaphthoyl)-propionic acid from acenaphthene. The condensation of aromatic compounds with phthalic anhydride to produce aroylbenzoic acids furnishes a convenient method of preparing crystalline derivatives for identification of compounds. Maleic anhydride can be employed in a similar manner to prepare β-aroyl-acrylic acids. An example is β-benzoylacrylic acid, obtained by the Friedel-Crafts reaction with benzene

$$C_6H_6 + \begin{array}{c} CHCO \\ \| \quad \diagdown O \\ \| \quad \diagup \\ CHCO \end{array} \xrightarrow{AlCl_3} C_6H_5COCH{=}CHCO_2H$$

However, it was recently observed that aluminum chloride and related Friedel-Crafts catalysts (such as stannic chloride, boron trifluoride, ferric chloride, and titanium chloride) also catalyze the Diels-Alder addition of maleic anhydride and related dienophiles to aromatic compounds (190,652).

c. Sulfocarboxylic acid anhydrides

"Mixed" anhydrides of sulfonic and carboxylic acids are known. Friedel-Crafts acylation of aromatics with o-sulfobenzoic anhydride, in the presence of aluminum chloride, proceeds similarly as does the like reaction with phthalic anhydride.   However, the product is a ketosulfonic acid instead of a ketocarboxylic acid (368)

As in the case of chlorosulfonylacyl halides, the acyl function is more reactive than the sulfonyl and ketones are formed in preference to sulfones.

Acetic sulfonic anhydrides, like acetic methanesulfonic anhydride, are obtainable either from the reaction of the corresponding sulfonyl halides with silver acetates (40,41),

$$RSO_2Cl + AgOOCCH_3 \rightarrow AgCl + RSO_2OOCCH_3$$

or result from the reaction of ketene with the appropriate sulfonic acid (474).

$$CH_3SO_3H + CH_2{=}C{=}O \rightarrow CH_3{-}SO_2{-}O{-}CO{-}CH_3$$

Acetic methanesulfonic anhydride, like the other mixed sulfo-carboxylic acid anhydrides, give ketones and not sulfones in Friedel-Crafts acylations.

$$C_6H_6 + CH_3COOSO_2CH_3 \xrightarrow[\text{HF}]{\text{AlCl}_3} C_6H_5COCH_3 + CH_3SO_3H$$

C. Carboxylic Acids

Although acyl halides and acid anhydrides are the most frequently employed acylating agents, it is also possible to apply the carboxylic acids themselves as acylating agents in Friedel-Crafts ketone syntheses.   To achieve the reaction milder catalysts are generally used. Aluminum chloride is of limited application (92,228,458) as there is evidence which strongly indicates that with its use the reaction is not direct acylation by the acid.   Instead there is an intermediate formation of acyl halide followed by the orthodox Friedel-Crafts acylation.   Thus more than two moles of aluminum halide are required for each mole of ketone formed.

$$RCOOH + 2AlCl_3 \rightarrow RCOCl{\cdot}AlCl_3 + AlOCl + HCl \xrightarrow{C_6H_6} C_6H_5COR + HCl$$

Zinc chloride is a catalyst frequently used in the acylation of aromatics with carboxylic acids (Nencki, 1881) (82,134,441).   Good results are obtained when it is the condensing agent in the acylation of resorcinol (or other oxybenzenes) (439).

However, the $ZnCl_2$ catalyzed acylation frequently also gives undesirable by-products.

As it is improbable that $ZnCl_2$ acts primarily as a halogenating agent to form the acyl halide, the reactions are indeed acylation by acids.

The theory that acids act as acylating agents *via* acyl chlorides, which are indeed formed by $AlCl_3$, cannot apply either to acylations catalyzed by BF_3, which is the catalyst most frequently used in Friedel-Crafts ketone synthesis when acids are the acylating agents (Meerwein, 1933) (414).

Phosphorus pentoxide, used as catalyst in the acylation of aromatic amines (183), is less favorable (363,418) than $ZnCl_2$ or BF_3.

Certain proton acids such as hydrogen fluoride (181) and polyphosphoric acid (PPA) (210,582) have also been applied successfully. In some cases phosphorus oxychloride (349,490) and trifluoroacetic anhydride (76) have been used as catalysts.

Phenols and heterocyclic aromatics, which are sufficiently reactive are acylated in good yield by carboxylic acids. The method is, however, difficult to apply to aromatic amines and with aromatic hydrocarbons the yields are generally low.

Aromatic acids can also act as acylating agents under Friedel-Crafts conditions. Polycarboxylic acids, such as pyromellitic acid, condense with benzene to yield keto acids (504).

D. Esters

Acylation of aromatics with esters has not been extensively studied.   It generally occurs with simultaneous alkylation

$$C_6H_6 + RCOOR' \rightarrow C_6H_5COR + R'OH \quad (C_6H_5R' + RCOOH)$$

Such substances as phenyl acetate, O-acetylsalicylic acid, or ethyl acetate will yield acetophenone with benzene in the presence of aluminum chloride (143,466).

O-acetylsalicylic acid gives a 70% yield of the ketone.   Dialkyl-anilines are acylated by diethyl oxalate to give ethyl p-dialkyl-aminophenyl glyoxylate in 60% yields (236).

$$C_6H_5N(CH_3)_2 + (CO_2C_2H_5)_2 \rightarrow (CH_3)_2N\!\!\left\langle\!\!\bigcirc\!\!\right\rangle\!\!-\!\!\underset{\underset{O}{\|}}{C}\!\!-\!\!\underset{\underset{O}{\|}}{C}\!\!-\!\!OC_2H_5$$

The acylation of phenols proceeds (202) through the transformation of phenol esters followed by rearrangement to the corresponding o- and p-hydroxyaryl ketones (Fries reaction).

E. Ketene, Carbon Suboxide

The similarity of the behavior of acyl halides and anhydrides in Friedel-Crafts reactions is easily understood when it is recalled that acyl halides themselves are actually mixed anhydrides (of hydrogen halides and carboxylic acids).   Ketene also behaves as an internal anhydride of acetic acid.

It was first shown by Hurd (286) that ketene would react with aromatic hydrocarbons in the presence of aluminum chloride to form acetophenones according to the equation:

$$C_6H_6 + H_2C\!\!=\!\!C\!\!=\!\!O \xrightarrow{\text{AlCl}_3} C_6H_5\!\!-\!\!\underset{\underset{O}{\|}}{C}\!\!-\!\!CH_3$$

Since Hurd's discovery several investigators (491,514,587,645) have studied ketene more extensively in an endeavor to increase the yield of the ketones and to identify the other products formed in the reactions.

Boese (66) has found that diketene would also react in a manner similar to ketene to form benzoylacetone with benzene in the presence of aluminum chloride.

$$C_6H_6 + \underset{\underset{CH_2-C=O}{|}}{CH_2\!\!=\!\!C\!\!-\!\!O} \xrightarrow{\text{AlCl}_3} C_6H_5\!\!-\!\!\underset{\underset{O}{\|}}{C}\!\!-\!\!CH_2\!\!-\!\!\underset{\underset{O}{\|}}{C}\!\!-\!\!CH_3$$

If carbon suboxide, the anhydride of malonic acid with a ketene type structure, should react with benzene in the presence of aluminum chloride similarly to ketene, then one would expect to obtain dibenzoylmethane. However, if only one of the two ketene functional groups participated in the Friedel-Crafts reaction, then benzoylacetic acid, or by its decarboxylation acetophenone would be anticipated.

$$ArH + O{=}C{=}C{=}C{=}O \xrightarrow{\text{AlCl}_3} \underset{O}{Ar{-}\overset{\displaystyle O}{\overset{\|}{C}}{-}CH{=}C{=}O} + H_2O$$

$$\underset{}{Ar{-}\overset{\displaystyle O}{\overset{\|}{C}}{-}CH_2COOH} \xrightarrow{-CO_2} Ar{-}\overset{\displaystyle O}{\overset{\|}{C}}{-}CH_3$$

Indeed, except for a large amount of polymerized carbon suboxide, the only product identified in the reaction of carbon suboxide with benzene, after hydrolysis by dilute acids, was acetophenone (54, 153). That acetophenone was formed by way of decarboxylation of benzoylacetic acid was substantiated (48) by the decarboxylation of the latter in the presence of dilute acids. In contrast dibenzoylmethane is stable in this medium.

F. Nitriles (Houben-Hoesch Synthesis)

A method of aromatic acylation closely related to the Gatterman aldehyde synthesis and particularly suited for the preparation of hydroxy ketones from phenols was discovered by Hoesch (265) and developed by Houben (281). It consists of the condensation of nitriles with phenols in the presence of hydrogen chloride and zinc chloride. An imide chloride is probably formed as an intermediate.

$$CH_3CN + HCl \rightarrow CH_3C\overset{\displaystyle NH}{\underset{\displaystyle Cl}{<}}$$

The reaction is general for phenols and nitriles and also has been used to characterize the latter (283).

When applied to trichloroacetonitrile the method serves to produce nitriles as well as ketones. The ketimine initially formed may be hydrolyzed to the ketone in acid media, but under the influence of alkali yields the nitrile.

$$
\text{ArH} + \text{Cl}_3\text{CCN} \xrightarrow{\text{AlCl}_3} \underset{\text{ArCCCl}_3}{\overset{\text{NH}}{\parallel}} \begin{cases} \nearrow & \underset{\text{ArCCCl}_3}{\overset{\text{O}}{\parallel}} \\ \searrow & \text{ArCN} + \text{CHCl}_3 \end{cases}
$$

Cyanation by this method affords good yields with many aromatic compounds, including toluene, xylenes, mesitylene, and tetralin. The yield of mesitonitrile is 68% (280).

G. Amides

Certain reactive aromatics can be acylated by anilides in the presence of phosphorus oxychloride (176). One of the most important applications of this method is the acylation of N,N-dialkyl aromatic amines (287,418,565).

The reaction can also be carried out by first converting the anilide to the corresponding imino chloride and then condensing the latter through the catalytic effect of aluminum chloride with the aromatic (564).

Zollinger developed the method of acylation with aliphatic amides and pointed out the close similarity with the Vilsmeier aldehyde synthesis (74).

N-methylacetamide and N,N-dimethylacetamide give acetophenones with reactive aromatic compounds when allowed to react in the presence of phosphorus oxychloride. Similarly N-methylbenzamide and N,N-dimethylbenzamide yield the corresponding benzophenones

3. Cycliacylation of Aromatic Compounds

A. Intramolecular

Cyclic ketones can be prepared by intramolecular Friedel-Crafts acylation involving an aromatic ring and an acyl halide group in an attached side chain. The method is used in the preparation of tetralones, hydrindones, chromones, xanthones, etc.

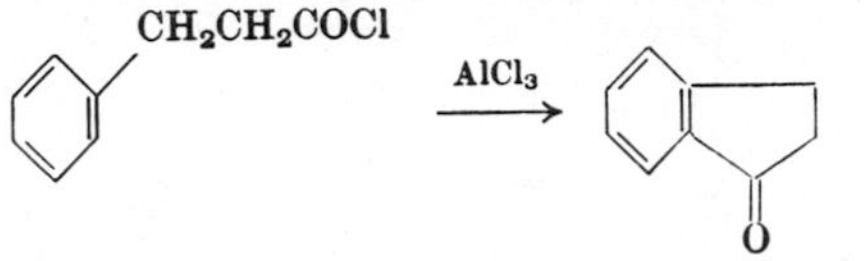

1-Hydrindone

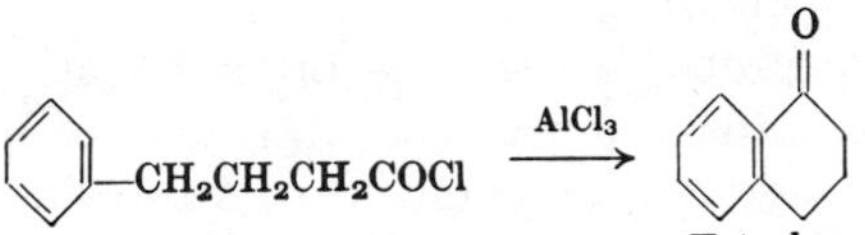

α-Tetralone

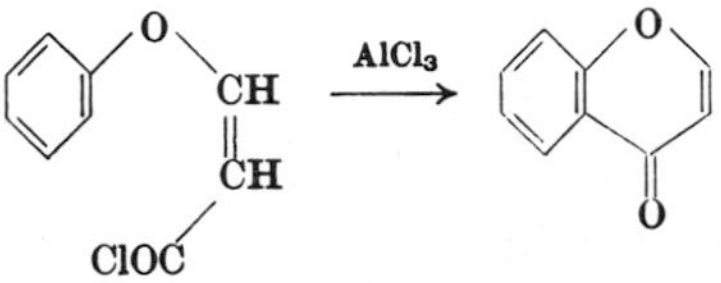

Chromone

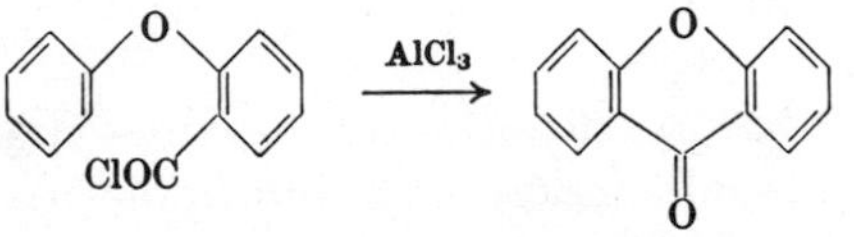

Xanthone

Cyclizations of ketoacyl chlorides leading to the formation of coumarandiones, chromandiones, indandiones, and anthraquinones are further examples of intramolecular cycliacylations.

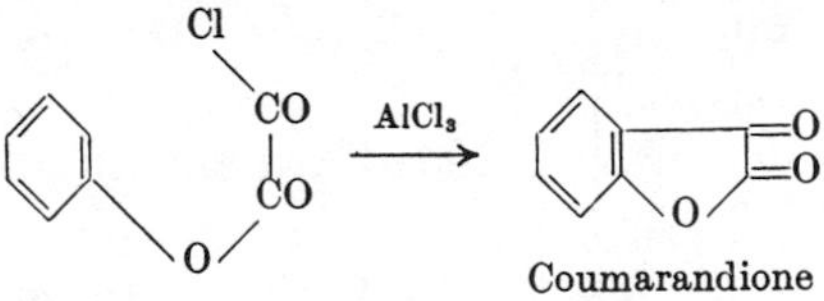

Coumarandione

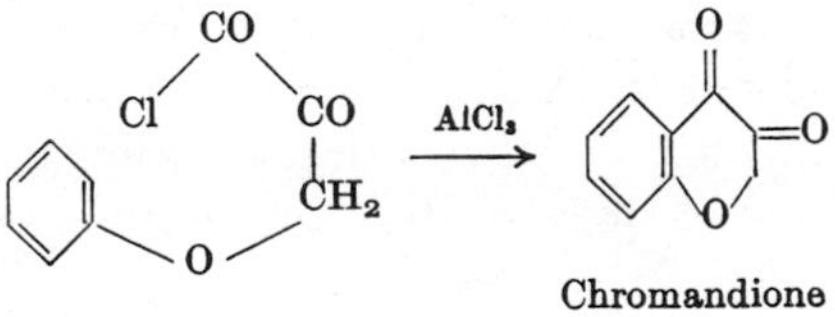

Chromandione

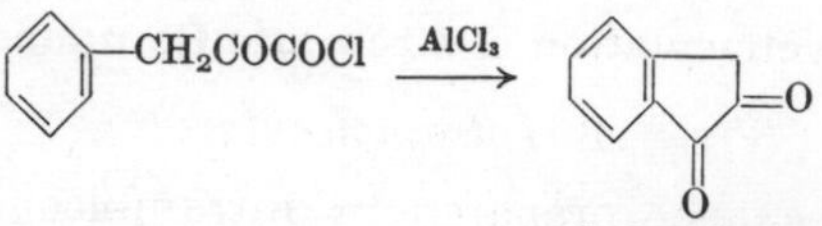

Indandione

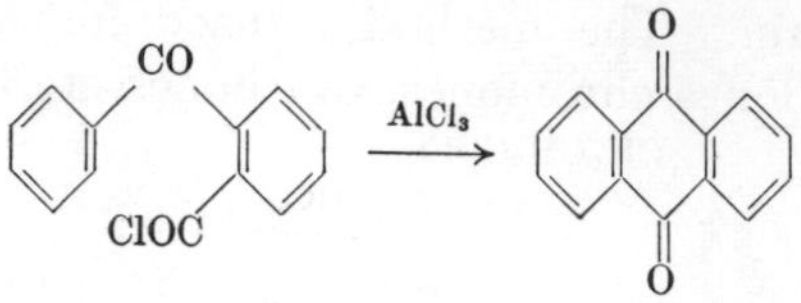

Anthraquinone

Five- and six-membered ring ketones are generally obtained in good yield by this method.

An alternative procedure that is often effective involves cycliacylation of the acids themselves catalyzed by hydrogen fluoride. γ-Phenylbutyric acid can be cyclized to α-tetralone.

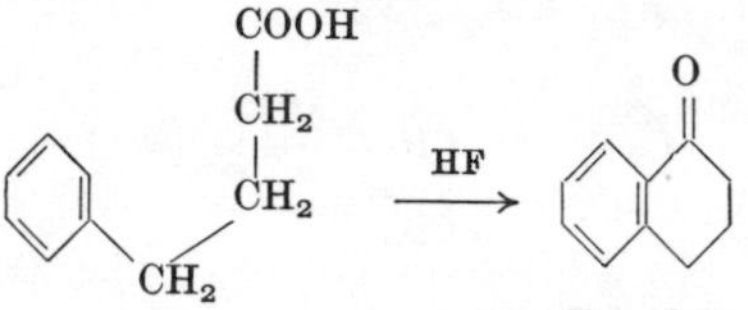

α-Tetralone

Polyphosphoric acid (PPA) is often employed as a catalyst for the cycliacylations.

Phenyl groups that are deactivated by an acyl group and thereby rendered incapable of undergoing intermolecular acylation may participate in intramolecular acylation (26).

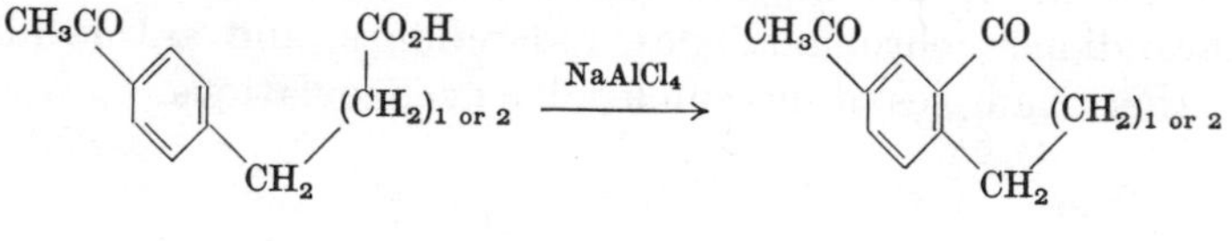

Similarly, o-aroylbenzoic acids afford anthraquinones.

B. Intermolecular, by Bifunctional Acylating Agents

Cycliacylations also take place readily in intermolecular acylations involving bifunctional acylating agents. Both functional groups may be acyl (as in the case of α,ω-diacyl halides) or one may be an alkylating group (as in unsaturated acyl halides or certain haloacyl halides) (192).

Diethylmalonyl   Diethylindandione
chloride

β-Chloropropionyl   1-Hydrindone
chloride

4. Aromatic Aldehyde Syntheses

Formylation, according to the general Friedel-Crafts acylation principle, would be expected to take place when formyl chloride or formic anhydride is allowed to react with an aromatic compound in the presence of aluminum chloride or other catalysts of the Friedel-Crafts type. However, neither the acid chloride nor anhydride of formic acid is known.

A. Gattermann-Koch Synthesis

In the presence of aluminum chloride and a small amount of cuprous halide as catalyst a mixture of hydrogen chloride and carbon monoxide serves as a formylating agent of aromatics (Gattermann and Koch) (213).

$$C_6H_5CH_3 + CO + HCl \xrightarrow[Cu_2Cl_2]{AlCl_3} CH_3-\!\!\bigcirc\!\!-CHO + HCl$$

Current theories suggest that the existence of formyl chloride is unnecessary since the actual alkylating agent, CHO^+, can be produced by protonation of carbon monoxide or its complexes. One of the practical difficulties involved in the laboratory use of this method is that of obtaining an equimolecular mixture of anhydrous hydrogen chloride and carbon monoxide. A suitable laboratory preparation involves the reaction of chlorosulfonic acid with formic acid (52),

$$HCO_2H + ClSO_3H \rightarrow H_2SO_4 + CO + HCl$$

or benzoyl chloride with formic acid (471).

$$HCO_2H + C_6H_5COCl \rightarrow C_6H_5CO_2H + CO + HCl$$

The Gattermann-Koch aldehyde synthesis is suitable for the preparation of simple aromatic aldehydes from benzene and its substituted derivatives, as well as from polycyclic aromatics. In the formylation reaction the *para*-isomers are produced preferentially. Aromatics with *meta*-directing substituents cannot be formylated.

Generally phenols and phenol ethers cannot be ring formylated by this reaction. Gattermann suggested that Cu_2Cl_2 is not soluble in phenolic solutions and this could be the cause of failure. However, it is more plausible that formylation of phenols first involves the formation of phenyl formates and these, instead of undergoing an acid-catalyzed Fries rearrangement, lose CO again very easily.

Investigation of the possible Fries rearrangement of phenyl formate fully supports this view. No *p*-hydroxybenzaldehyde is obtained even with relatively mild catalysts (HF, BF_3, PPA, etc.) (471).

The use of cuprous halides in the Gattermann-Koch reaction is necessary only if the reaction is carried at atmospheric pressure. In reactions carried out at higher pressures Cu_2Cl_2 is unnecessary. Thus the role of cuprous halides is mainly to complex with carbon monoxide to provide a sufficient carbon monoxide concentration even under atmospheric pressure and not a real co-catalytic effect.

A modification of the Gattermann-Koch reaction was introduced by Gresham and Tabet (226) by using BF_3 + HF (under pressure) instead of $AlCl_3$ + HCl

$$ArH + CO + HF + BF_3 \rightarrow ArCHO \cdot BF_3 + HF.$$

B. Formylation with Formyl Fluoride

The stability of formyl fluoride (452) allows a direct, simple Friedel-Crafts formylation reaction (Olah and Kuhn) (473,478).

$$ArH + FCHO \xrightarrow{BF_3} ArCHO + HF$$

Attempts to use the stable mixed anhydride, acetic formic anhydride in Friedel-Crafts type acylation with Lewis acid metal halide catalysts, result only in acetylation.

$$ArH + CH_3COOOCH \xrightarrow{AlCl_3} ArCOCH_3 + H_2O + CO$$

However, using anhydrous HF as catalyst a small amount of aldehyde is also formed. This is in accordance with the fact that acetic formic anhydride gives both acetyl and formyl fluoride with HF

$$2 \quad \begin{matrix} CH_3-C \diagdown \\ O \\ H-C \diagup \end{matrix} + 2HF \rightarrow CH_3COF + HCOF + CH_3CO_2H + HCO_2H$$

and with continuous removal of the low boiling HCOF the reaction can be directed to predominant formation of this compound (478).

C. Gattermann Reaction

The Gattermann-Koch reaction cannot generally be applied to formylation of phenolic compounds. To overcome this difficulty Gattermann (212) found an alternate method in which a mixture of hydrogen cyanide and hydrogen chloride reacts with phenols in the presence of zinc chloride or aluminum chloride.

$$\text{—OH} + HCN + HCl \xrightarrow[ZnCl_2]{AlCl_3} \text{—OH}$$

A useful modification of the method consists of using zinc cyanide (Adams). Hydrogen cyanide is generated *in situ* by addition of hydrogen chloride and at the same time zinc chloride, the required Friedel-Crafts catalyst, is produced.

$$Zn(CN)_2 + 2HCl \rightarrow ZnCl_2 + 2HCN$$

Similarly sodium cyanide is used in certain modifications of the reaction. This procedure has a great advantage in that it obviates the handling of hydrogen cyanide. If zinc chloride is not a sufficiently strong catalyst aluminum chloride is also added. Curiously, very pure zinc chloride will not bring about the reaction. However, small amounts of added metal halides, such as sodium or potassium chloride, will activate the catalyst.

A further modification of the synthesis of oxyaldehydes was introduced by Karrer (334), who used the more readily to handle cyanogen bromide in place of hydrogen cyanide.

$$BrCN + HCl \rightarrow Br—C{=}N—Cl$$
$$\underset{H}{\phantom{Br—C{=}N—}}$$

D. Vilsmeier Aldehyde Synthesis

The introduction of the aldehyde group into reactive aromatic and heterocyclic compounds through reaction with dialkyl- or alkylaryl-formamides in the presence of phosphorus oxychloride (45,331,330, 629,630,650) is called the Vilsmeier reaction.

Subsequently it was shown that arylalkylformamides are also able to act as formylating agents in the presence of phosgene (533).

That the Vilsmeier reaction is indeed an electrophilic substitution or a Friedel-Crafts type formylation was first pointed out by Wit-zinger (394,648), who suggested that the intermediate cation formed by the interaction of phosphorus oxychloride with the formamide is the electrophylic reagent. Arnold and Sorm (12) and Jutz (329) also suggest ionic addition compounds of the formamide with phosgene or phosphorus oxychloride.

E. Formylation with Orthoformic Acid Derivatives

H. Fischer (182) found in 1934 that porphyrin derivatives can be formylated with dichloromethyl ethyl ether in the presence of $SnBr_4$ as catalyst.

Recently (1960) Rieche and Gross (526) demonstrated that a general aromatic aldehyde synthesis can be achieved when dichloro-methyl methyl ether reacts with aromatics in the presence of a suitable Friedel-Crafts catalyst ($AlCl_3$, $TiCl_4$, $SnCl_4$) according to the equation:

$$ArH + Cl_2CHOCH_3 \xrightarrow{TiCl_4} [ArCH(Cl)OCH_3] \xrightarrow{-CH_3Cl} ArCHO.$$

In this reaction dichloromethyl methyl ether acts as a dihalide of methyl orthoformate.

$$H-\overset{\displaystyle OCH_3}{\underset{\displaystyle Cl}{C}}-Cl$$

Gross (229) was also able to show that the corresponding sulfide is another useful formylating agent

$$ArH + Cl_2CHSCH_3 \xrightarrow[AlCl_3]{-HCl} [ArCH(Cl)SCH_3] \xrightarrow{hydr.} ArCHO$$

In a simlar reaction phenols are ring formylated by trialkyl orthoformates in good yield:

$$C_6H_5OH + HC(OC_2H_5)_3 \xrightarrow{AlCl_3} HOC_6H_4CH(OC_2H_5)_2 \xrightarrow{hydr.} HOC_6H_4CHO$$

5. Syntheses of Aryl Nitriles and Aldoximes

A. Mercuric Fulminate

Mercuric fulminate reacts with benzene and aluminum chloride containing some hydrated salt to give benzaldoxime. It is assumed that the reaction proceeds through primary formation of fulminic acid by the action of HCl on the fulminate,

$$C_6H_6 + C{=}NOH \rightarrow C_6H_5CH{=}NOH.$$

Using freshly sublimed aluminum chloride, benzonitrile is formed. The formation of benzonitrile is probably due to subsequent formation of cyanogen chloride by action of the catalyst on fulminic acid and condensation with the hydrocarbon (557,558).

$$C{=}N-OH \xrightarrow{AlCl_3} C{=}N-Cl \longrightarrow Cl-C{\equiv}N \xrightarrow[-HCl]{C_6H_6} C_6H_5CN$$

B. Cyanogen Halides

Cyanogen bromide condenses with toluene in the presence of $AlCl_3$ to give p-tolunitrile. The reaction appears to be more satisfactory with phenol ethers (335,336).

$$CH_3C_6H_5 + BrCN \xrightarrow{AlCl_3} CH_3C_6H_4CN + HBr$$

C. Trichloroacetonitrile (Houben-Fischer Synthesis)

An alternative method described by Houben and Fischer (279, 280) consists of the reaction of trichloroacetonitrile, with a hydrocarbon and $AlCl_3$. A ketimine is formed which is hydrolyzed by treatment with alkali into the nitrile and chloroform. The reaction has been applied to toluene, some of its homologs and to phenols.

$$ArH + Cl_3CCN \xrightarrow{AlCl_3} [Ar-\overset{\overset{\displaystyle NH \cdot AlCl_3}{\|}}{C}-CCl_3] \xrightarrow{KOH} ArCN + CHCl_3$$

6. Syntheses of Arylcarboxylic Acids and Derivatives

The general Friedel-Crafts acylation principle can be successfully applied to the preparation of aromatic carboxylic acids.

A. Carbonyl Halides

Carbonyl halides (e.g., phosgene, carbonyl chloride fluoride, or carbonyl fluoride) are diacyl halides of carbonic acid.

Phosgene reacts with aromatic hydrocarbons in the presence of a Friedel-Crafts catalyst to give aroyl chlorides which yield acids on hydrolysis (201).

$$C_6H_6 + COCl_2 \xrightarrow[-HCl]{AlCl_3} C_6H_5COCl \xrightarrow{H_2O} C_6H_5CO_2H$$

However, the reaction generally does not stop at the formation of the aroyl chloride, but the latter readily reacts with excess of aromatics to give the corresponding ketones.

$$C_6H_5COCl + C_6H_6 \xrightarrow[-HCl]{AlCl_3} C_6H_5-CO-C_6H_5$$

The use of carbon disulfide as solvent helps the formation of the acid, as the intermediate complex formed ($C_6H_5COCl \cdot AlCl_3$) is insoluble in the solvent and therefore by precipitating helps to avoid secondary ketone formation.

Phosgene sometimes can be replaced advantageously by oxalyl chloride, which according to Liebermann (385) does not necessarily act as a primary source of phosgene. More probably it first acylates the aromatic compound to arylglyoxalyl chloride, which under the catalytic effect of aluminum chloride then readily loses CO.

$$ArH + ClOC-COCl \xrightarrow[-HCl]{AlCl_3} [ArCOCOCl] \xrightarrow{-CO} ArCOCl \xrightarrow{H_2O} ArCO_2H$$

Ketone formation is very much depressed in favor of acid formation in this modification of the reaction.

B. Carbamyl Chlorides (Gattermann Amide Synthesis)

Ketone formation can be avoided if one of the functional acyl halogens is "blocked" in phosgene. Gattermann (1888) (211) found that carbamyl chlorides, readily obtainable by the reaction of phosgene with ammonia or amines, are very suitable reagents for the preparation of amides in direct Friedel-Crafts acylation of aromatics. The formed amides can then be hydrolyzed to the acids (274).

$$ArH + ClOCNR_2 \xrightarrow{AlCl_3} ArCONR_2 \xrightarrow{hydr.} ArCO_2H$$

The Gattermann amide synthesis, although very versatile, has received relatively little attention. Recently its practical applicability was demonstrated in an industrial preparation of terephthalic acid (540).

The reaction succeeds well with substituted carbamyl chlorides such as $C_6H_5(C_2H_5)NCOCl$.

C. Isocyanates

Amides result from the reaction of aromatic hydrocarbons with isocyanates such as phenylisocyanate, in the presence of aluminum chloride.

$$C_6H_5NCO + C_6H_6 \xrightarrow{AlCl_3} C_6H_5NHCOC_6H_5$$

Phenyl isothiocyanate similarly gives thioanilides (389,390).

In the reactions, as correctly suggested (72,383,384) by Gattermann, the active acylating agent is indeed the carbamyl chloride, formed by interaction of isocyanate with HCl.

$$C_6H_5NCO + HCl \longrightarrow C_6H_5NHCOCl$$

D. Carbon Dioxide

Carbon dioxide can be looked upon as the acid anhydride of the dibasic carbonic acid. Accordingly it is not unexpected that it reacts with benzene to give benzoic acid and benzophenone in the presence of aluminum chloride (197,200,290,325,467).

$$ArH + CO_2 \xrightarrow{AlCl_3} ArCO_2AlCl_2 + HCl \xrightarrow{H_2O} ArCO_2H$$

With more reactive substances such as phenols, zinc or ferric chloride may be substituted as catalysts (429,467).

The reaction is, however, not easy to carry out in good yield and without side reactions. Elevated temperatures and high pressure are generally needed and only reactive substrates (such as phenols) react readily.

E. Miscellaneous

Alkyl chloroformates are easily decomposed by aluminum chloride (and related catalysts).

$$ROOCCl \xrightarrow{AlCl_3} RCl + CO_2$$

Therefore in Friedel-Crafts reactions they tend to give alkylated products instead of esters.

$$ArH + ClCOOR \xrightarrow{AlCl_3} ArR + HCl + CO_2$$

$$ArH + ClCOOR \rightarrow ArCOOR \xrightarrow{hydr.} ArCO_2H$$

Cyanogen chloride or bromide react with aromatics to give

$$Ar + BrCN \xrightarrow{AlCl_3} ArCN \xrightarrow{hydr.} ArCO_2H$$

nitriles, which can be hydrolyzed to the acids. Owing to side reactions the method has not received wide attention.

As mentioned, when trichloroacetonitrile is applied in the Houben-Hoesch synthesis, the intermediately formed ketimine can be hydrolyzed in acid media directly to the corresponding acid.

7. Aromatic Sulfonylation

Under Friedel-Crafts reaction conditions sulfonyl halides and sulfonic acid anhydrides sulfonylate aromatics, a reaction which can be considered as the analog of the related acylation with acyl halides and anhydrides. The products of the sulfonylations are sulfones, instead of ketones obtained in the acylation reactions. The first detailed investigation of the sulfonylation of aromatic compounds with sulfonyl chlorides was carried out by Olivier (489).

$$ArH + RSO_2Cl \xrightarrow{AlCl_3} ArSO_2R + HCl$$

Sulfonyl chlorides are the most frequently used reagents, although the bromides and fluorides also react. A summary of sulfonylation reactions up to 1943 was published by Suter (602).

The availability of sulfonic acid anhydrides, like methanesulfonic anhydride and benzenesulfonic anhydride, permits the use of these anhydrides in Friedel-Crafts sulfonylation reactions.

$$ArH + (CH_3SO_2)_2O \rightarrow ArSO_2CH_3 + CH_3SO_3H$$

Benzenesulfonic anhydride was claimed to be superior to benzene-sulfonyl chloride in Friedel-Crafts sulfonylation (180).

Catalysts used besides aluminum chloride are ferric chloride, antimony pentachloride, aluminum bromide, and boron trifluoride (618).

8. Aromatic Sulfonation

As in the case of nitration, the general Friedel-Crafts acylation principle can be applied in the case of sulfonation to the halides and anhydride of sulfuric acid (halosulfonic acids, sulfur trioxide). In a general sense sulfonations by sulfuric acid or oleum should also be considered in the scope of Friedel-Crafts acylations, the acids providing both sulfonating agent and catalyst. Aluminum chloride and boron trifluoride were found to be effective catalysts in certain sulfonations with halosulfonic acids.

The sulfonation of aromatic hydrocarbons with sulfuric acid is catalyzed by hydrogen fluoride. Similarly boron trifluoride has a pronounced catalytic effect (608) and in its presence reactions take place at lower temperatures. More uniform products are also obtained and considerably less sulfuric acid is needed in the sulfonation, probably due to the fact that BF_3 complexes with the water formed in the reaction and thus prohibits dilution of the sulfuric acid.

$$ArH + H_2SO_4 + BF_3 \rightarrow ArSO_3H + BF_3 \cdot H_2O$$

The migration of alkyl groups or halogen atoms during the sulfonation of polyalkyl (polyhalo) benzenes containing at least four substituents is known as the Jacobsen reaction (262,320,321,576). The product of sulfonation of a tetra- or pentamethylbenzene rearranges on contact with sulfuric acid towards a thermodynamically more stable isomer. Thus durenesulfonic acid is rearranged chiefly to prehnitenesulfonic acid.

9. Aromatic Sulfination

Benzene and homologs react with SO_2 in the presence of $AlCl_3$ and HCl to form sulfinic acids (351).

$$C_6H_6 + SO_2 \xrightarrow[\text{HCl}]{\text{AlCl}_3} C_6H_5SO_2H$$

10. Aromatic Nitration

The general Friedel-Crafts acylation principle can also be applied to inorganic acid halides and anhydrides. Consequently nitrations involving nitryl halides, dinitrogen tetroxide, and dinitrogen pentoxide (the halides and anhydrides of nitric acid) should be considered as Friedel-Crafts type reactions, as there obviously exists a very close analogy to the corresponding Friedel-Crafts ketone synthesis involving acyl halides and anhydrides.

A. Nitric Acid

In a more general sense nitric acid nitrations catalyzed by proton acids (H_2SO_4, $HClO_4$, HF, etc.) or by Lewis acid type metal halides (BF_3, $AlCl_3$, etc.) should also be considered as reactions of the Friedel-Crafts type, in analogy to the Friedel-Crafts ketone syntheses using carboxylic acids (acetic, propionic, etc.) as acylating agents. A more detailed discussion of the acid-catalyzed nitric acid nitration is, however, outside the scope of present treatment. An excellent summary of the reaction by de la Mare (155) is available.

B. Nitryl Halides

Friedel-Crafts nitration using nitryl chloride with aluminum chloride as catalyst has been reported by Price and Sears (520).

$$ArH + NO_2Cl \xrightarrow{AlCl_3} ArNO_2 + HCl$$

Aluminum chloride may cause reduction of nitro compounds giving β-phenylhydroxylamines which rearrange to p-aminophenol derivatives.

Tetramethylene sulfone was found to be a suitable solvent for carrying out homogenous Friedel-Crafts nitrations with nitryl halides and catalysts like $TiCl_4$ (370).

Anhydrous silver tetrafluoroborate is also a suitable cation-forming agent in these reactions (482).

$$ArH + NO_2Cl + AgBF_4 \to ArNO_2 + AgCl + HF + BF_3$$

C. Nitrogen Oxides

Dinitrogen tetroxide was found by Schaarschmidt (543) to be an effective Friedel-Crafts nitrating agent for aromatics in the presence of aluminum chloride or ferric chloride. Pinck (505) used sulfuric acid as catalyst in the same reaction.

Bachman (19) found that boron trifluoride was a suitable catalyst. He prepared the stable $N_2O_4 \cdot BF_3$ complex as a crystalline, insoluble salt which is suitable for nitration of aromatics. However, the suggested nitronium salt structure for the complex ($NO_2^+ BF_3 NO_2^-$) is not in agreement with later spectroscopic investigations. It was found from Raman spectroscopic investigations that the high melting, crystalline salt is a mixture of $NO^+ BF_4^-$ and $NO_2^+ BF_4^-$. $TiCl_4$,

BF_3, BCl_3, BF_5 and AsF_5 are effective catalysts for the N_2O_4 nitrations in homogenous media using tetramethylene sulfone as solvent.

Dinitrogen pentoxide, the anhydride of nitric acid, is a powerful nitrating agent, even in the absence of catalysts. Solid dinitrogen pentoxide is known to be the nitronium nitrate, $NO_2^+ NO_3^-$. Klemenz and Scholler (346) have shown that N_2O_5 nitrations are carried out preferably in sulfuric acid solutions. Bachman (18) reported the use of BF_3 as catalyst for the reaction. He isolated the stable, insoluble $N_2O_5 \cdot BF_3$ complex, which was found to be a stable nitronium salt. $TiCl_4$, $SnCl_4$, and SbF_5 are also effective catalysts in homogenous Friedel-Crafts nitrations with N_2O_5 (370).

D. Nitronium Salts

Stable nitronium salts such as $NO_2^+ BF_4^-$, $NO_2^+ AsF_6^-$, $NO_2^+ PF_6^-$, $(NO_2^+)_2^+ SiF_6^=$, $NO_2^+ ClO_4^-$, $NO_2^+ HS_2O_7^-$ etc. are extremely powerful Friedel-Crafts nitrating agents for aromatics (370,477).

$$ArH + NO_2^+ BF_4^- \rightarrow ArNO_2 + HF + BF_3$$

Nitronium salts are easily obtained either from N_2O_5, HF, and BF_3 (477)

$$N_2O_5 + 2HF + 3BF_3 = 2NO_2^+ BF^- + BF_3 \cdot H_2O$$

or from nitric acid, HF and BF_3 (370)

$$HNO_3 + HF + 2BF_3 = NO_2^+ BF_4^- + BF_3 \cdot H_2O$$

The nitrations can be carried out in homogenous media, using tetramethylenesulfone or nitromethane as solvent. A large variety of aromatic compounds have been nitrated with the nitronium salts in excellent yields. An advantage of the method is that it is carried out in non-aqueous media. Thus sensitive compounds, otherwise easily hydrolyzed or oxidized by nitric acid, can be nitrated without secondary effects.

11. Aromatic Amination

The direct introduction of the amino group into aromatic hydrocarbons was first reported by Graebe in 1901 (222). He treated hydroxylamine hydrochloride in the presence of aluminum chloride or ferric chloride, with benzene or toluene and obtained a small yield of aniline or toluidines. In the same year, independently, Jaubert (323) reported the same reaction.

$$C_6H_6 + NH_2OH \cdot HCl \xrightarrow[-H_2O]{AlCl_3} C_6H_5NH_2 \cdot HCl$$

Hydroxylamine sulfate was also used in a similar reaction (623, 624,625).

In subsequent investigations hydroxylamine-*o*-sulfonic acid and some of its derivatives were used as aminating agents (339,364).

$$ArH + H_2NOSO_3H \xrightarrow{AlCl_3} ArNH_2 + H_2SO_4$$

N-alkyl hydroxylamines have also been applied in amination of aromatics (365).

Aluminum chloride-catalyzed amination of aromatics with chloramines was recently extensively investigated by Kovacic and co-workers.

$$ArH + ClNH_2 \rightarrow ArNH_2 \cdot HCl$$

12. Aromatic Perchlorylation

The stability of perchloryl fluoride ($FClO_3$) recently made possible the development of a new type of aromatic substitution reaction, perchlorylation, which is closely related to the Friedel-Crafts acylation (296).

The general perchlorylation of reaction can be written

$$ArH + FClO_3 \xrightarrow{AlCl_3} ArClO_3 + HF$$

The fact that such a reaction can be achieved, despite the apparent oxidative nature of perchloryl fluoride, lies in its unexpected stability. Aluminum bromide was found to be an equally active catalyst for the reaction. Boron trifluoride is considerably less active than the aluminum halide catalyst, probably because of its limited solubility in aromatic hydrocarbons. The perchloryl aromatics obtained are interesting compounds, although they must be handled with care, owing to their explosive nature and sensitivity to mechanical shock and local overheating.

13. Aromatic Halogenation

The halogenation of a wide variety of aromatic compounds proceeds readily in the presence of aluminum chloride and related Friedel-Crafts catalysts. The halogenation of aromatics can be classified as a generalized acylation reaction. Halogenating agents used are elementary chlorine, bromine, or iodine (which can be considered as the halides of the corresponding hydrogen halide acids), and interhalogens (such as iodine monochloride and bromine monochloride).

A. Catalysts and Halogenating Agents

Anhydrous aluminum chloride, like ferric chloride and other Lewis acid metal chlorides, catalyzes nuclear chlorination and bromination

of aromatic compounds (73,117,324).   Iodine and iodine trichloride
are also effective ring halogenation catalysts.

$$Ar—H + X_2 \xrightarrow{\text{FeCl}_3} Ar—X + HX \quad (X = Cl, Br, I)$$

Although iodination can also be achieved, oxidative conditions
must be provided to remove HI formed in the reaction which otherwise tends to reduce the ring iodinated compounds (636).

The extent of halogenation is regulated by the amount of halogenating agents used and by reaction conditions.   Alkylated benzenes, phenols, phenol ethers, and polynuclear hydrocarbons undergo
ring halogenation with ease.

Iodine monochloride when used as halogenating agent under
Friedel-Crafts conditions preferentially gives ring iodination and
bromine monochloride results only in bromination.   Fluorine monochloride has not yet been reported as an aromatic halogenating
agent under Friedel-Crafts conditions, but would be expected to
give only ring chlorination.   In the presence of cation-forming silver
salts, such as anhydrous silver tetrafluoroborate, ring halogenations
with both elementary halogens and interhalogens take place readily.
Fluorine monochloride (FCl) was found to effect only ring chlorination with $AgBF_4$ (471).

Bromination can be conveniently effected by transfer of bromine
from one nucleus to another.   As the Friedel-Crafts isomerization of
bromoaromatic compounds generally takes place through an intermolecular mechanism, the migrating bromine atom serves as a
source of positive bromine, thus effecting ring brominations (487).
For example, bromobenzene is obtained from benzene by brominating with bromonaphthalene, tribromophenol, bromochlorophenol,
tetrabromocresol, or tribromocresol in the presence of aluminum
halides.   Monobromotoluene was prepared from toluene by brominating with the same brominating agents.   Tribromophenol was
prepared by brominating phenol with dibromobenzene.

Carbon tetrachloride and benzotrichloride are sometimes used as
chlorinating agents in the presence of aluminum chloride.

Paraffinic hydrocarbons having a polychloro-substituted carbon
atom can yield their halogen when heated with aromatic acid anhydrides (like phthalic anhydride) and aluminum chloride at high
temperatures.   For example, phthaloyl chloride has been prepared
in this manner from phthalic anhydride and carbon tetrachloride
(401).

When p-cresol is treated with carbon tetrachloride and aluminum chloride, 1-methyl-1-trichlormethyl-4-ketodihydrobenzene is obtained (657).

$$p\text{-}CH_3C_6H_4OH + CCl_4 \xrightarrow[-HCl]{AlCl_3}$$

This so-called Zincke-Suhl reaction was reinvestigated and its scope was established by Newman (456).

B. The Problem of Fluorination

Fluorine is not known to form either the cation F^+ or a positively polarized entity with acid catalysts. In the presence of Lewis acid fluorides chlorine trifluoride and bromine trifluoride form complex difluorohalonium cations F_2Cl^+, F_2Br^+, which under suitable conditions are able to give ring fluorinated products always formed together, however, with the corresponding chloro or bromo derivatives (481,494).

Ring fluorination with halogen fluorides, even in the presence of acidic fluoride catalysts, cannot be considered as electrophilic fluorination, but only as secondary reaction accompanying the interaction of the difluorohalonium salts

$$ArH + F_2Br^+SbF_6^- \rightarrow ArBrF_2 + HSbF_6$$
$$ArBrF_2 + ArH \rightarrow ArF + ArBr + HF$$

The fluorinating action of phenyliodofluoride, presumably in a radical type of substitution reaction, is known (Bockemüller).

Acidic fluoride catalysts (BF_3, SbF_5, etc.) are unable to effect heterolytic cleavage of elementary fluorine. Consequently no electrophilic fluorination with fluorine is known.

Aromatic ring bonded fluorine was found unable to undergo either intra- or intermolecular migration under Friedel-Crafts conditions (488). No trans-fluorination of aromatics with difluorobenzenes or fluorophenols could be achieved under acid catalyzed conditions.

C. Thiocyanation

Thiocyanogen [$(CNS)_2$] as a pseudo-halogen allows aromatic ring thiocyanation in the presence of aluminum chloride or aluminum thiocyanate [$Al(CNS)_3$] catalysts (583).

$$ArH + (CNS)_2 \xrightarrow[Al(CNS)_3 \cdot 2(C_2H_5)_2O]{AlCl_3} ArCNS + HCNS$$

14. Acylation of Aliphatic and Cycloaliphatic Compounds

The Friedel-Crafts syntheses are still mainly associated with the preparation of aromatic compounds. However, one of the most important fields of Friedel-Crafts syntheses is the acylation of aliphatic compounds.

A. Alkenes

a. With acyl halides

Olefins and cycloolefins react with acyl halides in the presence of aluminum chloride to form unsaturated ketones, the acyl group replacing a hydrogen atom in the olefinic group (Kondakoff, 1892; Darzens, 1910) (150,358,369).

Other acidic halides such as $ZnCl_2$, $SnCl_4$, $FeCl_3$, and $SbCl_3$ may be also used as catalysts (121–125,152,448).

Wieland and Bettag (1922) (642) realized first the reaction to be the addition of an acyl chloride to an olefinic double bond.

tetrahydroacetophenone

Saturated chloroketones are formed as intermediates in the reaction, a fact which has been established by isolating these compounds from the products when the reaction was carried out at a sufficiently low temperature.

$$CH_2=CH_2 + CH_3COCl \rightarrow [ClCH_2CH_2COCH_3] \rightarrow CH_2=CHCOCH_3 + HCl$$

Chloroketones are obtained when the olefin, acyl chloride, and aluminum chloride are added simultaneously to a carbon disulfide solution at 0° or below. However, even under these conditions unsaturated ketones are also formed in varying amounts (642).

The intermediate chloroketone eliminates HCl during prolonged heating with aluminum chloride and transforms to the unsaturated ketone.

Nenitzescu (446) found that if cyclohexane is used as a solvent for the reaction of olefins with acyl chlorides saturated ketones are formed. For example, when aluminum chloride was added to a mixture of cyclohexene and acetyl chloride in cyclohexane solution at about −10° and the reaction mixture was then warmed to 70°, hydrogen chloride was evolved and methyl cyclohexyl ketone was obtained.

$$\text{(cyclohexene)} + CH_3COCl \xrightarrow[AlCl_3]{C_6H_{12}} \text{(cyclohexane-COCH}_3) + HCl$$

Apparently the unsaturated ketone is hydrogenated at the expense of the solvent.

With cycloheptene acetyl chloride gives a mixture of acetylmethyl-cyclohexenes (204).

$$\text{(cycloheptene)} + CH_3COCl \xrightarrow[AlCl_3]{C_6H_{12}} \text{(acetylmethylcyclohexene)}$$

Nenitzescu (448) also found that when cyclohexene was added to acetyl chloride and aluminum chloride in carbon disulfide at $-15°$ and then benzene added, 1-phenyl-4-acetyl-cyclohexane was formed. The reaction probably involves a migration of the chlorine atom in the chloroketone followed by a Friedel-Crafts alkylation (327).

$$\text{(cyclohexene)} + CH_3COCl \rightarrow \underset{COCH_3}{\overset{Cl}{\bigcirc}} \rightleftharpoons \underset{COCH_3}{\overset{Cl}{\bigcirc}} \xrightarrow[-HCl]{C_6H_6} \underset{COCH_3}{\overset{C_6H_5}{\bigcirc}}$$

In a similar manner 4-phenyl-2-pentanone was prepared from propene.

$$CH_3CH{=}CH_2 \xrightarrow[C_6H_6, AlCl_3]{CH_3COCl} CH_3CH(C_6H_5)CH_2COCH_3$$

The same product is obtained when the unsaturated ketone reacts with benzene in the presence of aluminum chloride.

$$CH_3CH{=}CHCOCH_3 \xrightarrow[AlCl_3]{C_6H_6} CH_3CH(C_6H_5)CH_2COCH_3$$

Halogenated unsaturated ketones are secured by reaction of vinyl halides with acyl halides in the presence of aluminum chloride (46). Vinyl chloride and benzoyl chloride thus yield phenyl chlorovinyl ketone

$$ClCH{=}CH_2 + C_6H_5COCl \xrightarrow{AlCl_3} ClCH{=}CH{-}CO{-}C_6H_5$$

Analogously, acetyl chloride yields chlorovinyl methyl ketone, and isovaleroyl chloride yields isobutyl β-chlorovinyl ketone.

The reactions can be effected in carbon tetrachloride as solvent (292). That the condensation proceeds through intermediate formation of a saturated product, involving acylation of the alkenyl halide by the acyl halide

$$ClCH{=}CH_2 + C_6H_5COCl \rightarrow Cl_2CHCH_2{-}CO{-}C_6H_5$$

may be proved by isolation, under certain conditions, of the intermediate dihaloketone which tends to loose hydrogen chloride easily, even during distillation of the crude product under reduced pressure. Thus the acyl halide, showing a much higher reactivity than the vinylic halogen, react selectively.

The reaction of acyl halides with olefins does not always proceed as smoothly as with aromatic hydrocarbons and yields can be low. It is most conveniently effected by addition of the alkene to a solution of the acyl halide–aluminum halide complex in methylene or ethylene chloride, although excess acyl halide or nitromethane have been used successfully as solvents (30,111).

The acyl group combines with that carbon atom of the ethylenic bond which holds the smaller number of alkyl groups and the resulting cation may (a) combine with chloride ion, forming a β-chloroketone, (b) isomerize by transfer of hydride ion and subsequently combine with chloride ion, or (c) afford unsaturated ketone by loss of a proton. All these processes participate in the reaction of cyclohexene with acetyl chloride in the presence of aluminum chloride. Cleaner products and more reproducible results are obtained when the acyl halide is replaced by the anhydride of the corresponding monocarboxylic acid.

Balaban and Nenitzescu (33) observed that when branched-chain olefins react in the presence of aluminum chloride or stannic chloride catalyst with acyl chlorides, alkyl-substituted pyrylium salts are formed in addition to the chloroketones and unsaturated ketones.

In place of olefins the corresponding t-alkyl chlorides will also react with acyl chlorides. In the presence of Lewis acid halides they are in equilibrium with the olefins.

Tertiary alcohols can be also substituted for the branched olefins (and tertiary alkyl chlorides) and the acylating agent can be an acid anhydride (35,516). Useful catalysts include $SbCl_5$, $FeCl_3$, $TiCl_4$, $ZnCl_2$, $BF_3 \cdot Et_2O$, $NaClO_4$, 70% $HClO_4$, H_2SO_4.

Cyclic olefins such as methylene cycloalkanes

$$\underset{CH_2-C=CH_2}{\overset{[CH_2]_n}{\left[\rule{0pt}{18pt}\right.}} \quad (n=3,4)$$

are readily bis-acylated, yielding pyrylium salts (35). Allylbenzene similarly can be diacetylated (36).

An interesting ramification of the reaction is afforded by the decomposition of tertiary acyl halides in the presence of aluminum halides. Carbon monoxide is readily eliminated and a tertiary alkyl cation is formed which loses a proton and provides an olefin, which in turn combines with excess of the acyl halide. Decomposition of pivalyl chloride exemplifies the process.

$$(CH_3)_3CCOCl \cdot AlCl_3 \longrightarrow CO + (CH_3)_3C^+AlCl_4^- \longrightarrow$$
$$HCl + AlCl_3 + (CH_3)_2C=CH_2 \xrightarrow{(CH_3)_3CCOCl} (CH_3)_2C=CHCOC(CH_3)_3$$

Formation of an olefin from an alkyl halide in the presence of an acylating agent involves an alkyl cation as intermediate and may therefore incur isomerization: e.g., addition of cyclohexyl chloride or bromide to acetyl chloride–aluminum chloride in methylene chlorides gives methyl 2-methyl-1-cyclopentenyl ketone, and addition of n-butyl iodide gives 3-methyl-3-penten-2-one. Rearrangement of the carbon skeleton in reactions of this type can be prevented by addition of nitromethane, nitrobenzene, acyl halide, or another suitably basic compound which, while not preventing reaction, will assist removal of a proton from the alkyl cation and consequently enable the formation of olefin to forestall isomerization; thus interaction of cyclohexyl halide and acetylating agent in the presence of a nitro compound gives 1-cyclohexenyl methyl ketone (26).

b. With acid anhydrides

Olefins react with acid anhydrides in the presence of zinc chloride according to the general scheme (359,421):

$$\begin{array}{c} R-CH \\ \| \\ R-CH \end{array} + \begin{array}{c} R'CO \\ \diagdown \\ R'CO \end{array}O \xrightarrow{ZnCl_2} \begin{array}{c} R-CH-COR' \\ | \\ R-CH-OCOR' \end{array} \longrightarrow \begin{array}{c} R-C-COR' \\ \| \\ R-CH \end{array} + R'COOH$$

Stannic chloride is also recommended as a catalyst for the reaction (538)

$$\text{(cyclohexene)} + (CH_3CO)_2O \xrightarrow[25-35°]{SnCl_4} \text{(1-acetylcyclohexene, COCH}_3) + CH_3COOH$$

but the catalyst most frequently used is boron trifluoride.

It is probable that in the presence of BF_3 the reaction takes place in a similar way as in the case of $ZnCl_2$, but the keto esters first formed in the presence of BF_3 easily dissociate into the unsaturated ketones and the corresponding acids, and therefore are not observed during the reactions.   Cyclic- or iso-olefins react particularly readily with anhydrides of carboxylic acids containing 2–6 carbon atoms (104).

Diisobutylene, isononylene, and isodecylene react at room temperature.   Cyclohexene and acetic anhydride form tetrahydroacetophenone in good yield (416).

Anhydrides of dicarboxylic acids do not readily react with olefins. When milder catalysts (zinc chloride or stannic chloride) are used addition of hydrogen chloride to the olefin, giving alkyl or cycloalkyl chlorides may become an important side reaction.   This addition process is readily reversible in the presence of aluminum halides. Alkyl halides in the presence of this catalyst may therefore be used as a convenient source of olefin.

c. With carboxylic acids

The possibility of acylation of olefins by organic acids was suggested by Menshutkin (419,420) during the investigation of the dissociation of esters of tertiary alcohols into acids and olefins.   In practice the reaction was first carried out by Konovaloff (361) who used zinc chloride as catalyst.

A number of Lewis acid halides (including aluminum chloride) and proton acids (usually sulfuric acid) have been used as catalysts in the reaction in addition to zinc chloride.   The chief disadvantage of the acidic halide catalysts, however, is their low catalytic activity which causes the esters to be obtained in yields not exceeding 30%.

Boron trifluoride was first applied as a catalyst in the acylation of olefins with carboxylic acids by Nieuwland and co-workers (160). Propylene reacts in the presence of boron trifluoride with acetic, chloroacetic, dichloroacetic, and trichloroacetic acid to give the appropriate isopropyl esters.   Zavgorodnii (654) studied the reaction in detail and found that boron trifluoride is a general and very active catalyst.   Boron trifluoride can be used in the reaction as its coordination compound with acetic acid.   The most suitable catalyst, however, is boron fluoride–ethyl etherate which enables the addition of ethylene to acids to be carried out under relatively mild conditions which do not bring about side reactions (mainly the polymerization of olefins) and enables esters to be obtained in yield of 40–95%. Zavgorodnii also investigated the reaction of a large variety of other olefins and carboxylic acids.

Solutions of carboxylic acids in trifluoroacetic anhydrides have also been used for the acylation of olefins (256). The unsymmetrical anhydride formed adds across the unsaturated bond at room temperature. The initially formed acyl ester decomposes spontaneously to give the acylated olefin and trifluoroacetic acid.

$$RCOOOCCF_3 + R'CH{=}CHR'' \rightarrow RCOCHR'CHR''OOCCF_3 \rightarrow$$
$$RCOCR'{=}CHR'' + CF_3COOH$$

B. Alkynes

a. With acyl halides

Acyl halides react with acetylenic compounds in the presence of aluminum chloride. β-Chlorovinyl ketones are obtained as the product of the reaction of acetylene with acyl chlorides.

$$CH{\equiv}CH + CH_3COCl \rightarrow ClCH{=}CHCOCH_3$$

The reaction has been carried out with various homologs of acetylene and with acetyl, propionyl, butyryl, isovaleroyl, stearoyl, and benzoyl chlorides.

Phenylacetylene reacts in the presence of zinc or stannic chloride with acyl chlorides to give α-chloro-ω-acylstyrene.

b. With anhydrides

Phenylacetylene reacts with acetic anhydride in the presence of zinc chloride to give ω-acetyl-α-phenylethyl acetate, which loses acetic acid with ease and gives β-acetyl phenylacetylene.

c. With carboxylic acids

The addition of acetylene and its derivatives to carboxylic acid to form vinyl esters is catalyzed by boron trifluoride (635). Vinyl acetate was obtained from acetylene and acetic acid in 85% yield in the presence of boron trifluoride. Butyric, crotonic, and higher monobasic acids also add to acetylene in the presence of BF_3 and its compounds to form vinyl esters in high yields.

C. Alkanes

a. With acyl halides

Hopff (271) as well as Nenitzescu and co-workers (449) observed in 1931 that not only unsaturated but also saturated aliphatic hydrocarbons react with acyl chlorides in the presence of aluminum chloride to give ketones.

The saturated ketones obtained were derivatives, however, not of the starting hydrocarbons but of their isomerization products. As an example, cyclohexane and acetyl chloride gave 1-methyl-2-acetylcyclopentane. The isomeric six-membered ring ketone is not produced during the reaction even in traces (445).

Isomerization occurring during the reactions cannot be prevented by addition of nitro compounds or similar reagents since these, when present, prevent all reactions from occurring.

Nenitzescu and Cantuniari (444) were the first to show that a saturated hydrocarbon can hydrogenate the products of reaction of an olefin with an acylating agent. They obtained cyclohexyl methyl ketone in good yield by addition of cyclohexane to the reaction mixture of the acetylation of cyclohexene. The preparation of 3,4-dimethyl-2-pentanone and methyl 2-methylcyclopentyl ketone by reaction of acetylating agent with n-pentane and cyclohexane respectively exemplifies the process.

Hydride abstraction from a saturated hydrocarbon occurs most readily at a tertiary carbon atom. Thus decalin or methylcyclohexane react more readily with acetylating agents than does cyclohexane.

b. With carbon monoxide

Hopff (270) achieved condensation of carbon monoxide with saturated hydrocarbons according to the Gattermann-Koch reaction of aromatics. However, in the reaction not aldehydes but ketones were obtained. The carbon skeleton is split during the reaction, and in the cases where it is unbranched isomerization takes place. From n-butane and isobutane methyl isopropyl ketone is obtained, from n-pentane ethyl isopropyl ketone and from cyclohexane methyl cyclopentyl ketone. As by-products small amounts of acids and higher molecular weight ketones are formed. The reaction mechanism is generally interpreted in terms that involve primary formation of aldehydes with subsequent rearrangement to the corresponding ketone. This rearrangement was indeed carried out by Danilov (149).

In more detailed investigations Hopff and Nenitzescu (269,271, 277) found that in the reaction of CO with alkanes or cycloalkanes in the presence of aluminum chloride at slightly elevated temperatures and under pressure three products are always formed: (a) acid $C_zH_{y-1}COOH$, (b) ketone C_zH_yCO, and (c) ketone $(C_zH_{y-1})_2CO$. For example, in the reaction of isobutane with carbon monoxide pivalic acid, methyl isopropyl ketone, and isobutyl t-butyl ketone are formed.

Pines and Ipatieff (506) obtained 5% isobutyric acid from the reaction of propane and carbon monoxide at 80° together with 23% isopropyl isobutenyl ketone and 44% isobutyl isobutyrate. The latter was suggested to be formed by a Cannizzaro reaction of the intermediate isobutyraldehyde.

Hopff (32,273,274) using the complex catalyst $AlCl_3$, Cu_2Cl_2, HCl obtained isobutyric acid, methyl ethyl ketone, and isopropyl isobutyl ketone in the same reaction at lower reaction temperatures.

Balaban and Nenitzescu (32) investigated in detail the mechanism of the reaction of alkanes with carbon monoxide. They were unable to obtain proof of the intermediate formation of aldehydes suggested by Pines and Ipatieff. Instead they found evidence for the nucleophilic addition of CO to the intermediate carbonium ion, formed by hydride abstraction.

D. Aldehydes and Ketones

a. With acid anhydrides

Aliphatic compounds containing active methylene groups readily undergo Friedel-Crafts acylations. The condensation of ketones with acid anhydrides in the presence of boron trifluoride catalyst was first developed by Meerwein (416).

$$RCOCH_3 + (R'CO)_2O \xrightarrow{BF_3} RCOCH_2COR' + R'COOH$$

This method provides a very convenient synthesis of β-diketones and has achieved widespread application (249).

Besides aluminum chloride and boron fluoride, zinc chloride and ferric chloride are sometimes used as catalyst in the acylation of ketones with acid anhydrides (499).

An example is the acetylation of acetophenone, catalyzed by zinc chloride, to yield benzoylacetone.

$$C_6H_5COCH_3 + (CH_3CO)_2O \xrightarrow{ZnCl_2} C_6H_5COCH_2COCH_3 + CH_3CO_2H$$

Cycloaliphatic ketones such as cyclohexanone are also acylated in high yields by acid anhydrides in the presence of boron fluoride. Thus cyclohexanone gives acetylcyclohexanone with an 86% yield when acetylated with acetic anhydride in the presence of boron trifluoride (400).

b. With ketene

Ketene in the presence of boron trifluoride or its etherate is a particularly suitable acetylating agent in Friedel-Crafts reactions with aldehydes and ketones.

Ketones and aldehydes form lactones of the corresponding acids with ketene. In the presence of suitable catalysts ($ZnCl_2$, $AlCl_3$, etc.) ketene combines with many carbonyl compounds to produce β-lactones. An example is the formation of β-propiolactone from ketene and formaldehyde in the presence of zinc chloride (225, 237,373).

$$CH_2{=}C{=}O + CH_2O \xrightarrow{ZnCl_2} \begin{array}{c} CH_2{-}C{=}O \\ | \quad\quad | \\ CH_2{-}O \end{array}$$

Acetone yields dimethyl propiolactone.

$$CH_2{=}C{=}O + (CH_3)_2C{=}O \longrightarrow \begin{array}{c} CH_2{-}C{=}O \\ | \quad\quad | \\ (CH_3)_2C \quad O \end{array}$$

Acetone, ketene, and boron trifluoride etherate give β,β-dimethyl-acrylic acid and isopropenylacetic acid on subsequent hydrolysis of the primary addition products (67).

The β-lactone of 3-hydroxy-4-methyl-4-pentanoic acid is formed from ketene and methacraldehyde in the presence of boron trifluoride etherate in isopropyl ether solution (238).

Unsaturated ketones and ketene in the presence of boron fluoride give lactones of the corresponding unsaturated acids (653).

Diketones may enter into reaction with one or two molecules of ketene in the presence of boron fluoride etherate to form mono- and dilactones of the corresponding acids, which when decarboxylated form unsaturated ketones and dienic hydrocarbons (237).

Ketene and biacetyl, in the presence of boron trifluoride etherate, form a di-β-lactone.

$$2CH_2\!=\!C\!=\!O + \begin{array}{c} CH_3\!-\!C\!=\!O \\ | \\ CH_3\!-\!C\!=\!O \end{array} \longrightarrow \begin{array}{c} CH_2\!-\!C\!=\!O \\ | \qquad | \\ CH_3\!-\!C\!\!-\!\!-\!\!O \\ | \\ CH_3\!-\!C\!\!-\!\!-\!\!O \\ | \qquad | \\ CH_2\!-\!C\!=\!O \end{array}$$

Methylformal is acetylated by ketene to give methyl β-methoxy-propionate (85).

$$CH_2(OCH_3)_2 + CH_2\!=\!C\!=\!O \xrightarrow{\text{BF}_3} CH_3OCH_2CH_2COOCH_3$$

E. Alcohols

a. With acyl halides

An interesting reaction first described by Combes in 1886 (126,127, 128,129,235) but apparently little used since is that of the acyl chloride–aluminum chloride complexes with alcohols. In this early work it was observed that when an aliphatic acyl chloride is treated with aluminum chloride in carbon disulfide or chloroform solution, a solid crystalline compound is formed with evolution of hydrogen chloride. The claimed complex, $(CH_3CO)_2CHCCl_2 \cdot O \cdot AlCl_2$, was obtained by gradual addition of aluminum chloride to acetyl chloride dissolved in chloroform. The complex is violently decomposed upon addition of water with formation of acetylacetone. When it is added in small portions to well cooled ethyl alcohol, an energetic reaction takes place with evolution of hydrogen chloride. The isolated reaction products are ethyl acetate, ethyl acetoacetate, and ethyl diacetoacetate, $(CH_3CO)_2CHCOOC_2H_5$. The main product of the reaction is ethyl diacetoacetate. The readiness with which it splits up in the presence of alcohol is responsible for the formation of ethyl acetoacetate and ethyl acetate in this reaction.

In later work (475) it was observed that stable 1:1 addition compounds of acyl fluorides with boron trifluoride (and other Lewis acid fluorides such as SbF_5, PF_5, and AsF_5) which are acylium salts react easily and practically quantitatively with alcohols to form the corresponding esters.

$$RCO^+MF^-_{4,6} + R'OH \rightarrow RCOOR' + HF + MF_{3,5} \quad (M = B, Sb, As, P)$$

The ester formations therefore can be considered as O-acylation of alcohols by the Friedel-Crafts acylating agents.

b. With acid anhydrides

The esterification of alcohols with acid anhydrides is catalyzed by proton acids like sulfuric acid and trifluoroacetic acid, trifluoroacetic anhydride, or Lewis acids like boron trifluororide-etherate and other boron trifluoride complexes (415).

The reaction is so fast that in many instances it can be carried out in aqueous solution (using sulfuric acid as catalyst) because the esterification reaction exceeds in order of magnitude any hydrolysis of the anhydride which might otherwise take place in the aqueous media.

c. With carboxylic acids

The formation of an ester by the interaction of a carboxylic acid and an alcohol is generally carried out in the presence of a mineral acid catalyst. The reaction is a reversible process and can be made to proceed virtually to completion in either direction by the use of excess of alcohol or water.

$$RCOOH + R'OH \rightleftharpoons RCOOR' + H_2O$$

It is usual to employ one of the reactants, selected on the basis of availability and ease of recovery, in a large excess in order to convert the other as completely as possible to the ester. Since the rate of esterification, like the rates of many other reactions, is approximately doubled for every ten degree rise in temperature it is usual to esterify with application of heat. The rate of esterification is also increased enormously by the presence of strong mineral acids. In general practice hydrochloric acid and sulfuric acid are the ones most commonly used, the former being favored in the laboratory on account of its efficiency and the latter in the plant on account of its cheapness, and on account of its less corrosive effect on metals. In many cases a satisfactory yield can be obtained only by removal of one of the reactants during the process, generally the water.

That the esterification of alcohols by acids is indeed an acylation reaction and not O-alkylation of the acid by alcohol was proved by the reaction of $CH_3O^{18}H$ with benzoic acid. The heavy oxygen was retained completely in the ester.

$$C_6H_5COOH + CH_3O^{18}H \rightarrow C_6H_5COO^{18}CH_3 + H_2O$$

The esterification of highly hindered acids is carried out in 100% sulfuric acid. The reaction involves the formation of an acylium ion, which then interacts with the alcohol in a similar O-acylation as mentioned previously for acyl halide complexes.

$$RCOOH + HA \rightleftharpoons (RCO)^+ + H_2O + A^-$$
$$(RCO)^+ + ROH + A \rightleftharpoons RCOOR' + HA$$

Esters of mesitoic acid are formed in this way (454,614).

It has been pointed out previously that proton acids and Lewis acid halides show considerable similarities as catalysts. Therefore, although the treatment of the acid-catalyzed esterification itself is not considered in the scope of the present discussion, it is of some interest to point out that Lewis acid catalysts generally used in Friedel-Crafts systems (such as aluminum chloride or boron trifluoride) are highly effective catalysts in addition to proton acids in the achievement of esterification of alcohols by acids. The catalytic effect of aluminum chloride was investigated by Ishibashi and Tarama (317). The property of boron trifluoride of forming complex compounds with appreciably increased acidity with alcohols and carboxylic acids makes it a particularly good catalyst for esterification. If acetic acid and ethyl alcohol are heated under reflux, in the presence of 1–2% of the coordination compound of $BF_3 \cdot 2CH_3COOH$, ethylacetate is formed in good yield. Similarly, a series of other aliphatic alcohols and acids were found to form esters with high yield under the catalytic effect of boron trifluoride (264).

Boron trifluoride–ethyl etherate is also a very advantageous catalyst for the reaction (611).

Polyhydric alcohols and aliphatic acids containing not less than 12 carbon atoms in the molecule are reported to condense in the presence of boron trifluoride to the corresponding esters (542).

Substituted aromatic acids are appreciably more difficult to use in esterification of alcohols in the presence of boron trifluoride. As an example, salicylic acid forms esters with alchohols in the presence of BF_3. However, on prolonged reaction they isomerize to alkyl-substituted derivatives of salicylic acid and these are capable of entering into reaction with the second molecule of alcohol, finally giving alkyl-substituted salicylates (142).

Trifluoroacetic anhydride was introduced as a very effective esterification catalyst of general use (75). It combines the properties of providing a strong proton acid catalyst (CF_3CO_2H) with the dehydrating power of the anhydride, thus helping the reaction to go to completion by removal of water. Trifluoroacetates are generally less stable than normal carboxylic esters and this may explain their virtual absence from the reaction products.

d. With amides

The acetamide–boron trifluoride complex was found to be a good acylating agent by Nieuwland (585). It reacts readily with alcohols to form esters.

$$CH_3CONH_2 \cdot BF_3 + ROH \rightarrow CH_3CO_2R + BF_3 \cdot NH_3$$

$$(R = CH_3, C_2H_5, iso\text{-}C_3H_7, C_4H_9, t\text{-}C_4H_9, C_6H_5)$$

F. Ethers

a. With acyl halides

Ethers are cleaved by acyl halides in the presence of a Lewis acid halide catalyst, giving rise to alkyl halide and ester:

$$ROR + R'COX \xrightarrow{ZnCl_2} R'CO_2R + RX$$

The reaction was first observed by Descude in 1901 (156), who obtained good yields of ethyl chloride and ethyl acetate from the spontaneous reaction of ethyl ether with acetyl chloride in the presence of zinc chloride. The ability of various Lewis acids (such as $ZnCl_2$, $SnCl_4$, $TiCl_4$, $SbCl_5$, $FeCl_3$, $AlCl_3$, $SbCl_3$, BF_3) to catalyze the cleavage was investigated extensively by Meerwein and Maier-Hüser (415).

The preferred catalysts for the reaction are zinc chloride and stannic chloride. In some instances, notably in the case of cyclic ethers, titanium tetrachloride is as good as or superior to stannic chloride. In most instances, it would appear that only traces of the Lewis acid catalyst are required, despite the fact that many workers have used molecular equivalents of acyl and metal halides.

A few dibasic acid chlorides have been used. In the case of oxalyl chloride (627), ethyl ether, and zinc chloride, very poor yields of the expected product are obtained.

b. With acid anhydrides

The reaction of acid anhydrides with ethers in the presence of Lewis acid halide catalysts leads to a mixture of esters or in the case of cyclic ethers to α,ω-diacyloxyalkanes.

$$ROR' + (R''CO_2)O \rightarrow R''CO_2R + R''CO_2R'$$

The cleavage of ethers by means of acid anhydrides in the presence of Lewis acid catalysts was discovered by Knoevenagel (350). He obtained a small yield of ethyl acetate when ethyl ether was heated with acetic anhydride in the presence of ferric chloride for eight hours at 100°. Later workers obtained similar results with aluminum chloride. Cyclic ethers such as tetrahydrofuran and tetrahydropyran generally give much higher yields than do open chain ethers (493).

$$\text{[cyclic ether structure]} + (CH_3CO)_2O \xrightarrow[66\%]{ZnCl_2} CH_3COO(CH_2)_4OOCCH_3$$
$$\text{1,4-butanediol diacetate}$$

Stable, isolated alkyl oxocarbonium (acylium) salts acylate ethers readily, thus proving that the reactions are indeed typical Friedel-Crafts acylations (471).

c. With carboxylic acids

In place of alcohols ethers can also be esterified with carboxylic acids using boron trifluoride as catalyst. According to Hennion, Hinton, and Nieuwland (258) ethers react with aliphatic and aromatic carboxylic acids in the presence of boron trifluoride to give the corresponding esters with yields ranging from 21 to 54%.

G. Carboxylic Acids

A rather unusual application of aliphatic Friedel-Crafts acylation is the preparation of acid anhydrides by the reaction of an acyl chloride with the corresponding carboxylic acid in the presence of aluminum chloride (587).

H. Thiols

The acid-catalyzed reaction of mercaptans with carboxylic acids (with exception of formic acid) gives thiol esters.

$$RCOOH + RSH \rightarrow RCOSH + H_2O$$

Similarly to the reaction of acylium salts with alcohols, Olah and co-workers (478) have observed acylation of thiols to the corresponding thiol esters.

$$RCO^+MF_{4.6}^- + R'SH \rightarrow RCOSR' + HF + MF_{3.5}$$

15. Aliphatic Aldehyde Syntheses

Friedel-Crafts type aliphatic aldehyde syntheses are considerably less known than the corresponding aromatic aldehyde syntheses. However, it should be pointed out that the hydroformylation reaction of olefins (Roeben, 1938) (532) and the related oxo synthesis are effected by catalyst, which in nature is a strong acid, *e.g.*, cobalt tetracarbonyl hydride, $HCo(CO)_4$.

$$RCH{=}CH_2 + CO + H_2 \xrightarrow{HCo(CO)_4} \underset{\overset{|}{CHO}}{R{-}CH{-}CH_3} \text{ and } RCH_2CH_2{-}CHO \xrightarrow{H_2}$$

$$\underset{\overset{|}{CH_2OH}}{R{-}CH{-}CH_3} \text{ and } RCH_2CH_2CH_2OH$$

Although the treatment of these reactions is outside the scope of the Friedel-Crafts reactions there seems to be some inevitable correlation with the acid-catalyzed formylation of aromatics with carbon monoxide.

The Gattermann-Koch reaction when applied to alkenes or alkanes gives ketones or acids, but not aldehydes. However, application of the Vilsmeier aldehyde synthesis to aliphatic compounds is known.

1,2-Dialkoxyethylenes react with N-methylformanilide and $POCl_3$ to give alkoxymalondialdehydes (166).

$$\begin{array}{c} RO{-}CH \\ \| \\ RO{-}CH \end{array} \begin{array}{c} POCl_3 \\ \xrightarrow{\hspace{1cm}} \\ OHC{-}\underset{\overset{|}{CH_3}}{N}{-}C_6H_5 \end{array} \begin{array}{c} CHO \\ | \\ CH{-}OR \\ | \\ CHO \end{array}$$

Sorm (13) was able to obtain a series of malondialdehydes from alkoxyethylenes, dimethylformamide, and phosgene.

$$RO{-}CH{=}CH{-}R \xrightarrow[\text{(CH}_3)_2N{-}CHO]{COCl_2} \underset{\overset{|}{CHO}}{\overset{\overset{|}{CHO}}{R{-}CH}}$$

Vinyl alkyl ethers with orthoesters in the presence of boron trifluoride etherate are converted into acetals of dicarbonyl compounds (136). For example, vinyl methyl ether and trimethoxymethane react to give malondialdehyde dimethylacetal.

$$CH_3OCH{=}CH_2 + HC(OCH_3)_3 \xrightarrow{BF_3 \cdot O(C_2H_5)_2} (CH_3O)_2CHCH_2CH(OCH_3)_2.$$

16. Syntheses of Aliphatic Carboxylic Acids and Derivatives

Alkenes are carbonylated in the presence of acid catalysts at temperatures ranging from 75–110° and under high pressure (600–900 atm.) to give carboxylic acids (184,185,534).

$$(CH_3)_2C\!\!=\!\!CH_2 + CO + H_2O \xrightarrow[\text{BF}_3\cdot 2\text{H}_2\text{O}]{\text{H}_2\text{SO}_4} (CH_3)_3C\text{—COOH}$$

Koch (1958) was able to extend the scope of the acid-catalyzed olefin carbonylation when he found that olefins can be carbonylated in concentrated sulfuric acid at modest temperatures (0–40°) and low pressures with formic acid, which serves as the source of carbon monoxide (355).

$$\begin{array}{c} R_1 \qquad R_3 \\ \diagdown \quad \diagup \\ C\!\!=\!\!C \qquad + CO + H_2O \xrightarrow[\text{HF}\cdot \text{BF}_3]{\text{H}_2\text{SO}_4} \\ \diagup \quad \diagdown \\ R_2 \qquad R_4 \end{array} \qquad \begin{array}{c} R_1 \qquad R_3 \\ \diagdown \quad \diagup \\ CH\!\!-\!\!C\text{—COOH} \\ \diagup \quad \diagdown \\ R_2 \qquad R_4 \end{array}$$

$$\text{and } HOOC\text{—}\begin{array}{c} R_1 \qquad R_3 \\ \diagdown \quad \diagup \\ C\text{—CH} \\ \diagup \quad \diagdown \\ R_2 \qquad R_4 \end{array}$$

Liquid hydrogen fluoride, preferably in the presence of boron trifluoride, is an equally good catalyst and solvent system. Normal carboxylic acids are preferentially formed, the migration of alkyl groups and dimerization of olefins being suppressed if sufficient CO pressure is used (20–100 atm.). Alkenes and cycloalkenes readily undergo the reaction which is of broad scope (354).

The formation of carboxylic acids, accompanying ketone formation in reaction of alkanes with CO, was discussed previously in connection with the preparation of aliphatic ketones.

Carbon monoxide forms acids when it reacts under pressure in the presence of aluminum chloride with alkyl halides and the reaction product is hydrolyzed:

$$C_2H_5Cl + CO \xrightarrow{\text{AlCl}_3} C_2H_5COCl \xrightarrow[-\text{HCl}]{\text{H}_2\text{O}} C_2H_5CO_2H$$

Ketones are formed as by-products, together with higher homologous acids. t-Butyl chloride reacts very readily with CO (38) to give pivalic acid

$$(CH_3)_3CCl + CO \xrightarrow[\text{150 atm.}]{\text{FeCl}_3} (CH_3)_3CCOCl \xrightarrow{\text{H}_2\text{O}} (CH_3)_3CCO_2H$$

Carbon monoxide reacts with polychlorinated methanes such as carbon tetrachloride in the presence of aluminum chloride and under pressure to yield chloroacetyl chlorides (198).

$$CCl_4 + CO \xrightarrow{AlCl_3} CCl_3COCl$$

The high-pressure aluminum chloride catalyzed reaction of poly-chlorinated ethanes also yields chloroacyl chlorides. However, side reactions are more troublesome and yields are lower (293).

Carbon monoxide reacts with dialkyl ethers in the presence of aluminum chloride at elevated temperature and under high pressure to form aliphatic esters (393).

$$(C_2H_5)_2O \cdot AlCl_3 + CO \rightarrow C_2H_5COOC_2H_5 \cdot AlCl_3$$

Boron trifluoride in the presence of water is an especially useful catalyst for the reaction of ethers with carbon monoxide. For example methyl acetate is obtained from dimethyl ether and CO in good yield (161).

$$(CH_3)_2O + CO \xrightarrow{BF_3 \cdot H_2O} CH_3COOCH_3$$

Phosgene reacts with olefins according to its acid halide nature. Chloroacyl chlorides are obtained which can either be hydrolyzed to the corresponding acids or be esterified. Propylene and phosgene give β-chloroisobutyryl chloride in the presence of aluminum chloride, which can subsequently be esterified and dehydrohalo-genated to give methacrylates (523).

$$COCl_2 + CH_2{=}CH{-}CH_3 \xrightarrow{AlCl_3} \underset{\underset{CH_3}{|}}{CH_2Cl{-}CH{-}COCl}$$

$$\xrightarrow{ROH} \underset{\underset{CH_3}{|}}{\cdot CH_2Cl{-}CHCOOR} \xrightarrow{-HCl} \underset{\underset{CH_3}{|}}{CH_2{=}C{-}COOR}$$

Although the Gattermann amide synthesis has not yet been reported with olefins, it may represent a useful application:

$$R{-}CH{=}CH_2 + \underset{R'}{\overset{R'}{\diagdown}} N{-}COCl \cdot AlCl_3 \rightarrow \underset{\underset{Cl}{|}}{R{-}CH{-}CH_2{-}CONR'_2 \cdot AlCl_3} \xrightarrow{-HCl}$$

$$\rightarrow R{-}CH{=}CH{-}CONR'_2$$

Phosgene reacts in the presence of aluminum chloride with alkanes and cycloalkanes to give ketones (544). Although acyl chlorides must be intermediates in the reaction their isolation prior to ketone formation seems to be difficult.

Carbamyl chloride or its alkyl derivatives have been reported to react with alkanes and cycloalkanes to give acid amides (274,276). *n*-Pentane for example gives capryl amide with the molecular compound of carbamyl chloride and aluminum chloride.

$$n\text{-}C_5H_{12} + ClCONH_2\cdot AlCl_3 \rightarrow C_5H_{11}CONH_2\cdot AlCl_3 + HCl$$

17. Aliphatic Sulfonylation

The sulfonylation of aliphatic compounds has been less extensively investigated than that of aromatic compounds.

Arylsulfonyl chlorides add, in the presence of Friedel-Crafts catalyst, to alkenes (ethylene, propene, 1-butene, isobutylene) to give chloroalkyl aryl sulfones (227).

$$ArSO_2Cl + CH_2\!=\!CH_2 \xrightarrow[\text{FeCl}_3, \text{BF}_3]{\text{AlCl}_3} ArSO_2CH_2CH_2Cl$$

Subsequent dehydrochlorination could give vinyl aryl sulfones.

$$ArSO_2CH_2CH_2Cl \xrightarrow{-HCl} ArSO_2CH\!=\!CH_2$$

The reaction is the analog of the acylation of olefins with aranoyl halides.

No similar reactions of alkylsulfonyl halides or sulfonic acid anhydrides have been reported but there is no reason why they should not take place in reactions analogous to acylation with acyl halides and acid anhydrides.

$$RSO_2Cl + CH_2\!=\!CH_2 \rightarrow RSO_2\!-\!CH_2\!-\!CH_2\!-\!Cl \xrightarrow{-HCl} RSO_2CH\!=\!CH_2$$

$$(CH_3SO_2)_2O + CH_2\!=\!CH_2 \xrightarrow[\text{BF}_3, \text{FeCl}_3]{\text{AlCl}_3} \begin{array}{c} CH_2\!-\!CH_2\!-\!SO_2\!-\!CH_3 \rightarrow \\ | \\ OSO_2CH_3 \end{array}$$

$$\rightarrow CH_2\!=\!CH\!-\!SO_2\!-\!CH_3 + CH_3SO_3H$$

Although zinc chloride does not catalyze the cleavage of ethyl ether by benzenesulfonyl chloride or toluenesulfonyl chloride, the use of ferric chloride or aluminum chloride (415) leads to the formation of ethyl chloride and ethyl *p*-toluenesulfonate. Boron fluoride is ineffective as catalyst.

$$(C_2H_5)_2O + ClO_2S\!-\!\langle\ \rangle\!-\!CH_3 \xrightarrow{\text{AlCl}_3} C_2H_5OO_2S\!-\!\langle\ \rangle\!-\!CH_3 + C_2H_5Cl$$

The reaction is a close analog of the Friedel-Crafts acylation of ethers.

18. Aliphatic Nitration

Aliphatic nitration is generally considered a free radical type of reaction. There are certain instances, however, where acid-catalyzed ionic nitration has importance.

Investigation of the effect of nitronium salts such as $NO_2^+BF_4^-$ on alkenes has shown that the nitronium ion readily attacks the olefinic double bond, resulting in the formation of a carbonium ion which then initiates polymerization of the olefin (484).

$$CH_2{=}CH{-}R + NO_2^+ BF_4^- \rightarrow \left[O_2NCH_2{-}\overset{+}{C}H{-}R\right] BF_4^- \xrightarrow{nRCH{=}CH_2}$$

$$
\begin{array}{l}
NO_2 \\
| \\
CH_2 \\
| \qquad\quad\left[\begin{array}{c} R \\ | \\ CH_2{-}CH \end{array}\right] \qquad\quad \begin{array}{c} R \\ | \\ CH_2{-}CH \end{array} \xrightarrow{-H^+} \\
CH{-}\!\!\!\overline{} \qquad\qquad\qquad\qquad {}_{n-1} \qquad\qquad\qquad {}_{+} \\
| \\
R
\end{array}
$$

$$
\begin{array}{l}
NO_2 \\
| \\
CH_2 \\
| \qquad\quad\left[\begin{array}{c} R \\ | \\ CH_2{-}CH \end{array}\right] \\
CH{-}\!\!\!\overline{} \qquad\qquad\qquad {}_{n-1} \qquad CH{=}CH{-}R \\
| \\
R
\end{array}
$$

So far no systematic investigation of the nitration of alkenes with nitronium salts has been carried out. It has also been observed that NO_2^+ is able to effect hydride abstraction from tertiary and secondary carbon atoms in alkanes (see Chapter III for recent developments).

The addition of nitryl halides and nitrogen oxides to olefinic double bonds takes place readily and no catalyst is generally needed. The presence of Lewis acid catalyst, however, sometimes helps the reaction with insufficiently reactive olefins enabling subsequent dehydrohalogenation to nitroolefins. Tetramethylene sulfone or nitromethane are suitable solvents for the reaction

$$
\begin{array}{c}
R{-}CH{=}CH_2 + NO_2Cl \longrightarrow R{-}CH{-}CH_2{-}NO_2 \\
| \\
Cl
\end{array}
$$

19. Aliphatic Halogenation

Both addition halogenation of unsaturated compounds and substitution halogenation of saturated compounds can be achieved.

A. Addition to Alkenes and Alkynes

Halogenation of unsaturated hydrocarbons, particularly olefins such as ethylene or propylene, is promoted by the use of catalysts comprising of a complex cyanide and aluminum chloride (43,525). 1,2-Dichloroethane is formed by passing ethylene, together with the amount of chlorine required, over the catalyst.

The addition of chlorine to ethylene may be effected in the liquid phase and in the presence of a co-solvent or in the vapor phase in the presence of metal halide contact agents. The use of solvents ordinarily leads to considerable substitution as well as addition. According to Hammond the addition of relatively small amounts (0.05–0.25% weight) of anhydrous ferric chloride serves to depress substitution reactions (242).

Anhydrous ferric chloride has also been successfully applied as catalyst in the chlorination of trichloroethylene to pentachloroethane and in the chlorination of perchloroethylene to hexachloroethane.

The formation of alkyl halides by the addition of hydrogen halides to olefins in the presence of catalysts has been studied extensively since the products of such reactions are of industrial importance. The reaction of ethylene and hydrogen chloride to form ethyl chloride is exothermic. The two gases will not combine, however, when heated, but in the presence of aluminum chloride and similar metal halide catalysts at 100–180° ethyl chloride is formed (50,641).

Propene reacts with hydrogen chloride at 80° using aluminum chloride on silica-gel as catalyst to give isopropyl chloride in practically quantitative yield. There is no indication of the formation of n-propyl chloride but a small amount of higher boiling compounds, probably hexyl chloride, has been reported as by-product of the reaction (86).

The form of the catalyst is of importance. Whereas aluminum chloride on asbestos or on glass gives good results, pure aluminum chloride gives a mixture of hydrocarbons and organo-aluminum compounds (640).

The use of superatmospheric pressure permits lower reaction temperatures (145,598).

When ethyl chloride is used as solvent, the reaction of ethylene and hydrogen chloride in the presence of aluminum chloride can be carried out even at −78° with practically quantitative yield. Ethyl bromide can be prepared similarly (622).

With increasing temperature the conversion to polymeric products increases, consequently overall yield of ethyl halides is decreased.

Halogenated olefins react with hydrogen halides in a similar way. For example, 1,1-dichloroethane is obtained when vinyl chloride reacts with hydrogen chloride over aluminum chloride at 125°. Only a small amount of 1,2-dichloroethane is formed as by-product (118).

Chlorinated solvents, such as trichloroethane, acetylene tetrachloride, or 1,1-dichloroethane can be used to carry out the reaction.

The chlorination of acetylene in the absence of catalysts occurs with explosive violence. The reaction may be effected in the presence of aluminum chloride (431). However, in recent practice other catalysts have been found to be more advantageous. The production of tetrachloroethane by the chlorination of acetylene is probably best carried out using a large body of the final product, tetrachloroethane, as a diluent for the reactant gases. Anhydrous ferric chloride is the catalyst and the reaction is carried out under reduced pressure (56,179). Tetrachloroethane can be dehydrochlorinated to trichloroethylene, which in turn can be hydrolyzed in the presence of sulfuric acid to give chloroacetic acid.

B. Substitution of Alkanes

Saturated aliphatic hydrocarbons can be chlorinated in the presence of aluminum chloride and related catalysts. Products such as tetrachloroethane may be prepared using molten aluminum chloride–ferric chloride–sodium chloride catalysts at 250–500° (223).

Ethylene polymers with a considerable range of molecular weight either in the molten state or preferably in an inert solvent can be halogenated using aluminum chloride as catalyst to produce rubberlike or fibrous products, which are useful for electrical insulation and other applications (177).

C. Substitution of Ethers and Alcohols

Alkyl halides can be prepared by the reaction of an aliphatic ether with anhydrous hydrogen halides in the presence of aluminum chloride at a temperature below the boiling point of the alkyl halide (84). Heating of aliphatic ethers with aluminum chloride or other metal halide catalysts generally also results in the cleavage of the ether bond with subsequent alkyl halide formation.

The reaction of hydrogen halides with alcohols (a typical esterification) is catalyzed by aluminum chloride and related acid catalysts. Alcohols themselves react with aluminum chloride and other metal halide catalysts on heating, to give alkyl chlorides (464,465).

Friedel-Crafts halides and particularly zinc and aluminum chloride have been found effective dehydrochlorinating agents. When pentachloroethane is heated with aluminum chloride, tetrachloroethylene is obtained (647). Similar catalytic methods were applied earlier in the dehydrochlorination of 1,2-dichloroethane to vinyl chloride. This reaction is now usually carried out by thermal dehydrochlorination.

D. Fluorination

Bromination and iodination reactions generally take place easily and under similar conditions to chlorinations. Fluorination represents a separate problem. Elementary fluorine attacks all organic compounds with violence. This violent and disruptive activity is in harmony with theoretical considerations and thermodynamic data. Any halogen, X_2, may react either by an ionic (heterolytic) mechanism giving the active X^+ ion, or by an atomic (homolytic) reaction, giving the active $X\cdot$ radical. Fluorine, the most electronegative of all elements, reacts atomically rather than ionically, because complete removal of an electron to form the positive F^+ ion would be very difficult. Consequently in direct organic fluorinations fast radical chain reactions occur, generally with subsequent decompositions.

Of the organic fluorination methods the displacement process of halogens with an inorganic fluoride like HF or SbF_3 is the most commonly used. The Swarts exchange reaction (603) of organic chlorides or bromides with antimony trifluoride catalyzed by antimony pentachloride (or pentafluoride) is closely related to the Friedel-Crafts type. Probably the quinquevalent fluoride acts as a complexing agent, transforming the C–Cl bond to the highly polarized haloantimonate complex, the decomposition of which results in the halogen exchange. (The Sb–Cl bond, being stronger than the Sb–F bond, can be the driving force for the reaction.) Thus it can be said that in the Swarts reaction the quinquevalent antimony compound is the fluorine carrier. In this relationship the reaction is closely associated with the chlorine carrier capacity of antimony pentachloride.

Recently it was shown that boron trifluoride is also capable of catalyzing the SbF_3 fluorination of aliphatic chlorides (bromides) (434).

Addition of hydrogen fluoride to olefinic and acetylinic compounds generally takes place with ease (no catalyst is needed, HF being itself a sufficiently strong acid).

References

1. Adams, J. T., B. Abramowitch, and C. R. Hauser, *J. Am. Chem. Soc.*, **65**, 552 (1943).
2. Althan, J., *Ber.*, **53**, 78 (1940).
3. Allen, R. H., *J. Am. Chem. Soc.*, **82**, 4856 (1960).
4. Allen, R. H., A. Turner, Jr., and L. D. Yats, *J. Am. Chem. Soc.*, **81**, 42 (1959).
5. Allen, R. H., and L. D. Yats, *J. Am. Chem. Soc.*, **81**, 5289 (1959).
6. Allen, R. H., and L. D. Yats, *J. Am. Chem. Soc.*, **83**, 2799 (1961).

7. Allen, R. H., L. D. Yats, and D. S. Erley, *J. Am. Chem. Soc.*, **82**, 4853 (1960).
8. American Petroleum Institute Research Project 44 at the National Bureau of Standards. "Selected Values of Properties of Hydrocarbons." 1945.
9. Anschütz, R., *Ann.*, **235**, 189 (1886).
10. Anschütz, R., and H. Immendorff, *Ber.*, **18**, 657 (1885).
11. Arnold, R. T., J. S. Buckley, and J. Richter, *J. Am. Chem. Soc.*, **69**, 2322 (1947).
12. Arnold, Z., and F. Sorm, *Chem. Listy*, **51**, 1082 (1957).
13. Arnold, Z., and F. Sorm, *Collection Czech. Chem. Commun.*, **23**, 452 (1958).
14. Arundale, E., and L. A. Mikeska, *Chem. Rev.*, **51**, 505 (1952).
15. Aschan, O., *Ann.*, **324**, 23 (1902).
16. Aston, J. G., *Ind. Eng. Chem.*, **34**, 529 (1942).
17. Averson, M. H., U.S. Pat. 2,125,872 (1938); *C.A.*, **32**, 7610 (1938).
18. Bachman, G. B., and J. L. Dever, *J. Am. Chem. Soc.*, **80**, 5871 (1958).
19. Bachman. G. B., H. Fever, B. R. Blaustein, and C. N. Vogt, *J. Am. Chem. Soc.*, **77**, 6188 (1951).
20. Bachman, G. B., and H. F. Hellman, *J. Am. Chem. Soc.*, **60**, 1772 (1948).
21. Bachman, W. E., and F. Y. Wiselogle, *J. Am. Chem. Soc.*, **56**, 1559 (1934).
22. Baddeley, G., *J. Chem. Soc.*, 232 (1944).
23. Baddeley, G., *J. Chem. Soc.*, S 229 (1949).
24. Baddeley, G., *J. Chem. Soc.*, 599 (1949).
25. Baddeley, G., *J. Chem. Soc.*, 944 (1950).
26. Baddeley, G., *Quart. Rev. (London)*, **8**, 355 (1954).
27. Baddeley, G., G. Holt, and D. Voss, *J. Chem. Soc.*, 100 (1952).
28. Baddeley, G., and J. Kenner, *J. Chem. Soc.*, 303 (1935).
29. Baddeley, G., and A. G. Pendleton, *J. Chem. Soc.*, 807 (1952).
30. Baddeley, G., H. T. Taylor, and W. Picles, *J. Chem. Soc.*, 124 (1953).
31. Baddeley, G., and R. Williamson, *J. Chem. Soc.*, 4647 (1956).
32. Balaban, A. T., and C. D. Nenitzescu, *Ann.*, **625**, 66 (1959).
33. Balaban, A. T., and C. D. Nenitzescu, *Ann.*, **625**, 74 (1959).
34. Balaban, A. T., and C. D. Nenitzescu, *J. Chem. Soc.*, 3561 (1961).
35. Balaban, A. T., and C. D. Nenitzescu, *J. Chem. Soc.*, 3553 (1961).
36. Balaban, A. T., C. D. Nenitzescu, M. Gavat, and G. Matescu, *J. Chem. Soc.*, 3564 (1961).
37. Balsohn, M., *Bull. Soc. Chim. France*, (2) **31**, 539 (1879).
38. Balz, G., and G. Schiemann, *Ber.*, **60B**, 1186 (1927).
39. Barbier, H., *Helv. Chim. Acta*, **11**, 157 (1928).
40. Baroni, A., *Atti Accad. Nazl. Lincei*, **17**, 1081 (1933); through *C.A.*, **28**, 1661 (1934).
41. Baroni, A., G. Marrano, and Modigliani, *Gazz. Chim. Ital.*, **63**, 23 (1933); through *C.A.*, **27**, 3447 (1933).
41a. Barry A. J., J. W. Gilkey, and D. E. Hook in "Metal Organic Compounds," Advances in Chemistry Monograph 23, American Chemical Society, 1959.
42. Bartlett, P. D., F. E. Condon, and A. Schneider, *J. Am. Chem. Soc.*, **66**, 1531 (1944).

43. Bataafsche Petroleum Maatschappij N.V., Germ. Pat. 660,642 (1938), *C.A.*, **32**, 6666 (1938).

44. Baum, F., and V. Meyer, *Ber.*, **28**, 3212 (1895).

45. Bayer, O., in Houben-Weyl "Methoden Der Organischen Chemie." Vol. VII, pp. 29. IVth Edition, Stuttgart, 1954.

46. Bayer, O., and J. Nelles, U.S. Pat. 2,137,664 (1938), *C.A.*, **33**, 1758 (1939).

47. Becker, S. B., U.S. Pat. 2,433,020 (1947); *C.A.*, **42**, 2094 (1948).

48. Beckmann, E. and Paul, *Ann.*, **266**, 17 (1891).

49. Bergmann, E. D., *Chem. Rev.*, **29**, 529 (1941).

50. Berl, E., and J. Bitter, *Ber.*, **57**, 95 (1924).

51. Berry, T. M., and E. E. Reid, *J. Am. Chem. Soc.*, **49**, 3142 (1927).

52. Bert, L., *Compt. rend.*, **221**, 77 (1945).

53. Beyerstedt, J., U.S. Pat. 2,284,554 (1942); *C.A.*, **36**, 6547 (1942).

54. Billman, J. H., G. E. Tripp, and R. V. Cash, *J. Am. Chem. Soc.*, **62**, 770 (1940).

55. Biltz, H., *J. Prakt. Chem.*, **142**, 193 (1935).

56. B.I.O.S. final report *1056*, item 22.

57. Birch, S. F., A. E. Dunstan, F. A. Fidler, F. B. Pim, and T. Tait, *J. Inst. Petrol. Tech.*, **24**, 303 (1938).

58. Birch, S. F., A. E. Dunstan, F. A. Fidler, F. B. Pim, and T. Tait, *Ind. Eng. Chem.*, **31**, 884 (1939).

59. Bischler, A., and B. Napieralski, *Ber.*, **26**, 1903 (1893).

60. Blanc, G., *Bull. Soc. Chim. France*, (4) **33**, 313 (1923).

61. Bloch, H. S., U.S. Pat. 2,476,955 (1949).

62. Bloch, H. S., H. Pines, and L. Schmerling, *J. Am. Chem. Soc.*, **68**, 153 (1946).

63. Boedeker, E. R., and W. E. Erner, *J. Am. Chem. Soc.*, **76**, 3591 (1954).

64. Boedtker, E., *Bull. Soc. Chim. France*, (3) **35**, 835 (1906).

65. Boedtker, E., and O. M. Halse, *Bull. Soc. Chim. France*, (4) **19**, 444 (1916).

66. Boese, A. B. Jr., *Ind. Eng. Chem.*, **32**, 16 (1940).

67. Boese, A. B., Jr., U.S. Pat. 2,382,464 (1945); *C.A.*, **40**, 1867 (1946).

68. Boeseken, J., *Rec. Trav. Chim.*, **24**, 210 (1905).

69. Boeseken, J., *Rec. Trav. Chim.*, **27**, 10 (1908).

70. Boeseken, J., and J. H. Prins, *K. Akad. Wetenschappen*, **19**, 776 (1911).

71. Boeseken, M. J., and E. Arias, *Rec. Trav. Chim.*, **54**, 711 (1935).

72. Bogert, M. T., and M. Meyer, *J. Am. Chem. Soc.*, **44**, 1568 (1922).

73. Bornwater, J. T., and A. F. Holleman, *Rec. Trav. Chim.*, **31**, 396 (1912).

74. Bosshard, H. H., and H. Zollinger, *Helv. Chim. Acta*, **42**, 1659 (1959).

75. Bourne, E. J., M. Stacey, J. C. Tatlow, and J. M. Tedder, *J. Chem. Soc.*, 2976 (1949).

76. Bourne, E. J., M. Stacey, J. C. Tatlow, and J. M. Tedder, *J. Chem. Soc.*, 718 (1951).

77. Bouveault, L., *Bull. Soc. Chim. France*, (3) **17**, 1014 (1897).

78. Bowden, E., *J. Am. Chem. Soc.*, **60**, 645 (1938).

79. Bradsher, C. K., *Chem. Rev.*, **38**, 447 (1946).

80. Breslow, D. S., and C. R. Hauser, *J. Am. Chem. Soc.*, **62**, 2385 (1940).

81. Breslow, D. S., and C. R. Hauser, *J. Am. Chem. Soc.*, **62**, 2611 (1940).

82. Brewster, C. M., and G. G. Watters, *J. Am. Chem. Soc.*, **64**, 2579 (1942).

83. Brintzinger, H., and M. Langheck, *Ber.*, **86**, 557 (1953).
84. Brooks, B. T., U.S. Pat. 2,015,706 (1935); *C.A.*, **29**, 8004 (1935).
85. Brooks, R. E., U.S. Pat. 2,436,286 (1948); *C.A.*, **42**, 3431 (1948).
86. Brouwer, L. G., and J. P. Wibaut, *Rec. Trav. Chim.*, **53**, 1001 (1934).
87. Brown, H. C., and J. D. Brady, *J. Am. Chem. Soc.*, **74**, 3570 (1952).
88. Brown, H. C., and M. Grayson, *J. Am. Chem. Soc.*, **75**, 6285 (1953).
89. Brown, H. C., and H. Jungk, *J. Am. Chem. Soc.*, **77**, 5579 (1955).
90. Brown, H. C., and K. L. Nelson, *J. Am. Chem. Soc.*, **75**, 6292 (1953).
91. Brown, H. C., and C. R. Smoot, *J. Am. Chem. Soc.*, **78**, 6255 (1956).
92. Bruce, D. B., A. J. S. Sorrie, and R. H. Thompson, *J. Chem. Soc.*, 2403 (1953).
93. Bruner, F. H., U.S. Pat. 2,363,116; *C.A.*, **39**, 3300 (1944).
94. Bruner, F. H., L. A. Clarke, and R. L. Sawyer, U.S. Pat. 2,345,095; *C.A.*, **38**, 3663 (1944).
95. Bruson, H. A., and J. W. Kroeger, *J. Am. Chem. Soc.*, **62**, 36 (1940).
96. Bujwid, Z. J., W. Gerrard, and M. F. Lappert, *Chem. and Ind.*, 1091 (1959).
97. Burwell, R. L., Jr., and S. Archer, *J. Am. Chem. Soc.*, **64**, 1032 (1942).
98. Burwell, R. J., Jr., and L. M. Elkin, *J. Am. Chem. Soc.*, **73**, 502 (1951).
99. Burwell, R. L., and A. D. Shields, *J. Am. Chem. Soc.*, **77**, 2766 (1955).
100. Butlerov, A. M., *J. Russ. Chem. Soc.*, **9**, 38 (1877).
101. Butlerov, A. M., and V. Goryainov, *J. Russ. Chem. Soc.*, **5**, 302 (1873).
102. Buu-Hoï, N. G., and P. Cagncunt, *Bull. Soc. Chim. France*, (5) **9** 887 (1942).
103. Buu-Hoï, N. P., N. Hoan, and N. D. Xuong, *J. Chem. Soc.*, 3499 (1951).
104. Byrns, A. C., U.S. Pat. 2,463,742; *C.A.*, **43**, 4685 (1949).
105. Calcott, W. S., J. M. Thinker, and V. Weinmayr, *J. Am. Chem. Soc.*, **61**, 1010 (1939).
106. Calloway, N. O., *Chem. Rev.*, **17**, 327 (1935).
107. Calloway, N. O., *J. Am. Chem. Soc.*, **59**, 1474 (1937).
108. Calloway, N. O., and L. D. Green, *J. Am. Chem. Soc.*, **59**, 809 (1937).
109. Capeller, R. de, *Helv. Chim. Acta*, **11**, 426 (1928).
110. Carruthers, W., *J. Chem. Soc.*, 3486 (1953).
111. Catch, J. R., D. F., Elliott, D. H. Hey, and E. R. H. Jones, *J. Chem. Soc.*, 278 (1948).
112. Ceasar, P. D., and A. W. Francis, *Ind. Eng. Chem.*, **33**, 1426 (1941).
113. Chalmers, W., *Can. J. Res.*, **7**, 113, 472 (1932); *J. Am. Chem. Soc.*, **56**, 912 (1934).
114. Chattaway, F. D., *J. Chem. Soc.*, **63**, 1885 (1893).
115. Chattaway, F. D., and W. H. Lewis, *J. Chem. Soc.*, **65**, 869 (1894).
116. Clar, E., *Ber.*, **65**, 846 (1932).
117. Cohen, J. B., and H. D. Dakin, *J. Chem. Soc.*, **79**, 1117 (1901).
118. Coleman, G. H., U.S. Pat. 1,900,276 (1933); through *C.A.*, **27**, 2965 (1933).
119. Collet, A., *Bull. Soc. Chim. France*, (3) **17**, 66, 506 (1897); *Compt. rend.*, **125**, 717 (1897); **126**, 1577 (1898).
120. Collet, A., *Compt. rend.*, **125**, 305 (1897); *Bull. Soc. Chim. France*, (3) **17**, 506 (1897).
121. Colonge, J., and J. Chambion, *Compt. rend.*, **224**, 128 (1947).

122. Colonge, J., and J. Chambion, *Bull. Soc. Chim. France*, 999, 1002, 1006 (1947).
123. Colonge, J., and P. Dumont, *Bull. Soc. Chim. France*, 38 (1947).
124. Colonge, J., and K. Mostafavi, *Bull. Soc. Chim. France*, (5) **6**, 335, 342 (1939).
125. Colonge, J., and E. Puroux, *Bull. Soc. Chim. France*, (5) **7**, 459 (1940).
126. Combes, A., *Compt. rend.*, **103**, 814 (1886).
127. Combes, A., *Compt. rend.*, **104**, 855 (1887).
128. Combes, A., *Ann. Chim. (Paris)*, (6), **12**, 205 (1887).
129. Combes, A., *J. Chem. Soc.*, **52**, 127 (1887).
130. Condon, F. E., *J. Am. Chem. Soc.*, **70**, 2265 (1948).
131. Condon, F. E., *J. Am. Chem. Soc.*, **71**, 3544 (1949).
132. Condon, F. E., *J. Am. Chem. Soc.*, **73**, 3938 (1951).
133. Condon, F. E., in "Catalysis," ed. P. H. Emmett.   Vol. VI, Reinhold Publishing Corp., New York, 1958.
134. Cooper, S. R., *Org. Syn.*, **21**, 103 (1941).
135. Copenhaver, J. W., U.S. Pat. 2,487,525 (1949); *C.A.*, **44**, 3011 (1950).
136. Copenhaver, J. W., U.S. Pat. 2,527,533 (1950); *C.A.*, **45**, 1623 (1951).
137. Cox, M. V., *Bull. Soc. Chim. France*, **37**, 1549 (1925).
138. Cristol, S. J., and H. L. Haller, *Ind. Eng. News*, **23**, 2070 (1945).
139. Croxall, W. J., and H. T. Neher, U.S. Pat. 2,448,660 (1948); *C.A.*, **42**, 8816 (1948); U.S. Pat. 2,446,171 (1948); *C.A.*, **43**, 1433 (1949).
140. Croxall, W. J., and H. T. Neher, U.S. Pat. 2,473,014 (1949); *C.A.*, **43**, (6649).
141. Croxall, W. J., and H. T. Neher, U.S. Pat. 2,490,337 (1949); *C.A.*, **44**, 7867 (1950).
142. Croxall, W. J., F. J. Sowa, and J. A. Nieuwland, *J. Org. Chem.*, **2**, 253 (1937).
143. Cryer, J., *Trans. Roy. Soc. Can., Sec. III*, (3), **19**, 29 (1925).
144. Cullinane, N. M., and D. M. Leyshon, *J. Chem. Soc.*, 2942 (1954).
145. Curme, G. O., Jr., U.S. Pat. 1,518,182 (1924); *C.A.*, **19**, 523 (1925).
146. Dainton, F. S., and G. B. B. M. Southerland, *J. Polymer Sci.*, **4**, 37 (1949).
147. Dainton, F. S., and R. H. Tomlinson, *J. Chem. Soc.*, 151 (1953).
148. Dainton, F. S., R. H. Tomlinson, and T. L. Batte, in "Cationic Polymerization", ed. P. H. Plesch, Academic Press, New York, 1953, p. 86.
149. Daniloff, S., and E. Venus-Danilova, *Ber.*, **59**, 377 (1926).
150. Darzens, G., *Compt. rend.*, **150**, 707 (1910).
151. Darzens, G., *Compt. rend.*, **189**, 766 (1929).
152. Darzens, G., and H. Rost, *Compt. rend.*, **151**, 758 (1910).
153. Dashkevich, L. B., and L. G. Izrailev, *J. Gen. Chem. U.S.S.R.*, **30**, 3060 (1960).
154. Davies, N. C., *J. Chem. Soc.*, 462 (1935).
155. De la Mare, P. B. D., and J. H. Ridd, "Aromatic Substitution; Nitration and Halogenation", Academic Press, New York, 1959.
156. Descude, M., *Compt. rend.*, **132**, 1129 (1901).
157. Deville, H. St. C., *Ann. Chim. (Paris)*, **75**, 66 (1839).
158. Dixon, Y. K., and K. W. Saunders, *Ind. Eng. Chem.*, **46**, 652 (1954).
159. Dominion Tar and Chemical Co., Ltd., Brit. Pat. 616,751 (1949); through *C.A.*, **43**, 5636 (1949).

160. Dorris, T. B., F. J. Sowa, and G. A. Nieuwland, *J. Am. Chem. Soc.*, **56**, 2689 (1934).
161. Du Pont, E. I. de Nemours and Co., Brit. Pat. 486,560 (1938); *C.A.*, **32**, 8438 (1938).
162. Egloff, G., U.S. Pat. 2,010,948 (1935); U.S. Pat. 2,010,949 (1935).
163. Egloff, G., *Ind. Eng. Chem.*, **28**, 1461 (1936).
164. Egloff, G., and G. Hulla, *Chem. Rev.*, **37**, 323 (1945).
165. Egloff, G., and P. C. Weinert, *Proc. 3rd World Petroleum Congress*, **4**, 201 (1951).
166. Eistert, B., and F. Haupter, *Ber.*, **92**, 1921 (1959).
167. Elbs, K., *J. Prakt. Chem.*, (2), **35**, 503 (1887).
168. Evans, A. G., B. Holden, P. Plesch, M. Polanyi, H. A. Skinner, and M. A. Weinberger, *Nature*, **157**, 102 (1946).
169. Evans, A. G., and G. W. Meadows, *J. Polymer Sci.*, **4**, 359 (1949).
170. Evans, A. G., G. W. Meadows, and M. Polanyi, *Nature*, **158**, 94 (1946).
171. Evans, A. G., and M. Polanyi, *J. Chem. Soc.*, 252 (1947).
172. Evans, T. W., and K. R. Edlund, *Ind. Eng. Chem.*, **28**, 1186 (1936).
173. Evering, B. L., E. L. d'Ouville, A. P. Lien, and R. C. Waugh, *Ind. Eng. Chem.*, **45**, 582 (1953).
174. Evering, B. L., and A. K. Roebuck, *J. Am. Chem. Soc.*, **75**, 1631 (1953).
175. Evering, B. L., and A. K. Roebuck, U.S. Pat. 2,897,248 (1959); U.S. Pat. 2,975,223 (1961).
176. Farbwerke vorm. Meister, Lucius, and Bruning, Germ. Pat. 41,751 (1887); through *Z.*, 564 (1888).
177. Fawcett, E. Wm., Brit. Pat. 481,515 (1936); *C.A.*, **33**, 6365 (1938).
178. Feasley, C. F., U.S. Pat. 2,502,000 (1950); *C.A.*, **44**, 6616 (1950).
179. F.I.A.T. final report 843.
180. Field, L., *J. Am. Chem. Soc.*, **74**, 394 (1952).
181. Fieser, L. F., and E. B. Hershberg, *J. Am. Chem. Soc.*, **61**, 1272 (1939).
182. Fischer, H., and A. Schwarz, *Ann.*, **512**, 239 (1934).
183. Fischer, O., *Ann.*, **206**, 88 (1881).
184. Ford, T. A., U.S. Pat. 2,419,131 (1947).
185. Ford, T. A., H. W. Jacobsen, and F. C. McGrew, *J. Am. Chem. Soc.*, **70**, 3793 (1948).
186. Francis, A. W., *Chem. Rev.*, **43**, 257 (1948).
187. Francis, A. W., unpublished work, quoted in ref. (186).
188. Francis, A. W., and E. E. Reid, *Ind. Eng. Chem.*, **38**, 1194 (1946).
189. Frank, C. E., A. T. Hallowell, C. W. Theobald, and G. T. Vaala, *Ind. Eng.*, **41**, 2061 (1949).
190. Fray, G. I., and R. Robinson, *J. Am. Chem. Soc.*, **83**, 249 (1961).
191. Freund, M., *Monatsh.*, **17**, 399 (1896).
192. Freund, M., and K. Gleischer, *Ann.*, **373**, 291 (1910).
193. Friedel, C., and J. M. Crafts, *Compt. rend.*, **84**, 1392 (1887).
194. Friedel, C., and J. M. Crafts, *Compt. rend.*, **84**, 1450 (1877).
195. Friedel, C., and J. M. Crafts, *Compt. rend.*, **85**, 74 (1877).
196. Friedel, C., and J. M. Crafts, *Bull. Soc. Chim. France*, (2) **30**, 146 (1878).
197. Friedel, C., and J. M. Crafts, *Compt. rend.*, **86**, 1368 (1878).
198. Friedel, C., and J. M. Crafts, *Ann. Chim. (Paris)*, (6), **1**, 507 (1884)
199. Friedel, C., and J. M. Crafts, *Compt. rend.*, **100**, 692 (1885).
200. Friedel, C., and J. M. Crafts, *Ann. Chim. (Paris)*, (6), **14**, 433 (1888).

201. Friedel, C., J. M. Crafts, and E. Ador, *Compt. rend.*, **85**, 673 (1877).
202. Fries, K., and G. Finck, *Ber.*, **41**, 4271 (1908).
203. Fries, K., and W. Pfaffendorf, *Ber.*, **43**, 212 (1910).
204. Friess, S. L., and R. Pinson, Jr., *J. Am. Chem. Soc.*, **73**, 3512 (1951).
205. Froning, J. F., and G. F. Hennion, *J. Am. Chem. Soc.*, **62**, 653 (1940).
206. Fuson, R. C., and C. H. McKeever, *J. Am. Chem. Soc.*, **62**, 2088 (1940).
207. Fuson, R. C., and C. H. McKeever, *Org. Reactions*, **1**, 63.   John Wiley and Sons, New York, 1942.
208. Galle, K., *Ber.*, **16**, 1747 (1883).
209. Gannasch, P., and A. Rathjen, *Ber.*, **32**, 2391 (1899).
210. Gardner, P. D., *J. Am. Chem. Soc.*, **76**, 4550 (1954).
211. Gattermann, L., *Ann.*, **244**, 29 (1888); *Ber.*, **32**, 1117 (1899).
212. Gattermann, L., *Ber.*, **31**, 1149 (1898); *Ann.*, **357**, 313 (1907).
213. Gattermann, L., and J. A. Koch, *Ber.*, **30**, 1622 (1897).
214. Gerard, W., and E. F. Mooney, *Chem. and Ind.*, 1259 (1958).
215. Gilman, H., and G. E. Dunn, *Chem. Rev.*, **52**, 77 (1953).
216. Given, P. H., and D. L. Hammick, *J. Chem. Soc.*, 928 (1947).
217. Given, P. H., and D. L. Hammick, *J. Chem. Soc.*, 1237 (1947).
218. Goldsby, R., U.S. Pat. 2,383,056; *C.A.*, **39**, 5466 (1945).
219. Goldschmidt, S., and R. R. Wolff, U.S. Pat. 2,060,195 (1937); *C.A.*, **31**, 420.
220. Gorin, M. H., C. S. Kuhn, Jr., and C. B. Miles, *Ind. Eng. Chem.*, **38**, 795 (1946).
221. Gottschalk, M., *Ber.*, **20**, 3287 (1887).
222. Graebe, C., *Ber.*, **34**, 1778 (1901).
223. Grebe, J. J., J. H. Reilly, and R. M. Wiley, U.S. Pat. 2,034,292 (1935); *C.A.*, **30**, 3178 (1936).
224. Greenhalg, R., U.S. Pat. 1,977,516 (1935); *C.A.*, **29**, 181 (1935); Brit. Pat. 395,732 (1934); *C.A.*, **28**, 492 (1934),
225. Gresham, T. L., J. E. Jansen, and F. W. Shaver, *J. Am. Chem. Soc.*, **70**, 998 (1948).
226. Gresham, W. F., and G. E. Tabet, U.S. Pat. 2,485,237·(1946).
227. Greune, H., and L. Bicker, Germ. Pat. 926,965 (1955); *C.A.*, **50**, 2668 (1956).
228. Groggins, P. H., R. H. Nagel, and A. J. Stirton, *Ind. Eng. Chem.*, **26**, 1317 (1934).
229. Gross, H., personal communication.
230. Grosse, A. V., and V. N. Ipatieff, *J. Am. Chem. Soc.*, **57**, 2415 (1935).
231. Grosse, A. V., and V. N. Ipatieff, *J. Org. Chem.*, **2**, 447 (1937).
232. Grosse, A. V., J. M. Mavity, and V. N. Ipatieff, *J. Org. Chem.*, **3**, 137 (1938).
233. Guenther, F., U.S. Pat. 1,670,505 (1928); *C.A.*, **22**, 2378 (1928).
234. Gustavson, J. *Bull. Soc. Chim. France*, **34**, 322 (1880).
235. Gustavson, G., *J. Prakt. Chem.*, (2), **37**, 109 (1888).
236. Guyot, A., *Compt. rend.*, **144**, 1651 (1907); **144**, 1120 (1908).
237. Hagenmeyer, H. J., Jr., *Ind. Eng. Chem.*, **41**, 765 (1949).
238. Hagenmeyer, H. J., Jr., U.S. Pat. 2,478,388 (1949); *C.A.*, **44**, 1133a (1950).
239. Hahn, W., R. Seydel, and R. Stroh, *Angew. Chem.*, **69**, 699 (1957).
240. Halle, W. J., and E. C. Britton, *J. Am. Chem. Soc.*, **41**, 841 (1919).

241. Hamblet, C. H., and A. McAlevy, U.S. Pat. 2,426,017 (1947); *C.A.*, **42**, 206 (1948).
242. Hammond, J. H. S., U.S. Pat. 2,393,367 (1946); *C.A.*, **40**, 2162 (1946).
243. Hansford, R. C., C. G. Myers, and A. N. Sachanen, *Ind. Eng. Chem.*, **37**, 671 (1945).
244. Haoglin, R. E., and D. H. Hirsh, *J. Am. Chem. Soc.*, **71**, 3468 (1949).
245. Hartmann, C., and L. Gattermann, *Ber.*, **25**, 3531 (1892).
246. Hartough, H. D., and A. I. Kosak, *J. Am. Chem. Soc.*, **69**, 3098 (1947).
247. Hauser, C. R., and J. T. Adams, *J. Am. Chem. Soc.*, **64**, 728 (1942).
248. Hauser, C. R., and D. S. Breslow, *J. Am. Chem. Soc.*, **62**, 2389 (1940).
249. Hauser, C. R., F. W. Swamer, and J. T. Adams, *Org. Reactions*, **8**, 59. John Wiley and Sons, New York, 1954.
250. Heid, J. V., and R. Levine, *J. Org. Chem.*, **13**, 409 (1948).
251. Heise, R., *Ber.*, **24**, 768 (1891).
252. Heise, R., and A. Tohl, *Ann.*, **270**, 155 (1892).
253. Heldmann, J. D., *J. Am. Chem. Soc.*, **66**, 1791 (1944).
254. Heller, G., *Ber.*, **45**, 665 (1912).
255. Helse, R., and A. Tolh, *Ann.*, **279**, 155 (1892).
256. Henne, A. L., and J. M. Tedder, *J. Chem. Soc.*, 3627 (1953).
257. Hennion, G. F., and L. A. Auspos, *J. Am. Chem. Soc.*, **65**, 1603 (1943).
258. Hennion, G. F., H. D. Hinton, and J. A. Nieuwland, *J. Am. Chem. Soc.*, **55**, 2857 (1933).
259. Hennion, G. F., and R. A. Kurtz, *J. Am. Chem. Soc.*, **65**, 1001 (1943).
260. Hennion, G. F., and S. F. de C. McLeese, *J. Am. Chem. Soc.*, **64**, 2421 (1942).
261. Herbert, R. H., *J. Am. Chem. Soc.*, **80**, 5080 (1958).
262. Herzig, J., *Ber.*, **14**, 1205 (1881).
263. Hinton, H. D., and J. A. Nieuwland, *J. Am. Chem. Soc.*, **52**, 2893 (1930).
264. Hinton, H. D., and J. A. Nieuwland, *J. Am. Chem. Soc.*, **54**, 2017 (1932).
265. Hoesch, K., *Ber.*, **48**, 1122 (1915).
266. Hoffman, E., U.S. Pat. 2,538,001, *C.A.*, **45**, 4031 (1931).
267. Hoogzand, C., and W. Hubel, *Angew. Chem.*, **73**, 680 (1961).
268. Hoogzand, C., and W. Hubel, *Tetrahedron Letters*, 637 (1961).
269. Hopff, H., Germ. Pat. 512,718 (1927).
270. Hopff, H., *Ber.*, **64**, 2739 (1931).
271. Hopff, H., *Ber.*, **65**, 482 (1932).
272. Hopff, H., Germ. Pat. 594,968 (1934).
273. Hopff., H., *Angew. Chem.*, **60**, 218 (1948).
274. Hopff, H., *Angew. Chem.*, **60**, 245 (1948).
275. Hopff, H., *Helv. Chim. Acta*, **44**, 19 (1961).
276. Hopff, H., H. Kellermann, and A. Freytag, U.S. Pat. 2,168,161 (1959).
277. Hopff, H., C. C. Nenitzescu, D. A. Isacescu, and I. P. Cantuniari, *Ber.*, **69**, 2244 (1936)
278. Hopff, H., and A. K. Wiek, *Helv. Chim. Acta*, **43**, 1473 (1960).
279. Houben, J., and W. Fischer, *Ber.*, **63**, 2464 (1930).
280. Houben, J., and W. Fischer, *Ber.*, **66**, 339 (1933).
281. Houben, J., and T. Weyl, "Die Methoden der Organischen Chemie", 2nd ed., Thieme, Verl., Leipzig, 1923, vol. **3**, p. 391.

282. Houston, R. C., R. L. Guile, J. J. Sculati, and W. N. Wasson, *J. Org. Chem.*, **6**, 252 (1941).

283. Howells, H. P., and J. G. Little, *J. Am. Chem. Soc.*, **54**, 2451 (1932).

284. Hubel, W., and C. Hoogzand, *Ber.*, **93**, 103 (1960).

285. Humblet, C. H., and A. McElvey, Brit. Pat. 590,571 (1947); *C.A.*, **42**, 589 (1948).

286. Hurd, C. D., *J. Am. Chem. Soc.*, **47**, 2777 (1925).

287. Hurd, C. D., and C. M. Webb, *Org. Syn. Coll.*, Vol. I, 217 (1941).

288. Ide, W. S., and Y. S. Buck, *Org. Reactions*, **4**, Ch. 5.   John Wiley and Sons, New York, 1948.

289. I.G. Farbenindustrie, A.G., Brit. Pat. 354,992; *C.A.*, **26**, 5574 (1930).

290. I.G. Farbenindustrie, A.G., Germ. Pat. 524,186 (1931).

291. I.G. Farbenindustrie, A.G., French Pat. 716,604; *C.A.*, **26**, 2198.

292. I.G. Farbenindustrie, A.G., Brit. Pat. 466,891 (1937); *C.A.*, **31**, 7881.

293. Imperial Chemical Industries Limited, Brit. Pat. 604,579 (1948); *C.A.*, **43**, 1434.

294. Ingold, C. K., C. G. Raisin, and C. L. Wilson, *Nature*, **134**, 847 (1934)

295. Ingold, C. K., C. G. Raisin, and C. L. Wilson, *J. Chem. Soc.*, 915 (1936)

296. Inmann, C. E., R. E. Oesterling, and E A. Tyczkowsky, *J. Am. Chem. Soc.*, **80**, 5286 (1958).

297. Ipatieff, V. N., U.S. Pat. 2,006,695 (1935); 2,039,798 (1936).

298. Ipatieff, V. N., U.S. Pat. 2,639,798 (1937).

299. Ipatieff, V. N., B. B. Corson, and G. Egloff, *Ind. Eng. Chem.*, **27**, 1077 (1935).

300. Ipatieff, V. N., B. B. Corson, and H. Pines, *J. Am. Chem. Soc.*, **58**, 919 (1936).

301. Ipatieff, V. N., and A. V. Grosse, *J. Am. Chem. Soc.*, **57**, 1616 (1935).

302. Ipatieff, V. N., and A. V. Grosse, *J. Am. Chem. Soc.*, **57**, 2415 (1935).

303. Ipatieff, V. N., and A. V. Grosse, *J. Am. Chem. Soc.*, **58**, 915 (1936).

304. Ipatieff, V. N., and A. V. Grosse, *Ind. Eng. Chem.*, **28**, 461 (1936).

305. Ipatieff, V. N., A. V. Grosse, H. Pines, and V. I. Komarewski, *J. Am. Chem. Soc.*, **58**, 913 (1936).

306. Ipatieff, V. N., and H. Pines, *Ind. Eng. Chem.*, **27**, 1364 (1935).

307. Ipatieff, V. N., and H. Pines, *Ind. Eng. Chem.*, **28**, 684 (1936).

308. Ipatieff, V. N., and H. Pines, *J. Am. Chem. Soc.*, **58**, 1026 (1936).

309. Ipatieff, V. N., and H. Pines, *J. Am. Chem. Soc.*, **59**, 56 (1937).

310. Ipatieff, V. N., and H. Pines, *Ind. Eng. Chem.*, **37**, 362 (1945).

311. Ipatieff, V. N., H. Pines, and R. C. Olberg, *J. Am. Chem. Soc.*, **70**, 2123 (1948).

312. Ipatieff, V. N., H. Pines, and L. Schmerling, *J. Org. Chem.*, **5**, 253 (1940).

313. Ipatieff, V. N., and O. Rutal, *Ber.*, **46**, 1748 (1913).

314. Ipatieff, V. N., and L. Schmerling, *Ind. Eng. Chem.*, **40**, 2354 (1948).

315. Ipatieff, V. N., and L. Schmerling, "Advances in Catalysis," Vol. 1, p. 27, Academic Press, New York, 1948.

316. Isagulyants, V. I., and N. A. Slavskaya, *J. Appl. Chem. U.S.S.R.*, **33**, 949 (1960).

317. Ishibashi, T., and K. Tarama, *J. Chem. Soc. Japan, Ind. Chem. Sect.*, **55**, 214 (1952); through *C.A.*, **47**, 10977 (1953).

318. Jacobsen, O., *Ber.*, **14**, 2624 (1881).

319. Jacobsen, O., *Ber.*, **18**, 338 (1885).
320. Jacobsen, O., *Ber.*, **19**, 1209 (1886).
321. Jacobsen, O., *Ber.*, **21**, 2814 (1888).
322. Jacobsen, R. A., H. B. Dykstra, and W. H. Carothers, *J. Am. Chem. Soc.*, **56**, 1169 (1934).
323. Jaubert, G. F., *Compt. rend.*, **132**, 841 (1901).
324. Jenkins, R. L., R. McCullough, and C. F. Booth, *Ind. Eng. Chem.*, **22**, 31 (1930).
325. Johnson, J. Y., Brit. Pat. 307,223 (1929).
326. Johnson, T. B., and F. W. Lane, *J. Am. Chem. Soc.*, **43**, 348 (1921).
327. Johnson, W. S., and R. D. Offenhauer, *J. Am. Chem. Soc.*, **67**, 1045 (1945).
328. Johnstone, W. W., U.S. Pat. 2,470,894; *C.A.*, **43**, 5788h (1949); U.S. Pat. 2,491,496; *C.A.*, **45**, 10614 (1949).
329. Jutz, C., *Ber.*, **91**, 850 (1958).
330. Kalischer, G., and K. Keller, Germ. Pat. 519,806 (1927); *C.A.*, **25**, 3012.
331. Kalischer, G., H. Scheyer, and K. Keller, Germ. Pat. 514,415; *C.A.*, **25**, 1536 (1927).
332. Kamarewsky, V. I., and N. Balai, *Ind. Eng. Chem.*, **30**, 1051 (1938).
333. Kane, H. L., and A. Lowy, *J. Am. Chem. Soc.*, **58**, 2605 (1936).
334. Karrer, P., *Helv. Chim. Acta*, **2**, 89 (1919).
335. Karrer, P., and E. Zeller, *Helv. Chim. Acta*, **2**, 482 (1919).
336. Karrer, P., and E. Zeller, *Helv. Chim. Acta*, **3**, 361 (1920).
337. Kaschtanov, L. I., *J. Gen. Chem. U.S.S.R.*, **2**, 515 (1932).
338. Kastner, D., "Neuere Methoden der Preparativen Organischen Chemie," Vol. I, p. 413, Berlin, 1943.
339. Keller, R. N., and P. A. S. Smith, *J. Am. Chem. Soc.*, **66**, 1122 (1944); **68**, 899 (1946).
340. Kennedy, J. P., and R. M. Thomas, *J. Polymer Sci.*, **45**, 229 (1960).
341. Kennedy, R. M., in "Catalysis," Vol. VI, p. 1, edited by P. H. Emmett, Reinhold Publishing Corp., New York, 1948.
342. Kennedy, R. M., and A. Schneider, U.S. Pat. 2,626,966; *C.A.*, **47**, 6644 (1953).
343. Kenner, J., M. Polanyi, and P. Szego, *Nature*, **135**, 267 (1935).
344. Kinckell, *Ber.*, **38**, 2609 (1905).
345. Kinney, R. E., and L. A. Hamilton, *J. Am. Chem. Soc.*, **76**, 786 (1954).
346. Klemenz, A., and K. Scholler, *Z. anorg. allgem. Chem.*, **141**, 231 (1927).
347. Klieger, A., and H. Huber, *Ber.*, **53**, 1646 (1920).
348. Klit, A., and A. Langseth, *Z. Physik. Chem.*, **A176**, 65 (1936).
349. Klos, J., *Arch. Pharm.*, **288**, 48 (1955).
350. Knoevenagel, E., *Ann.*, **402**, 133 (1913).
351. Knoevenagel, E., and J. Kenner, *Ber.*, **41**, 3315 (1908).
352. Knowles, C. L., *J. Am. Chem. Soc.*, **43**, 896 (1921).
353. Kobe, K. A., and T. F. Doumani, *Ind. Eng. Chem.*, **31**, 257 (1939).
354. Koch, H., *Brennstoff-Chem.*, **36**, 321 (1955).
355. Koch, H., and W. Haaf, *Angew. Chem.*, **70**, 311 (1958); *Ann.*, **618**, 251 (1958).
356. Kohler, E. P., *J. Am. Chem. Soc.*, **42**, 375 (1909).
357. Kohler, E. P., G. L. Hertage, and M. C. Burnley, *J. Am. Chem. Soc.*, **44**, 60 (1910).
358. Kondakov, I. L., *Bull. Soc. Chim. France*, (3) **7**, 576 (1892).

359. Kondakov, I. L., *J. Russ. Chem. Soc.*, **26**, 232 (1894).
360. Konigs, and J. U. Nef, *Ber.*, **19**, 2431 (1886).
361. Konowaloff, M., *J. Russ. Phys. Chem. Soc.*, **17**, 459 (1885); *ibid.*, **19**, 916 (1887).
362. Konowaloff, M., *J. Russ. Phys. Chem. Soc.*, **27**, 457 (1895).
363. Kosolapoff, G. M., *J. Am. Chem. Soc.*, **69**, 1652 (1947).
364. Kovacic, P., and R. E. Bennett, *J. Am. Chem. Soc.*, **83**, 221 (1961).
365. Kovacic, P., and G. L. Foote, *J. Am. Chem. Soc.*, **83**, 743 (1961).
366. Kraft, F., and G. Konig, *Ber.*, **23**, 2382 (1890).
367. Kraft, F., and A. von Hansen, *Ber.* **22**, 803 (1889).
368. Krannich, C., *Ber.*, **33**, 3485 (1900).
369. Krapivin, S., *Bull. Soc. Naturalistes, Moscow*, **1** (1908); *C.A.*, **5**, 1281 (1911).
370. Kuhn, S. J., and G. A. Olah, *J. Am. Chem. Soc.*, **83**, 4564 (1961).
371. Kunckell, F., *Ber.*, **33**, 2641 (1900).
372. Kunckell, F., *Ber. Deut. Pharm. Ges.*, **23**, 188 (1913); **22**, 180 (1912).
373. Kung, F. E., U.S. Pat. 2,356,459
374. Kursanov, D. N., and R. R. Zelvin, *J. Gen. Chem. U.S.S.R.*, **9**, 2173 (1939).
375. Kynaston, W., B. E. Larcombe, and H. S. Turner, *J. Chem. Soc.*, 1772 (1960).
376. Kynaston, W., and H. S. Turner, *Proc. Chem. Soc.*, 304 (1958).
377. La Coste, W., and A. Michaelis, *Ann.*, **201**, 184 (1880).
378. Lappert, M. F., *Proc. Chem. Soc.*, 121 (1957).
379. Lavaux, Y., and M. Lombard, *Bull. Soc. Chim. France*, (4) **7**, 913 (1910).
380. Lee, R. J., H. M. Knight, and J. T. Kelly, *Ind. Eng. Chem.*, **50**, 1001 (1958).
381. Lee, S. W., and G. Dougherty, *J. Org. Chem.*, **4**, 48 (1939).
382. Leiserson, J. L., and A. Weissberger, *Org. Syn. Coll.*, **3**, 183, John Wiley and Sons, New York, 1955.
383. Leuckart, R., *Ber.*, **18**, 873 (1885).
384. Leuckart, R., *J. Prakt. Chem.*, **41**, 301 (1890).
385. Liebermann, C., *Ber.*, **45**, 1187 (1912).
386. Lien, A. P., U.S. Pat. 2,399,126 (1946), *C.A.*, **40**, 4076 (1946).
387. Lien, A. P., and D. A. McCaulay, *J. Am. Chem. Soc.*, **75**, 2407 (1953).
388. Lien, A. P., and B. H. Shoemaker, U.S. Pat. 2,430,516 (1947); *C.A.*, **42**, 1048g (1948).
389. Linn, C. B., and V. N. Ipatieff, Preprints of Papers Presented before the Division of Petroleum Chemistry, Spring Meeting, A.C.S., 1949.
390. Linner, F., U.S. Pat. 1,892,990 (1933); *C.A.*, **27**, 2319 (1933).
391. Linstead, R. P., *Ann. Rept. Progr. Chem. (Chem. Soc. London)*, **34**, 259 (1937).
392. Lock, G., and W. Clemens, *Monatsh.*, **74**, 77 (1943).
393. Loder, D. J., U.S. Pat. 2,135,449 (1938); *C.A.*, **33**, 992 (1939).
394. Lorenz, H., and R. Witzinger, *Helv. Chim. Acta*, **28**, 600 (1945).
395. Mackor, E. L., A. Hofstra, and J. H. Van der Waals, *Trans. Faraday Soc.*, **54**, 186 (1958).
396. Makarova, I. G., and E. A. Gribchenko, *Bull. Acad. Sci. U.S.S.R., Div. Chem. Sci.*, 693 (1958).

397. Makarova, L. G., and M. K. Matveeva, *Bull. Acad. Sci. U.S.S.R.*, *Div. Chem. Sci.*, 565 (1958).
398. Makarova, L. G., M. K. Matveeva, and E. A. Gribchenko, *Bull. Acad. Sci. U.S.S.R.*, *Div. Chem. Sci.*, 1399 (1958).
399. Malson, P. E., and J. H. Gardner, Paper presented at the A.C.S. Meeting, Baltimore, 1939.
400. Manyik, R. M., F. C. Frostick, J. J. Sanderson, and C. R. Hauser, *J. Am. Chem. Soc.*, **75**, 5030 (1953).
401. Mares, J. R., U.S. Pat. 2,051,096 (1936).
402. Marschner, R. F., and D. R. Carmody, *J. Am. Chem. Soc.*, **73**, 604 (1951).
403. Mattox, W. J., U.S. Pat. 2,360,814 (1944).
404. Mavity, J. M., H. Pines, R. C. Wackher, and J. A. Brooks, *Ind. Eng. Chem.*, **40**, 2374 (1948).
405. McCaulay, D. A., *J. Am. Chem. Soc.*, **81**, 6437 (1959).
406. McCaulay, D. A., and A. P. Lien, *J. Am. Chem. Soc.*, **74**, 6246 (1952).
407. McCaulay, D. A., and A. P. Lien, *J. Am. Chem. Soc.*, **75**, 2411 (1953).
408. McCaulay, D. A., A. P. Lien, and P. J. Launer, *J. Am. Chem. Soc.*, **76**, 2354 (1954).
409. McCaulay, D. A., B. H. Shoemaker, and A. P. Lien, *Ind. Eng. Chem.*, **42**, 2103 (1950).
410. McKenna, J. F., and F. J. Sowa, *J. Am. Chem. Soc.*, **59**, 470 (1937).
411. McKenna, J. F., and F. J. Sowa, *J. Am. Chem. Soc.*, **59**, 1204 (1937).
412. McKenna, J. F., and F. J. Sowa, *J. Am. Chem. Soc.*, **60**, 124 (1938).
413. Meadow, J. R., U.S. Pat. 2,403,013 (1945).
414. Meerwein, H., *Ber.*, **66**, 411 (1933).
415. Meerwein, H., and H. Maier-Huser, *J. Prakt. Chem.*, **134**, 51 (1932).
416. Meerwein, H., and D. Vossen, *J. Prakt. Chem.*, **141**, 149 (1934).
417. Meinert, R. N., U.S. Pat. 2,348,637; *C.A.*, **39**, 527 (1944).
418. Meisenheimer, J., E. von Budkewicz, and G. Kananow, *Ann.*, **423**, 75 (1921).
419. Menshutkin, N. A., *J. Russ. Chem. Soc.*, **10**, 367 (1878); *ibid.*, **14**, 292 (1882).
420. Menshutkin, N. A., and D. P. Konovaloff, *Ber.*, **17**, 1361 (1884).
421. Mesheryakov, A. P., and L. V. Petrova, *Bull. Acad. Sci. U.S.S.R.*, *Div. Chem. Sci.*, 576 (1951).
422. Meyer, H., and K. Bernhauer, *Monatsh.*, **53**, 721 (1929).
423. Michaelis, A., *Ann.*, **293**, 197 (1896).
424. Midland Silicones Ltd., Brit. Pat. 751,370 (1956); *C.A.*, **51**, 5828 (1957).
425. Milas, N. A. *et al.*, *J. Am. Chem. Soc.*, **70**, 1597 (1948).
426. Miron, S., and R. J. Lee, Preprints of Papers presented before the Division of Petroleum Chemistry, A.C.S. Boston Meeting, April, 1959.
427. Mohlau, R., and R. Berger, *Ber.*, **26**, 1994, 1196 (1893).
428. Monacelli, W. J., and G. F. Hennion, *J. Am. Chem. Soc.*, **63**, 1722 (1941).
429. Morgan, G. T., and D. D. Pratt, Brit. Pat. 353,464 (1930); *C.A.*, **26**, 5308 (1938).
430. Mosby, W. L., *J. Am. Chem. Soc.*, **74**, 2564 (1952).
431. Mouneyrat, A., *Compt. rend.*, **126**, 1805 (1898).
432. Moyle, C. L., and L. I., Smith, *J. Org. Chem.*, **2**, 122 (1937).
433. Muetterties, E. L., *J. Am. Chem. Soc.*, **81**, 2597 (1959).

434. Müller, R., and C. Dathe, *J. Prakt. Chem.*, **13**, 306 (1961).
435. Mundici, C. M., *Gazz. Chim. Ital.*, **34**, 114 (1904).
436. Murrow, *Bull. Soc. Chim. France*, (3) **9**, 568 (1893).
437. Nametkin, S. S., and A. E. Abakumovskaya, *J. Gen. Chem. U.S.S.R.*, **6**, 1166 (1936).
438. Nazarova, Z. N., and I. P. Tsukervanik, *J. Gen. Chem. U.S.S.R.*, **10**, 1151 (1940).
439. Nencki, M., *J. Prakt. Chem.*, **25**, 273 (1882).
440. Nencki, M., *Ber.*, **30**, 1766 (1897).
441. Nencki, M., and N. Sieber, *J. Prakt. Chem.*, (2), **23**, 147 (1881).
442. Nenitzescu, C. D., *Experientia*, **16**, 332 (1960).
443. Nenitzescu, C. D., M. Avran, and E. Sliam, *Bull. Soc. Chim. France*, 1266 (1955).
444. Nenitzescu, C. D., and I. P. Cantuniari, *Ann.*, **510**, 269 (1934).
445. Nenitzescu, C. D., and I. Chicos, *Ber.*, **66**, 969 (1933).
446. Nenitzescu, C. D., and E. Cioranescu, *Ber.*, **69**, 1820 (1936).
447. Nenitzescu, C. D., E. Cioranescu, and M. Maican, *Ber.*, **74**, 687 (1941).
448. Nenitzescu, C. D., and I. G. Gavat, *Ann.*, **519**, 260 (1936).
449. Nenitzescu, C. D., and C. N. Ionescu, *Ann.*, **491**, 189 (1931).
450. Nenitzescu, C. D., E. Necsoiu, A. Glatz, and M. Zalman, *Ber.*, **92**, 10 (1959).
451. Nenitzescu, C. D., and V. Przemetzky, *Ber.*, **74**, 676 (1941).
452. Nesmeyanov, A. N., and E. J. Kahn, *Ber.*, **67**, 370 (1934).
453. Nesmeyanov, A. N., and L. G. Makarova, *Bull. Acad. Sci. U.S.S.R., Div. Chem. Sci.*, 213 (1947).
454. Newman, M. S., *J. Am. Chem. Soc.*, **63**, 2531 (1941).
455. Newman, M. S., *J. Am. Chem. Soc.*, **67**, 345 (1945).
456. Newman, M. S., and A. G. Pinckus, *J. Org. Chem.*, **19**, 978 (1954).
457. Newton, A., *J. Am. Chem. Soc.*, **65**, 2434 (1943).
458. Newton, H. P., and P. H. Groggins, *Ind. Eng. Chem.*, **27**, 1397 (1935).
459. Nightingale, D. V., *Chem. Rev.*, **25**, 329 (1939).
460. Nightingale, D. V., *Chem. Rev.*, **40**, 117 (1947).
461. Nightingale, D. V., H. B. Hucker, and O. L. Wright, *J. Org. Chem.*, **18**, 244 (1942).
462. Norris, J. F., *Org. Syn.*, *Coll.*, **1**, 548. Second Edition, John Wiley and Sons, New York, 1941.
463. Norris, J. F., and P. Arthur, *J. Am. Chem. Soc.*, **62**, 874 (1940).
464. Norris, J. F., and J. N. Ingraham, *J. Am. Chem. Soc.*, **69**, 1421 (1938).
465. Norris, J. F., and B. M. Sturgis, *J. Am. Chem. Soc.*, **61**, 1413 (1939).
466. Norris, J. F., and B. M. Sturgis, *J. Am. Chem. Soc.*, **61**, 1461 (1939).
467. Norris, J. F., and J. E. Wood, *J. Am. Chem. Soc.*, **62**, 1428 (1940).
468. Norrish, R. G. W., and K. E. Russel, *Trans. Faraday Soc.*, **48**, 91 (1952).
469. Oblad, A. G., G. A. Mills, and H. Heinemann, in "Catalysis," Vol. VI, p. 341, edited by P. H. Emmett, Reinhold Publishing Corporation, New York, 1958.
470. O'Connor, M. J. and F. J. Sowa, *J. Am. Chem. Soc.*, **60**, 125 (1938).
471. Olah, G. A., unpublished results.
472. Olah, G., and S. Kuhn, *Ber.*, **89**, 866 (1956).
473. Olah, G. A., and S. J. Kuhn, *J. Am. Chem. Soc.*, **82**, 2380 (1960).

474. Olah, G. A., and S. J. Kuhn, *J. Org. Chem.*, **27**, 2667 (1962).
475. Olah, G., S. Kuhn and S. Beke, *Ber.*, **89**, 862 (1956).
476. Olah, G. A., S. J. Kuhn, and S. H. Flood, *J. Am. Chem. Soc.*, **84**, 1688 (1962).
477. Olah, G. A., and S. J. Kuhn, and A. Mlinko, *J. Chem. Soc.*, 4257 (1956).
478. Olah, G. A., and S. J. Kuhn, *J. Org. Chem.*, **26**, 237 (1961).
479. Olah, G. A., and M. Meyer, *J. Org. Chem.*, **27**, 3464 (1962).
480. Olah, G. A., and A. Pavlath, *Acta Acad. Sci. Hung.*, **3**, 203 (1953).
481. Olah, G. A., and A. E. Pavlath, Hung. Pat. 145,691, 145,788, 145,579 (1957).
482. Olah, G. A., A. E. Pavlath, and S. J. Kuhn, *Chem. and Ind.*, 50 (1957).
483. Olah, G. A., A. Pavlath, and J. Olah, *J. Chem. Soc.*, 2174 (1957).
484. Olah, G. A., H. W. Quinn, and S. J. Kuhn, *J. Am. Chem. Soc.*, **82**, 426 (1960).
485. Olah, G. A., and W. S. Tolgyesi, *J. Org. Chem.*, **26**, 2053 (1961).
486. Olah, G. A., and W. S. Tolgyesi, *J. Org. Chem.*, **26**, 2319 (1961).
487. Olah, G. A., W. S. Tolgyesi, and R. E. A. Dear, *J. Org. Chem.*, **27**, 3449, 3455 (1962).
488. Olah, G. A., Abstracts of Papers 12M, 140th Meeting, ACS, Chicago, Ill., September, 1961; G. A. Olah, W. S. Tolgyesi, and R. E. A. Dear, *J. Org. Chem.*, **27**, 3441 (1962).
489. Olivier, S. C. J., *Rev. Trav. Chim.*, **33**, 91, 163, 244 (1914).
490. Ott, E., *Ber.*, **59**, 1071 (1926).
491. Packendorff, K., N. D. Zelinsky, and L. Leder-Packendorff, *Ber.*, **66**, 1069 (1933).
492. Papa, D., E. Schwenk, and A. Klingsberg, *J. Am. Chem. Soc.*, **68**, 2133 (1946).
493. Paul, R., *Bull. Soc. Chim. France*, (5) **6**, 1162 (1939).
494. Pavlath, A. E., U.S. Pat. 2,993,937 (1961).
495. Payne, R. E., *Petrol Ref.*, **37**, 316 (1958).
496. Pepper, D. C., *Trans. Faraday Soc.*, **45**, 397 (1949).
497. Pepper, D. C., *Sci. Proc. Roy. Dublin Soc.*, **25**, 131 (1950).
498. Pepper, J. M., B. P. Robinson, and G. W. Schwanbeck, *Can. J. Chem.*, **40**, 122 (1962).
499. Perfetti, B. M., and R. Levine, *J. Am. Chem. Soc.*, **75**, 626 (1953).
500. Perrier, G., *Ber.*, **33**, 815 (1900).
501. Perrier, G., *Bull. Soc. Chim. France*, (3) **31**, 859 (1903); (3) **31**, 859 (1904).
502. Pevere, E. F., L. A. Clarke, G. B. Hatch, and F. H. Bruner, U.S. Pat. 2,346,452 (1944); *C.A.*, **38**, 5393 (1944).
503. Philips, M., *J. Am. Chem. Soc.*, **49**, 473 (1927).
504. Phillipi, E., *Monatsh.*, **32**, 631 (1911).
505. Pinck, L. A., *J. Am. Chem. Soc.*, **49**, 2536 (1927).
506. Pines, H., and V. N. Ipatieff, *J. Am. Chem. Soc.*, **69**, 1337 (1947).
507. Pines, H., F. J. Pavik, and V. N. Ipatieff, *J. Am. Chem. Soc.*, **74**, 5544 (1952).
508. Pines, H., L. Schmerling, and V. N. Ipatieff, *J. Am. Chem. Soc.*, **62**, 2901 (1940).
509. Pines, H., and R. C. Wackher, *J. Am. Chem. Soc.*, **68**, 595, 1642, 2518 (1946).

510. Plant, S. G. P., K. M. Rogers, and S. B. C. Williams, *J. Chem. Soc.*, 741 (1935).
511. Plesch, P. H., "Cationic Polymerization and Related Complexes," Academic Press, New York, 1953.
512. Plesch, P. H., *Research (London)*, **2**, 267 (1949).
513. Plesch, P. H., M. Polanyi, and H. A. Skinner, *J. Chem. Soc.*, 257 (1947).
514. Ploeg, W., *Rec. Trav. Chim.*, **45**, 342 (1926).
515. Posner, T., *Ber.*, **38**, 646 (1905).
516. Praill, P. F. G., and A. L. Whitear, *J. Chem. Soc.*, 3573 (1961).
517. Price, C. C., *Chem. Rev.*, **29**, 37 (1941).
518. Price, C. C., *Org. Reactions*, **3**, 1 (1946); John Wiley and Sons, New York.
519. Price, C. C., and N. Lund, *J. Am. Chem. Soc.*, **62**, 3105 (1940).
520. Price, C. C., and C. A. Sears, *J. Am. Chem. Soc.*, **75**, 3276 (1953).
521. Prins, H. J., *Chem. Weekblad*, **16**, 1510 (1919).
522. Pritzker, G. K., *Petroleum Processing*, **1**, 58 (1946).
523. Reid, E. E., U.S. Pat. 2,028,012 (1936).
524. Reppe, W., *et al.*, *Ann.*, **596**, 1 (1955).
525. Reynhart, A. F. A., U.S. Pat. 2,043,932 (1935); *C.A.*, **30**, 5235 (1936).
526. Rieche, A., H. Gross, and E. Hoft, *Ber.*, **93**, 88 (1960).
527. Rinehart, K. L., and D. H. Gustafson, *J. Org. Chem.*, **26**, 1836 (1960).
528. Risi, J., and D. Gauvin, *Can. J. Res.*, **B14**, 255 (1936).
529. Roberts, R. M., and S. G. Brandenberger, *J. Am. Chem. Soc.*, **79**, 5484 (1957).
530. Roblin, R. O., D. Davidson, and M. T. Bogert, *J. Am. Chem. Soc.*, **57**, 151 (1935).
531. Roe, A., *Org. Reactions*, **6**, 193 (1949); John Wiley and Sons, New York.
532. Roehen, Germ. Pat. 849,548 (1938).
533. Roh, N., and G. Kochendorfer, Germ. Pat. 677,207 (1939); *C.A.*, **33**, 6880 (1939).
534. Roland, J. R., J. D. C. Wilson, and W. E. Hanford, *J. Am. Chem. Soc.*, **72**, 2122 (1950).
535. Rotenberg, I. A., and M. A. Favorskaya, *J. Gen. Chem. U.S.S.R.*, **6**, 185 (1936).
536. Rosen, R., and E. Arundale, U.S. Pat. 2,368,494; *C.A.*, **39**, 4529 (1945).
537. Rothstein, E., and R. W. Saville, *J. Chem. Soc.*, 1946 (1949).
538. Royals, E. E., and C. M. Hendry, *J. Org. Chem.*, **15**, 1149 (1950).
539. Rudkovskii, D. M., A. G. Trifel, and A. W. Frost, *Ukrain. J. Chem.*, **10**, 277 (1935); *Z.*, I, 3667 (1936).
540. Runge, F., H. Reinhardt, and G. Kühnhanss, *Chem. Techn. Berlin*, **8**, 644 (1956).
541. Russell, G. A., *J. Am. Chem. Soc.*, **81**, 4831 (1959).
542. Saunders, S. L. M., Brit. Pat. 498,414 (1939); *C.A.*, **33**, 4348 (1939).
543. Schaarschmidt, A., *Ber.*, **57**, 2065 (1924); *Angew. Chem.*, **39**, 1457 (1926).
544. Schaarschmidt, A., U.S. Pat. 1,995,752 (1935); *C.A.*, **29**, 3349.
545. Schaarschmidt, A., L. Hermann, and B. Szemzo, *Ber.*, **58**, 1914 (1925).
546. Schering-Kahlbaum, A. G., Brit. Pat. 397,505 (1933); *C.A.*, **28**, 1048 (1933).
547. Schmerling, L., in "The Chemistry of Petroleum Hydrocarbons," Vol. 3, Reinhold Publishing Corp., 1955.

548. Schmerling, L., *J. Am. Chem. Soc.*, **66**, 1442 (1944); *ibid.*, **67**, 1152, 1178 (1945); **68**, 275 (1946); *Ind. Eng. Chem.*, **40**, 1072 (1948).

549. Schmerling, L., *J. Am. Chem. Soc.*, **67**, 1778 (1945); **68**, 275 (1946).

550. Schmerling, L., *J. Am. Chem. Soc.*, **69**, 1121 (1947).

551. Schmerling, L., *J. Am. Chem. Soc.*, **71**, 701 (1949).

552. Schmerling, L., and J. D. West, *J. Am. Chem. Soc.*, **75**, 4275 (1953).

553. Schneider, A., and R. M. Kennedy, *J. Am. Chem. Soc.*, **73**, 5017 (1951).

554. Schneider, A., and R. M. Kennedy, *J. Am. Chem. Soc.*, **73**, 5013, 5024 (1951).

555. Schneider, A., *J. Am. Chem. Soc.*, **74**, 2553 (1952).

556. Schneider, H. G., U.S. Pat. 2,197,023 (1940); *C.A.*, **34**, 5465 (1940).

557. Scholl, R., *Ber.*, **32**, 3492 (1899).

558. Scholl, R., and F. Kacer, *Ber.*, **36**, 322 (1903).

559. Schorger, A. W., *J. Am. Chem. Soc.*, **39**, 2671 (1917).

560. Schultze, W. A., U.S. Patents, 2,382,505 and 2,382,506 (1945).

561. Seel, F., *Z. anorg. allgem. Chem.*, **250**, 331 (1943).

562. Sensel, E. E., and A. R. Goldsby, *Ind. Eng. Chem.*, **44**, 2716 (1952).

563. Serniuk, G. E., and B. M. Vanderbilt, U.S. Pat. 2,534,304 (1950); *C.A.*, **45**, 3861 (1951).

564. Shah, R. C., and J. S. Chaubal, *J. Chem. Soc.*, 650 (1932).

565. Shah, R. C., R. K. Deshpande, and J. S. Chaubal, *J. Chem. Soc.*, 642, 650 (1932).

566. Shoppee, C. W., and D. A. Prins, *Helv. Chim. Acta*, **26**, 201 (1943).

567. Silver, S. L., and A. Lowy, *J. Am. Chem. Soc.*, **56**, 2429 (1934).

568. Simons, J. H., and S. Archer, *J. Am. Chem. Soc.*, **62**, 1623 (1940).

569. Simons, J. H., and G. C. Bassler, *J. Am. Chem. Soc.*, **63**, 880 (1941).

570. Simons, J. H., and H. Hart, *J. Am. Chem. Soc.*, **66**, 1309 (1944).

571. Simonis, H., and S. Danischewski, *Ber.*, **59**, 2914 (1926).

572. Simonis, H., and C. Lear, *Ber.*, **59**, 2908 (1926).

573. Simpson, J. C. E., C. M. Atkinson, K. Schofield, and O. Stephenson, *J. Chem. Soc.*, 646 (1945).

574. Sisido, K., Y. Udo, T. Nakamura, and H. Nozaki, *J. Org. Chem.*, **26**, 1369 (1961).

575. Slamina, S. J., F. J. Sowa, and J. A. Nieuwland, *J. Am. Chem. Soc.*, **57**, 1547 (1935).

576. Smith, L. I., *Org. Reactions*, **1**, 370; John Wiley and Sons, New York (1942).

577. Smith, L. I., and C. O. Guss, *J. Am. Chem. Soc.*, **62**, 2635 (1940).

578. Smith, L. I., and S. A. Harris, *J. Am. Chem. Soc.*, 1289 (1935).

579. Smith, L. I., and J. W. Horner, *J. Am. Chem. Soc.*, **62**, 1349 (1940).

580. Smith, L. I., F. L. Taylor, and I. M. Webster, *J. Am. Chem. Soc.*, **59**, 1082 (1937).

581. Smith, L. I., and D. Tenenbaum, *J. Am. Chem. Soc.*, **57**, 1293 (1935).

582. Snyder, H. R., and C. T. Elston, *J. Am. Chem. Soc.*, **77**, 364 (1955).

583. Soderback, E., *Acta Chim. Scand.*, **8**, 1851 (1954).

584. Sommer, L. H., and G. A. Bougham, *J. Am. Chem. Soc.*, **83**, 3346 (1961).

585. Sowa, F. J., and J. A. Nieuwland, *J. Am. Chem. Soc.*, **55**, 5052 (1933).

586. Spoerri, P. E., and M. J. Rosen, *J. Am. Chem. Soc.*, **72**, 4918 (1950).

587. Spring, F. S., and T. Vickerstaff, *J. Chem. Soc.*, 1873 (1935).

588. Standard Oil Development Co., French Pat. 822,405 (1937); *C.A.*, **32**, 4173 (1938).

589. Standard Oil Development Co., Brit. Pat. 546,406 (1942); *C.A.*, **37**, 4074 (1943).

590. Standard Oil Development Co., Brit. Pat. 550,711 (1943); *C.A.*, **38**, 2460 (1944).

591. Standard Oil Development Co., Brit. Pat. 558,487 (1944); *C.A.*, **39**, 4463 (1945).

592. Standard Oil Development Co., Brit. Pat. 577,747 (1945); *C.A.*, **41**, 2569 (1946).

593. Staudinger, H. P., and K. H. W. Tuerck, Brit. Pat. 539,163 (1941); *C.A.*, **36**, 3509.

594. Staudinger, H. P., and K. H. W. Tuerck, U.S. Pat. 2,316,465 (1943); *C.A.*, **37**, 5734.

595. Stevenson, D. P., and J. H. Morgan, *J. Am. Chem. Soc.*, **70**, 2773 (1948).

596. Stockhausen, F., and L. Gattermann, *Ber.*, **25**, 3521 (1892).

597. Strosacker, C. J., and C. C. Schwegler, U.S. Pat. 1,713,104 (1929); *C.A.*, **23**, 3235 (1929).

598. Suida, H., Germ. Pat. 485, 434 (1924); *C.A.*, **24**, 866 (1924).

599. Sulivan, F. W., "Science of Petroleum," Vol. IV, p. 2664, Oxford University Press, 1938.

600. Sullivan, F. W., V. Voorhees, A. W. Neeley, and R. V. Shankland, *Ind. Eng. Chem.*, **23**, 604 (1931).

601. Sullivan, R. F., C. J. Egan, G. E. Langlois, and R. P. Sieg, *J. Am. Chem. Soc.*, **83**, 1156 (1961).

602. Suter, C. M., "The Organic Chemistry of Sulfur," John Wiley and Sons, New York, 1944, p. 675.

603. Swarts, F., *Acad. Roy. Belg.*, **24**, 309,474 (1892).

604. Taylor, W. J., D. D. Wagman, M. G. Williams, K. S. Pitzer, and F. D. Rossini, *J. Res. Nat. Bur. Std.*, **37**, 95 (1946).

605. Texaco Development Corp., Brit. Pat. 540,459 (1941); *C.A.*, **36**, 4127 (1942).

606. Theobald, R. S., and K. Schofield, *Chem. Rev.*, **46**, 171 (1950).

607. Thomas, C. A., "Anhydrous Aluminum Chloride in Organic Chemistry," Reinhold Publ. Corp., New York, 1941.

608. Thomas, R. J., W. F. Anzilotti, and G. F. Hennion, *Ind. Eng. Chem.*, **32**, 408 (1940).

609. Thomas, R. M., W. J. Sparks, and P. K. Frolich, *J. Am. Chem. Soc.*, **62**, 276 (1940).

610. Tomita, M., *J. Pharm. Soc. Japan*, **56**, 906 (1936).

611. Toole, S. G., and F. J. Sowa, *J. Am. Chem. Soc.*, **59**, 1971 (1937).

612. Topchiev, A. V., G. N. Yegorova, and V. N. Vasilieva, *Proc. Acad. Sci., U.S.S.R., Chem. Sect.*, **67**, 475 (1949).

613. Toussaint, N. F., and G. F. Hennion, *J. Am. Chem. Soc.*, **62**, 1145 (1940).

614. Treffers, H. P., and L. P. Hammett, *J. Am. Chem. Soc.*, **59**, 1708 (1937).

615. Trenkin, E. I., A. A. Prokhorova, Y. M. Paushkin, and A. V. Topchiev, *Bull. Acad. Sci. U.S.S.R., Div. Chem. Sci.*, 1507 (1960).

616. Truce, W. E., and F. P. Hoerger, *J. Am. Chem. Soc.*, **76**, 5357 (1954).

617. Truce, W. E., and C. E. Olson, *J. Am. Chem. Soc.*, **74**, 4721 (1952).

618. Truce, W. E., and C. W. Vriesen, *J. Am. Chem. Soc.*, **75**, 5032 (1953).

619. Tsukervanik, I. P., and N. I. Bogdanova, *J. Gen. Chem. U.S.S.R.*, **22**, 410 (1953).
620. Tsukervanik, I. P., and K. Tokareva, *J. Gen. Chem. U.S.S.R.*, **5**, 764 (1935); *C.A.*, **30**, 442 (1936).
621. Tsukervanik, I. P. and G. Vikhrova, *J. Gen. Chem. U.S.S.R.*, **7**, 632 (1937); *C.A.*, **31**, 5779 (1937).
622. Tulleners, A. J., M. C. Tuyn, and H. I. Waterman, *Rec. Trav. Chim.*, **53**, 544 (1934).
623. de Turski, J. F., Germ. Pat. 287,756 (1914); *C.A.*, **10**, 2128 (1914).
624. Turski, J. S. F., Brit. Pat. 626,661 (1949); *C.A.*, **44**, 2761 (1950).
625. Turski, S., U.S. Pat. 2,585,355; *C.A.*, **47**, 875.
626. Ulich, H., and G. Heyne, *Z. Electrochem.*, **41**, 509 (1935).
627. Underwood, H. W., and R. L. Wakeman, *J. Am. Chem. Soc.*, **52**, 387 (1930).
628. Vierling, K., U.S. Pat. 2,198,046 (1940); *C.A.*, **34**, 5465 (1940).
629. Vilsmeier, A., *Chem. Ztg.*, **75**, 133 (1951).
630. Vilsmeier, A., and A. Haack, *Ber.*, **60**, 119 (1927).
631. Vollman, H., and H. Becker, U.S. Pat. 2,126,360 (1938); *C.A.*, **32**, 7478 (1938).
632. Waddington, T. C., and F. Klanberg, *Naturwissenschaften*, **20**, 528 (1959).
633. Waddington, T. C., and J. A. White, *Proc. Chem. Soc.*, 85 (1960).
634. Waldmann, H., *J. Prakt. Chem.*, (2), **131**, 71 (1931).
635. Weibezahn, W., U.S. Pat. 1,912,608, Germ. Pat. 582,544; *C.A.*, **27**, 4243 (1933).
636. Wertyporoch, E., *Ann.*, **493**, 153 (1932).
637. Weston, A. W., and C. M. Suter, *J. Am. Chem. Soc.*, **61**, 2556 (1939).
638. Whaley, W. M., and Govindachuri, T. R., *Org. Reactions*, **6**, 74 (1951); John Wiley and Sons, New York.
639. Whitmore, F. C., *J. Am. Chem. Soc.*, **54**, 3274 (1932); *Ind. Eng. Chem.*, **26**, 94 (1934).
640. Wibaut, J. P., *Z. Elektrochem.*, **35**, 602 (1929).
641. Wibaut, J. P., J. J. Diekmann, and A. J. Rutgers, *Rec. Trav. Chim.*, **47**, 477 (1928).
642. Wieland, H., and L. Bettag, *Ber.*, **55**, 2246 (1922).
643. Wieland, H., and A. Kulenkampff, *Ann.*, **431**, 30 (1923).
644. Williams, J. W., Y. J. Dickert, and J. A. Krynitsky, *J. Am. Chem. Soc.*, **63**, 2510 (1941).
645. Williams, J. W., and J. M. Osborn, *J. Am. Chem. Soc.*, **61**, 3438 (1939).
646. Willstatter, R., and H. Kubli, *Ber.*, **42**, 4163 (1910).
647. Wimmer, J., and M. Mugdan, U.S. Pat. 2,249,512 (1941).
648. Witzinger, R., *J. Prakt. Chem.*, **154**, 25 (1939).
649. Wohl, A., and E. Wertyporoch, *Ber.*, 64, 1360 (1931).
650. Wolff, P., Germ. Pat. 614,325 (1935); *C.A.*, **29**, 5861 (1935); Germ. Pat. 615,130 (1933); *C.A.*, **29**, 6248 (1935).
651. Yamase, Y., *Bull. Chem. Soc. Japan*, **34**, 480 (1961).
652. Yates, P., and P. Eaton, *J. Am. Chem. Soc.*, **82**, 4436 (1960).
653. Young, F. G., *J. Am. Chem. Soc.*, **71**, 1346 (1949).

654. Zavgorodnii, S. V., *J. Gen. Chem. U.S.S.R.*, **14**, 270 (1944).
655. Zeavin, J. M., and A. M. Fischer, *J. Am. Chem. Soc.*, **54**, 3738 (1932).
656. Zeide, O. A., and B. M. Dubinin, *J. Gen. Chem. U.S.S.R.*, **2**, (64), 455 (1932).
657. Zincke, T., and R. S. Suhl, *Ber.*, **39**, 4148 (1906).

CHAPTER III

Recent Advances 1965–1972*

I. Introduction

The rapid development of chemistry in our times makes monographs and reviews obsolete soon after they are published. Preceding Chapter II, reproduced from the 1963–1965 edition of *Friedel-Crafts and Related Reactions* (1), covered the scope and general aspects of the reactions from the original discovery in 1877 until the publication date. The present shorter version, *Friedel-Crafts Chemistry*, is intended not only to provide in a single accessible volume the more general aspects of the original compendium but also to give an up-to-date review of this extensive subject. This chapter attempts to review the major advances in the field during the period 1965–1972. No attempt is made at an encyclopedic coverage, instead major developments are reviewed, and inevitable preference is given to the reviewers' own interests.

Two reviews have been published on Friedel-Crafts chemistry since 1965: by Roberts (1a) and by Olah and Cupas (1b). They are, however, of a more general type and not intended to be of detailed nature.

II. Alkylation (Including Isomerization and Polymerization)

1. Alkylation of Aromatic Compounds

A. Alkylating Agents

a. Haloalkanes and dialkylhalonium ions

Gallium chloride was used as a catalyst for the methylation of benzene by methyl chloride. In an excess of methyl chloride, the rate equation is

$$\text{Rate} = k[\text{PhH}][\text{GaCl}_3]^2$$

* Written in cooperation with Dr. R. H. Schlosberg, Department of Chemistry, Wisconsin State University, Clearwater, Wisconsin.

and hexadeuterated benzene is methylated at the same rate as isotopically normal benzene. DeHaan, Brown, and Hill (2) discussed the probable mechanism in detail and concluded that the slow step is an attack of the complex $(MeCl)_2GaCl_2^+GaCl_4^-$ on benzene to form a σ-complex. Similar results were obtained for the methylation of toluene and dimethylbenzene (3). Isotopic exchange occurs betweeen methyl chloride and gallium chloride, but the two reactions involve different intermediates. In the latter case a dimethylchloronium ion is formed (4). This ion was directly observed and prepared by Olah and DeMember (5). With ethyl chloride, exchange involves bascially the same mechanism as methylation (6).

Olah and his co-workers (7) continued their study of alkyl halide–antimony pentafluoride systems. Most alkyl fluorides ionize to the corresponding alkylcarbenium fluoroantimonates in the powerful ionizing agent antimony pentafluoride. Alkylcarbenium ions were found to be remarkably stable in SbF_5 solutions (neat or diluted with SO_2 or SO_2ClF).

Methyl fluoride and ethyl fluoride form stable addition complexes with antimony pentafluoride ($CH_3F–SbF_5$ and $CH_5F–SbF_5$), which are remarkably powerful methylating and ethylating agents, respectively (8). They surpass as alkylating agents all previously known systems and indeed act as carbenium ion type reagents. Not only n- and π-donor substrates (including aromatic hydrocarbons) are alkylated, but also saturated σ-donor systems.

In their study of alkyl halide–antimony pentafluoride systems, Olah and DeMember (5, 9) found that open-chain dialkylhalonium ions $R\overset{+}{X}R$ are formed when 2 moles (with some excess) of an alkyl halide (except fluoride), are reacted with a sulfur dioxide solution of $SbF_5–SO_2$. Anhydrous, complex fluoro silver salts gave similar results in metathetic reactions.

$$2RX + SbF_5–SO_2 \xrightarrow{\ SO_2\ } R\overset{+}{X}R\ SbF_5X^-$$

$$2RX + AgSbF_6\ (or\ AgBF_4) \xrightarrow{\ SO_2\ } R\overset{+}{X}R\ SbF_6^-(BF_4^-) + AgX$$

Prepared dialkylhalonium ions include

$$R = CH_3,\ C_2H_5,\ i\text{-}C_3H_7,\ C_4H_9$$

$$X = Cl,\ Br,\ I$$

Methyl, alkylhalonium and ethyl, alkylhalonium ions are prepared by methylating (ethylating) alkyl halides (other than fluorides) with the powerful CH_3F–SbF_5 and C_2H_5F–SbF_5 complexes, respectively.

$$RX + CH_3F\text{–}SbF_5(C_2H_5F\text{–}SbF_5) \rightarrow R\overset{+}{X}CH_3(C_2H_5)SbF_6^-$$

$$R = CH_3, C_2H_5, i\text{-}C_3H_7 \quad X = Cl, Br, I$$

The latter method allows the preparation of mixed dialkylhalonium ions.

Dialkylhalonium ions react with aromatic hydrocarbons as rather selective alkylating agents (10). The reactions are considered as nucleophilic displacements.

$$ArH + CH_3\overset{+}{X}CH_3\,SbF_6^- \rightleftharpoons \left[ArH\text{---}\overset{\overset{H\quad H}{\diagdown\!\diagup}}{\underset{H}{C}}\text{---}XCH_3 \right]^+ SbF_6^-$$

$$ArCH_3 + HF + SbF_5 \longleftarrow Ar\overset{+}{H}CH_3\,SbF_6^- + CH_3X$$

For steric reasons, as well because of high selectivity, *para*-substitution in these reactions is more predominant than in more conventional alkylations. There is, however, close similarity to some alkylations, particularly with alkyl iodides, indicating that dialkylhalonium (particularly iodonium) ion formation can play a role in general Friedel-Crafts alkylations. It should be remembered that excess alkyl halide used in the reactions can always act as an *n*-base and thus be alkylated (in competition with π-aromatics).

Nakane, Kurihara, and Natsubori (11) reported that in the $AlBr_3$-catalyzed ethylation of benzene by ethyl-2-C^{14} iodide migration of radioactivity from the β- to the α-atom occurs. These investigators suggested that the reagent is either a carbenium ion or a polarized complex formed between ethyl halide and the metallic halide, the former occurring in non-polar and the latter in more basic solvents. Toluene, ethyl fluoride, and BF_3 form a complex which was thought to be the σ-complex occurring as an intermediate in Friedel-Crafts alkylations. However, Nakane (12) suggested, from a study of visible, infra-red, and n.m.r. spectra and isotope effects, that it is in fact an oriented π-complex.

The relative reactivities of substituted benzenes in Friedel-Crafts alkylations do not always follow an obvious order. In ethylation by ethyl fluoride and BF_3, toluene is *less* reactive than benzene (11). The normal order in electrophilic substitution is Me > Et > i-Pr > t-Bu, and steric hindrance to solvation, rather than hyperconjugation, has been suggested as being responsible for it.

Myhre, Rieger, and Stone (13) reported a high yield in the synthesis of 1,3,5-tri-t-butylbenzene from t-butylbenzene with t-butyl chloride at −10°. p-Di-t-butylbenzene formed predominantly in the monoalkylation step is obviously isomerized by $AlCl_3$ to the *meta*-isomer, which is then further alkylated.

Heating a mixture of HF with a 10:1 molar mixture of benzene–chloroalkane (2HF–hydrocarbon mixture) led to the production of 90% alkylate (14). The straight-chain alkylbenzenes so formed are useful as detergents.

Alkylation of diphenylalkanes (15) using a C_8–C_{16} haloalkane (or olefin) in the presence of aluminum chloride at 50° followed by treatment with oleum (23% SO_3) produces intermediates suitable for detergent preparation.

Alkylation of anisole by isopropyl chloride (and propylene) in the presence of $AlCl_2$–H_2PO_4 or CH_3NO_2–$AlCl_3$ has been studied along with the disproportionation reaction is isopropylanisole (16).

b. Dihaloalkanes

In the alkylation of benzene by dichloroalkanes in the presence of boron trifluoride and hydrogen fluoride, the rate increases as the distance between the two chlorines increases, owing to the inductive effect of the second chlorine on the developing positive charge produced by partial ionization of the first (17).

A second substituent affects the rate of reaction between halogenooctanes and benzene catalyzed by HF–BF_3, even if separated by seven methylene groups (18).

2,4-Dichloropentane alkylated benzene with an aluminum chloride catalyst under two sets of conditions (19). Using 0.5 M AlCl$_3$ at 0° for 10 hours led to a 47% yield of monophenyl and an 18% yield of diphenyl products. With 0.16 M AlCl$_3$ at 25° for 3 hours, a 32% yield of monophenyl products and a 23% yield of diphenyl products were obtained.

The Friedel-Crafts alkylation of benzene with (CH$_3$)$_2$C(Br)CH(Br)-CH$_3$, CH$_3$CH(Br)CH(CH$_3$CH(Br)CH$_3$, (CH$_3$)$_2$C(ϕ)CH$_2$CHBrCH$_3$, and related compounds has been reported (20).

The formation of the triphenylpropionic acid from β-phenyl, α,β-dibromopropionic acid and benzene has been found to proceed in a stepwise fashion (21).

$$C_6H_5CH-CH-COOH \quad \xrightarrow{C_6H_6/AlCl_3} \quad (C_6H_5)_2CH-CH-COOH$$
$$\underset{Br}{|} \quad \underset{Br}{|} \qquad\qquad\qquad\qquad\qquad \underset{Br}{|}$$

$$\downarrow C_6H_6/AlCl_3$$

$$(C_6H_5)_2CH-CH(C_6H_5)-COOH$$

The condensation of p-bis(chloromethyl)benzene with benzene under the influence of SnCl$_4$ leads to the expected condensation products (22).

c. *Haloalkenes*

A study of the condensation of benzene with allyl bromide in the presence of sulfuric acid led to the observation of seven products: 1-bromo-2-phenylpropane (a), α-methylbibenzyl (b), 2-bromo-1-propylbenzene (c), allylbenzene (d), 1,2-dibromopropane (e), α-methylstyrene (f), and propenylbenzene (g). Two pathways were proposed to account for the formation of the observed products (23).

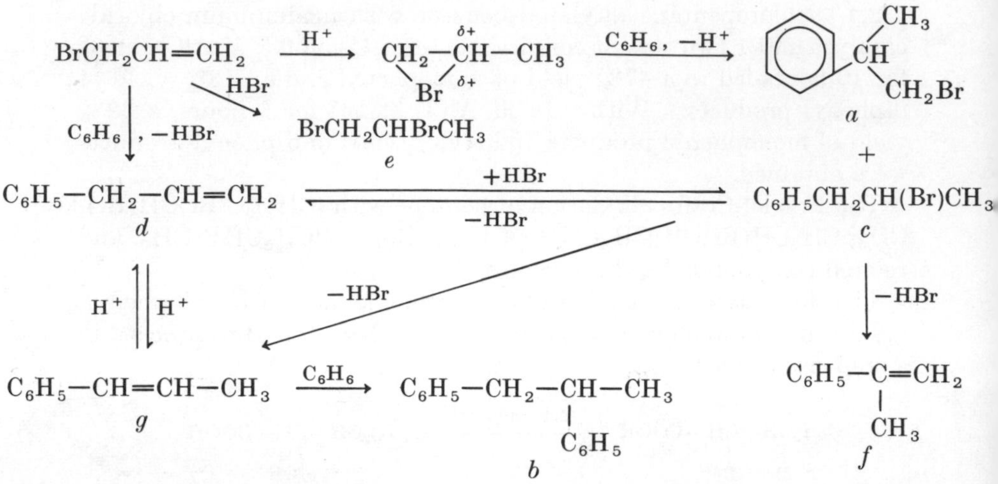
BrCH2CH=CH2
H+
CH2—CH—CH3
Br
δ+
C6H6, −H+
CH3
CH
CH2Br
a
HBr
C6H6, −HBr
BrCH2CHBrCH3
e
+HBr
C6H5—CH2—CH=CH2
d
−HBr
+
C6H5CH2CH(Br)CH3
c
H+
H+
−HBr
−HBr
C6H5—CH=CH—CH3
g
C6H6
C6H5—CH2—CH—CH3
C6H5
b
C6H5—C=CH2
CH3
f

The action of propargyl chlorides in the Friedel-Crafts reaction has been investigated (24).

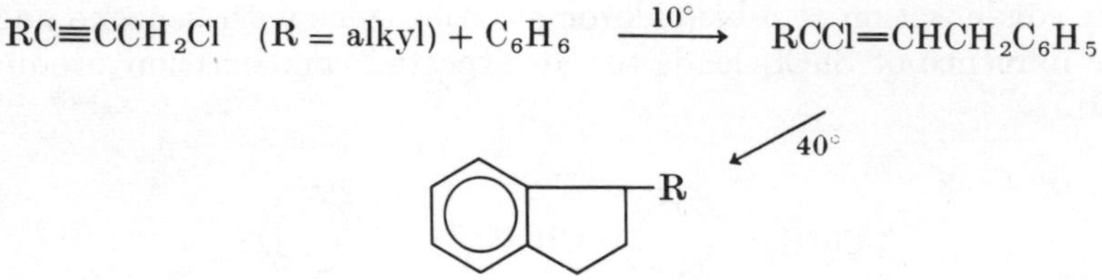
RC≡CCH2Cl (R = alkyl) + C6H6
10°
RCCl=CHCH2C6H5
40°
R

$$C_6H_5CH_2C{\equiv}CCH_2Cl + C_6H_6 \longrightarrow \text{variety of products}$$

Ried and Lantzsch (25) have reported the alkylation of benzene with a chlorinated cyclobutenone as the alkylating agent.

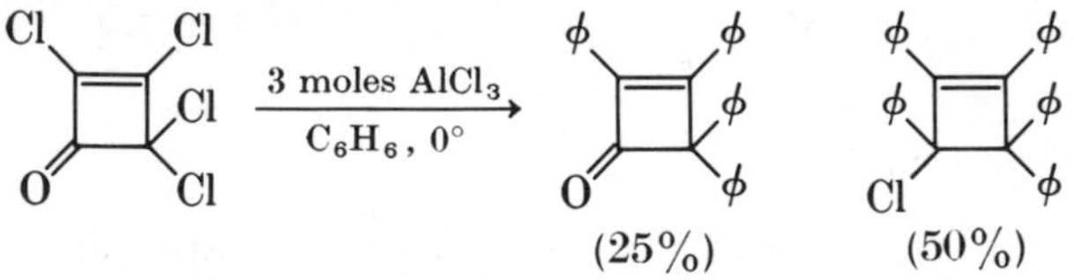
Cl
Cl
Cl
Cl
O
3 moles AlCl3
C6H6, 0°
O
Cl
(25%)
(50%)

d. Alkenes

Alkenes are the most frequently employed industrial alkylating agents because of their low cost and ready availability.

Sulfur-free benzene when treated with propene (free of allene and methylacetylene) in the presence of aluminum chloride at 70–80° for 1 hour was reported to lead to an improved preparation of cumene (26). Results are optimum when a 3:1 ratio of benzene to propene is employed.

The Friedel-Crafts alkylation of benzene with butadiene has been reported (27). The three isomeric diphenylbutanes formed when excess aromatic was employed were 23–50% 1,1-, 29–31% 1,2-, and 19–47% 1,3-isomers.

In a brief note it was reported that even with relatively bulky olefin–catalyst complexes (e.g., BF_3–H_3PO_4–trans-2-butene) the 1-isomer is the kinetically favored product (58% of the 1-isomer and 23% of the 2-isomer being formed) in the Friedel-Crafts alkylation of naphthalene (28).

In the presence of $AlCl_3$, 1-dodecene and trans-6-dodecene alkylate benzene to give phenyldodecanes with a similar isomer distribution (29). Styrene reacts almost quantitatively with benzene in the presence of palladium acetate to give trans-stilbene (30).

Anisole is alkylated by propene catalyzed by $AlCl_3$ or H_2SO_4 and eventually gives the triisopropyl derivative (31).

e. Alcohols

Alkylation of aromatic hydrocarbons can be carried out with primary, secondary, or tertiary alcohols by using the same catalysts employed in alkylations with alkyl halides or with alkenes. Benzylations using benzyl alcohols and CH_3NO_2–$AlCl_3$ were carried out (32). In order to maximize yields of diphenylmethane products, 0.9 mole of catalyst per mole of alcohol was employed. The relative rates of benzylation of benzene using p-methylbenzyl alcohol, benzyl alcohol, and p-chlorobenzyl alcohol were found to be 1.5:1.0:1.0. Benzyl alcohol–aluminum chloride is an effective reagent for benzylation (33). Naphthalene and methanol react over an alumina catalyst to give C-2 and C-4 ring-methylated products (34).

Aluminum chloride-catalyzed alkylations of benzene with 1-propanol, cyclobutylcarbinol, cyclopentylcarbinol, and cyclohexylcarbinol have been reported to take place with rearrangement of the alkylating agent and products (35).

Alkylations with alcohols are in fact consistent in terms of mechanism with other aluminum chloride-catalyzed alkylations with alkyl derivatives, alkylations in which reaction conditions play a very significant role in product distribution.

Alkylations with isobutyl alcohol (and isobutyl chloride) were attempted under severe reaction conditions (36). An excess of aluminum chloride catalyst (versus the alkylate) was employed in all cases. Products obtained again reflected extensive isomerizations.

f. Isocyanates and sulfinylamines

A review of the condensation of phenylisocyanate and phenyliso-thiocyanate with benzene, naphthalene, anthracene, and phenan-threne in the presence of aluminum chloride has appeared (37).

A novel alkylation procedure was reported by Olah and associates (38). This work demonstrated that Friedel-Crafts type alkylation occurs smoothly with either N-alkylsulfinylamines or alkyliso-cyanates in the presence of nitrosonium salts. Yields are of the

$$ArH + RNSO + NO^+SbF_6^- \rightarrow ArR + N_2 + SO_2 + HSbF_6$$
$$(PF_6^-, BF_4^-) \qquad\qquad\qquad\qquad\qquad\qquad (HPF_6, HBF_4)$$

$$ArH + RNCO + NO^+SbF_6^- \rightarrow ArR + N_2 + CO_2 + HSbF_6$$

$$ArH + RCONCO + NO^+SbF_6^- \rightarrow ArCOR + N_2 + CO_2 + HSbF_6$$

$$ArH + RCONSO + NO^+SbF_6^- \rightarrow ArCOR + N_2 + SO_2 + HSbF_6$$

order of 30–75%, and isomer distributions are in line with other alkylation results (*para > ortho > meta*). Further studies of arene alkylation by amines under diazotization conditions gave results which show that substrate reactivities ($k_{toluene}/k_{benzene} = 1.5$) are quite similar to those for Friedel-Crafts reactions (40). Absolute rate determinations have been obtained for these reactions, and the results are in good agreement with corresponding data obtained from competitive kinetic experiments (41).

g. Heterocyclic alkylating agents

Propyleneimine was reacted with substituted arenes in the pres-ence of aluminum chloride (42). Electron donors increase the α- to β-methyl-β-phenylethylamine ratio, while increasing the reaction temperature has the opposite effect. Chlorobenzene and toluene was reacted with aziridines. In these reactions the nature of the sub-stituents had little effect on the *ortho/para* ratio. In a related study (43) benzene was reacted with propylene oxide under the influence of Friedel-Crafts catalysts, the product being 2-phenyl-1-propanol.

A report has appeared on the Friedel-Crafts alkylation of benzene with a variety of oxiranes and oxetanes (44), catalyzed by aluminum chloride or stannic chloride.

B. Activity of Catalysts

The different Friedel-Crafts catalysts vary in activity in alkylation reactions (as in other reactions). The Friedel-Crafts self-conden-sation of benzyl chloride has been chosen as a means of evaluating relative catalytic activities of a large number of Lewis acid halides. The lowest temperature at which reaction occurs was chosen as an activity index (45).   The observed order of activity was in agreement with earlier observations.   A comparative study of about 40 catalysts in benzylation reactions with methylbenzyl chlorides lead to the suggestion of four categories of catalyst actvity, *i.e.*, type *A*, very active, causing both alkylation and isomerization; *B*, moderately active, effective in alkylation but causing little or no isomerization; *C*, weak, give low yields in alkylation and no isomerization; *D*, very weak or inactive (46).

C. Effect of Solvent

The solvent continues to appear to be an important factor controlling both substrate and positional selectivity of Friedel-Crafts alkylations.

Nakane, Natsubori, and Kurihava (47), in a series of important papers, showed the effect of solvent on selectivity in alkylation reactions.

D. Orientation

In Friedel-Crafts benzylation, Olah, Tashiro, and Kobayashi (46, 48) recently found that it is possible to vary in a systematic way the electrophilicity of the alkylating agent by using suitably substituted benzyl halides. In this way a regular change in the nature of the transition state of the reactions can be observed from late ones resembling σ-complexes to early ones of a π-complex nature. The system studied was the titanium tetrachloride-catalyzed benzylation of toluene and benzene with substituted benzyl chlorides.

Data obtained showed the effect of substituents on isomer distributions, particularly on *ortho/para* 'ratios. Electron-donating substituents *ortho* and *para* to the benzylic center decrease the *ortho/para* isomer ratio in the methyldiphenylmethanes formed, *i.e.*, *para*-substitution becomes predominant. In contrast, electron-withdrawing substituents increase the *ortho/para* isomer ratio.

A neopentyl and a methyl substituent were found to have but slightly differing orienting effects (49). A cyclopropyl substituent activates the benzene ring appreciably more than does a cyclobutyl or isopropyl substituent (50).

Kovacic and Hiller (51) have introduced the concept of "linear coordination" to explain unusually high *ortho/para* ratios observed in the alkylation of chlorobenzene and anisole. The concept refers to coordination of the reagent at the Lewis base substituent, followed by transfer to the *ortho*-position *via* an unspecified bridging mechanism presumed to involve the π-electrons of the ring. It is thus distinguished from *ortho*-substitution directed by chelation to the substituent already present. Linear coordination is also suggested to explain the high ratio of *ortho*-nitration of anisole by nitronium fluoroborate.

E. Dealkylation and Transalkylation

Friedel-Crafts catalysts are not restricted to the introduction of alkyl groups into aromatics. They can also remove alkyl groups and transfer them from one nucleus to another. Ethylbenzene and biphenyl were reported to react under the influence of aluminum chloride to provide a greater than 80% yield of 3- and 4-ethylbiphenyl (52).

F. Isomerization in Aromatic Alkylations

A well-known characteristic of Friedel-Crafts alkylations is the tendency toward rearrangement of alkyl groups during the alkylation process.

Rearrangement of alkylating agents

Alkylation of benzene with 1,1-dideuterio-1-chloropropane occurred with only 5% (maximum) of the possible scrambling of deuterium atoms

$$C_6H_6 + CH_3CH_2CD_2Cl$$

$$\xrightarrow{AlCl_3} C_6H_5CD_2CH_2CH_3 + C_6H_5CH(CH_3)CD_2H$$

as measured by mass spectroscopy and n.m.r. spectroscopy (53). This result indicates that protonated cyclopropanes did not intervene as intermediates in the major pathway involved in this Friedel-Crafts alkylation reaction, which can be considered as an S_N2 type displacement.

G. Rearrangement of Alkylaromatics

Olah, Carlson, and Lapierre (54) reported their findings in a study of the water-promoted, aluminum chloride-catalyzed isomerization of *o*-, *m*-, and *p*-di-*t*-butylbenzene. The equilibrium mixture consisted of 52% *m*- and 48% *p*-di-*t*-butylbenzene, no *ortho*-isomer being present. The isomerization of *o*-di-*t*-butylbenzene is very rapid, aided in large part by relief of steric strain.

Isomerization of the three *t*-butyltoluenes was carried out with water-promoted aluminum chloride in excess aromatic (heterogeneous system) and in nitromethane (homogeneous system) (55). In both systems, and starting from any of the *t*-butyltoluenes, the equilibrium isomer distribution was found to be 64% *m*- and 36% *p*-*t*-butyltoluene. Similar isomerization of isomeric diethylbenzenes (56) gave the equilibrium isomer mixture; 3% *o*-, 69% *m*-, and 28% *p*-diethylbenzene. The diisopropylbenzenes equilibrated to 68% *m*- and 32% *p*-diisopropylbenzene (57).

Friedel-Crafts isomerization of diethyltetramethylbenzenes and dineopentyltetramethylbenzenes have been observed under acid catalysis (58). The mere fact that intramolecular isomerization and not dealkylation occurs in these highly crowded molecules is of interest.

A study of the isomerization of di- and monomethylnaphthalenes in hydrogen fluoride–boron trifluoride was undertaken (59).

An elegant study by Roberts, Khalaf, and Douglass (60) showed that the principal reactions of *n*-alkyltoluenes and xylenes with aluminum chloride at room temperature were isomerisations to *m*-*n*-alkyltoluenes and *m*-*n*-alkylxylenes. At higher temperatures both isomerisations and *n*-alkyl group rearrangements competed. The results of this study led to the conclusion that the susceptibility toward aluminum chloride-induced rearrangement of *n*-propyl side chains decreased in the series *n*-propylbenzene, *n*-propyltoluene, and *n*-propylxylene. This is consistent with the fact that respectively more aluminum chloride is tied up in the lower energy reorientation (to the *meta*-product) process in going fron *n*-propylbenzene to toluene to the xylenes.

The various aluminum chloride-catalyzed reactions of *sec*-butylbenzene have been studied (61). It has been found that transalkylation occurs most readily, followed by rearrangement to *t*-butylbenzene. The slowest reaction observed was the dealkylation–alkylation to produce butane and 2,2-diphenylbutane.

Priddy (62) alkylated biphenyl under conditions that lead to the smallest amount of isomerization from the kinetically controlled products (*ortho* and *para*) to the thermodynamically more stable *meta*-product.

Alkylation of toluene with ethylene under the influence of aluminum chloride has been investigated to see what effect the addition of nitromethane to the reaction mixture would produce (63). The results from this study support the assumption that the CH_3NO_2–$AlCl_3$ system prevents isomerization and disproportionation of the primary products formed in the Friedel-Crafts alkylation.

H. Positional Rearrangement during Substitution Reactions

Hart and Janssen (64) studied the mechanism of the Jacobsen rearrangement (65). Two most likely mechanisms proposed for the Jacobsen rearrangement are:

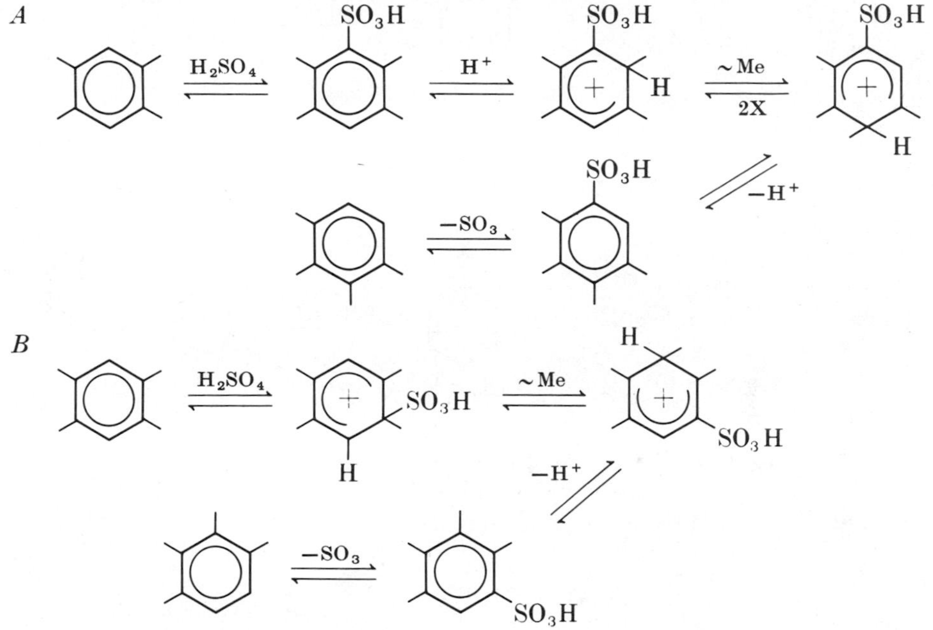

A study of the aluminum chloride-catalyzed reaction of carbon tetrachloride with durene and mesitylene gave the following results:

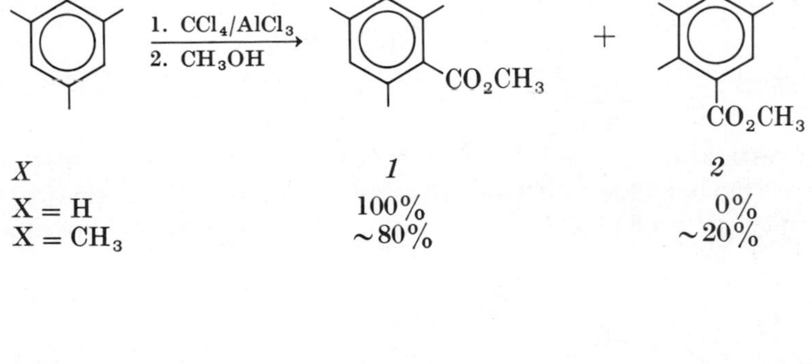

X	1	2
X = H	100%	0%
X = CH$_3$	~80%	~20%

The proposed mechanism for the reaction is similar to the mechanism *B* above offered for the Jacobsen rearrangement

Koeberg-Telder, Prinsen, and Cerfontain (66) studied the kinetics of the isomerization of *o*-xylenesulfonic acids in 74% H_2SO_4 at 141° and obtained the equilibrium mixture:

Under identical conditions the *meta*-derivative gave

$(0.0 \pm 0.5\%)$ $(17.5 \pm 1.5\%)$ $(79 \pm 2\%)$ $(3.5 \pm 1.5\%)$

Investigation (67) of the Friedel-Crafts *t*-butylation of methyl 2-pyrrole carboxylate and 2-acetylpyrrole to produce the 5- (major) and 4-*t*-butyl products has been carried out.   Isomerization experiments involving these products showed that the 4-*t*-butyl isomer undergoes rearrangement to the more stable 5-position very readily.

I. Alkylation by Metathetic Reactions of Halides with Silver Salts

In an example of a somewhat unusual Friedel-Crafts alkylation reaction, the similarities between alkyl halide- and alkyl chloroformate–silver ion-initiated Friedel-Crafts alkylations have been reported (68). The similarities between these reactions are demonstrated by the observation that reaction of cyclohexyl chloroformate and cyclohexyl chloride both yield *o*-chlorocyclohexylbenzene (33%) and *p*-chlorocyclohexylbenzene (44%) upon treatment with an equimolar amount of silver tetrafluoroborate in chlorobenzene at room temperature.

J. Stereoselective Alkylations

An interesting example of stereoselectivity was observed in the Friedel-Crafts alkylation of benzene by optically active 4-valerolactone, catalyzed by aluminum chloride (69). After allowance for the slow racemization of the valerolactone under the reaction conditions, the actual substitution step was found to proceed with nearly 50% net inversion of configuration. The mechanism of this process was envisaged as:

The internal ion pair complex of lactone with aluminum chloride can racemize by bond rotation and subsequently revert to starting material (to give racemized lactone), or it can attack a benzene molecule.   Consistent with this type of mechanism was the observation that dilution with carbon disulfide led to increased racemization and the reaction between benzene and (S)-$(+)$-2-methyl tetrahydrofuran proceeds with inversion of configuration (70).

The alkylation of benzene by 3-chlorobutanoic acid is stereospecific (43% inversion), and it appears that benzene displaces the cyclic intermediate as a nucleophile (71).

K. Intermediate Complexes

Olah and his group (72) continued their study of alkyl halide–Lewis acid halide complexes, primarily those with antimony pentafluoride.   The methyl fluoride–SbF_5 and ethyl fluoride–SbF_5 complexes were studied in detail (73).

Friedel-Crafts reactions are the most complex of the common examples of electrophilic aromatic substitution.   One factor is the number of intermediates that may be formed between the different reactants.   A termolecular π-complex of ethyl fluoride, boron trifluoride, and toluene has been detected spectrophotometrically as a reversible species in the course of ethylation of toluene (74). Stable σ-complexes were obtained by Effenberger from tripyrrolidinobenzene and alkyl halides (75).

2. Cyclialkylation of Aromatic Compounds

A. Cyclialkylation of Aralkyl Halides

Barclay and Stanford (76,77) reported a continued study of reactions of phenylalkyl chlorides. It was shown that Friedel-Crafts cyclialkylation of a series of dideuterated primary phenylpentyl chlorides yielded specifically labeled 1-methyltetralins. Determination of the location of the deuterium labels indicated that rearrangement occurs more rapidly than cyclialkylation and thus precedes it. Several phenylalkyl chlorides were synthesized for a comparative study of Friedel-Crafts cyclialkylations and competing reactions involving hydride exchange.

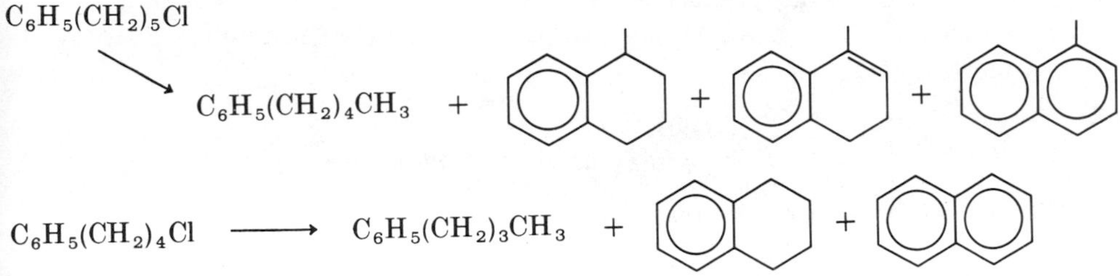

$C_6H_5(CH_2)_5Cl$

$C_6H_5(CH_2)_4CH_3$ +

$C_6H_5(CH_2)_4Cl \longrightarrow C_6H_5(CH_2)_3CH_3$ +

$C_6H_5(CH_2)_3Cl \longrightarrow$ mostly propylbenzene

Khalaf and Roberts (78) reported on cyclialkylation and competing Friedel-Crafts reactions. They observed cyclialkylation and rearrangements of both primary and secondary phenylalkyl chlorides. The extent of rearrangement in these systems was less than in comparable intermolecular alkylations only in those cases in which a six-membered ring (tetralin system) was formed directly. They also found that five-membered ring formation is facilitated in phenylalkyl halides by having the halogen on a tertiary carbon in the γ-position.

In cyclialkylation processes compounds of general formula $Ar(CH_2)_nCl$ cyclize preferentially to give six-membered rather than five- or seven-membered rings when aluminum chloride is used as catalyst. For example, five-membered rings are formed only from tertiary carbonium ions,

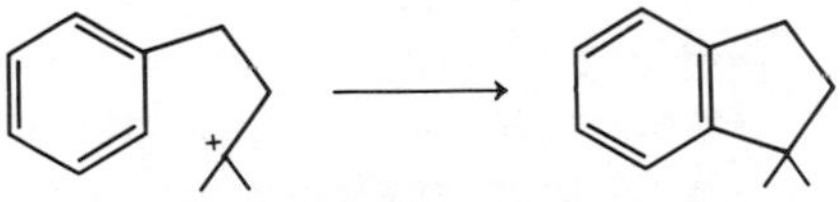

whereas six-membered ones can be obtained from primary carbonium ions (accompanied by rearrangement products). An example of this appears to be the conversion of *a* into *b* by $AlCl_3$–HCl. Polyphosphoric acid (PPA) gives *c*, thought to be the kinetically controlled product.

b a c

The use of low-activity Lewis acid catalysts, such as ferric chloride, allowed ring-closure reactions without extensive hydride exchange, and side-chain rearrangements were less extensive than previously reported (79).

B. Cyclialkylation of Arylalkenes

The ring closure of cinnamanilide

is retarded by the presence of *ortho*-substituents in the N-phenyl nucleus and is prevented entirely by a *para*-nitro group in the C-aryl nucleus (80). In related work (81) it was found that isomerization in the H_3PO_4-catalyzed cyclization of cinnamanilide to 4-phenyl-3,4-dihydrocarbostryril is prevented by nitro and severely retarded by other substituents *ortho* to the NH group.

C. Cyclialkylation of Arylaliphatic Alcohols

Tertiary 3- and 4-phenylalkanols undergo cyclialkylation readily with sulfuric acid. Primary and most secondary 3-phenylalkanols do not cyclialkylate even at high temperature with phosphoric acid. The requirement for a stable secondary or tertiary carbenium ion intermediate is implied by these results (82).

3. Alkylation of Aliphatic Compounds

A. Alkylation of Alkanes

Despite frequent literature references to electrophilic alkylation of alkanes by olefins, from a mechanistic point of view these reactions must be considered alkylations of the olefin by the carbenium ion derived from the isoalkane by intermolecular hydride transfer (83). The suggested reaction pathway is reflected in the products from

$$R_2'C = CH_2 \xrightarrow{\;H^+\;} R_2\overset{+}{C}CH_3 + R_3CH \longrightarrow R_2'CHCH_3 + R_3C^+$$

$$R_3C^+ + R_2C = CH_2' \longrightarrow R_2'C^+ - CH_2'CR_3$$

$$\xrightarrow{\;H^-\;} R_2'CHCH_2CR_3$$

propylene and isobutylene with isobutane. Products do not contain 2,2,3-trimethylbutane or 2,2,3,3-tetramethylbutane, which would be expected as the primary alkylation products in direct alkylation of isobutane with propylene and isobutylene, respectively.

Recently, the Olahs (84) carried out the first alkylation of alkanes under conditions where no olefin formation is involved.

The reaction of stable alkylcarbenium ions with alkanes was studied in sulfuryl chloride fluoride solution at low temperature (generally as low as $-78°$). A small amount (2% of the C_8 fraction) of 2,2,3,3-tetramethylbutane was detected in the reaction of t-butyl

$$(CH_3)_3C^+ + HC(CH_3)_3 \rightleftharpoons (CH_3)_3C - C(CH_3)_3$$

fluoroantimonate with isobutane. This observation is of substantial importance since no other way than direct alkylation of the C—H bond of isobutane by the t-butyl cation can account for its formation. That the direct alkylation of alkanes can indeed be carried out with ease is further shown when isobutane is reacted with the less bulky isopropyl fluoroantimonate, or when propane is reacted with t-butyl fluoroantimonate, giving the primary alkylation product 2,2,3-trimethylbutane in yields of about 15%. As intermolecular hydrogen transfer is much faster than alkylation, and the isopropyl cation more reactive than the t-butyl cation, the alkylation reaction in both systems is considered to be mainly the propylation of isobutane.

$$(CH_3)_2CH_2 + \overset{+}{C}(CH_3)_3 \rightleftharpoons (CH_3)_2CH^+ + HC(CH_3)_3$$

$$(CH_3)_3CH + H\overset{+}{C}(CH_3)_2 \rightleftharpoons (CH_3)_3CCH(CH_3)_2$$

The alkylation products obtained were not only those derived from the starting alkane and carbenium ion, but also those from the alkanes and carbenium ions formed by intermolecular hydrogen transfer, which is a faster reaction than alkylation. Since carbocations also undergo isomerization, and since intermolecular hydrogen transfer produces new alkanes from these ions, alkylation products obviously are increasingly complex. Generally, the bulkier tertiary carbenium ions can attack the more shielded tertiary C–H bonds only with great difficulty. The processes mentioned, however, produce with ease secondary and incipient primary ions in the systems, which then alkylate C–H and C–C bonds (*alkylolysis*). The alkylation reactions of alkanes with stable carbenium ion salts in low-nucleophilicity solvents produce a variety of products which, however, can be quantitatively analyzed with today's methods (gas chromatography, mass spectrometry). Since olefins are formed under the reaction conditions only to a minimal degree, if at all, and at low temperatures ($-78°$) and with short reaction times (<30 seconds), isomerizations are considered of lesser importance, and for the first time *direct alkylation of alkanes* was observed. Alkylation of alkanes can also be effected with the $CH_3F–SbF_5$ and $C_2H_5–SbF_5$ complexes (8) (which show incipient methyl and ethyl cation properties) and with dialkylhalonium ions. It is also reasonable to suggest, that, in more conventional Friedel-Crafts systems, direct alkylation of alkanes can play a role, although in systems in which olefin formation is possible, alkene reactivity vastly excedes that of alkanes.

B. *Alkylation of Alkenes and Alkynes*

Alkylation of alkenes continues to attract widespread interest, mostly in the petrochemical field, referred to extensively in the patent literature, which is, however, considered to be outside the scope of this review.

In the benzylation reaction of hexyne-3 reported by Martens and Hoornaert (85), it was found that in high-polarity solvents the yields

$$\phi\text{-}CH_2Cl + C_2H_5C{\equiv}CC_2H_5 \xrightarrow[CH_2Cl_2]{AlCl_3}$$

2.3-diethylindenone

+

$$\begin{array}{c} \phi CH_2 \\ \diagdown \\ \\ CH_3CH_2 \diagup \end{array} C{=}C(Cl)CH_2CH_3$$

cis and *trans*

of the *trans*-addition compound and of the indenone were increased. Thus an ionic mechanism apparently produces the *trans*-addition compound, while a more covalent complex or ion pair leads to the formation of the *cis*-isomer.

4. Isomerization of Aliphatic Compounds

A. Initiation by Carbocations

Isomerization of *n*-alkanes has assumed great importance in recent years, particularly because of the increased need for high-octane branched hydrocarbons (which require less or no lead additives).

It has been found (86) that hydrogen transfer reactions between alkanes containing tertiary carbon–hydrogen bonds and tertiary cations occur readily over a water-promoted aluminum bromide catalyst in 1,2,4-trichlorobenzene. Normal alkanes appear to react much more slowly under these conditions, probably because both the rates of formation and the rates of rearrangement of secondary ions are slow. Ionic rearrangements leading to similarly branched species are fast, and it was not possible to distinguish between these rates and those of hydride transfer.

A detailed investigation of the aluminum chloride-catalyzed isomerization of cyclopropylcarbinyl chloride and cyclobutyl chloride has been carried out (87). The original study of these systems was that of Robert's group (88) in the early 1950's. The final product from the aluminum chloride-catalyzed isomerization of cyclopropylcarbinyl chloride and cyclobutyl chloride was always exclusively allylcarbinyl chloride. No formation of cyclopropylcarbinyl chloride from the latter was observed under the reaction conditions.

B. Side Reactions Accompanying Isomerization

7,7-Dichloronorcarane when treated at 7–8° with aluminum chloride gave the following products (89).

(35% of total)

+ higher molecular weight hydrocarbons (40–45%)

When 7,7-dichloronorcarane was used to alkylate benzene in the presence of aluminum chloride, a mixture of cyclohexylbenzenes was produced (90). The expected 1-chloro-6-phenylcyclohexene and 1,6-diphenylcyclohexene were not observed.

5. Polymerization

Excellent reviews have appeared on many aspects of cationic polymerizations of the Friedel-Crafts type (91). Therefore only a few representative examples of the progress in this area of Friedel-Crafts chemistry are given, instead of attempting a complete review.

The present status of cationic polymerization has been discussed with emphasis on the nature of the catalysts, the counterions, and medium effects, and the question of "living" cationic polymers has been raised by Kennedy (91a). A detailed summary by Plesch of the chemistry of cationic polymerization has been published (91b).

The polymerization and copolymerization of olefins with alkyl-aluminum halides and related metal alkyl catalyst systems (in the presence of co-catalysts) has gained substantial importance (92). The use of stable carbocations (particularly triphenylmethyl and cycloheptatrienyl cation salts) as polymerization catalysts was reviewed in the polymerization of vinyl alkyl ethers, N-vinylcarbazol, and tetrahydrofuran (91j).

Weichman and Fierce (93) polymerized butadiene with benzyl chloride in the presence of zinc chloride at about 75°.

By means of preparative gel chromatography, the early products (molecular weight < 800) of the $SnCl_4$-catalyzed polycondensation of p-bis(chloromethyl)benzene with benzene have been separated and identified (94). Evidence was provided indicating the major pathway of the polycondensation and the relationship between thermal stability and chemical structure.

p-Bis(chloromethyl)benzene and diphenylmethane were polycondensed under the influence of $SnCl_4$. Products with molecular weight < 1000 were separated and identified and rate curves were obtained for the formation of each product (94).

A solvent dependence study of cyclicopolymerization has been reported by Gaylord and Svestka (95). Isoprene was cyclicopolymerized at 20° with $TiCl_4$ (88%), PF_5 (82%), and $SbCl_5$ (35%) in nitrobenzene. Using solvents such as n-heptane, benzene, acetonitrile, or nitromethane led to only trace formation of polymer.

Using a liquid polymerization medium and water-saturated propylene in the presence of 0.5 mole % BF_3 catalyst for 5 minutes at 30° under 28 atm. pressure led to the formation of propylene oligomers averaging C_{24} (96). The oligomers are useful for the alkylation of benzene, o-xylene, and other aromatic hydrocarbons to produce compounds suitable for preparing detergents by means of subsequent sulfonation.

Detergent precursors have been prepared by combining diphenylmethane with a twofold excess of 1-tetradecene and 1/6 mole aluminum chloride at 50° (97). This reaction produces a 34% yield of alkylated tetralins and a 40% yield of dialkylbenzenes. The reaction product is heated to 65° with a fourfold excess of dodecane and a fourfold excess of oleum to produce an oil-soluble sulfonated layer which is used as detergent feedstock.

Polycondensation of arenesulfonyl chlorides in nitrobenzene using ferric chloride as a catalyst has been described (98). Polycondensation proceeds smoothly at 120–140° to yield a high polymer.

Isobutylene is polymerized under the influence of $AlBr_3$ or AlI_3 with a co-catalyst of $TiCl_4$, $SnCl_4$, or BCl_3 (99), with the most active catalyst system being $AlBr_3$–$TiCl_4$ (100). The amount of conversion and the molecular weight of the polymer depend on the $AlBr_3$ concentration but are nearly independent of the $TiCl_4$ concentration. Of the three cocatalysts rate enhancement of polymerization is greatest with $TiCl_4$ followed by $SnCl_4$ (101).

An attempted polymerization of acetone using Friedel-Crafts catalysts was carried out at -78, 23, and 56° (102). In all cases the major product was that of an aldol condensation followed by dehydration giving mesityl oxide.

III. Acylation

1. Aromatic Acylation

A. General Scope

Olah and Kobayashi (103) studied an extensive series of substituted acetyl and benzoyl chlorides in the Friedel-Crafts acylation of benzene and toluene.

Olah and his co-workers (104) also introduced a series of stable oxocarbenium (acyl) hexafluoro- and hexachloroantimonate salts as powerful acylating agents.

A reexamination of the Freidel-Crafts acetylation of 2-bromonaphthalene was reported (105).

Data show that 2-bromonaphthalene is one-seventh as reactive as naphthalene. The 8-position of 2-bromonaphthalene is 0.58 times as reactive and the 6-position is 0.63 times as reactive as the α- and β-positions, respectively, of naphthalene. Bisanz and Prejzner (106) published their study on orientation in the $ZnCl_2$-catalyzed Friedel-Crafts acylation of 1- and 2-naphthol derivatives.

Phenanthrene has been acetylated under a wide variety of conditions (107) to give 1-, 2-, 3-, 4-, and 9-acetylphenanthrenes. In ethylene dichloride the 9-isomer predominates, and in $CHCl_3$ equimolar amounts of the 3- and 9-isomers (37% each) are produced, while in nitrobenzene, nitromethane, benzene, and carbon disulfide, the 3-isomer is the major product. The relative reactivities for various ring positions were also determined from competitive studies in chloroform at room temperature. The results are: 1-naphthyl, 1.00; 3-phenanthryl, 0.64; 9-phenanthryl, 0.62; 1-phenanthryl, 0.29; 2-naphthyl, 0.28; 2-phenanthryl, 0.12; and 4-phenanthryl, 0.009.

The ratio of isomers derived from the Friedel-Crafts acetylation (acetyl chloride–aluminum chloride) of hemimellitene (1,2,3-trimethylbenzene) is dependent upon: the mode of addition of reagents, the solvent, the catalyst, and the concentration of reactants. Under carefully arranged conditions either the 4- or 5-isomer can be prepared in better than 95% yield (108).

Mono-, di-, and trialkylcymantrenes were subjected to Friedel-Crafts acylation (109) and substitution at the β-position was favored.

The acylation of 1-(X)-cymantrenes (where X = F, Cl, Br, I, $NHCOOCH_2\phi$, or NHAc), and of 1,2-diiodocymantrene, has been reported (110). The acylating agent was acetyl chloride–aluminum chloride, and in all cases the β-isomer predominated in the product mixture.

A further study of the Friedel-Crafts acylation of acephenanthrene has been described (111). Acetyl chloride–aluminum chloride acetylation in nitrobenzene led to two products, while in all other acylations only a single product (the 9-derivative) was obtained.

The acylation of aceanthrene (ethylene-1,10-anthracene) has also been described (112).

The benzoylation of stilbene oxide has been reported (113).

$$\phi CH \overset{\displaystyle}{\underset{O}{\text{—}}} CH\phi + C_6H_5COCl \xrightarrow[\text{AlCl}_3]{\phi NO_2} \begin{array}{c} \text{2-benzoyldiphenylene} \\ \text{oxide (77\%)} \\ + \\ \text{3-benzoyldiphenylene oxide} \end{array}$$

Acylation of p-methylacetanilide with acetyl chloride and aluminum chloride in carbon disulfide solvent led to the formation of three products: 2'-amino-5'-methylacetophenone (8%), 5'-amino-2'-methylacetophenone (54%), and 2,6-dimethyl-4(1H)-quinoline (24%) (114).

The Friedel-Crafts acylation of 1-methyl-2-(1H)-quinoline produces the 3- and 6-acyl derivatives (115).

The acetylation of methyl-2,3-dimethyl-4- and 7-benzofuran carboxylates with acetyl chloride–aluminum chloride in carbon

disulfide at 25° was carried out (116).   The major product was the
6-acetyl compound.

Capeletti, DeCasablanca, and Mantovani (117) have described the
monoacetylation of various polyalkylbenzenes.

In the acylation of phenylcyclopropane by acetyl chloride, there is
no attack on the cyclopropane ring (118), and the minor product
reported by Hart and Levitt (119) is in fact 1-(4'-acetyl)phenyl-2-
chloropropane.

The Friedel-Crafts benzoylation (using benzoyl chloride–aluminum
chloride) of the isomeric dichlorobenzenes has been studied (120).
By using the Perrier procedure, the reactions were run in nitrobenzene solution at 100° for about 10 hours to promote completion of
reaction.

Ferrocene has been acylated at room temperature with $(CF_3CO)_2O$,
$(C_2F_5CO)_2O$, $(n\text{-}C_3F_7CO)_2O$, $n\text{-}C_3F_7COCl$, and $CHF_2(CF_2)_3COCl$,
with $AlCl_3$ as the catalyst (121,122).   The absence of diacetyl
derivatives is ascribed to the powerful deactivation of the system by
the fluorinated acyl groups.

Toma and Kaluzayova (123) investigated the Friedel-Crafts
acetylation of ferrocene analogs of benzophenone, diphenylmethane,
and stilbene.

B. Role of Catalyst

The effect of various catalysts on the benzoylation of anisole to give *p*- and *o*-benzoylanisoles, with the *para*-isomer predominating by a ratio of 31:1 (124), was studied. The relative percent yields were determined to be: $ZnCl_2$, 93; $SnCl_4$, 90; $FeCl_3$, 83; $MoCl_5$, 80; $NbCl_5$, 54; $AlCl_3$ and $TaCl_5$, 52; $BiCl_3$, 49; WCl_6, 47; $SbCl_5$, 43; $ZrCl_4$, 39; $TiCl_4$, 35; $TiCl_2$, 32; VCl_3, 10. The effect of metal halides on various acylation reactions has been described in numerous articles (124–126). *O*-Benzoylation was found to predominate in the reaction of 2-methylnaphthoresorcinol monobenzoate with benzoyl chloride when zinc chloride was used as the catalyst (122).

C. Role of Solvent

Several new reports of the effect of solvent, notably nitrobenzene, on isomer distribution in Friedel-Crafts acylations have been published (127).

The effect of solvent as an important factor controlling positional selectivity in the Friedel-Crafts acylation of phenanthrene (48), 1,3,5-triphenylbenzene (49), 2-methoxynaphthalene (50), and 2-bromonaphthalene (51) was studied.

D. Directing Effects and Selectivity

The effect of ring substituents in benzoyl chloride on the rate of benzoylation of toluene, correlates roughly with σ constants, although *meta*- and *para*-substituents lie on two different lines (132). There is, however, a better correlation with σ^+ constants. Rates of the $AlCl_3$-catalyzed aroylation of toluene in chlorobenzene have been measured for a series of substituted aroyl chlorides. These were shown to exhibit isokinetic behavior and to give a fairly good σ^+ correlation consistent with $ArCO^+$ being the reactive species (133). Variation in reactivities of compounds RCOCl in acylation, as a function of R, have been measured for several widely differing acyl groups, and a range of mechanistic behavior with varying extents of ionization ($RCOCl-AlCl_3$ to $RCO^+AlCl)_4^-$ indicated (134).

Olah and Kobayashi (107) carried out a systematic study of the effect of substituents in acylating agents (acyl chlorides and fluorides) on the selectivity of Friedel-Crafts acylations.

Friedel-Crafts acylation of aromatic hydrocarbons such as toluene and benzene is characterized by generally high selectivity of the reactions reflected in high substrate rate ratios (usually $k_{toluene}/k_{benzene} > 100$) and related predominant *para*-substitution. The latter was long considered to be a consequence of the steric requirements of the acylating agents, causing large steric *ortho*-hindrance. Studies now show that substituents in the acylating agents can cause a change in the transition state from a σ- to a π-complex nature and consequently substantially effect isomer distribution of the acylation reactions, particularly the *ortho/para* isomer ratio.

When acetylation and benzoylation of toluene and benzene with acyl and substituted acyl halides were studied, the results clearly proved the importance of substituents in the electrophilic substituting agent, influencing both positional as well as substrate selectivity. Aluminum chloride-catalyzed acetylation of tolene with acetyl chloride gives 2.5% 2- and 95.5% 4-methyl acetophenone, whereas a similar reaction with dichloroacetyl chloride yields 17.3% *ortho*- and 79.5% *para*-product. Negative substituents increase the electron deficiency of the reagent, causing the transition state to be "early" on the reaction coordinate, resembling starting aromatics more than intermediate σ-complexes (which is the case in "late" transition state reactions). The effect is even better shown in benzoylations. Benzoyl chloride in reaction with toluene ($AlCl_3$ catalyst) yields 8% 2- and 91% 4-methylbenzophenone. In contrast, 2,4-dinitrobenzoyl chloride gives 42% *ortho*- and 55% *para*-product.

Steric effects, although obviously of importance, consequently cannot be considered the only reason for observed directing effects in aromatic acylations. It has become obvious that the selectivity of the reagents plays an important role.

Changing not only the electrophilicity of the reagent but also the nucleophilicity of the aromatic substrate can cause the same effect. With increasingly more basic aromatics, even a relatively weak electrophile results in the transition state of highest energy becoming relatively early, resembling starting aromatics more than intermediates.

Jensen and Smart (135) reported that the $AlCl_3$-catalyzed benzoylation of phenylnorbornane is faster than that of isopropylbenzene or even toluene, although the norbornyl group is far bulkier than any simple alkyl group. Hyperconjugation must therefore be important, but as the fastest reaction is that of 1-phenylnorborane, in which there are no α-hydrogens, C–C hyperconjugation must be as important as or more important than C–H hyperconjugation in toluene.

E. Intermediate Complexes

X-ray studies

In the crystal it has been found that a 1:1 complex of aluminum chloride and benzoyl chloride is coordinated through oxygen and not through chlorine (136).

An X-ray study of the benzoyl chloride–antimony pentachloride complex also shows that coordination is through oxygen (137).

The structure of acetyl fluoroantimonate ($CH_3CO^+SbF_6^-$), obtained from acetyl fluoride and antimony pentafluoride, was confirmed by Boer (138) by a detailed X-ray structural study to be that of a linear ion salt.

F. Kinetic Isotope Effects

Jensen observed (139) kinetic hydrogen isotope effects of intermediate magnitude in the Friedel-Crafts benzoylation, as well acetylation, of naphthalene. Details of these investigations, however, were not published.

Competitive acetylation of benzene and benzene-d_6 was carried out by Olah et al., with $CH_3CO^+SbF_6^-$ and $CH_3CO^+SbCl_6^-$ in nitromethane solution at 24° (140a) and gave the kinetic isotope effects:

	$CH_3CO^+SbCl_6^-$	$CH_3CO^+SbF_6^-$
$k_H/k_D =$	2.15	2.22

The same kinetic isotope effect was also obtained from the competitive acetylation of toluene, and benzene as compared with toluene and benzene-d_6. The observed kinetic isotope effect is greater for ring-deuterated toluene than for benzene ($k_H/k_D = 3.2$).

The larger isotope effect in the case of toluene is probably a consequence of the increased conjugative stabilization of the σ-complex type transition state by the p-methyl group.

In the case of the acetylation of mesitylene, the steric hindrance by a pair of flanking o-methyl groups may interfere with the conjugative stabilization of the benzenium ion. Therefore the effect may be a combination of electronic and steric effects (k_H/k_D is 1.9).

The primary kinetic isotope effects indicate that proton elimination in the acetylation of benzene, toluene, and mesitylene is at least partially rate-determining. Isotopic substitution of the methyl group in toluene gives rise to secondary kinetic isotope effect. The acetylation of $CD_3C_6H_5$, $CD_3C_6H_5$, and $CH_3C_6D_5$ was compared with that of toluene, in competitive acetylations with benzene, showing this effect using $CH_3CO^+SbF_6^-$ in CH_3NO_2 solution as the acetylating agent.

$$k_{tolune}/k_{benzene} = 125$$

$$k_{toluene-\alpha,\alpha,\alpha-d_3}/k_{benzene} = 118 \quad k_{\alpha H}/k_{\alpha D} = 1.06$$

$$k_{toluene-d_5}/k_{benzene} = 38.6$$

$$k_{toluene-d_8}/k_{benzene} = 37 \quad k_{\alpha H}/k_{\alpha D} = 1.04$$

Propionylation and benzoylation of benzene and toluene with $CH_3CH_2CO^+SbF_6^-$ and $C_6H_5CO^+SbF_6^-$, respectively, also show primary hydrogen isotope effects (140b). A primary isotope effect has been detected in the benzoylation of toluene with $AlCl_3$ as catalyst (141).

There is a small isotope effect ($k_H/k_D = 1.2$) in the cyclization of o-benzoylbenzoic acid to anthraquinone (142).

G. Diacylation

Gore and Hoskins (143) studied in detail the Friedel-Crafts acetylation of mesitylene. In order to obtain high yields of diacetylmesitylene, a sixfold excess of catalyst and the Elbs addition sequence were employed.

Fletcher and Marlow (144) reported the following anomalous Friedel-Crafts acylation.

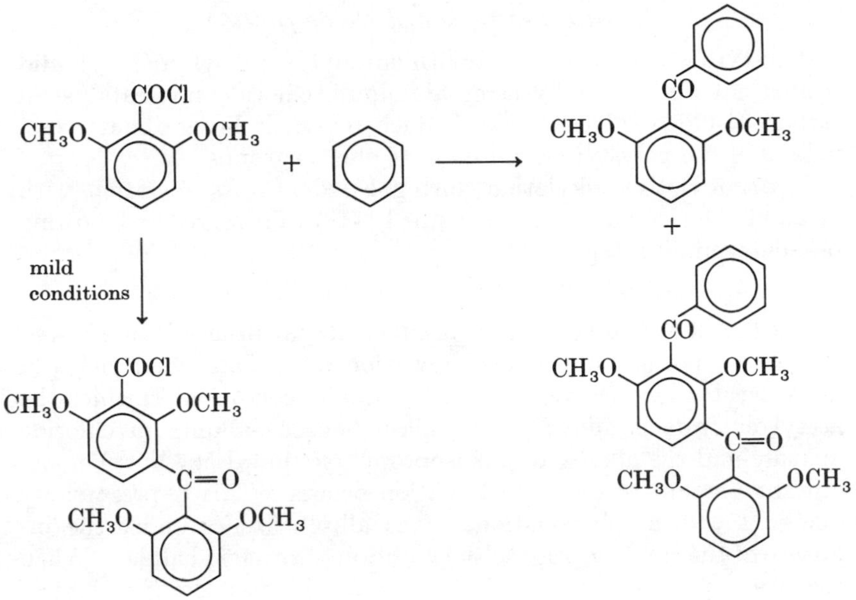

COCl

CH₃O   OCH₃

$+$

$\longrightarrow$

CO

CH₃O   OCH₃

$+$

mild
conditions

COCl

CH₃O   OCH₃

C=O

CH₃O   OCH₃

CO

CH₃O   OCH₃

C=O

CH₃O   OCH₃

Buu-Hoi, Mabille, and Do-Cao-Thang (145), in their series on the Friedel-Crafts acylation reactions of polycyclic aromatic hydrocarbons, reported that the acetylation of fluoranthrene leads in a 63% conversion to the diacylated product.

The acylation (and alkylation) of 1,3-dimethoxybenzene in polyphosphoric acid has been reported (146). Polyphosphoric acid, unlike aluminum chloride, effects Friedel-Crafts acylations without bringing about arylalkyl ether cleavage. It was shown that the products of the reaction of 3-halo- and 3-ethoxypropionic acids with 1,3-dimethoxybenzene in polyphosphoric acid are:

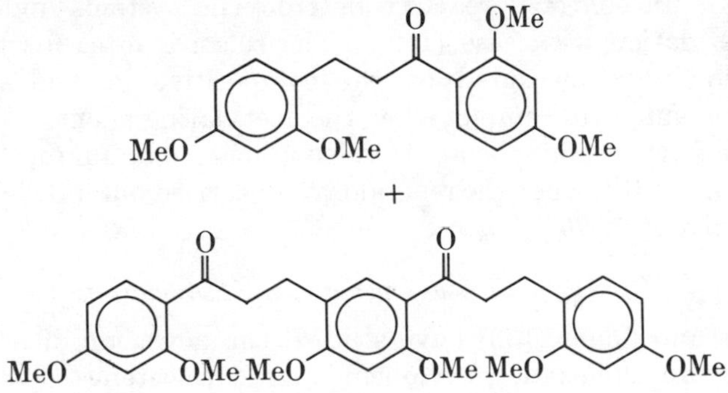

O   OMe

MeO   OMe   OMe

$+$

O   O

MeO   OMe MeO   OMe MeO   OMe

H. Side Reactions and Decomposition

Polystyrene can be acylated with adipoyl, sebacoyl, malonyl, and related chlorides (147) by using aluminum chloride as a catalyst in carbon disulfide solution at 0°. Such reactions bring about cross-linking of the polystyrene through its phenyl groups.

Decarbonylative alkylation during Friedel-Crafts acylation with pivaloyl chloride has been determined (148) to involve the following rate-determining step.

$$(CH_3)_3CO^+ \; AlCl_4^- \quad \longrightarrow \quad CO + (CH_3)_3C^+ AlCl_4^-$$

A study of the acetylation of primary alkylbenzenes (149) showed that under normal and inverse acylation conditions, good yields of p-alkylacetophenones were obtained. Under conditions in which the acetyl chloride is added to the alkyl benzene–aluminum chloride mixture and the alkyl group is isopropyl, sec-butyl or t-butyl, a significant amount of disproportionation occurs. This is presumably due to the disproportionation of the alkylbenzenes under the influence of the catalyst prior to the addition of the acid halide. When the order of addition was such as to avoid this problem (i.e., hydrocarbon added to a acetyl chloride–aluminum chloride complex) the yield of p-acetyl-sec-butylbenzene rose from 37% to 93%.

I. Haloacylations

The relative rates of acylation of aromatic hydrocarbons with $ClCH_2CH_2COCl$ at 25° in CH_2ClCH_2Cl are reported (150) to be: benzene 1; toluene, 103; ethylbenzene, 102; cumene, 102; t-butylbenzene, 90.5; and biphenyl, 47.

A thorough study has appeared (151) of the Friedel-Crafts acylation using polyflyoroaromatic acid halides and ferric or aluminum chloride as the catalyst.

Trifluoroacetylation of benzenoid aromatics takes place only with difficulty. In contrast, reactive heterocyclic systems undergo trifluoroacetylation with ease (152). The relative rates for trifluoroacetylation were measured by the competitive method with trifluoracetic anhydride employed as the acetylating agent. The rates of reaction were: thiophene, 1; selenophene, 6.5; furan, 140; and pyrrole, 5.3×10^7; when the reactions were carried out in 1,2-dichloroethane solvent at 75°.

2. Aliphatic and Cycloaliphatic Acylation

Groves and Jones (153) have studied the acylation of a series of substituted cyclohexanes, cyclopentes and cycloctenes.

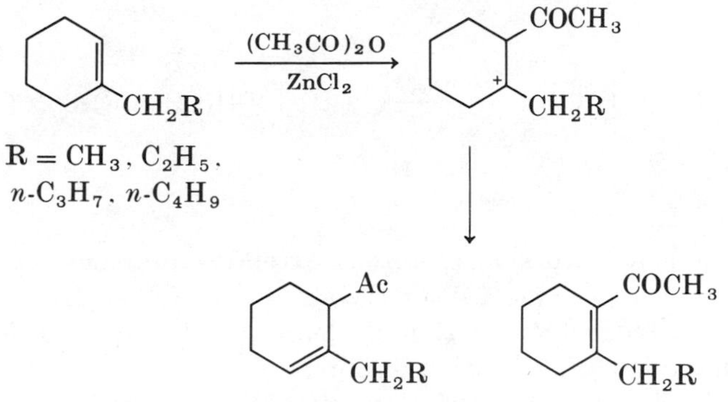

R = CH$_3$, C$_2$H$_5$.
n-C$_3$H$_7$, n-C$_4$H$_9$

(major, 50–70%)   (minor)

Acetyl chloride and styrene, when added at -5 to $0°$ to a benzene–aluminum chloride suspension, produced a 62% yield of ϕ_2-CHCH$_2$-COCH$_3$ after a reaction time of 96 hours (154).

A series of bicyclo(3.3.1)nonane deivatives was prepared by a Friedel-Crafts reaction of 3-(cyclohexen-1-yl)propionyl chloride under various conditions (155):

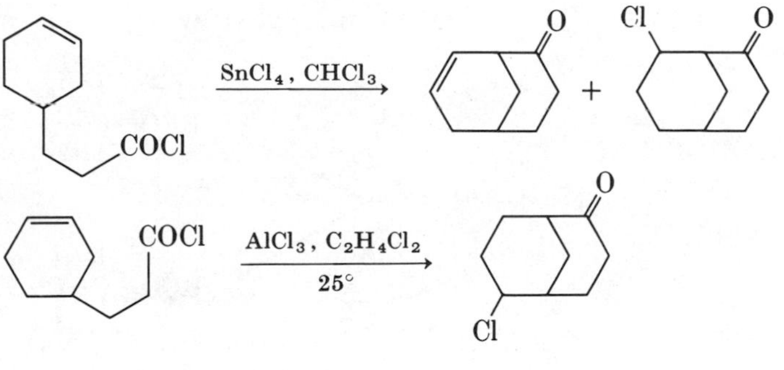

Studies on the effect of sequence of addition and solvent on product distribution were carried out by Hart and Schlosberg (156,157) in their study on the acylation of cyclopropane.

The results of the acylation of t-amyl chloride with acetyl chloride have been reported (158).

$$CH_3CH_2-\underset{\underset{CH_3}{|}}{\overset{\overset{CH_3}{|}}{C}}-Cl + CH_3\overset{\cdot}{C}OCl \xrightarrow{AlCl_3} CH_3COCH(CH_3)C(CH_3)_2Cl +$$

$AlCl_4^-$

Smit and his co-workers (159,160) extended previous studies by Olah, Quinn, and Kuhn (161) on the acyl cation ($RCO^+BF_4^-$ or RCO^+ SbF_6^-)-initiated polymerization of olefins and found conditions that result in monomeric acylation products.

1,5-Polyenes react by initial attack of the acyl cation on the terminal (2,3) double bond followed either by ring formation due to nucleophilic participation of the second (6,7) double bond (path A), or by C-1 proton loss with the regeneration of a new double bond (path B).

$$\text{RO} \quad \xleftarrow[\text{path } A]{RCO^+BF_4^-} \quad \overset{3}{\underset{4}{}}\overset{1}{\underset{5}{}}\overset{2}{\underset{6}{}}\overset{7}{} R \quad \xrightarrow[\text{path } B]{RCO^+BF_4^-} \quad \text{ROC} \quad R$$

In continuation of the work at low temperature (0–70°), monomeric products, according to an addition-elimination mechanism were found (path C), or when a nucleophile was present path D resulted.

$$\begin{array}{c} RCO^+BF_4^- \\ \text{or} \\ RCO^+SbF_6^- \end{array} + \overset{}{\underset{}{}}C=\overset{|}{C}-CH_2- \begin{array}{c} \xrightarrow[\text{path } C]{-H^+} \quad RCO-\overset{|}{C}-\overset{|}{C}=CH- \\ \\ \xrightarrow[\text{path } D]{Nu^-} \quad RCO-\underset{\underset{Nu}{|}}{\overset{|}{C}}-\overset{|}{\underset{|}{C}}-CH_2- \end{array}$$

3. Biologically Related Synthesis

Biologically important syntheses involving a Friedel-Crafts type acylation reaction have been reported (162). 9-β-D-Ribofuranosyladenine was synthesized by condensation of N^6-octanoyladenine with 1-O-acetyl-2,3,5-tri-O-benzoyl-β-D-ribofuranose or the tetraacetylribofuranose in sym-dichloroethane or chlorobenzene in the presence of Friedel-Crafts catalysts. The condensation when followed by hydrolysis of the acyl groups, led to a product free from contamination of the corresponding anomer.

A general synthesis of pyrimidine nucleosides using Friedel-Crafts catalysts has been described (163). Thus, for example, 1-*O*-acetyl-2,3,5-tri-*O*-benzoyl-D-ribofuranose was reacted with the 2,4-bisilyl derivative of 6-azauracil in 1,2-dichloroethane in the presence of stannic chloride to give a 92% yield of 2′,3′,5′-tri-*O*-benzoyl-6-azauridine.

IV. Amination

In a series of articles, Kovacic and co-workers (164–169) examined the trichloramine–aluminum chloride amination of aromatic and aliphatic hydrocarbons. Since in aromatic amination in the case of toluene predominantly *meta*-substitution was observed, a novel mechanism, called σ-substitution, was invoked to explain the results (*o*- and *p*-chlorotoluenes are major side products in the reaction).

The *m*-toluidine is considered to arise by σ-substitution which involves nucleophilic attack of NCl_3 on the *p*-chlorotoluenium ion (σ-complex) formed by electrophilic chlorination of toluene in the system.

Aminobiphenyls and diphenylamines are produced in modest yield from the reaction of HBF_4 with phenylhydroxylamine in sulfolane in the presence of aromatic compounds (170). A neighboring-group effect is postulated to be responsible for *ortho*-substitution in anisole.

V. Hydroxylation

Aromatic hydroxylation by titanium(III) and hydrogen peroxide is electrophilic in character. Oxygenation of aromatic compounds by isopropyl peroxydicarbonate catalyzed by aluminium chloride gives moderate yields of products (15–70%) showing a characteristic electrophilic substitution pattern (171) A possible mechanistic scheme is:

$$
\underset{\text{i-PrOCOOCOPr-i}}{\overset{\displaystyle O \quad O-AlCl_3}{\underset{\displaystyle \|\quad\|}{}}} \quad \xrightarrow{\text{ArH}} \quad \begin{cases} \text{i-PrOCOOCOAlCl}_2 + \text{Ar-i Pr} + \text{HCL} \\[2ex] \text{ArOCO}_2\text{i-Pr} + \text{i-PrOCO}_3\text{AlCl}_2 + \text{HCl} \end{cases}
$$

$$
\text{ArOCO}_2\text{-i Pr} \quad \xrightarrow[\text{AlCl}_3]{\text{ArH}} \quad \text{Ar-i Pr} + \text{ArOCO}_2\text{AlCl}_2 + \text{HCl}
$$

$$
\left. \begin{array}{l} \text{ArOCO}_2\text{-i Pr} \\ \text{ArOCO}_2\text{AlCl}_2 \end{array} \right\} \quad \xrightarrow[\text{work-up}]{\text{acid}} \quad \text{ArOH}
$$

Trifluoroperacetic acid was used in hydroxylating aromatic compounds (172). Relative reactivities and isomer distributions indicate that hydroxylation of aromatic compounds by trifluoroperacetic acid in methylene chloride is clearly an electrophilic process (173).

In the aluminum chloride-catalyzed oxygenation of aromatic compounds with peresters, the initial formation of mixed carbonates is

involved in the case of secondary alkyl peroxydicarbonates (174,175), and initial *t*-butoxylation is considered responsible in the reaction with *t*-butylpercarbonates (176).

Rather different isomer ratios, although again characteristically electrophilic, are obtained in oxygenations catalyzed by iron (177) and copper salts (178). In these instances acyloxy radical intermediates are, it is tentatively suggested, oxidized to acyloxy cations which then give rise to substitution products.

$$\underset{R-C-O}{\overset{O}{\parallel}} \quad \xrightarrow{Cu^{++}} \quad \underset{R-C-O^+ + Cu^+}{\overset{O}{\parallel}}$$

Liquid hydrogen fluoride is an effective catalyst for hydroxylation of aromatic compounds with aqueous hydrogen peroxide at temperatures ranging from -30 to $50°$ (179). Toluene is converted to o- and p-cresol, and isomeric xylenes to xylenols. The hydroxyl group is oriented principally *ortho* over *para* (ratio $>2 : 1$), with little *meta*-isomer formed. A surprisingly high conversion of benzene to phenol, hydroquinone, and catechol does occur when the reaction is carried out in the presence of carbon dioxide (under pressure).

An effective hydroxylation of aromatics (anisole, toluene, chlorobenzene, o-xylene, mesitylene, and even benzene) was described using hydrogen peroxide (90 or 30%) in the presence of aluminum chloride (180). Yields of 10–70% were observed. $k_{toluene}/k_{benzene} = 19$ and *ortho/para* directing effects (60% o-, 8% m-, 32% p-cresol) are indicative of the reactions.

VI. Nitration

A. Aromatic Nitration

Coe, Jukes, and Tatlow (181) have reported the use of boron trifluoride as a catalyst in the nitric acid nitration (with tetramethylene sulfone as solvent) of some polyfluoroaromatic compounds. Pentafluorobenzene gave pentafluoronitrobenzene in 82% yield unaccompanied by oxidation. Other compounds successfully nitrated were 1,2,3,4-tetrafluorobenzene, 1,2,3,5-tetrafluorobenzene, and 2H-nonafluorobiphenyl.

Taylor (182) published his results on the nitration of biphenyl. The low *ortho/para* ratio (*ca.* 0.6) in HNO_3–H_2SO_4 is due to the reaction being heterogeneous. When a homogeneous reaction mixture was used, the *ortho/para* ratio was 1.5. It was proposed that nitration with preformed nitronium ions gives rise to normal *ortho/para* ratios, whereas nitration with *all* species of the type NO_2X gives enhanced ratios because of an additional mode of *ortho*-substitution involving, in the case of biphenyl, nucleophilic displacement of X (or X^-) by the π-electrons of the unsubstituted ring to give a π-complex involving NO_2^+. This π-complex goes over to the most accessible σ-complex which, because of the non-coplanarity of the phenyl rings, is the one formed at the *ortho*-position of the other ring.

Kovacic (183) presented data supporting the existence of a linear coordination effect in electrophilic aromatic substitution of n-donor-substituted aromatics. While nitration of toluene or chlorobenzene gives little variation in the *ortho/para* ratio with different nitrating agents, the *ortho/para* nitration ratio with anisole can be substantially altered by appropriate choice of the reagent. This was suggested to be the result of the proposed linear coordination mechanism, involving initial nitration of the n-donor oxygen atom of the methoxy group with subsequent migration of the nitro group to the ring (preferring the *ortho*-position).

Myhre and Beug (184) studied the nitration of 2,4,6-tri-t-butyl-nitrobenzene and -toluene with the following results.

They proposed a mechanism to account for the products formed, involving nitrodealkylation.

Nitration of a series of 1-X-2,4,6-tri-t-butylbenzenes (X = H, F, NO$_2$, CH$_3$) led to results (186) that are consistent with a two-step electrophilic substitution mechanism. A demonstrable primary isotope effect (k_H/k_D = 3.8 for X = CH$_3$), provided the first evidence that aromatic nitration can occur with rate-limiting proton transfer (185).

B. Aliphatic Nitration

Based on the general concept of reactivity of single bonds (σ-donors), Olah and Lin (187) demonstrated typical electrophilic nitration of alkanes with nitronium salts. To avoid the possibility of free-radical reaction resulting from the use of nitric acid and to avoid acid cleavage of reaction products, nitrations were carried out with stable nitronium salts, such as $NO_2^+ PF_6^-$, in aprotic solvents such as methylene chloride–sulfolane. In the case of methane (in which case the nitration product–nitromethane—is not acid-sensitive), anhydrous HF or FSO$_3$H were also used as solvents.

$$R\text{—}H \xrightarrow[\substack{CH_2Cl_2\text{–sulfolane,}\\ 25°,\\ \text{dark}}]{NO_2^- PH_6^+} R\text{—}NO_2 + H^+$$

Aliphatic nitro compounds, particularly tertiary and secondary ones, are acid-sensitive and undergo protolytic cleavage reactions with ease. This significant difference from their aromatic counterparts necessitates specific conditions, such as the use of acid-free nitronium salts, to obtain products such as 2-methyl-2-nitropropane ((CH$_3$)$_3$CNO$_2$).

Acknowledgment

We gratefully acknowledge the assistance of Professor E. M. Arnett and the Pittsburgh Chemical Information Center, University of Pittsburgh, for carrying out a computer literature search profile.

References

1a. Roberts, R. M., *Chem. Eng. News,* January 25, 1965, p. 96.
1b. Olah, G. A., and C. A. Cupas, in "Encyclopedia of Chemical Technology," R. E. Kirk and D. F. Othmer, eds. 2nd Ed., Vol. II, Wiley-Interscience, New York, 1966, pp. 135–166.
2. DeHaan, F. P., and H. C. Brown, *J. Am. Chem. Soc.,* **91**, 4844 (1969).
3. DeHaan, F. P., H. C. Brown, and J. C. Hill, *J. Am. Chem. Soc.,* **91**, 4850 (1969).
4. DeHaan, F. P., H. C. Brown, D. C. Conway. and M. G. Gibby, *J. Am. Chem. Soc.,* **91** 4854 (1969).
5. Olah, G. A., and J. R. DeMember, *J. Am. Chem. Soc.,* **91**, 2113 (1969).
6. DeHaan, F. P., M. G. Gibby, and D. R. Aebersold, *J. Am. Chem. Soc.,* **91**, 4860 (1969).
7. For comprehensive reviews and references given therein see: Olah, G. A., *Chem. Ing. News,* **45**, March 27, 1967, pp. 76–88; G. A. Olah and J. A. Olah, in "Carbonium Ions," Vol. II, G. A. Olah and P. v. R. Schleyer, eds., Wiley-Interscience, New York, 1970, 715–782; G. A. Olah, *Science,* **168**, 1798 (1970).
8. Olah, G. A., J. R. DeMember, and R. H. Schlosberg, *J. Am. Chem. Soc.,* **91**, 2112 (1969); G. A. Olah, J. R. DeMember, R. H. Schlosberg, and Y. Halpern, *ibid.,* **94**, 156 (1972).
9. Olah, G. A., and J. R. DeMember, *J. Am. Chem. Soc.,* **92**, 718, 2562 (1970).
10. Olah, G. A., J. R. DeMember, P. Schilling, and J. A. Olah, *J. Am. Chem. Soc.,* in press.
11. Nakane, R., O. Kurihara, and A. Natsubori, *J. Am. Chem. Soc.,* **91**, 4528 (1969).
12. Nakane, R., *J. Chem. Soc. Japan,* **90**, 17 (1969); *Chem. Abstr.,* **70**, 86873 (1969).
13. Myhre, P. C., T. Rieger, and J. T. Stone, *J. Org. Chem.,* **31**, 3425 (1966).
14. Vives, V. V., and C. W. Kruse, U.S. Patent 3,413,359; *Chem. Abstr.,* **70**, 47054 (1969).
15. Kerfoot, O. C., and D. D. Krehbiel, Brit. Patent 1,156,110; *Chem. Abstr.,* **71**, 101502 (1969).
16. Beginina, M. S., E. P. Babin, V. P. Marshtupa, and N. L. Zotova, *Zh. Prikl. Khim.,* **41**, 370 (1968); *Chem. Abstr.,* **69**, 35109 (1968).
17. Ransley, D. L., *J. Org. Chem.,* **33**, 1517 (1968).
18. Ransley, D. L., *J. Org. Chem.,* **34**, 2618 (1969).
19. Gelin, R., S. Gelin, and B. Chantegrel, *Compt. Rend., Ser. C,* **266**, 813 (1968).
20. Gelin, R., B. Chantegrel, and S. Gelin, *Compt. Rend., Ser. C,* **270**, 1123 (1970).
21. Barnes, R. A., and A. Costa Neto, *J. Org. Chem.,* **35**, 4259 (1970).

22. Grassie, N., and I. G. Meldrum, *Europ. Polymer J.*, **5**, 195 (1969); *Chem. Abstr.*, **70**, 97225 (1969).

23. Bodrikov, I. V., and I. S. Okrokova, *J. Org. Chem.* (*USSR*), **3**, 1663 (1967); translated from *Zh. Org. Kh.*, **3**, 1706 (1967).

24. Gelin, R., S. Gelin, D. Pigasse, *Compt. Rend. Ser. C*, **270**, 1875 (1970).

25. Reid, W., and R. Lantzsch, *Synthesis*, **2**, 303 (1970).

26. Dalin, M. A., V. V. Lobkina, A. M. Bakhshizade, and S. L. Plaksunova, *Neftepererab. Neftekhim* (*Moscow*), **21** (1968); *Chem. Abstr.*, **70**, 30601 (1969).

27. Inukai, T., *J. Org. Chem.*, **31**, 1124 (1966).

28. Friedman, H. M., and A. L. Nelson, *J. Org. Chem.*, **34**, 3211 (1969).

29. Alul, H. R., *J. Org. Chem.*, **33**, 1522 (1968).

30. Fujiwara, I., I. Moritani, M. Matsuda, and S. Teranishi, *Tetrahedron Letters*, 633 (1968).

31. Babin, E. P., I. G. Gakh, M. S. Berginina, and L. G. Gakh, *Zh. Prikl. Khim.*, **41**, 342 (1968); *Chem. Abstr.*, **69**, 66606 (1968).

32. Tsuge, O., M. Tashiro, and A. Torii, *Kogyo Kagaku Zasshi*, **71**, 229 (1968); *Chem. Abstr.* **69**, 18370 (1968).

33. Tsuge, O., M. Tashiro, and A. Torii, *Kogyo Kagaku Zasshi*, **70**, 2287 (1967); *Chem. Abstr.* **68**, 104300 (1968).

34. Klemm, L. H., J. Shabtai, and D. R. Taylor, *J. Org. Chem.*, **33**, 1480, 1489, 1494 (1968).

35. Roberts, R. M., Y. -T. Lin, and G. P. Anderson, Jr., *Tetrahedron*, **25**, 4173 (1969).

36. Roberts, R. M., G. P. Anderson, Jr., and S. E. McGuire, *Tetrahedron*, **25**, 4523 (1969).

37. Desai, R. D., *J. Indian Chem. Soc.*, **45**, 193 (1968).

38. Olah, G. A., N. Friedman, J. M. Bollinger, and J. Lukas, *J. Am. Chem. Soc.*, **88**, 5328 (1966).

39. Jurewicz, A. T., J. H. Bayless, and L. Friedman, *J. Am. Chem. Soc.*, **87**, 5788 (1965).

40. Olah, G. A., N. A. Overchuk, and J. C. Lapierre, *J. Am. Chem. Soc.*, **87**, 5785 (1965).

41. Olah, G. A., and N. A. Overchuk, *J. Am. Chem. Soc.*, **87**, 5786 (1965).

42. Milstein, N., *J. Heterocyclic Chem.*, **5**, 339 (1968).

43. Milstein, N., *J. Heterocyclic Chem.*, **5**, 337 (1968).

44. Nakamoto, Y., T. Nakajuma, and S. Suga, *Kogyo Kagaku Zasshi*, **72**, 2594 (1969); *Chem. Abstr.*, **72**, 100192 (1970).

45. Asaoka, T., C. Shimasaki, K. Taki, M. Funayama, M. Sakano, and Y. Kamimura, *Yuki Gosei Kagaku Kyokai Shi*, **27**, 783 (1969); *Chem. Abstr.*, **72**, 2920 (1970).

46. Olah, G. A., S. Kobayashi, M. Tashiro, *J. Am. Chem. Soc.*, **94**, 7448 (1972).

47. Nakane, R., A. Natsubori, and O. Kurihava, *J. Am. Chem. Soc.*, **87**, 3597 (1965); **88**, 3011 (1966); R. Nakane, T. Oyama, and A. Natsubori, *J. Org. Chem.*, **33**, 275 (1968); **34**, 949 (1969); and references therein.

48. Olah, G. A., M. Tashiro, and S. Kobayashi, *J. Am. Chem. Soc.*, **92**, 6369 (1970).

49. Gurka, D. F., and W. M. Schubert, *J. Org. Chem.*, **31**, 3416 (1966).

50. Shabarov, Y. S., N. A. Donskaya, L. D. Sychkova, and R. Y. Levina, *Vestn. Mosk. Univ., Ser. II, Khim.*, **20**, 73 (1965); *Chem. Abstr.*, **64**, 4890.

51. Kovacic, P., and J. J. Hiller, *J. Org. Chem.*, **30**, 1581, 2871 (1965).
52. Inoue, H., K. Fujimoto, and Y. Suenaga, Ger. Offen., 1,962,689; *Chem. Abstr.*, **73**, 120280 (1970).
53. Lee, C. C., and D. J. Woodcock, *Can. J. Chem.*, **48**, 858 (1970).
54. Olah, G. A., C. G. Carlson, and J. C. Lapierre, *J. Org. Chem.*, **29**, 2687 (1964).
55. Olah, G. A., M. W. Meyer, and N. A. Overchuk, *J. Org. Chem.*, **29**, 2310 (1964).
56. Olah, G. A., M. W. Meyer, and N. A. Overchuk, *J. Org. Chem.*, **29**, 2313 (1964).
57. Olah, G. A., M. W. Meyer, and N. A. Overchuk, *J. Org. Chem.*, **29**, 2315 (1964).
58. Olah, G. A., and R. H. Schlosberg, unpublished results.
59. Suld, G., and A. P. Stuart, *J. Org. Chem.*, **29**, 2939 (1964).
60. Roberts, R. M., A. A. Khalaf, and J. E. Douglass, *J. Org. Chem.*, **29**, 1511 (1964).
61. Itoh, K., G. Yamada, S. Hamanaka, M. Ogawa, *Bull. Chem. Soc. Japan*, **41**, 2504 (1968); *Chem. Abstr.*, **70**, 28488 (1969).
62. Priddy, E. B., *Ind. Eng. Chem.*, *Prod. Res. Develop.*, **8**, 239 (1969).
63. Itoh, M., I. Takeuchi, K. Matsuura, A. Arase, and A. Suzuki, *Aromatik-kushu*, **20**, 129 (1968); *Chem. Abstr.*, **69**, 106022 (1968).
64. Hart, H. and J. F. Janssen, *J. Org. Chem.*, **35**, 3637 (1970).
65. Jacobsen, O., *Ber.*, **19**, 1209 (1886).
66. Koeberg-Telder, A., A. J. Prinsen, and H. Cerfontain, *J. Chem. Soc.*, *B*, 1004 (1969).
67. Anderson, H. J. and C. W. Huang, *Can. J. Chem.*, **48**, 1550 (1970).
68. Beak, P., R. J. Trancik, J. B. Mooberry, and P. Y. Johnson, *J. Am. Chem. Soc.*, **88**, 4288 (1966).
69. Brauman, J. I., and A. J. Pandell, *J. Am. Chem. Soc.*, **89**, 5421 (1967).
70. Brauman, J. I., and A. Solladie-Cavallo, *Chem. Commun.*, 1124 (1968).
71. Suga, S., T. Nakajima, Y. Nakamoto, and K. Matsumoto, *Tetrahedron Letters*, 3283 (1969).
72. For a review see: Olah, G. A., in "Organic Reaction Mechanism," Special Publ., No. 19, The Chemical Society, London, 1965, pp. 21–69.
73. Olah, G. A., J. R. DeMember, R. H. Schlosberg, and Y. Halpern, *J. Am. Chem. Soc.*, **94**, 156 (1972).
74. Nakane, R., T. Oyama, and A. Natsubori, *J. Org. Chem.*, **33**, 275 (1968).
75. Niess, R., K. Nagel, and F. Effenberger, *Tetrahedron Letters*, 4265 (1968).
76. Barclay, L. R. C., and E. C. Stanford, *Can. J. Chem.*, **46**, 3315 (1968).
77. Barclay, L. R. C. and E. C. Stanford, *Can. J. Chem.*, **46**, 3325 (1968).
78. Khalaf, A. A., and R. M. Roberts, *J. Org. Chem.*, **31**, 89 (1966).
79. Baddeley, G., and R. Williamson, *J. Chem. Soc.*, 4647 (1956); L. R. C. Barclay, in "Friedel-Crafts and Related Reactions," Vol. II, G. A. Olah, ed., Wiley-Interscience, New York, 1964, p. 786.
80. Johnston, K. M., *Tetrahedron*, **24**, 5595 (1968).
81. Johnston, K. M., *J. Heterocyclic Chem.*, **6**, 847 (1969).
82. Khalaf, A. A., and R. M. Roberts, *J. Org. Chem.*, **34**, 3571 (1969).
83. For a summary see: Schmerling, L., in "Friedel-Crafts and Related Reactions," G. A. Olah, ed., Vol. II, Wiley-Interscience, New York, 1969, pp. 1075–1131, and references therein.

84. Olah, G. A., and J. A. Olah, *J. Am. Chem. Soc.*, **93**, 1256 (1970).

85. Martens, H., and G. Hoornaert, *Tetrahedron Letters*, 1821 (1970).

86. Kramer, G. M., *J. Org. Chem.*, **34**, 2919 (1969).

87. Olah, G. A., and C. H. Lin, *J. Am. Chem. Soc.*, **90**, 6468 (1969).

88. Roberts, J. D., and R. H. Mazur, *J. Am. Chem. Soc.*, **73**, 2509 (1951).

89. Nefedov, O. M., E. S. Agavelyan, and O. S. Chizhov., *Izv. Akad. Nauk. SSSR, Ser. Khim.*, 2084 (1969), *Chem. Abstr.*, **72**, 12887 (1970).

90. Sandler, R. S., *Chem. Ind. (London)*, 565 (1970).

91a. Kennedy, J. P., "Polymer Preprints," **12**, No. 2, p. 6, American Chemical Society, September 1971.

91b. Plesch, P. H., "The Chemistry of Cationic Polymerization," Pergamon Press, Oxford, 1963.

91c. Zlamal, Z., in "Kinetics and Mechanism of Polymerization," G. E. Ham, ed., Dekker, New York, 1969, Chapter 6, p. 231.

91d. Higashimura, T., in "Structure and Mechanism of Vinyl Polymerization," T. Tsuruta and K. F. O'Driscoll, eds., Dekker, New York, 1970, Chapter 10, p. 313.

91e. Kennedy, F. P., in "Polymer Chemistry of Synthetic Elastomers," Vol. I, J. P. Kennedy and E. Tornquist, eds., Wiley-Interscience, New York, 1968, Chapter 5, p. 291.

91f. Gandini, A., and P. H. Plesch, "The Chemistry of Polymerization Processes," Society of Chemical Industry, London, 1966, pp. 107–114, 122–129.

91g. Kennedy, J. P., A. Shinkawa, and F. Williams, *J. Polymer. Sci.*, **9**, 1551 (1971).

91h. Yamashita, Y., *Advan. Chem. Ser.*, **91**, 350 (1969).

91i. Dreyfuss, P., and M. P. Dreyfuss, *Advan. Polymer. Sci.*, **4**, 528 (1967).

91j. Ledwith, A., *Advan. Chem. Ser.*, **91**, 317 (1969).

92.a. Kennedy, J. P., and G. E. Milliman, in "Addition and Polymerization Processes," *Advan. Chem. Ser.*, **91**, 287 (1969).

92b. Saegusa, T., in "Structure and Mechanism of Vinyl Polymerization," T. Tsuruta and K. F. O'Driscoll, eds., Dekker, New York, 1969.

93. Weichman, R. L., and W. L. Fierce, U.S. Patent 3,442,882; *Chem. Abstr.*, **71**, 13707 (1969).

94. Grassie, M., and I. G. Meldrum, *Europ. Polymer. J.*, 499, 513 (1970); *Chem. Abstr.*, **73**, 4213, 15320 (1970).

95. Gaylord, N. G., and Svestka, *J. Polymer Sci.*, *Part B*, **7**, 55 (1969).

96. Robert, M., and R. Perrus, Fr. Patent 1,500,178; *Chem. Abstr.*, **69**, 58933 (1968).

97. Kerfoot, O. C., D. D. Krehbiel, U.S. Patent 3,401,208 (1968).; *Chem Abstr.*, **69**, 86581 (1968).

98. Cudby, M. E. A., R. G. Gaskin, M. E. B. Jones, and J. B. Rose, *J. Polymer Sci.*, *Part C*, **22**, 747 (1967).

99. Lopour, P., M. Marek, *Makromol. Chem.*, 134 (1970).

100. Chmelir, M., M. Marek, *J. Polymer Sci.*, *Part C*, 177 (1967).

101. Marek, M., M. Chmelir, *J. Polymer Sci.*, *Part C.* 223 (1968).

102. Guidry, C. L., M. A. F. Walker, *Polymer*, **11**, 548 (1970); *Chem. Abstr.*, **73**, 131419 (1970).

103. Olah, G. A., and S. Kobayahi, *J. Am. Chem. Soc.*, **93**, 6964 (1971).

104. Olah, G. A., S. J. Kuhn, S. H. Flood, and B. A. Hardie, *J. Am. Chem. Soc.*, **86**, 2203 (1964).
105. Cirdler, R. B., P. H. Gore, and J. A. Hoskins, *J. Chem. Soc. (Org.)*, 518 (1966).
106. Bisanz, T., and J. Prejzner, *Roczniki Chem.*, **42**, 803 (1968); *Chem. Abstr.*, **69**, 96313 (1968).
107. Girdler, R. B., P. H. Gore, and C. K. Thadani, *J. Chem. Soc.*, *C*, 2619 (1967).
108. Friedman, L., and R. J. Honour, *J. Am. Chem. Soc.*, **91**, 6344 (1969).
109. Egger, H., and A. Nikiforov, *Montash. Chem.*, **99**, 2311 (1968).
110. Egger, H., A. Nikiforov, *Montash. Chem.*, **99**, 296 (1968).
111. Hoffinger, J. P., P. Jacquignon, and N. P. Buu-Hoi, *Bull. Soc. Chim. France*, 974 (1970).
112. Hoffinger, J. P., P. Jacquignon, and N. P. Buu-Hoi, *Bull. Soc. Chim. France*, 2534 (1970).
113. Keumi, T., K. Kitagawa, and Y. Oshima, *Kogyo Kagaku Zasshi*, **73**, 536 (1970); *Chem. Abstr.*, **73**, 45232 (1970).
114. Thomsen, A. D., and H. Lund, *Acta Chem. Scand.*, **23**, 2930 (1969).
115. Tomisawa, H., M. Watanabe, R. Fujita, and H. Hongo, *Chem. Pharm. Bull.*, **18**, 919 (1970).
116. Kawase, Y., T. Okada, and T. Miwa, *Bull. Chem. Soc., Japan*, **43**, 2884 (1970); *Chem. Abstr.*, 73, 109597 (1970).
117. Capeletti, R., A. L. DeCasablanca, and V. Mantovani, *Rev. Fac. Ing. Quim., Univ. Nac. Litoral*, **37**, 119 (1969); *Chem. Abstr.*, **73**, 87588 (1970).
118. Hart, H., R. H. Schossberg, and R. K. Murray, *J. Org. Chem.*, **33**, 3800 (1968).
119. Hart, H., and G. Levitt, *J. Org. Chem.*, **24**, 1261 (1959).
120. Goodman, P. A., and P. H. Gore, *J. Chem. Soc.*, *C*, 2452 (1968).
121. Sokolova, E. B., G. P. Chalykh, and A. P. Suslov, *Zh. Obshch. Khim.*, **38**, 537 (1968); *Chem. Abstr.*, **69**, 44004 (1968).
122. Bisanz, T., and J. Prejzner, *Roczniki Chem.*, **44**, 1041 (1970); *Chem. Abstr.*, **73**, 76925 (1970).
123. Toma, S., and E. Kaluzayova, *Chem. Zvesti*, **23**, 540 (1969); *Chem. Abstr.*, **73**, 14960 (1970).
124. Tsukervanik, I. P., and N. V. Veber, *Dokl. Akad. Nauk, SSR*, **180**, 892 (1968).
125. Leont'eva, L. I., and I. P. Tsukervanik, *Uzbeksk. Khim. Zh.*, **11**, 44 (1967); *Chem. Abstr.*, **68**, 86960 (1968).
126. Tsukervanik, I. P., K. A. Suerbaer, and A. A. Azizov, *Uzbeksk. Khim. Zh.*, **11**, 20 (1967); *Chem. Abstr.*, **68**, 68599 (1968).
127. Girdler, R. B., P. H. Gore, and J. A. Hoskins, *J. Chem. Soc.*, *C*, 181, 518 (1966).
128. Buu-Hoi, N. P., P. Mabille, and Do-Cao-Thang, *Bull. Soc. Chim., France*, 180 (1966).
129. Lewis, G. E., *J. Org. Chem.*, **31**, 749 (1966).
130. Girdler, R. B., P. H. Gore, and J. A. Hoskins, *J. Chem. Soc.*, *C*, 181 (1966).
131. Girdler, R. B., P. H. Gore, and J. A. Hoskins, *J. Chem. Soc.*, *C*, 518 (1966).

132. Slootmaekers, P. J., and R. Verbeest, *Bull. Soc. Chim., Belges,* **77**, 273 (1968).

133. Slootmaekers, P. J., A. Rasschaert, and W. Janssens, *Bull. Soc. Chim. Belges,* **75**, 199 (1966).

134. Gore, P. H., and J. A. Hoskins, *Chem. Commun.,* 835 (1966).

135. Jensen, F. R., and B. E. Smart, *J. Am. Chem. Soc.,* **91**, 5686 (1969).

136. Rasmussen, S. E., and N. C. Broch, *Chem. Commun.,* 289 (1965).

137. Weiss, R., and B. Chevrier, *Chem. Commun.,* 145 (1967).

138. Boer, F. P., *J. Am. Chem. Soc.,* **90**, 6706 (1968).

139. Jensen, F. R., Dissertation, Purdue University, 1955; F. R. Jensen, 9th Conference on Reaction Mechanism. Sept., 1962, Brookhaven, N.Y.

140a. Olah, G. A., S. J. Kuhn, S. H. Flood, and B. A. Hardie, *J. Am. Chem. Soc.,* 86, 2203 (1964).

140b. Olah, G. A., J. Lukas, and E. Lukas, *J. Am. Chem. Soc.,* **91**, 5319 (1969).

141. Vanhaverbeke, D., *Meded. Vlaam. Chem. Ver.,* **29**, 110 (1967); *Chem. Abstr.,* **68**, 86460 (1968).

142. Noyce, D. S., P. A. Kittle, and E. H. Banitt, *J. Org. Chem.* **33**, 1500 (1968).

143. Gore, P. H., and J. A. Hoskins, *J. Chem. Soc., C,* 517 (1970).

144. Fletcher, P., and W. Marlow, *J. Chem. Soc., C,* 937 (1970).

145. Buu-Hoi, N. P., P. Mabille, Do-Cao-Thang, *Bull. Soc. Chim. France,* 981 (1968).

146. Kasturi, T. R., and K. M. Damodaran, *Can. J. Chem.,* **47**, 1529 (1969).

147. Wolf, F., K. Frederich, and G. Schwachula, *Plaste Kaut.,* **16**, 727 (1969); *Chem. Abstr.,* **71**, 113565 (1969).

148. Stepanov, B. I., and V. F. Traven, *Zh. Org. Khim.,* **4**, 1067 (1968); *Chem. Abstr.,* **69**, 51271 (1968).

149. Olafsson, P. G., *J. Org. Chem.,* **35**, 4257 (1970).

150. Ruotsalainsen, H., L. A. Kumpulainen, and P. O. I. Virtanen, *Suomen Kemistilehti, B.,* **43**, 91 (1970); *Chem. Abstr.,* **73**, 13984 (1970).

151. Anichkina, S. A., V. A. Barkhash, and N. N. Vorozhtsov, Jr., *Zh. Obshch. Khim.,* **38**, 2493 (1968); *Chem. Abstr.,* **70**, 57349 (1969).

152. Clementi, S., and G. Marino, *Tetrahedron,* **25**, 4599 (1969).

153. Groves, J. K., and N. Jones, *J. Chem. Soc.* C.2215, 2898 (1968); 608, 1718, 2350 (1969).

154. Zaitseva, N. T., and A. D. Grebenyuk, *Zh. Org. Khim.,* **5**, 904 (1969); *Chem. Abstr.,* **71**, 38521 (1969).

155. Marvell, E. N., R. S. Knutson, T. McEwen, D. Sturma, W. Frederici, and K. Salisbury, *J. Org. Chem.,* **35**, 391 (1970).

156. Hart, H., and R. H. Schlosberg, *J. Am. Chem. Soc.,* **88**, 5030 (1966).

157. Hart, H., and R. H., Schlosberg, *J. Am. Chem. Soc.,* **90**, 5189 (1968).

158. Khayat, M. A. R., Fifth Arab Science, Congress, Baghdad (1966); *Chem. Abstr.,* **70**, 11481 (1969).

159. Smit, W. A., A. V. Semenowsky, *Tetrahedron Letters,* 3651 (1965).

160. Smit, W. A., A. V. Semenowsky, V. F. Kurchenov, T. N. Clerneva, M. Z. Krivoes, and O. V. Lubinskaya, *Tetrahedron Letters,* 3101 (1971).

161. Olah, G. A., H. Quinn, and S. Kuhn, *J. Am. Chem. Soc.,* **82**, 426 (1960).

162. Furukawa, Y., and M. Honjo, *Chem. Pharm. Bull. (Tokyo),* **16**, 1076 (1968).

163. Neidball, U., and H. Vorbrueggen, *Angew. Chem., Intern. Ed. Engl.,* **9** 461 (1970); *Chem. Abstr.,* **73**, 66854 (1970).

164. Kovacic, P., and R. M. Lange, *J. Org, Chem.*, **28**, 968 (1963).
165. Kovacic, P., J. J. Hiller, Jr., J. F. Gormish, and J. A. Levitsky, *Chem. Commun.*, 580 (1965).
166. Kovacic, P., R. M. Lange, J. L. Foote, C. T. Goralski, J. J. Hiller, Jr., and J. A. Levitsky, *J. Am. Chem. Soc.*, **86**, 1650 (1964).
167. Kovacic, P., C. T. Goralski, J. J., Hiller, J. A. Levisky, and R. M. Lange, *J. Am. Chem. Soc.*, **87**, 1262 (1965).
168. Kovacic P., and J. A. Levitsky, *J. Am. Chem. Soc.*, **88**, 1000 (1966).
169. Kovacic, P., and J. F. Gormish, *J. Am. Chem. Soc.*, **88**, 3819 (1966).
170. Parish, J. H., and M. C. Whiting, *J. Chem. Soc.*, 4713, (1964).
171. Kovacic, P., and S. T. Morneweck, *J. Am. Chem. Soc.*, **87**, 1566 (1965); P. Kovacic, and M. E. Kurz, *ibid.*, 4811.
172. Hart, H., and C. A. Buecher, *J. Org. Chem.*, **29**, 2397 (1964).
173. Davidson, A. J., and R. O. C. Norman, *J. Chem. Soc.*, 5405, (1964).
174. Razuvaev, G. A., N. A. Kartashova, and L. S. Boguslavskaya, *Zh. Organ. Khim.*, **1**, 1927 (1965); *Chem. Abstr.* **64**, 9544.
175. Kovacic, P., and M. E. Kurz, *J. Org. Chem.*, **31**, 2011 (1966).
176. Kovacic, P., and M. E. Kurz, *J. Org. Chem.*, **31**, 2459 (1966).
177. Kovacic, P., and M. E. Kurz, *Chem. Commun.*, 321, 1966.
178. Kovacic, P., and M. E. Kurz, *Tetrahedron Letters*, 2689, 1966, *J. Am. Chem. Soc.*, **88**, 2068 (1966).
179. Vesely, J. A., and L. Schwedling, *J. Org. Chem.*, **35**, 4028 (1970).
180. Kurz, M. E., and G. J. Johnson, *J. Org. Chem.*, **36**, 3184 (1971).
181. Coe, P. L., A. E. Jukes, and J. C. Tatlow, *J. Chem. Soc.*, *C*, 2323 (1966).
182. Taylor, R., *J. Chem. Soc.*, *B.* 727 (1966).
183. Kovacic, P., and J. J. Hiller, *J. Org. Chem.*, **30**, 1581, 2871 (1965).
184. Myhre, P. C., and M. Beug, *J. Am. Chem. Soc.*, **88**, 1568 (1966).
185. Myhre, P. C., and M. Beug, *J. Am. Chem. Soc.*, **88**, 1569 (1966).
186. Myhre, P. C., M. Beug, and L. L. James, *J. Am. Chem. Soc.*, **90**, 2105 (1968).
187. Olah, G. A., and H. C. Lin, *J. Am. Chem. Soc.*, **93**, 1259 (1971).

CHAPTER IV

Catalysts and Solvents

I. The Nature of Catalysts

The catalysts concerned with the initiation of reactions of the Friedel-Crafts type are all electron acceptors, falling into the general class of acids as defined by G. N. Lewis (355).

Acidic catalysts capable of effecting reactions of the Friedel-Crafts type can be divided into the following groups on the basis of chemical constitution:

(a) acidic halides (Lewis acids)
(b) metal alkyls and alkoxides
(c) proton acids (Brønsted acids)
(d) acidic oxides and sulfides (acidic chalcides)
(e) cation exchange resins
(f) metathetic cation-forming agents
(g) stable carbonium and related complexes.

1. Acidic Halides (Lewis Acids)

Friedel and Crafts, as well as the generation of chemists who followed in their footsteps, concerned themselves primarily with the class of *acidic halides* as reaction catalysts.   These *metal halide type Lewis acid catalysts*, of which aluminum chloride and bromide are the most frequently used, are generally known as Friedel-Crafts catalysts.   Other frequently used active metal halide catalysts are, among others, $BeCl_2$, $CdCl_2$, $ZnCl_2$, BF_3, BCl_3, BBr_3, $GaCl_3$, $GaBr_3$, $TiCl_4$, $TiBr_4$, $ZrCl_4$, $SnCl_4$, $SnBr_4$, $SbCl_5$, $SbCl_3$, $BiCl_3$, $FeCl_3$, and UCl_4.

The common feature of these acidic halides is that they have an electron-deficient central metal atom capable of electron acceptance from the basic reagents.   They are spread throughout the Periodic Table and are commonly chlorides or bromides.

Iodides also show good catalytic activity but their general use is excluded by their tendency to decompose.

Boron fluoride is a very active catalyst but fluorides of the other elements are used less, mainly because of their highly ionic nature. Fairbrother (176) reported the mixed pentafluorides of Nb and Ta to be good catalysts. PF_5, SbF_5, and AsF_5 were also found to be effective catalysts for Friedel-Crafts reactions (337).

If a metal has two or more states of oxidation the highest usually seems to be the catalytically most active, e.g., Sb^v, Sn^{iv}, Fe^{iii}, etc.

Lewis acid type metal halide catalysts can be used to effect Friedel-Crafts reactions not only in a previously prepared form but also in situ. In reactions involving reactive organic halides the use of metals (Al, Al + Cu, Al + Hg, Al + Zn, Cu, Bi, Cd, Hg, Mg, Mo, Zn, etc.) as catalysts is frequently useful. In the course of the reaction the organic halide reacts with the metal forming the corresponding metal halide, which is then responsible for furthering the reaction.

The acidic halide catalysts can be used in the presence of solvents or carriers, with which they may be partially or wholly complexed.

It has been amply demonstrated that a source of protons or other cations, such as water, hydrogen halide, or alkyl halides must be present, even if in comparatively small amounts, for the acidic halides to show catalytic activity.

2. Metal Alkyls and Alkoxides

Acidic halide catalysts effect Friedel-Crafts reactions through their Lewis acid activity. Consequently it is not surprising to find that Lewis acids like metal alkyls and metal alkoxides also possess catalytic activity in Friedel-Crafts reactions.

Metal alkyls (e.g., aluminum and boron alkyls) are fairly active catalysts in accordance with their Lewis acidity. Hyperconjugation with the electron-deficient metal atom, however, tends to decrease the electron deficiency (a similar back donating effect is also observed in the halides, especially in the case of fluorides). The effect is even stronger in alkoxides, which are therefore generally fairly weak Lewis acids.

3. Proton Acids (Brønsted Acids)

The realization that many Friedel-Crafts reactions carried out with metal halide catalysts are indeed proton catalyzed reactions, as shown by the necessity of a third substance (water, hydrogen halide, alkyl halide, etc.) to activate the reaction by the formation of a conjugate proton acid shows the very close relationship of these reactions to those carried out with *proton acids* (*Brønsted acids*). Sulfuric acid is typical and is the most widely used proton acid catalyst, but anhydrous hydrogen fluoride also enjoys wide application. Further proton acids used in effecting reactions of the Friedel-Crafts type are: phosphoric acid, polyphosphoric acid, perchloric acid, chlorosulfonic acid, fluorosulfonic acid, alkane sulfonic acids (such as ethanesulfonic acid), *p*-toluenesulfonic acid, and related aromatic sulfonic acids, chloroacetic acids, trifluoroacetic acid, etc.

The proton acid catalysts (as well as the promoted Lewis acid type halides) effect their activity through formation of cations. Basically their activity is measured by their ability to protonate a suitable substrate or to abstract hydride ion.

4. Acidic Oxides and Sulfides (Acidic Chalcides)

The acidic mixed oxide and sulfide catalysts (acidic chalcides) deserve specific attention as they seem to be in some respects the most attractive catalysts of the future.

Chalcide catalysts include a great variety of solid oxides and sulfides. The most widely used comprise alumina, silica, and mixtures of alumina and silica, either natural or synthetic, in which other oxides such as chromia, magnesia, molybdena, thoria, tungstic oxide, and zirconia may be also present, as well as certain sulfides, such as sulfides of molybdenum. A large variety of compositions is possible. The composition and structure of many catalysts of this type, including different types of bauxites, floridin, Georgia clay, and other natural aluminosilicates, are still not well known. Some synthetic catalysts, other than silica–alumina compositions, representative of the acidic chalcides are BeO, Cr_2O_3, P_2O_5, TiO_2, ThO_2, $Al_2(SO_4)_3$ which may be regarded as $Al_2O_3 \cdot 3SO_3$, $Al_2O_3 \cdot xCr_2O_3$, $Al_2O_3 \cdot Fe_2O_3$, $Al_2O_3 \cdot CoO$, $Al_2O_3 \cdot MnO$, $Al_2O_3 \cdot Mo_2O_3$, $Al_2O_3 \cdot V_2O_3$, $Cr_2O_3 \cdot Fe_2O_3$, MoS_2, MoS_3.

In contrast to sulfuric acid, which may be regarded as a fully hydrated chalcide, the chalcides of this group are seldom very highly hydrated under conditions of use.

Of the acidic mixed oxide catalysts silica–alumina has been studied most extensively. The acidity could rest in Lewis acid or Brønsted acid sites. Dehydrated silica–alumina is inactive as isomerization catalysts, but addition of water increases activity until a maximum is reached. Additional water then decreases activity. The effect of water suggests that Brønsted acidity is responsible for catalyst activity (125,430).

At any rate silica–alumina is quantitatively at least as acidic as 90% sulfuric acid (512).

Absorbed protons seem to be essential to the catalytic activity of all acidic chalcide catalysts. These catalysts possess properties of being physically and chemically stable and catalytically active at temperatures approaching the threshold for thermal decomposition of hydrocarbons, yet their acidity is not sufficient to lead them to form stable complexes with unsaturated hydrocarbons and other donor reagent molecules, as do the aluminum halides for example. For these reasons they are frequently used at high temperatures. They are particularly preferred for the isomerization of unsaturated hydrocarbons, which are polymerized by strongly acidic catalysts at low temperatures.

5. Cation Exchange Resins

The Brønsted acid catalytic activity in Friedel-Crafts reactions has resulted in the successful use of "solid acids" of the acidic oxide and sulfide type in carrying out a variety of reactions. Based on a similar principle it is easily understandable that cation exchange resins which are also solid acids are active catalysts in alkylations, acylation, esterifications, etc. The use of this type of catalyst may provide ample possibility in future work in many other reactions.

6. Metathetic Cation-forming Agents

There are a number of substances capable of forming cations in a way other than protonation or hydride abstraction. These are effective "catalysts" in Friedel-Crafts type reactions, although their activity in the real sense is not catalytic, because they are generally consumed in the cation-forming reaction.

Anhydrous silver salts such as $AgClO_4$, $AgBF_4$, $AgSbF_6$, $AgPF_6$, $AgAsF_6$, Ag_3PO_4, etc., are representatives of this class of compounds. Owing to the close resemblance of their ability to initiate reactions of the Friedel-Crafts type, it is customary to include them in the class of catalysts related to the Friedel-Crafts type (463).

II. The Role of Co-catalysts and the General Acid Nature of the Catalysts

1. The Question of "Anhydrous" Aluminum Chloride and Related Halide Catalysts

Although Friedel and Crafts in their original work described investigations with "anhydrous aluminum chloride" in organic reactions, it is necessary to stress at this point that conditions allowing absolutely "anhydrous" aluminum chloride to be used are very rarely encountered.

As we now know it is rather difficult to obtain a Lewis acid type metal halide, either aluminum chloride, ferric chloride, boron trifluoride, or any other catalytic metal halide, in a really anhydrous state. Utmost care and precaution must be applied, even if the dry salt is obtained, during its application to a reaction mixture if the claim is to be made that the reaction was carried out in the absence of moisture or other impurities. The reaction conditions used in the Sorbonne laboratory by Charles Friedel and James Mason Crafts during the pioneering epoch of their work were quite obviously, to present eyes, rather crude. As pointed out in Chapter I dealing with the history of the reaction, only the enthusiasm and genius of the workers enabled them to overcome these difficulties and to establish in an astonishingly short period of time the scope of the reaction to the extent that they did. The equipment used (retorts, flasks, etc., with cork connections) was quite obviously unsuitable for the maintenance of strictly anhydrous conditions. Neither did the quality of their aluminum chloride conform to our present-day standards of an anhydrous salt. To prepare really anhydrous aluminum chloride is quite difficult. Dehydration of aluminum chloride (or bromide) by repeated sublimation is not entirely suitable, as minor amounts of basic aluminum halides may co-sublime without decomposition. The only suitable method of obtaining anhydrous aluminum halides of sufficient purity is to react high purity aluminum (99.99%) in a combustion tube attached to a vacuum line system with dry, pure halogen (206).

A suggested purification of commercially available aluminum chloride giving a product closest to the pure anhydrous salt consists of repeated sublimation followed by addition of aluminum curls and 1–2% weight sodium chloride (in a current of dry nitrogen) followed by gentle heating of the closed system (214,484).

All handlings of the dry salt must be carried out in closed vacuum system. As the reaction of moisture with aluminum chloride does not take place at once, small amounts of moisture may change the whole of a stored sample of aluminum halide. If the surface of aluminum chloride is allowed to "glass" with a coating of aluminum chloride hydrate ($AlCl_3 \cdot 6H_2O$) and the compound is then sealed, the water of the hydrate slowly diffuses into the remaining anhydrous material and reacts by releasing HCl. Pressure build-up may cause explosion of the container (505).

Not only aluminum chloride, but other Lewis acid halides present similar problems concerning the preparation of anhydrous salts. Boron trifluoride, a low-boiling gas which should be more readily obtained in an anhydrous state than solid salts, was shown by Evans and Polanyi (171) to contain small amounts of bonded moisture and purification to obtain the anhydrous compound involved months of work. At last repeated low-temperature distillation in a vacuum system from sodium resulted in the preparation of the anhydrous compound.

2. Promoter Effect of Water and Other Impurities

In view of previous facts it is obvious that neither the original inventors, Friedel and Crafts, nor the thousands of subsequent researchers, who carried out most successful work with aluminum chloride and related catalyst systems, worked under truly anhydrous conditions. Thomas, when he published his excellent compilation of "Anhydrous aluminum chloride in organic chemistry" in 1941 (623), reported many thousands of investigations and reactions carried out under the catalytic effect of anhydrous aluminum chloride. However, impurities such as water, oxygen, hydrogen halides, organic halides, etc. (generally called promoters or co-catalysts) were present in almost all cases.

In spite of the fact that many of the original workers publishing in the Friedel-Crafts field advised "complete exclusion" of moisture in aluminum chloride reactions, the presence of traces of moisture has fortunately been found to accelerate rather than hinder the reactions. It must be emphasized that the conditions in most of the work carried out till the late forties and early fifties were not suitable for the fulfillment of strictly anhydrous conditions. This means that in most of the investigated preparative work appreciable amounts of moisture were present, detectable not only with more refined physical methods, but observable to the average organic chemist. The beneficiary action of traces of moisture has been observed especially in reactions involving additions to the olefinic double bond. Here a small amount of water or hydrogen chloride (resulting from the reaction of anhydrous aluminum chloride with water, which was long considered to be the real source of co-catalyst) initiates the reaction. The accelerating effect of moisture had likewise been noted in studying the isomerization and cracking of hydrocarbons. Small amounts of moisture may also facilitate reaction by causing greater solubility of aluminum chloride in the reaction media, thus providing more efficient contact of catalyst with reactant. Dry pentane undergoes little or no change in the presence of freshly sublimed aluminum chloride. Addition of anhydrous hydrogen bromide, hydrogen chloride, water, hydrated aluminum chloride, or alkyl chloride results in immediate reaction (213).

It was also observed that excess water weakened the catalyst by decomposing it. Moreover, it is obvious that in reactions involving easily hydrolyzible reactants like acyl halides, reasonably anhydrous conditions must be observed.

3. Inactivity of Pure Anhydrous Lewis Acid Halides in Isomerization, Polymerization and Alkylation

It was Nenitzescu who first called attention to the fact that anhydrous, freshly sublimed, pure aluminum chloride fails to initiate certain Friedel-Crafts reactions, but these reactions take place if the aluminum chloride is activated by water (HCl, HBr, etc., *e.g.*, co-catalysts). Aschan (14) observed that cyclohexane isomerizes under the catalytic effect of aluminum chloride to methyl cyclopentane. However, Grignard and Stratford (232) were unable to reproduce this reaction. Nenitzescu and Cantuniari (411) have shown that cyclohexane, even at reflux temperature, remains completely unchanged if it is in contact with pure, freshly sublimed aluminum chloride in the absence of moisture. However, by addition of water isomerization takes place rapidly. Similar results have been obtained by adding HCl or HBr to the reaction mixture. From these results Nenitzescu concluded that the substance added (water, HCl, HBr) acts in such a way as to activate aluminum chloride.

The inactivity of pure anhydrous Lewis acid halides in Friedel-Crafts polymerization of olefins was first demonstrated by Ipatieff and Grosse in 1936 (281). They found that pure, dry aluminum chloride does not react with ethylene; however, the necessity of an additional component, the so-called co-catalyst, was not discussed in the literature until ten years later. Up until this time it was believed that the mechanism of the Friedel-Crafts polymerization of α-olefins involved the direct interaction of a Lewis acid type catalyst with the corresponding olefinic double bond (113). Krauss and Grund (333) demonstrated the inactivity of boron trifluoride and other Lewis acid halides towards olefinic double bonds in the absence of co-catalysts in alkylation reactions. They found that co-catalysts, like acetyl chloride or HCl, are needed with certain metal halides to cause the usual Friedel-Crafts reactions of olefins.

4. Necessity of Co-catalysts

Evans, Holden, Meadows, Plesch, Polanyi, and Skinner (69) have demonstrated first that boron trifluoride alone does not cause the polymerization of isobutene, either at room temperature or at low temperatures. They expressed their belief that this means that BF_3 does not interact with the olefinic double bond. It could be argued, however, that even if BF_3 interacted with the double bond to form a dipole

$$F_3\bar{B}-CH_2-\overset{+}{C}(CH_3)_2$$

the further reaction of this dipole with the olefin to give a polymer would be prevented by the fact that the energy required to separate the positive charge at the end of the growing chain from the negative charge located on the BF_3 would be prohibitive in the hydrocarbon medium. They noted that there was no color formed in the system prior to introduction of water vapor which then caused a colored ionic species. Evans, Polanyi, and their co-workers subsequently demonstrated the necessity for the presence of proton acids in the BF_3 and $TiCl_4$ system for forming reactive complexes, i.e., complexes which will cause polymerization. They postulated carbonium ion pairs as the active species (174,482). In their investigations they also demonstrated that boron trifluoride and isobutylene do not interact when kept absolutely dry in vacuum systems, but form polymers upon admission of water. The active species of polymerization is $BF_3 \cdot H_2O$, i.e., the complex acid $H^+BF_3OH^-$. In subsequent work Evans and Hammann (167) demonstrated that 1,1,3,3,-tetraphenyl-1-butene does not polymerize with boron trifluoride in the absence of a co-catalyst. Clark (118) has similarly shown that styrene is not polymerized by boron fluoride in carbon tetrachloride solution at temperatures of 0–25° if the system is kept dry. Addition of a small amount of water vapor, however, causes vigorous polymerization. The question of co-catalyst in Friedel-Crafts polymerizations was discussed in considerable detail at the *Cationic Polymerization Symposium* held in 1952 (481) and also in a number of summary surveys (243).

Concerning the polymerization of olefins with Friedel-Crafts catalysts, the necessity of co-catalyst is sometimes difficult to establish although it has been demonstrated in quite a number of cases. If the Friedel-Crafts catalysts alone cannot polymerize the olefin, then repeated purification of the catalyst and the monomer should stop the reaction completely and the necessity for a co-catalyst is thereby demonstrated. This has been done for some polymerization reactions. If, in certain cases, it were possible for the Friedel-Crafts catalyst alone to polymerize the olefin (as it was claimed for certain cases with aluminum halide) then however much these two components were purified, polymerization would still proceed. However, there would be some doubt as to whether the reaction would have occurred if the purification had been carried out even more rigorously. Therefore the claim that a Friedel-Crafts catalyst alone can cause the polymerization of an olefin should be made only if the effect on the reaction of extreme purification of both monomer and catalyst has been examined.

Kennedy and co-workers (150,540) have shown the need for a co-catalyst, such as an alkyl fluoride, for the boron trifluoride promoted isomerization, disproportionation, and self-alkylation of saturated isoalkenes, the conditions being such that in the absence of co-catalyst boron trifluoride has no effect.

Jordan and Treloar (300) recently reinvestigated the polymerization of styrene by aluminum chloride. Previously Jordan and Mathieson (299) claimed that styrene is polymerized by aluminum chloride in carbon tetrachloride solution in the absence of a co-catalyst. This was against the proton initiation theory (168) which has been found to hold for other metal halide systems. Jordan and Mathieson claimed that the system was anhydrous and hence that initiation occurred through a π-complex between the olefin and aluminum chloride. However, reinvestigation of this system under conditions designed to give more stringent drying than previously obtained, in order to eliminate the possible co-catalytic action of traces of water, has proved that water is necessary for the polymerization. This is in agreement with previous kinetic results on the system styrene–aluminum chloride–carbon tetrachloride, in which the drying of reagents, although not sufficient to stop polymerization, showed that the polymerization rate was below that obtained by Jordan and Mathieson. The orange color on addition of water is due to the formation of 1-phenylethyl ion and the growing polymer ion (300).

It is clear that polymerization in this system too must be initiated by proton addition to the styrene and hence water is a co-catalyst.

Excess water discharges the color because of the reactivity of the carbonium ions. Thus the system aluminum chloride–styrene–carbon tetrachloride is also in conformity with other polymerizations catalyzed by metal halides, since in all systems studied up to the present a co-catalyst is essential for the reaction to proceed.

5. Nature of Co-catalysts

Following the work of Nenitzescu, Glasebrook, Phillips, and Lovell (213), as well as Schuit, Hoog, and Verhens (549), expressed the opinion that the activator in Friedel-Crafts type reactions is hydrogen chloride formed by the reaction of water with aluminum chloride. Egloff, Hulla, and Komarewsky (159) suggested that the active catalyst indeed is the complex acid $HAlCl_4$. Leighton and Heldman (350) observed that aluminum bromide does not isomerize n-butane at 85°, but a small amount of HBr initiates the reaction. They suggested that the active catalyst is $HAlBr_4$. Heldman (258) has shown that aluminum bromide is activated in isomerization reactions of hydrocarbons by substances like NaBr, NaCl, CH_3Br, CH_2Br_2, C_4H_9Br, and BF_3. All these compounds with the exception of boron fluoride form complexes of the type $R^+(AlX_4)^-$ with aluminum halides. Heldman concluded that the active catalyst in all these reactions should be the complex anion AlX_4^-, but in our present-day interpretation this suggestion has no proof.

Nenitzescu (409,410,414) has shown that the active species in the isomerization of saturated hydrocarbons with aluminum chloride initiated by water is the complex acid $H(AlCl_3OH)$. Thus water is a specific activator and its effect is not that of HCl formed by reaction with $AlCl_3$.

6. Conjugate Friedel-Crafts Acids of the Type $HMX_{4,6}$

Experiments directed toward preparation of compounds between aluminum halides and hydrogen halides, such as $HAlX_4$, provided evidence that such compounds are not formed in the absence of proton acceptor "bases."

Similarly the investigation of the system BF_3–HF in the temperature range $-100°$ to $50°$ provided no evidence of compound formation (HBF_4) (378,379,436,444).

Pines demonstrated the need for an activator for the aluminum halide catalyzed isomerization of hydrocarbons (475–479).

In his investigations he used high purity materials and developed the necessary vacuum line techniques and exact experimental methods. He was also able to show using heavy water as an activator, that HBr or HCl (DBr and DCl respectively) are not the real activators in the water promoted aluminum halide reactions. In accordance with Nenitzescu's suggestion in similar water promoted systems, the reaction

$$AlBr_3 + D_2O \rightarrow DOAlBr_2 + DBr$$

takes place when the DBr formed is pumped from the system. The active promoter species is $DOAlBr_2$. It was shown that in isomerization experiments this catalyst causes hydrogen–deuterium exchange.

It has been suggested that the catalytic activity of HCl–$AlCl_3$ and HBr–$AlBr_3$ is due to the formation of the unstable conjugate acids, $HAlCl_4$ and $HAlBr_4$, which presumably function as exceedingly strong acids (350,475,476,490,623).

However, a careful investigation by H. C. Brown of the hydrogen chloride–aluminum chloride system under a variety of conditions including temperature as low as $-120°$ yielded no evidence indicating any combination of the two components (82).

In the case of the hydrogen bromide–aluminum bromide system the solubility of the salt in liquid hydrogen bromide at $-80°$ was found to be 1.77 g./mole of solvent. From the vapor pressure lowering the aluminum bromide is present as dimeric Al_2Br_6 which again points to the absence of complex formation (87).

The same conclusion was also reached by Fontana and Harold as a result of their study of the solubility of hydrogen bromide in aliphatic hydrocarbons containing dissolved aluminum bromide (193).   Likewise McCaulay, Shoemaker, and Lien (378) as well as Olah and Kuhn (436) found no evidence of compound formation in solutions of boron fluoride in dry liquid hydrogen fluoride even at temperatures as low as $-100°$. It must therefore be concluded, according to Brown and Nelson (407), that "the postulated substances HMX_4 must be considered as hypothetical acids, whose salts are stable, but which do not themselves exist in other than possibly minute concentration."

In contrast complexes of the types $R^+AlX_4^-$ are formed by the interaction of alkyl and acyl halides (RX) with aluminum halides and are discussed in Chapter VII.

The stability of the conjugate acid HBF_4 was established in solvents like anhydrous hydrogen fluoride (where it is present as

$HF_2^+ BF_4^-$ according to Kilpatrick (316)), nitromethane and tetra-methylene sulfone, as well as in basic aromatic hydrocarbons where stable σ-complexes or benzenonium ions are formed. The acidity of $HF + 7$ wt.% BF_3 was derived from data of Mackor (364) by McCaulay (375). H_0 (Hammett Acidity Function) was found to be -16.6, compared with about -11 for 100 per cent sulfuric acid and -10 for anhydrous HF. Thus the conjugate acid HBF_4, which presumably has an acidity similar to that of $HAlCl_4$ and $HAlBr_4$, is more acidic than sulfuric acid by a factor of 10^5. The difference in acidity can account for the greater catalytic reactivity.

Katz, Kilpatrick, and co-workers (272) were recently able to determine H_0 values for SbF_5 and NbF_5 in anhydrous hydrogen fluoride

$$H_0$$

	H_0
$HF + 0.02M$ NbF_5 (0.4 weight %)	-12.5
$HF + 0.36M$ NbF_5 (6.7 weight %)	-13.5
$HF + 0.36M$ SbF_5 (6 weight %)	-14.3
$HF + 3.0M$ SbF_5 (60 weight %)	$-15.2.$

These data also show that Lewis acid fluorides are co-acids in hydrogen fluoride, increasing the acidity considerably.

In the presence of a proton acceptor, "B", "salts" of the hypothetical strong acids $HAlX_4$, HBX_4, etc., are formed with the structure $BH^+ (MX_4)^-$ (82,83,84,86,444,435,447).

Pines (477) was able to show that hydrochloric and hydrobromic acids alone, in the absence of other substances, do not activate aluminum halides in the isomerization of n-butane. In contrast with this inactivity of the hydrogen halides, water acts as an independent activator for aluminum halides.

Pines (475,478,479) has also demonstrated that small amounts of alkenes are extremely active activators for systems composed of $AlCl_3 + HCl$ or $AlBr_3 + HBr$. The presence of 0.01 mol. % of ethylenic hydrocarbon is sufficient to initiate the isomerization of iso-butane. It is very probable that in these systems the hydrogen halide adds to the alkene forming an alkyl halide and these compounds are then the real activators for aluminum halides. Thus there is a definite difference in activating power of water, hydrogen halides, and alkyl halides in Friedel-Crafts systems.

Friedel-Crafts catalysts of the Brønsted-Lowry type (e.g., H_2SO_4, HF, $HClO_4$, CF_3COOH, etc.) function through their ability to supply protons. They differ from the acidic halide Lewis acid type Friedel-Crafts catalysts in that in this case the acceptor properties are due to a positively charged entity, the proton. Any strong acid, HX, generated in Friedel-Crafts systems, might be expected to behave in a similar way.   Thus the strong Friedel-Crafts acids arising from the interaction of Lewis acid halides with co-catalysts or reagents, as for example, $BF_3 \cdot H_2O \rightleftharpoons H^+BF_3OH^-$, do not differ in nature from other Brønsted-Lowry acid catalysts.   Friedel-Crafts acids can be expressed, according to Burton and Praill (93a) by the general formulae:

$H^+(MX_3Y)^-$ where X = halogen, and Y = halogen, OH^-, OR^-, OAc^-, etc.
$\qquad\qquad$ M = B,Al,Ga,Fe,Sb,Bi
$H^+(MX_5Y)^-$, M = As,Sb,P,Nb,Ta
$(H^+{}_2)(MX_2Y_2)^{--}$, M = Zn,Hg
$(H^+{}_2)(MX_4Y_2)^{--}$, M = Ti,Zr,Sn.

Their properties are obviously dependent on the nature of the central atom (M), the groups X and Y and on the relative strengths of the bonds M–X and M–Y.   Some measure of the relative stabilities of the complex ions can be obtained in terms of electron affinities of the groups MX_n.   This has been done for ions of the type BF_3Y^- by Skinner (575a).   The heat of addition (q) of various ions to BF_3, as expressed by the equation

$$BF_3 + Y^- \rightarrow BF_3Y^- + q$$

gives the values: F^- 75 kcal.; OH^- 79 kcal.; Cl^- 25 kcal.; Br^- 11 kcal.   The larger size of the aluminum halide complexes might be expected to lead to a decrease in electron affinity, but this would be compensated somewhat by less steric repulsion than for the corresponding boron compounds.   However, solution of this problem for aluminum and other metal halide complexes awaits further data.

7. Effect of Oxygen as Co-catalyst

Pines and Wackher (476) have shown that oxygen from the air can act as an activator for aluminum bromide catalyzed reaction giving rise to bromine

$$AlX_3 + \tfrac{1}{2}O_2 \rightleftharpoons AlOX + X_2$$

which reacts then with the hydrocarbon liberating HBr

$$C_4H_{10} + X_2 \rightarrow CH_3-CH_2-\underset{\underset{X}{|}}{CH}-CH_3 + HX.$$

This reaction proceeds much faster in the presence of sunlight. The consumption of halogen causes the shift of the equilibrium toward the right.   The alkyl halide formed acts, in the presence of catalyst, as a chain starter for the isomerization of alkanes (44).

The use of aluminum chloride as an oxidation catalyst was noted by Friedel and Crafts (199), who found that upon passing moist air through a mixture of benzene and aluminum chloride, some oxygen is fixed resulting in the formation of phenol and other oxygenated hydrocarbons. However, this explanation is difficult to apply to aluminum chloride reactions, as the oxidation of aluminum chloride by air oxygen to liberate chlorine seems quite improbable. Another interpretation of the co-catalytic action of oxygen may be that the hydrocarbons themselves could be oxidized first to an alcohol,

$$2C_4H_{10} + O_2 \longrightarrow 2C_4H_9OH$$

which then reacts with aluminum chloride to an alkoxyaluminum chloride and hydrogen halide. Part of the hydrogen halide could also transform the alcohol into an alkyl halide which may be the real co-catalyst or activator. Nenitzescu observed in his experiments that oxygen does not act as an activator in aluminum chloride catalyzed isomerization.

8. Division and Mode of Action of Catalysts and Co-catalysts

In summary two main groups of co-catalysts can be defined in Friedel-Crafts systems:

(a) Proton-releasing substances. In this class belongs first of all hydroxy compounds (water, alcohols) and proton acids (hydrogen chloride, sulfuric acid, phosphoric acid, organic acids, etc.).

(b) Cation-forming substances (other than proton). Co-catalysis by alkyl and acyl halides, as well as by a variety of other donor substances (O, S, N, halogen, etc., donors) leading to cations other than proton is summarized in this class. Besides carbonium ions formed with alkyl and acyl halides, oxonium, sulfonium, halonium, etc., complexes can act as the activators in different Friedel-Crafts reactions.

Concerning the mode of action of the catalyst we again must differentiate between two quite well defined separate fields in the very large area of Friedel-Crafts type reactions. In the first a positively polarized entity or carbonium ion is produced by the interaction of the catalyst with the reagent containing a functional group with a donor atom having unshared electron pairs (like halogen, oxygen, nitrogen, sulfur, etc.). This electrophile then reacts with an unsaturated carbon atom (aromatic, olefinic sp^2 carbon atoms or acetylenic sp carbon atoms). This reaction, of course, can take place under absolutely anhydrous conditions and in the absence of any co-catalyst, as the Lewis acid catalysts themselves are suitable acceptors to interact with the aforesaid donors.   Lewis acids, however, are generally insufficient acceptors in reactions of olefins or acetylenes or isomerization of hydrocarbons.   These reactions generally require a co-catalyst to form a strong conjugate acid or carbonium ion with the Lewis acid, which then initiates the reactions. The second main type of Friedel-Crafts reactions is that in which the catalyst extracts a hydride ion from a saturated carbon atom. This type of reaction takes place in alkylations with alkanes, in isomerization of saturated hydrocarbons, etc.   In this type of reaction the Lewis acid is, so far, not known to be able to abstract hydride ions to provide the active species for subsequent Friedel-Crafts type reactions.   A co-catalyst is needed to form a more powerful conjugate acid with the Lewis acid capable of initiating the reaction, thus hydride abstraction is effected by the proton or carbonium ion.

III. Acidic Halide Lewis Acid Catalysts

1. Group IB Halides

Copper, silver, and gold all have in the valence shell the $(n-1)d^{10}ns^1$ electron configuration.   The electron configuration of Cu is $3d^{10}4s^1$, of Ag, $4d^{10}5s^1$, and that of Au, $5d^{10}6s^1$.   They show considerable tendency to form covalent compounds.   The d-electrons frequently participate in bonding and the ions M^+, M^{2+}, and, with Cu and Au, M^{3+} are recognized in various complexes.   Copper appears with charge $+3$ in complexes such as CuF_6^{3-}, where paramagnetism indicates two singly occupied d-orbitals with an octahedral shape.   Gold, with its higher electronegativity, can obtain an adequate share of electrons from a smaller number of ligands and occurs more commonly in 4 coordination complexes like $AuCl_4^-$ and $AuBr_4^-$ which are square planar.

From the elements of Group IB of the Periodic Table copper and gold halides show Lewis acid catalytic activity in certain Friedel-Crafts reactions.

A. Copper Halides

a. Copper(I) halides

Cuprous fluoride appears to be unstable at ordinary temperatures, as it disproportionates very easily $2CuF = CuF_2 + Cu$ (127,242). Cuprous halides all have the zinc blende structure, showing a predominantly covalent character.

Cuprous chloride, m.p. 430°, b.p. 1367°, a colorless substance, forms double molecules $(CuCl)_2$ with the structure in the vapor phase

$$Cu \diagup \overset{Cl}{\underset{Cl}{}} \diagdown Cu$$

Molecular weight determinations in donor solvents (pyridine) have given values closer to the unimolecular formula CuCl. (It is suggested that in this case a coordination complex $C_5H_5N \to CuCl$ is already formed.)

Cuprous bromide, a pale greenish-yellow compound, melts at 483°, boils at 1345°, and *cuprous iodide*, a pure white substance, melts at 588° and boils at 1293°. Both have covalent structures similar to the chloride.

The copper(I) halides are generally mild Friedel-Crafts type catalysts in alkylation, acylation, isomerization and polymerization reactions and are useful in cases where stronger catalysts such as aluminum halides may cause decomposition (anthracene compounds, certain polycyclic compounds).

In reactions with halide reagents complexes of the type $R^+CuX_2^-$ are formed intermediately.

b. Copper(II) halides

Cupric fluoride is a high-melting (950°) ionic substance, with a fluoride type lattice. Accordingly it has no catalytic activity.

Cupric chloride forms dark brown crystals, melting at 498°. The crystal lattice similarly to that of *cupric bromide* shows flat chains in which each Cu atom has four halogen atoms arranged around it.

$$Cu \diagup \overset{Cl}{\underset{Cl}{}} \diagdown Cu \diagup \overset{Cl}{\underset{Cl}{}} \diagdown Cu \diagup \overset{Cl}{\underset{Cl}{}} \diagdown Cu$$

Cupric iodide is exceedingly unstable and almost as soon as it forms breaks up into cuprous iodide and iodine. Cupric chloride and bromide show Friedel-Crafts catalytic activity especially in halogenation and isomerization reactions.

Cupric halides very easily form coordination compounds with halogen ions and to some extent also with organic halides. The most frequent types being $M^+ (CuCl_3)^-, M^{2+}(CuCl_4)^{2-}$.

B. Silver Halides

Silver(I) halides all have ionic structures and their "acidity" is that of Ag^+.

Complex halo silver salts ($AgBF_4$, $AgClO_4$, $AgSbF_6$, $AgAsF_6$, $AgPF_6$) behave similarly.

The designation of silver perchlorate as a Lewis acid (355) because of the tendency of the silver to complete a stable shell of two electron pairs as in the formation of $Ag(NH_3)_2^+$ has given rise to the idea that silver perchlorate can act as a Friedel-Crafts catalyst. The catalysts normally used in Friedel-Crafts reactions (e.g., Lewis acid metal halides, sulfuric acid, HF, other proton acids) are true catalysts since they can be recovered unchanged except for halogen transfer, provided that they do not combine with the end product. When secondary reactions are excluded the catalysts are able to give a stoichiometrically greater yield than would be possible if each molecule gave rise to only one molecule of product. The reason is the constant regeneration of the catalyst. The process by which this could occur with silver perchlorate is not easy to envisage and one cannot accept it as a catalyst in the above sense. The alleged polymerizing effect of silver perchlorate (162,532) may be explained by the presence of perchloric acid or traces of alkyl halide or other halide type reagent (giving rise to powerful carbonium perchlorate), thus acting as co-catalysts (111).

Burton and Praill (93) indeed were unable to catalyze the polymerization of styrene with the pure, dry $AgClO_4$ and concluded that the polymerization reported previously by Eley and Salamon must have been initiated by traces of perchloric acid present as impurity.

Complex silver salts are powerful halide acceptors and widely used in metathetic reactions as cation-forming agents in reactions with organic halides.

With the exception of AgF_2 silver does not form divalent halides.

C. Gold Halides

Aurous halides are fairly unstable and no investigation of their Friedel-Crafts catalytic ability has been reported.

Auric halides. The trivalent halides of gold, of which the chloride, bromide, and iodide are known, have the covalent dimeric form Au_2X_6, with the bridged structure

Whenever molecular weights have been determined they have been found to correspond to dimeric molecules, Au_2X_6. *Auric fluoride*, however, may be ionic, although no definite compound has yet been obtained.

Auric chloride, $AuCl_3$, is a brown, crystalline substance, which melts under chlorine pressure at 288°. Auric chloride is perceptibly volatile at temperatures above 180°. Fischer (188) has measured the vapor pressure of auric chloride in excess of chlorine by a differential method and found it to be 3.5 mm. at 250° and 7.5 mm. at 263°. He also determined the molecular weight in the vapor phase and found it in close agreement with the dimeric formula Au_2Cl_6.

Auric bromide, $AuBr_3$, is a dark brown substance. It is converted into aurous bromide and bromine by heating to 160°–170°. Its molecular weight has been found by the elevation of the boiling point in bromine to correspond to Au_2Br_6 (90).

The coordinating tendency of trivalent gold is great. Therefore, auric halides have potentialities as powerful Lewis acid type catalysts. They easily form AuX_4^- complexes, but their expensive nature has so far prohibited detailed investigation of their catalytic abilities. Woolf (677) reported the coordination of $AuCl_3$ with ClCN and BrCN and suggested the possibility of using it similarly to $AlCl_3$ and other Lewis acid halides to introduce the nitrile group into aromatic molecules

$$ArH + ClCN \xrightarrow{\;AuCl_3\;} ArCN + HCl.$$

$AuCl_3$ and $AuBr_3$ were found to be strong Friedel-Crafts catalysts in alkylations and acylation with organic halides (434).

2. Group IIA Halides

The trend of properties throughout this group is similar to that of Group IA, but the greater nuclear charges make the atoms smaller. The electronic configurations of the valence shells are Be: $1s^22s^2$, Mg: $2p^63s^2$, Ca: $3p^64s^2$, Sr: $4p^65s^2$, and Ba: $5p^66s^2$.

A. Beryllium Halides

The promotion of $Be[1s^22s^2] \rightarrow Be[1s^2\,2s^1\,2p^1]$ occurs easily giving rise through sp hybridization to two largely covalent bonds. Provided electron-rich ligands are available, the number of valencies can be increased to four by donation, *e.g.*, using sp^3 hybrids and making the s- and p-orbitals available for covalent bond formation. Thus beryllium halides are fairly active Lewis acid type Friedel-Crafts catalysts.

Beryllium fluoride, BeF_2, is only feebly ionic.   It has not yet been obtained in a definite crystalline state, but forms a glassy mass whose X-ray diffraction pattern is that of "random network" (667) in which there are definite interatomic distances within the particles, but a random orientation of the particles themselves.   It softens at about 800° and begins to sublime at that temperature.   The fused salt is a very bad conductor of electricity (416).

With fluorides it easily forms fluoroberyllates of the type $Me_2^I(BeF_4)$.

BeF_2 appears to be one of the few metal fluorides which are not completely ionized in solution.

A slight Friedel-Crafts catalytic activity was found for BeF_2 although it is less active than $BeCl_2$ (66,238).

Beryllium chloride, $BeCl_2$, is a white crystalline substance.   It has no dipole moment, its molecule being linear corresponding to Be in the *sp* hybrid valence state.   Some dimeric Be_2Cl_4 molecules are present in the vapor.   The melting and boiling (or subliming) points and the percentage of the dimer Be_2X_4 in the vapor phase (at 564°) are given for the three beryllium halides in Table I (502).

TABLE I.   Melting and boiling points of beryllium halides

	M.p. °C	B.p. °C	Subl.p. °C	%Be_2X_4
$BeCl_2$	405	488	—	23
$BeBr_2$	488	—	473	34
BeI_2	480	488	—	—

According to Nilson and Pettersson (422) the chloride is completely monomeric at 745°.   In pyridine solution it is monomolecular, but here it has already formed a coordination compound (516).

The specific conductivity of fused beryllium chloride is very low (40,43,263,661).   According to Hevesy, the conductivity is about a thousandth of that of a purely ionized salt, such as sodium chloride and so indicates that in fused beryllium chloride about one molecule in a thousand is ionized.   In the solid state we may suppose that there are even fewer ions.

$BeCl_2$ is quite soluble in many organic solvents.   It is a fairly effective Lewis acid type Friedel-Crafts catalyst similar to aluminum chloride, but not quite as efficient.   Rather high reaction temperatures are needed, especially to break up the intermediate beryllium complexes (67,238).

Aldehydes, ketones, ethers, etc., readily form coordination complexes with $BeCl_2$ (similar to $BeBr_2$ and BeI_2) of the type

The complexes involve the use of sp^3 orbitals giving a tetrahedral disposition of the four bonds around the Be atoms.

The properties of *beryllium bromide* are almost identical to those of the chloride. It melts at 488° and sublimes below this temperature, having a vapor pressure of 1 atm. at 473° (502). Thus it is more volatile than the chloride. The vapor at 566° contains 34% of dimeric Be_2Br_4 molecules. The fused substance is practically a non-conductor (345). It is an effective Friedel-Crafts catalyst, catalyzing for example bromination of benzene (617). It also has an increased solubility over beryllium chloride.

Beryllium iodide is similar in general properties to the chloride and bromide, but much less stable. The fused substance does not conduct electricity. It is much more hygroscopic than the bromide.

B. Magnesium Halides

Magnesium fluoride has a rutile lattice, with a Mg–F distance of 2.05 Å (calculated values are for the ionic link 2.11 Å, for the covalent 2.00 Å). It melts at 1260° and is distinguished from the other halides by its very low solubility in water, acids, or any other solvent. Owing to its practically ionic nature no catalytic activity is observed.

Magnesium chloride. Anhydrous magnesium chloride melts at 712° (550 degrees lower than the fluoride) and has a vapor pressure of 25 mm. at 1000°. Its extrapolated boiling point is 1410° and it can be distilled in a current of hydrogen. Although in the crystalline state it is still highly ionic, its lower melting point and substantial volatility indicate that in the liquid and vapor phases some covalent character must be present.

It is quite soluble in a number of organic solvents and forms complexes with ethers, aldehydes, and ketones similar to those of $BeCl_2$.

Although for long it was stated to be inactive as a Friedel-Crafts catalyst, recently Bryce-Smith (46,76) found freshly prepared anhydrous $MgCl_2$ (as well as $MgBr_2$) to be a fairly active alkylation catalyst.

Anhydrous *magnesium bromide* melts at 711°. It has a layer type crystal lattice and a considerably increased solubility in many organic solvents, like alcohols, ketones, ethers, with many of which it forms solid addition compounds.

Menshutkin (391) reported its catalytic activity in Friedel-Crafts acylation (34).

Anhydrous *magnesium iodide* melts under hydrogen at 650°. It is even more soluble than the bromide in various oxygenated organic solvents (alcohols, acids, amides, ethers, ketones, etc.). Its crystal structure is that of the magnesium bromide. Its Friedel-Crafts catalytic activity has not been reported.

C. Alkaline-Earth Halides

The fluorides of the alkaline-earth metals are practically ionic, insoluble crystalline compounds. Calcium fluoride, CaF_2, occurs as the mineral fluorspar. It melts at 1330°. Strontium fluoride, SrF_2, has also the fluorite type crystal lattice and melts above 1450° and boils at 2489°. Barium fluoride, BaF_2, has the fluorite lattice, melts at 1285° and boils at 2137°. Owing to their completely ionic nature none of these fluorides has any proven catalytic activity.

The claims that fluorspar is an alkylation catalyst at higher temperatures (432) does not necessarily mean that CaF_2 is the active catalyst. Impurities of other metal halides in fluorspar may be responsible for the catalytic activity, or hydrolysis with moisture in the system (either effecting CaF_2 itself or, even more plausibly, Na_2SiF_6 present in fluorspar) may give rise to catalytically active hydrogen fluoride. The same considerations must be viewed for similar claims on BaF_2 and also to a certain degree on the activity of the corresponding chlorides and bromides (where of course HCl or HBr may be formed).

The melting points of the alkaline-earth metal chlorides, bromides, and iodides, as shown in Table II (622), indicate that they are mostly ionic compounds, although the considerably decreased melting points related to the fluorides indicate that at least in the molten state some covalent contribution to the structure is to be expected. The relative volatility of the bromides and iodides also indicates some covalent nature in these compounds.

TABLE II. Melting and boiling points of alkaline-earth halides

	M.p. °C	B.p. °C
$CaCl_2$	772	> 1600
$CaBr_2$	765	810
CaI_2	575	718
$SrCl_2$	872	
$SrBr_2$	643	
SrI_2	402	
$BaCl_2$	962	
$BaBr_2$	847	
BaI_2	740 (dec.)	

3. Group IIB Halides

The elements which comprise this group: zinc, cadmium, and mercury occur at the end of the three short transition series of the Periodic Table. All possess in their outer valency electron shells the $(n - 1)d^{10}ns^2$ configuration (Zn: $3d^{10}4s^2$, Cd: $4d^{10}5s^2$, Hg: $5d^{10}6s^2$).

Unlike copper, silver, and gold which immediately precede them and which, despite the possession of completed levels, can use one or more d-electrons in chemical interactions, the members of Group IIB utilize only the outer s-electrons. No oxidation states higher than $+2$ are known.

Melting and boiling points of Group IIB metal halides are summarized in Table III.

TABLE III. Melting and boiling points of Group IIB halides

	M.p. °C	B.p. °C
ZnF_2	872	1500
$ZnCl_2$	275	732
$ZnBr_2$	394	650
ZnI_2	446	624
CdF_2	1049	1748
$CdCl_2$	568	964
$CdBr_2$	580	963
CdI_2	388	715
HgF_2	645	650
$HgCl_2$	280	302 (subl.)
$HgBr_2$	236	322
HgI_2	140 (subl.)	> 310 (dec.)

The zinc, cadmium, and mercury halides (with the exception of the fluorides) display a considerable amount of covalent character.

The fluorides are highly ionic compounds, practically insoluble in most organic solvents, and consequently show no catalytic activity. This is the case even with mercuric fluoride, which forms colorless octahedra, melting at 645° and boiling a few degrees higher. It has a fluorite lattice with the Hg–F distance 2.40 Å. This is an ionic lattice and the distance is that expected for the ions. It is clear that with mercury as with aluminum or tin, the fluoride is far more ionized than the other halides. Mercuric fluoride is also a strong electrolyte.

A. Zinc Halides

Zinc fluoride as a Friedel-Crafts catalyst has been investigated by Calloway (103), who found it to be a relatively good catalyst for the reaction of *t*-butyl chloride with anisole. This reaction, in which both reactants are very active, requires only a feeble catalyst and in some cases (at higher temperatures) can be carried out even in the absence of a metal halide catalyst. On the other hand, as zinc fluoride is a proven fluorinating agent capable of exchanging chlorides to fluorides,

$$2R - Cl + ZnF_2 \rightarrow 2R - F + ZnCl_2$$

the reaction observed could have been the zinc chloride catalyzed alkylation of anisole by *t*-butyl fluoride. Therefore, it is doubtful that the highly ionic zinc fluoride really has any appreciable Friedel-Crafts catalytic activity.

The chlorides, bromides, and iodides of zinc and cadmium are considerably more soluble than the fluorides, and tend to form auto-complexes. Although it has been assumed that this tendency is much stronger with cadmium than with zinc, recent work makes this uncertain and suggests that the greater abnormality of the cadmium halides in solution is due at least in part to their having the tendency of the mercuric halides to go into a non-complex dicovalent form. The low melting (275°) and boiling (732°) points of *zinc chloride* together with the observed remarkable small hydrolysis in aqueous solutions point to its considerably covalent nature in solution. Its vapor density at 900° corresponds to monomeric $ZnCl_2$. Molten zinc chloride is a relatively poor electrical conductor. At 319° its specific conductance is only 2×10^{-4} ohm^{-1} cm.$^{-1}$. Zinc chloride was used by Friedel and Crafts as catalyst for the reaction of benzyl chloride with benzene (672). Wertyporoch did not find that it catalyzed this reaction or that of chloroform with benzene. It has been used in the reaction of alkyl chloride with benzene (412) and in polymerization (89) and isomerization (467) of hydrocarbons.

$ZnCl_2$ is widely used as catalyst in chloromethylation and other haloalkylation reactions, in alkylations of olefins with polyhalides, and in aralkylation of olefins with α-haloaralkanes (437).

As a fairly selective and mild catalyst its use is advantageous in many reactions where halides or alcohols are required to react selectively with olefinic double bonds. Aluminum chloride and other more powerful catalysts react with all functional groups present and show no selectivity and are consequently less suitable in these reactions.

Zinc bromide is very similar to the chloride, with an even larger solubility in many solvents. *Zinc iodide* is the most soluble of the halides.

The most common coordination number of both zinc(II) and cadmium(II) in their complexes is four, although six is shown in a few cases. Since the inner *d*-orbitals are completely filled in the metal ions, the four-coordinated complexes would be expected to be sp^3 hybrids and thus have a tetrahedral structure. This arrangement has been demonstrated by X-ray studies in a number of complexes, including the halide complexes R_2ZnX_4, R_2CdX_4.

B. Cadmium Halides

Although the zinc halides doubtless have stronger covalent characters than the corresponding cadmium compounds, the latter are still able to form halo complexes and thus show some Friedel-Crafts catalytic activity.

The halides of cadmium demonstrate the effect on structure of the easier polarization of an anion by a smaller cation. CdF_2 has the cubic fluorite lattice, but the chloride, bromide, and iodide form hexagonal crystals based on layer lattices.

Cadmium halides, with exception of the fluoride, which is a practically ionic, high-melting (1100°) solid, show considerable covalent nature.

Cadmium chloride melts at 568° and boils at 964°. In the molten state $CdCl_2$ shows good electric conductivity. However, it is somewhat soluble in organic solvents, like acetonitrile and has a tendency to form complexes of the type CdX_3^- and CdX_4^{2-}.

Cadmium bromide (m.p. 580°, b.p. 963°) and *cadmium iodide* (m.p. 388°, b.p. 715°) are similar in nature to $CdCl_2$, with somewhat increased covalent character and complex-forming ability.

Cadmium halides are rather weak Friedel-Crafts catalysts with some activity reported in polymerization and halogenation reactions (72,294,333).

C. Mercury Halides

With the exception of the fluoride, the mercury(II) halides exhibit essentially covalent bonding.

Mercuric chloride, bromide, and iodide resemble one another very closely and differ markedly from the fluoride. They are almost wholly covalent. Their covalency is shown by their relatively low sublimation, melting, and boiling points (see Table III).

The covalency of the compounds is also shown by the small conductivity in the fused state, the specific conductivity values just above the melting points being: $HgCl_2$ 0.82×10^{-4} cm.$^{-1}$ Ω^{-1} at 272°, $HgBr_2$ 15×10^{-4} cm.$^{-1}$ Ω^{-1} at 238° (39,40,262) and by Raman spectra of the halides both in the vapor (669) and in the solid and liquid state, which indicates a symmetrical, linear molecule, X–Hg–X. The linear shape of the molecule has been established for all three dihalides also by electron diffraction in the vapor phase (65,231) and by X-ray analysis for the chloride and bromide (64,544).

Mercuric fluoride has an ionic (fluorite type) lattice. It melts at 645° and does not show Friedel-Crafts catalytic activity. Mercuric chloride forms molecular crystals with linear Cl–Hg–Cl molecules packed together. The structure of mercuric bromide differs from that of $HgCl_2$ being intermediate between a molecular and a layer type of crystal. Mercuric iodide consists of infinite square network layers in which each mercury atom is surrounded tetrahedrally by four iodine atoms.

Mercuric iodide is so covalent that it is much more soluble in benzene than in water.

With the exception of the fluoride, mercuric halides readily form halide complexes of the type $M^I HgX_3$ or $M_2^I HgX_4$.

Mercuric chloride and bromide are very mild Friedel-Crafts catalysts in certain alkylations, acylations, polymerizations, and isomerizations. They are applied with some success in cases where aluminum chloride causes secondary reactions, as in the decomposition of sensitive molecules (50,547,615).

4. Group III Halides

The elements of Group III of the Periodic Table (aluminum family) are boron, aluminum, gallium, indium, and thallium. They all have the electron configuration of the valence shell: ns^2np^1

$$B:2s^2\,2p^1, \quad Al:3s^2\,3p^1, \quad Ga:4s^2\,4p^1, \quad In:5s^2\,5p^1, \quad Tl:6s^2\,6p^1.$$

Electronically these elements may be subdivided into two classes: one, consisting of boron and aluminum, possessing an inert gas kernel and the other (gallium, indium, and thallium) having completed s-, p-, and d-orbitals underlying the three (s^2p) valence electrons.

A. General Consideration and Comparison of Boron and Aluminum Halides

Boron and aluminum are effectively trivalent, as promotion of $ns^2np^1 \to ns^1np^2$ occurs very easily. They form cations with an inert-gas structure much less readily than the elements of Group II and their bonding is predominantly covalent. Like atoms of Group II they are electron-deficient since an octet is not normally present. There are only three electron pairs in the valence shell, instead of the four characteristic of Groups IV and VII and electron-

TABLE IV. Melting and boiling points of boron and aluminum halides

	Boron		Aluminum	
	M.p. °C	B.p. °C	M.p. °C[a]	B.p. °C
Fluoride	− 130	− 101		1291 (subl.)
Chloride	− 107	+ 12	192	180 (subl.)
Bromide	− 46	+ 91	97	255
Iodide	+ 43	+ 210	180	381

[a] In sealed tube under pressure.

pair repulsions are accordingly smaller than usual, so the atoms tend to be electron "acceptors." Simple molecules with an incomplete octet around the metal (like halides or alkyls of Al and B) invariably contain sp^2 hybrid bonds lying in a plane with the electron pairs as far apart as possible. But the tendency to complete an octet is shown by the existence of tetrahedral compounds, such as BH_3CO or in complexes of the type AlX_4^-, BX_4^-, in which sp^3 hybridization is involved.

The boron trihalides are all covalent and monomeric. Their melting and boiling points are in accordance with the covalent monomeric structures (Table IV).

The covalent, monomeric structures were proven in the vapor and liquid phase as well as for solutions, such as BCl_3 in benzene. By contrast AlF_3 is an ionic crystalline solid of high melting point. The more volatile aluminum chloride and bromide exist as dimers, both in the vapor phase and in non-polar solvents, in which the halogen atoms are tetrahedrally arranged about each aluminum atom.

The elements of Group III exhibit a common oxidation state of $+3$. The small calculated ionic radii of the tripositive ions indicate there is little tendency for the formation of simple ions in this oxidation state. Thus we find practically all simple compounds corresponding to the $+3$ state largely covalent in nature. The melting points of the trichlorides (as compared with those of Group II chlorides) serve to illustrate this point (Table V).

TABLE V. Comparison of melting points of Group II and III halides

	M.p. °C		M.p. °C
$BeCl_2$	440	BCl_3	-107
$MgCl_2$	708	$AlCl_3$	192.6 (1700 mm.)
		$GaCl_3$	78
$CaCl_2$	772	$InCl_3$	498 (subl.)
$SrCl_2$	873	$TlCl_3$	155
$BaCl_2$	963		

The boron trihalides, in which the boron atom has a sextet of valence electrons, have a great tendency to act as acceptor molecules with the formation of a variety of complexes and in general act as active Lewis acid type catalysts in Friedel-Crafts reactions.

B. Boron Trifluoride

a. Properties and structure

Boron trifluoride is a colorless gas. It boils at $-101°$ and melts at $-130.7°$. The Trouton constant of 26.1 is very high for the non-associated liquid (but that of SiF_4 is 25.1). The vapor density of boron trifluoride gives no indication of dimeric molecules (190).

Electron diffraction shows that the molecule in the vapor has a planar structure with angles of $120°$. The B–F distance being 1.31 Å (theoretical for B–F linkage 1.52 Å, but obviously shortened by back coordination from the halogen electron pairs to the boron). The structure is supported by the infra-red and Raman spectra (202,682).

A recent infra-red investigation of crystalline boron trifluoride at

83° K, together with isotopically substituted boron trifluorides, has shown that the crystal structure of the solid boron trifluoride at low temperature is not isomorphous with that of boron trichloride and tribromide and the data obtained give considerable evidence of a more specific type of intermolecular attraction, perhaps verging on association through weak bonds in the solid state (154).

The prediction of the possibility of association is not new and has its analogy in the dimeric Al_2Cl_6 and Al_2Br_6 and cross-linked SbF_5. It was suggested that when the crystal structure of BF_3 is studied, it will be found that there is at least a tendency toward formation of bridged B–F–B bonds, arising from an interaction of the fluorine electrons with the unused p-orbital of the boron of another molecule.

However, in the temperature range where boron halides are used as catalysts it is safe to say that they are always present in the monomeric form.

b. Catalytic activity

Our original knowledge on the catalytic activity of boron trifluoride is mainly due to the work of Meerwein (380) in Germany and of Nieuwland (261,264,586,589) in the United States. As Meerwein (381) has pointed out its behavior is very like that of aluminum chloride in the Friedel-Crafts reactions, except that in place of organic halides it generally involves the use of oxygen compounds (alcohols, ethers, acids, etc.) and eliminates water (with which it coordinates) instead of hydrogen chloride. The resemblance between the action of the two halides undoubtedly goes very far. With both the catalytic power is primarily due to the loosening of atomic linkages in the organic molecule through coordination with the Lewis acid. Examples of the catalytic power of boron trifluoride are very numerous and have been summarized many times (56, 306,635).

Boron trifluoride catalyzed reactions cover most of the wide field of Friedel-Crafts type reactions, including alkylations and acylations (both aromatic and aliphatic), different types of cyclizations, addition reactions, isomerizations, polymerizations, and many other reactions.

Some of the major fields of applications of BF_3 as a catalyst in reactions related to the Friedel-Crafts type are

Alkylation with
  olefins
  alkyl fluorides
  alcohols

Isomerization

Disproportionation of isoparaffins
Isomerization of paraffinic and unsaturated hydrocarbons
Isomerization and disproportionation of alkylbenzenes
Stereospecific isomerization of olefins

Acylation

Aromatics
Aliphatics (including active methylene compounds)
Nitration, sulfonation, etc.

Polymerization (or Copolymerization)

Olefins and diolefins
Styrene or derivatives with isoprene, butadiene, dimethylbuta-
diene, piperylene, etc.
Vinyl ethers and esters
Unsaturated heterocyclics (indene, coumarone, etc.)
Unsaturated acids and esters (included drying oils)
Terpenes and derivatives.

Rearrangements

Fries, Beckmann, Benzidine, etc.

One of the chief advantages in the use of boron fluoride as a Lewis
acid type Friedel-Crafts catalyst is the fact that being a low-boiling
gas it is easy to handle and to remove (or recover) from the reaction
mixtures. A further advantage is the fact that tarry and undesirable by-products are usually not obtained (307).

For a long time it was thought that boron trifluoride was a far
more effective Friedel-Crafts catalyst than the other boron halides
(probably due to the fact that it was the most investigated boron
halide, which was known to form many stable complexes). It is
now well established that this is not the case. Back donation to
the boron from the unshared halogen pairs decreases the electron
deficiency to a larger degree in BF_3 than in BCl_3, BBr_3, or BI_3.
Therefore, disregarding steric factors, the decreasing Lewis acid
strength of the boron halides is expected to be

$$BI_3 > BBr_3 > BCl_3 > BF_3.$$

Indeed this sequence was observed not only from the heat of

formation of different complexes, but was also found in the effectiveness of boron halide catalysts in certain Friedel-Crafts reactions (alkylation, acylation). For a more detailed discussion of the question of relative catalyst action of boron halides see Chapter VI.

c. Coordination compounds as catalysts

Boron trifluoride can be used in catalytic reactions not only neat, but also in the form of many of its coordination compounds. Depending on the type of reaction, the products desired and reaction conditions involved, a wide choice of stable coordination compounds of boron trifluoride is available.

Only the more common coordination compound catalysts will be mentioned here.

Frequently boron trifluoride is used as a catalyst in the presence of moisture, because under anhydrous conditions it will not catalyze certain reactions (171).

$BF_3 \cdot H_2O$ and $BF_3 \cdot 2H_2O$, which are formed when BF_3 is contacted with water are effective catalysts in reactions such as polymerization, alkylations, depolyalkylation, etc. Boron trifluoride hydrates are strong proton acids of the hydroxyfluoroboric or dihydroxyfluoroboric acid type and consequently will be discussed in connection with the proton acid catalysts.

Frequently boron trifluoride is used in the presence of anhydrous hydrogen fluoride. Although fluoroboric acid (HBF_4) does not exist in the anhydrous state, it is stable in the presence of proton acceptors (which can be HF, H_2O, nitromethane, tetramethylene sulfone, aromatic hydrocarbons, or many other solvents) and is a powerful Brønsted acid catalyst.

$BF_3 \cdot H_3PO_4$ is also a very effective catalyst system, as is $BF_3 \cdot H_2SO_4$.

Boron trifluoride addition compounds with organic oxygen-donor molecules (bases) are frequently used as catalysts.

Boron trifluoride generally forms complexes with oxygen containing compounds in the ratio of 1 : 1 or 1 : 2. From conductivity and electrolysis experiments it has been shown that the liquid (molten) complexes formed between boron trifluoride and water (220), alcohols (225), and carboxylic acids (225) are fairly good electrolytes, whilst complexes with ethers (224) and esters (221) are only slightly ionized.

The complexes are envisaged as being in equilibrium between an ionized and unionized form, represented for the 1 : 1 and 1 : 2 complexes respectively in the following way :

$$F_3B \leftarrow \overset{\overset{\displaystyle H}{|}}{O} - R \; \rightleftharpoons [H]^+[BF_3OR]^-$$

$$\underset{BF_3}{\overset{R}{\diagdown}} O - H \cdots \overset{H}{\underset{R}{\diagup}} O \quad \rightleftharpoons [ROH_2]^+[BF_3OR]^-.$$

The proton in the ionized structure of the 1 : 1 complex is probably further solvated.

As pointed out by Sharp (552) physical properties of the complexes do not completely confirm these hypotheses but, in general do not disprove them either.

X-ray powder investigations (318) support the $H_3O^+BF_3OH^-$ structure for $BF_3 \cdot 2H_2O$. Nuclear magnetic resonance spectroscopy (194), however, suggests that, in the slowly cooled specimens at least, there is no ionization.

The n.m.r. results for the complexes formed with alcohols are not at variance with the above structural ideas (140).

Some of the catalytically active boron trifluoride–oxygen donor complexes, together with their postulated ionized form, are summarized in Table VI according to Sharp (552).

It should be pointed out that the available evidence is in favor of only limited ionization for the boron trifluoride complexes considered. Whether the reactions of these derivatives take place through the ionized form is not certain, but there is strong evidence that the adducts themselves are the active catalysts in many reactions in which boron trifluoride is used. Thus the polymerization of isobutylene occurs only in the presence of a co-catalyst such as an alcohol or an acid (166,169,170,172,173,174); the same is true for the polymerization of stilbenes (62), alkylations (94), and the cis–trans isomerization of 2-butene (158). The existence of oxonium ions as reacting entities seems established from the direction of addition of alcohols to olefinic ethers under the influence of boron trifluoride as a catalyst (466). The postulation of alkyl carbonium ions in boron trifluoride–ether adducts is supported by the alkylation of aromatic hydrocarbons by s-butyl methyl ether (95), there being almost complete racemization during the alkylation step. On the other hand there appears to be no relation between electrical conductivity

of the complex and catalytic power (633) although the fall in conductivity during the course of a reaction does parallel the fall in
catalytic activity of the mixture.   Conductivities are such complex
phenomena that this argument may not be valid.   However, the
work of Clayton and Eastham (119) on the relation between the rate
of isomerization of 2-butene and the amount of water present in the

TABLE VI.   Catalytic boron trifluoride–oxygen donor complexes

Complex	Postulated ionization	Ref.
BF_3H_2O	$H^+[BF_3OH]^-$	222
$BF_3 \cdot 2H_2O$	$[H_3O]^+[BF_3OH]^-$	20,222
$BF_3 \cdot H_3PO_4$		632,633,634
$BF_3 \cdot H_4P_2O_7$		633
$BF_3 \cdot MeOH$	$H^+[BF_2OR]^-$	222
$BF_3 \cdot 2MeOH$	$[ROH_2]^+[BF_3OR]^-$	222
$BF_3 \cdot EtOH$		633
$BF_3 \cdot 2EtOH$		223
$BF_3 \cdot 2Pr^nOH$		
$BF_3 \cdot PhOH$		633
$BF_3 \cdot 2PhOH$		588
$BF_3 \cdot CH_3COOH$	$H^+[BF_3OOCCH_3]^-$	219,225
$BF_3 \cdot 2CH_3COOH$	$[CH_3COOH_2]^+[BF_3OOCCH_3]^-$	225
$BF_3 \cdot Et_2O$	$Et^+[BF_3OEt]^-$	295
	or $H^+\begin{bmatrix} BF_3O-C_2H_4 \\ \quad\quad\quad\searrow \\ \quad\quad\quad\quad Et \end{bmatrix}^-$	224
$BF_3 \cdot C_4H_8O_2$		223
$BF_3 \cdot 2C_4H_8O_2$		223
$BF_3 \cdot CH_3COOMe$		221,368
$BF_3 \cdot CH_3COOEt$	$H^+[BF_3 \cdot MeCO_2C_2H_4]^-$	
	or $[MeCO]^+[BF_3OEt]^-$	
$BF_3 \cdot CH_3COOPr^n$		
$BF_3 \cdot CH_3COOBu^n$		
$BF_3 \cdot (Bu^nO)_3B \cdot B(OH)_3$	$[(BuO)_3BH]^+[BF_3OBu]^-$	342
	or $[(BuO)_3B \cdot Bu^n]^+[BF_3OH]^-$	

boron trifluoride catalyst does cast considerable doubt on previous
ideas.   It is found that the maximum catalysis occurs when the
ratio $BF_3 : H_2O$ is $2 : 1$.   There is no correlation with the concentration of $BF_3 \cdot H_2O$, the entity previously suggested as the true
catalyst (174).   Since $2BF_3 \cdot H_2O$ is not a known adduct an adequate
explanation is difficult; it has been tentatively suggested that BF_3
may decompose an addition compound between butene and
$BF_3 \cdot H_2O$.

Some of the more important catalytically active individual boron trifluoride complexes are:

Boron trifluoride–diethyl ether, $BF_3 \cdot O(C_2H_5)_2$. It is a colorless to light straw colored liquid. B.p. 125°, f.p. lower than $-60°$. It darkens on standing and fumes on exposure to air.

Boron trifluoride etherate, with approximately 58% boron trifluoride content, being a liquid at room temperature represents a convenient way of using boron trifluoride as a catalyst in the liquid phase.

In general, the uses of boron trifluoride etherate are similar to those of boron trifluoride gas. However, in some cases it may perform while the gas would not and *vice versa*. Some uses of the etherate are in alkylations, polymerizations, cyclizations, acylations, and esterifications.

Generally boron trifluoride etherate is a weaker catalyst than BF_3 itself. This is obviously due to the fact, as in the case of all boron trifluoride coordination complex catalysts, that the equilibrium

$$BF_3\text{–base} \rightleftharpoons BF_3 + \text{base}$$

plays an important role in the catalyst activity. In order that BF_3 may effect catalytic activity in reactions involving alcohols, olefins, halides, etc., it must first be dissociated from the ether complex to be able to coordinate with the reagent (or co-catalyst). Of course it is not necessary that pre-dissociation from the ether complex should be complete before coordination with the reagent could start. An intermediate complex

$$\text{base} \ldots BF_3 \ldots \text{reagent}$$

can explain most of the reactivity. The donor–acceptor bond strengths in the ether–BF_3 and reagent–BF_3 complexes will determine the reactivity. Weak donor reagents are unable to pull BF_3 away from the strong ether coordination bonds, thus $BF_3 \cdot Et_2O$, will be a useless or weak catalyst in these systems. Other reagents or co-catalyst (such as H_2O) are however sufficiently strong "bases" and accordingly catalytic activity is found.

Boron trifluoride–dimethyl ether, $BF_3 \cdot O(CH_3)_2$, b.p. 126°, and *boron trifluoride-di-n-butyl ether*, $BF_3 \cdot O(n\text{-}C_4H_9)_2$, b.p. 69°/4 mm, are similar in their properties to the diethyl ether complex. *Boron trifluoride–anisole* complex is also frequently used.

Alcohols and phenols form stable alcoholates and dialcoholates (phenolates) of the formula $BF_3 \cdot ROH$ and $BF_3 \cdot 2ROH$, with boron trifluoride. The *boron trifluoride–phenol* complex, $BF_3 \cdot 2C_6H_5OH$, and, similarly, the *boron trifluoride–cresol* complexes are liquids at room temperature. When heated slowly BF_3 is released, starting at about 50°. Practically no BF_3 remains at the boiling point of phenol (cresols).

The $BF_3 \cdot 2C_6H_5OH$ complex contains about 25% BF_3 and represents a convenient means of introducing boron trifluoride into reactions in which boron trifluoride is a catalyst and phenol is one of the reactants. It is used in the preparation of some synthetic resins and the alkylation of phenol.

Boron trifluoride phenolate conceivably also can be used in other cases where the phenol of the complex would be desirable, or at least not objectionable, such as in the polymerization of unsaturated acids, esters, and terpenes.

The behavior of alcohols, phenols, and acids toward boron trifluoride is analogous to that of water, *i.e.*, they yield strong fluoroboric acid derivatives. Thus the compound types with empirical formula $BF_3 \cdot 2ROH$ and $BF_3 \cdot ROH$ in the ionized form are derivatives of $H^+BF_3OH^-$ and may be formulated as $H^+BF_3OR^-$ and $H \cdot ROH^+BF_3OR^-$ respectively. $BF_3 \cdot 2CH_3OH$ possesses a high electrical conductivity.

In the preparation of synthetic resins and in some acylation and polymerization systems useful catalysts are the addition compounds of boron trifluoride and organic acids.

Boron trifluoride–diacetic acid, $BF_3 \cdot 2CH_3COOH$ is a colorless liquid, b.p. 50°/2 mm. Similar 1 : 2 complexes with haloacetic acids (chloroacetic, trichloroacetic, trifluoroacetic) are also frequently used.

Boron trifluoride is also used with "promoters" such as phosphorus pentoxide, sulfuric acid, phosphoric acid, and benzenesulfonic acid in the alkylation of aromatics with alcohols, in alkylation with olefins or in isomerizations. Fluorosulfonic acid is sometimes also an effective promoter. In reactions where water is eliminated strong dehydrating agents, of course, help the reaction. As BF_3 itself is a good dehydrating agent, they may be used to keep the amount of BF_3 needed as low as possible. By complexing with BF_3 they form strong conjugate acids, which in organic media may be considerably stronger than H_2SO_4, HF, H_3PO_4, etc., themselves. This should be considered as another explanation of their increased catalytic effect.

Boron trifluoride forms more stable coordination complexes with nitrogen donors than with oxygen donors. These complexes are generally too stable to be of use as Friedel-Crafts catalyst in alkylation and acylation reactions (where oxygen or halogen donor reagents are not strong enough to compete with the nitrogen donor complexing agents). Complexes like *boron trifluoride–urea, boron trifluoride–triethanolamine, boron trifluoride–piperidine,* and *boron trifluoride–monoethylamine* are used as latent catalysts in certain polymerization reactions.

C. Boron Trichloride

a. Structure and properties

Boron trichloride has a planar, monomeric structure similar to that of boron trifluoride.

The crystal structures of *boron trichloride* and *boron tribromide* may be best described as hexagonal close-packed layers of halogen atoms with boron atoms inserted in appropriate interstitial positions, to produce the trihalide molecules (15,514).

There is no indication in the crystal structure of strong association tendencies, nor are their entropies of vaporization unusual. Although the crystal structure of boron trifluoride is not yet known, there is definite indication of strong association in the liquid (and presumably in the solid state also) from the high value (25 e.u.) of the entropy of vaporization. The infra-red spectrum of crystalline boron trifluoride also points to some degree of association not found in the case of boron trichloride or boron tribromide.

Boron trichloride, as well as the tribromide and triodide, is vigorously hydrolyzed by water to boric acid. However, boron trifluoride is quite soluble in water and like silicon tetrafluoride is only partially hydrolyzed, forming fluoroboric acid.

The inability of boron trichloride (as well as the bromide and iodide) to form stable hydroxyhaloboric acids or to form haloboric acids (by coordination with hydrogen halides) explains the inactivity of these catalysts in olefin alkylations and their weak activity in polymerizations.

The molecular weights of the boron trihalides are normal in the vapor phase and also, in the investigated cases of boron chloride, bromide, and iodide (by the freezing-point method), in benzene solutions.

b. Catalytic activity

Boron trichloride has been reported to polymerize styrene, yielding polymer of three times the molecular weight produced with other cationic catalysts. It is also used in copolymerizations (31,583,675).

It is a Friedel-Crafts type alkylation catalyst in the preparation of halogenated silanes (24), and a catalyst in alkylations of hydrocarbons (363).

As recently established, boron trichloride is a very active general Friedel-Crafts alkylation and acylation catalyst in reactions with organic fluoride reagents (439). BCl_3 is also used as a depolymerization catalyst (30) and as a catalyst in certain rearrangements (209).

D. Boron Tribromide and Triiodide

a. Structure

From the similarity of the pure nuclear quadrupole spectrum of iodine in boron triiodide to that of chlorine in boron trichloride and that of bromine in bromine tribromide, Laurita and Koski inferred that the structure of boron triiodide is similar to that of the other known boron trihalides (344a).

Recently the crystal structure of boron triiodide has been determined by the powder method (511a). The structure is hexagonal with two molecules per cell and the B–I distance of 2.10 Å.

Nuclear quadrupole data can be used to obtain information on π-bonding in boron trihalides. This method gives 6, 12, and 16% as the halogen double bond character in BCl_3 (113a), BBr_3 (113a), and BI_3 (344a).

When the boron–halogen π-bond character increases from 0 to 33%, the hybridization changes from trigonal to tetragonal and accordingly the boron covalent radius increases from 0.79 to 0.89 Å.

b. Catalytic activity

Boron tribromide has been reported as a polymerization catalyst (684). It was found to be a very active alkylation and acylation catalyst, together with *boron triiodide* in Friedel-Crafts reactions involving organic fluorides (439). However, the tendency to form coordination compounds is somewhat decreased in BI_3, probably owing to steric factors. The stability of BI_4^- in the recently isolated pyridinium tetraiodoborate, $C_5H_5NH^+BI_4^-$, proves, however, that steric hindrance does not prohibit formation of sp^3 hybridized tetrahedral complexes (663).

Although boron trifluoride was considered to be the most powerful acceptor molecule known (56) and consequently the strongest boron halide catalyst, the acceptor power of boron trichloride, boron tribromide, and boron triiodide in reality surpasses that of the fluoride (81,226). The observed acceptor power of boron trihalides decreases in the order $BI_3 > BBr_3 > BCl_3 > BF_3$. The reason for this order must be that the back donation from the unshared electron pairs of the halogens to the boron, which tend to decrease the acceptor power of the coordinatively unsaturated boron atom, is larger in the case of fluorine than from the considerably more shielded chlorine, bromine, or iodine atoms.   However, as the boron trihalides show their acceptor properties always against specific donor molecules, the nature of these latter influence the relative acceptor (and catalyst) strengths.   Consequently no general sequence of catalyst activity can be established.   It must be pointed out that in all of their catalytic actions Lewis acid halides, as boron trihalides, always act in a donor–acceptor system, thus we cannot speak about the acceptor strength (*e.g.*, acid strength) without stating what the specific reference base (*e.g.*, donor reagent) is.   It was shown that with proper organic reagent "bases," such as organic fluorides (439), the catalytic activity in Friedel-Crafts reactions parallels the acceptor strengths, being

$$BI_3 \gtrsim BBr_3 > BCl_3 > BF_3$$

For a more detailed discussion of this problem see Chapter VI.

Aluminum Halides

Aluminum is between those elements forming wholly ionic or wholly covalent compounds.   It will form compounds of one or the other type according to the conditions, particularly the deformability of the atoms to which it is attached.   Of the halides, the fluoride is the most readily ionized, and the other halides increasingly tend to go into the covalent form.

E. Aluminum Trifluoride

Anhydrous *aluminum fluoride* is a non-volatile, white crystalline compound with a density of 3.10.   It is practically insoluble in all known solvents.   X-ray analysis has shown that in aluminum fluoride (as well as in its fluoroaluminate complexes) aluminum always has a coordination number of 6.   In aluminum fluoride each fluorine atom is shared between two AlF_6 octahedra.   The structure is that of a continuous ionic network.   The salt character of the fluoride is obvious from the high sublimation point (1291°) and from its insolubility in non-dissociating solvents.   Mercuric fluoride and to a lesser degree stannic fluoride have the same ionic structure. Aluminum fluoride in all states is practically ionic, which easily explains its general inactivity as a Lewis acid catalyst.

Very recently Porter and Zeller (488) found that the vapor of aluminum trifluoride contains some dimer $(AlF_3)_2$ and the heat of dimerization

$$2AlF_3 \rightleftharpoons Al_2F_6$$

was determined (-48 kcal./mole dimer).

Benzene and anisole were not acylated in the presence of aluminum fluoride (103). AlF_3 does not polymerize cyclopentadiene (604).

A slight catalytic activity is sometimes observed in reactions at higher temperatures, as in acetylation of benzene with acetyl fluoride at 240° and t-butylation with t-butyl fluoride at 200°. This probably arises not from AlF_3 used as a "catalyst" in these systems, but because substitution reactions occur under identical conditions in the *absence* of aluminum fluoride, or any other catalyst (439).

Probably at the higher temperatures used either a homolytic reaction takes place or even more plausibly HF is eliminated in the reaction systems which acts as a cationic, Friedel-Crafts catalyst. In any case, the Lewis acid character of aluminum fluoride is considerably less than that of the other aluminum halides, mostly because the possibilities for Lewis acid sites exist only on the surface of the polymeric aluminum fluoride particles, where the electron deficiency of the aluminum atom is least satisfied by the surrounding fluorine atoms of the continuing crystal lattice. Since AlF_3 is not the strong Lewis acid required for most uses, the application of this halide to hydrocarbon catalysis has been relatively limited. Furthermore, in contrast to the other halides of aluminum, which must be essentially anhydrous in order to be catalytically active, strong evidence exists in the patent literature that only aluminum fluoride in the hydrated form, particularly the hemihydrate form, is catalytically active for hydrocarbon conversion processes, such as polymerization, hydration, reforming, isomerization, cracking, or the refining of hydrocarbon oils.

No systematic study of catalysis by AlF_3 has been presented outside the patent literature, although the surface aluminum atoms of AlF_3 could be Lewis acid sites. Kaiser, Moore, and Odioso (302a) found that an active and selective catalyst for the skeletal isomerization of olefins may be prepared by the addition of certain metals or metal oxides to anhydrous aluminum fluoride. This catalyst system can give equilibrium distributions of C_5 or C_6 olefins at 510°C, atmospheric pressure, with a relatively insignificant amount of undesired side reactions. It must be noted that the catalytic activity of aluminum fluoride is that of a solid acid with a limited amount of active acid sites on the surface. Consequently the necessary reaction conditions (high temperatures) correspond to those used with acidic oxides and differ basically from those of the strong Lewis acids, $AlCl_3$ or $AlBr_3$.

F. Aluminum Trichloride

a. Properties and structure

Anhydrous *aluminum chloride* is a colorless crystalline compound with a density of 2.44.

Commercial aluminum chloride may vary in color from very light yellow or white to green and gray (mostly due to impurities, such as ferric chloride). It forms deliquescent crystals having a strong hydrogen chloride odor. It reacts vigorously with water and because of its affinity for water it must be kept dry.

It is perceptibly volatile at ordinary temperature and sublimes at 183°. The melting point (192.6°) can be reached only under pressure.

Although conventionally we think of aluminum chloride as $AlCl_3$, it is in fact the dimer Al_2Cl_6 at ordinary temperatures. Up to a temperature of 440° the dimer Al_2Cl_6 prevails whether in the liquid

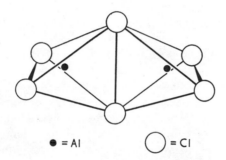

● = Al   ◯ = Cl

Fig. 1.   Structure of Al_2Cl_6 molecule.

or gas phase. Between 440° and 800° it is an equilibrium mixture of monomer and dimer. Between 800° and 1000° only the monomer exists, and above 1000° some ionic dissociation takes place.*

The dimeric molecule, Al_2Cl_6, has been found from electron-diffraction measurements on gaseous aluminum chloride (457) to

* Although anhydrous aluminum chloride, in its uncomplexed form, is thus dimeric Al_2Cl_6 at the generally used temperatures of the Friedel-Crafts type reactions in the liquid (or gas) phase, the catalytically active species—as will be discussed in more detail—is always the monomeric Lewis acid $AlCl_3$. Therefore it is customary to use in connection with discussion of Friedel-Crafts reactions the monomeric formula, $AlCl_3$, for the catalyst. The same practice is also used with similar other aluminum, ferric, gallium, etc., halides.

have the "bridged" structure, shown in Fig. 1, in which two tetrahedral aluminum atoms are bridged by chlorine atoms.   Below 400°
where only Al_2Cl_6 molecules are present, it has been shown that two
deformed $AlCl_4$ tetrahedra are joined along an edge.   Discussion of
the Raman spectra of aluminum halides in the liquid and crystalline
state had led to the same conclusions (207,208,322).

The double molecules have the symmetry D_{2h}.   The observation
is noteworthy that Al_2Br_6 and Al_2I_6 each respectively possesses the
same Raman spectrum in the liquid and crystalline state, while
with aluminum chloride considerable shifts and splitting of the
frequencies occur on passing from the liquid to the crystalline state.
X-ray analysis of solid aluminum chloride by Ketelaar (314) has
shown that no double molecules are present in its lattice.   The
structure is that of layers similar to the monoclinic deformed form
of the chromium trichloride lattice, which is like that of an in-
volatile compound with an apparently strongly polar bonding
character.   That aluminum chloride and chromic chloride possess
related structures was already put forward by Nowacki (429).

Previously suggested structures in which crystalline aluminum
chloride was assumed to be present as a pure ionic lattice has thereby
been superseded to a certain degree.   Biltz (41) postulated that
crystalline aluminum chloride is made up of ions and the liquid of
double molecules.   The ions in the lattice must already be deformed
to such a degree that a transformation to the less polar bonding of the
dimeric Al_2Cl_6 molecule does not require the expenditure of a large
amount of energy.   Consequently the lattice, as shown by the low
sublimation point of 183°, is decidedly labile.   That the interaction
between Al and Cl atoms or the corresponding deformed ions in
the crystal must be quite different from that in the melt, which can
be obtained under pressure, follows from the unusually large change
in volume on melting.   Aluminum bromide and iodide do not
exhibit such increases in volume nor, in contrast to the chloride, do
they possess any appreciable conductivity in the crystalline state
(41).   For the crystalline bromide a molecular lattice with Al_2Br_6
molecules has been ascertained by X-ray analysis (509).

Thus the heavier aluminum halides have the same type of bridged
structure as aluminum chloride, but in the solid state.   The chloride,
bromide, and iodide are strong Lewis acids like the halides of boron.
The melting and boiling points of the aluminum trihalides are shown
in Table VII.   These data also point to the fact that with the exception
of the fluoride the other aluminum halides are increasingly covalent
in character.   Indeed the chloride, bromide, and iodide of aluminum.

at least in the liquid state, are covalent and this is confirmed by the fact that they are extremely poor conductors of electric current at temperatures just above their melting point.

The covalent structure of aluminum halides is supported by the Raman spectra of the liquids (208,581). The bromide and iodide are clearly covalent in the solid state as well. But it is remarkable

TABLE VII.   Melting and boiling points of aluminum trihalides

	M.p °C	B.p C°
AlF_3	1290	sublimes at 1291
$AlCl_3$	192.6 (at 1700 mm.)	sublimes at 180
$AlBr_3$	97.5	255
AlI_3	179.5	381

that solid aluminum chloride is fairly ionized. This is established by the abnormally small molecular volume, while the expansion on heating and the heat of fusion are abnormally high (42).

The ionic nature of solid aluminum chloride is supported by conductivity measurements (40,41). The specific conductivity of solid aluminum chloride, which is zero at room temperature, increases very steeply just below the melting point, reaching about $5 \times 10^6 \ \Omega^{-1}$ cm.$^{-1}$, and then as it melts drops almost to zero, rising slowly to about $1 \times 10^6 \ \Omega^{-1}$ cm.$^{-1}$ at some 50° higher. The bromide and iodide show no such anomaly. Their conductivity in the solid state is zero, and in the liquid just above the melting point is of the same order of magnitude as that of the liquid aluminum chloride.

The bridged dimeric structure of Al_2Cl_6 was used by the adherents of constant valency long before the structure was proved by physical methods. They formulated these compounds in accordance with

Table VIII.   State of aluminum trihalides in liquid and solid phases

	AlF_3	$AlCl_3$	$AlBr_3$	AlI_3
Solid	ionic	ionic	covalent	covalent
Liquid	ionic	covalent	covalent	covalent

observations on the abnormal, twofold molecular weights determined from the vapor pressures measured up to 500°, as dimers with tetravalent metal atoms. For example, in all the work of Friedel and Crafts, aluminum chloride was consistently written as Al_2Cl_6.

To summarize, Table VIII shows the ionic or covalent state of the aluminum halides in the solid and liquid phases.

Aluminum chloride is generally insoluble, or only slightly soluble, in hydrocarbons; it is soluble in most alkyl and acyl halides (however, with complex formation and frequently direct reaction). In addition, aluminum chloride forms complexes and is soluble in phosgene, carbon disulfide, sulfur chlorides, sulfur dioxide, hydrogen cyanide, and other inorganic materials, as well as in many organic compounds.

The molecular weight in ethereal solution has been found to correspond to the monomeric formula $AlCl_3$. This is readily understood because in its coordination compounds aluminum shows a coordination number of 4, thus satisfying the electron octet demand of the aluminum. Coordination complexes of this type, $M^I(AlCl_4)$ as well as $M^I(AlBr_4)$ play an important part in Friedel-Crafts addition compounds of aluminum chloride and bromide. Complex iodides of the type $M^I(AlI_4)$ are also known with acyl or alkyl halides.

Monomeric $AlCl_3$ exerts a very strong polarizing effect. Its dipole moment is about 5.3 D. This accounts for its strong tendency to form addition compounds and also underlies its catalytic action, since it strongly modifies the binding forces within the molecule with which it forms addition compounds.

Pure, molten aluminum chloride, as mentioned previously, is a very poor conductor of the electric current. According to Biltz (41) solid crystalline aluminum chloride is a better conductor. As the temperature is raised to its melting point, its conductivity instantly vanishes. This is very striking, since as a rule molten salts are better conductors than the crystalline material. Biltz has correctly assumed, therefore, that solid aluminum chloride has a different structure from the liquid in that the former is composed of ions, the latter of molecules. This would also account for abnormally great increase in volume (almost twofold), which takes place when aluminum chloride is melted.

If it is necessary to obtain very powerful reaction conditions and no suitable solvent can be found, it is possible to carry out the reactions in a melt of aluminum chloride with alkali metal chlorides.

b. Catalytic activity

Aluminum chloride is the most widely used catalyst in all reactions of the Friedel-Crafts type. Its use as a catalyst was summarized previously in the books of Thomas (623) and Kränzlein (332) and is amply discussed in the present book in individual chapters. Therefore no detailed discussion of catalytic activity in individual reactions is necessary at this point.

All forms of aluminum chloride are not adapted for use as a catalyst. The method of preparation and its physical state has an effect on the nature of its action. It was found (53) that with aluminum chloride prepared from hydrogen chloride and aluminum the reaction of benzene and acetylene proceeded slowly, but with that prepared from aluminum and chlorine the reaction was rapid. Freshly prepared aluminum chloride (126) was found to be too active for the same reaction and any degree of activity could be obtained by aging for various periods of time. More liquid products were obtained when freshly prepared aluminum chloride was used in the polymerization of ethylene (287,506).

Impurities markedly alter both the yield and direction of the reaction catalyzed by aluminum chloride. The impurity which is most often present in aluminum chloride is ferric chloride. As an example, in proportion to its amount ferric chloride decreases the yield of product in reactions of benzene with acyl chlorides and with an excess of carbon tetrachloride (510,511).

Small amounts of impurities, however, frequently are beneficial and promote the action of aluminum chloride.

Differences in activity of aluminum chlorides may be due to the physical state of individual samples and impurities other than to the way of preparation. Storage time may have effect if moisture or other impurities can get to the dry salt.

The reactions of hydrocarbons in the presence of aluminum chloride are usually characterized by the formation of colored reaction mixtures (5). The color usually appears when aluminum chloride comes in contact with the hydrocarbons. Color is a characteristic property of certain complex compounds and since aluminum chloride readily forms complexes there is most probably a close relationship. The nature of various complexes formed between aluminum chloride and related Lewis acid catalysts and hydrocarbons, as well as a variety of reagents and reaction products is to be treated in Chapter IV.

Friedel-Crafts reactions are generally not induced by aluminum chloride hydrates. Its water of crystallization is held strongly enough to shield any electron attraction of the aluminum, which in the hydrates is coordinatively saturated and no longer acting as a Lewis acid. However, recently a number of Friedel-Crafts type alkylations were observed with aluminum chloride hydrates, even in aqueous media. It must be assumed that catalytic activity in these cases is due to Brønsted and not Lewis acidity of the catalyst.

c. Soluble aluminum chloride catalysts

A number of soluble aluminum chloride–solvent complexes are frequently used as Friedel-Crafts catalysts in different reactions. It must be realized that many of these addition complexes are distinct chemical compounds with specific properties, and in them the presence of aluminum chloride in a coordinated form effects a quite different, generally more selective catalyst activity than uncomplexed aluminum chloride.

Certain properties of anhydrous aluminum chloride decrease its usefulness as a catalyst for hydrocarbon conversions. It is often excessively active initially and catalyzes undesirable side reactions such as cracking or auto-destructive alkylations. It forms addition compounds (lower-layer sludge) with aromatic and olefinic hydrocarbons (present as reactants or formed during the reaction) which usually result in a decrease in the activity and life of the catalyst and also in excessive isomerization and disproportionations. Frequently it is advantageous both in industrial processes and in laboratory syntheses to use a catalyst which has uniform activity and a constant physical state. Fluid aluminum chloride catalyst or soluble aluminum chloride solvent complexes becomes a logical choice because of loss of the original crystalline form apparently inherent in all but a few reactions. Furthermore the liquid catalyst permits more efficient utilization of the aluminum chloride. There is no loss in activity because of the coating of the catalyst particles with sludge. Most liquid aluminum chloride catalysts described in the literature consist of lower-layer complexes, formed in the reactions and added uncombined "make-up" aluminum chloride. Mixtures of aluminum chloride and certain metal chlorides, *e.g.*, antimony trichloride, in the liquid state are also active catalysts. Nitrobenzene has been used as solvent for aluminum chloride in a number of cases. Other organic compounds such as chloroalkanes (methyl chloride and tetrachloroethane) and carbon disulfide have been used but have the disadvantage that they dissolve only a small amount of the metal halide. Thus the use of large proportion of solvent is necessary. On the other hand aluminum chloride is highly soluble in ethers, ketones, and alcohols but the resulting solutions are catalytically inactive at least for the alkylation of hydrocarbons. Nitroparaffins have high solvent power for anhydrous aluminum chloride. Solutions containing more than 50% by weight of the metal halide are obtainable. In the investigations of Schmerling (536) it was discovered that unlike the ether, ketone, and alcohol solutions, the nitroparaffin solutions of aluminum

chloride are excellent catalysts, particularly for alkylation reactions.

d. Aluminum chloride "sludges"

In order to develop processes for carrying out aluminum chloride
catalyzed Friedel-Crafts reactions on a commercial scale in continuous flow operations, liquid aluminum chloride sludges consisting
of aluminum chloride dissolved or dispersed in complexes of aluminum chloride with hydrocarbons were found to be particularly
suitable.   These sludges are mainly used in alkylation reactions.
The sludges are insoluble in the alkylate, settle rapidly and may be
separated from it very readily and recycled.   They are prepared by
a number of methods.   They may be formed in situ by the alkylation reaction, or be produced by the action of aluminum chloride on
olefins, or such substances as iso-octane, kerosene or similar hydrocarbon fractions (267,626,627).

An important advantage of the sludge catalysts is that they can
be maintained at a constant uniform level of activity by the addition
of small amounts of fresh catalyst.   When preformed sludge, such
as that formed by the action of aluminum chloride on alkylate
bottoms, is used, the make-up catalyst may be pumped into the reaction mixture, with an equivalent amount of spent catalyst being
removed at the same time.   When the sludge is formed in situ,
fresh aluminum chloride may be added as such at spaced time inter-
vals or continuously as a solution in the paraffin or aromatic hydrocarbons.   Generally fewer side reactions occur with sludge catalysts
than with unmodified aluminum chloride.

e. Aluminum chloriden–monomethanolate

The addition complex of equimolar amounts of aluminum chloride
and methanol is a crystalline compound which is stable at temperatures below 70°.   At higher temperatures it decomposes,
forming hydrogen chloride and methoxyaluminum dichloride,
further heating of which yields methyl chloride and oxyaluminum
chloride (427).   It very readily catalyzes the alkylation of benzene
in which it is soluble and of isobutane, in which it is not (534).   On
the other hand, the product of the reaction of aluminum chloride
with two molecular proportions of methanol (i.e., $AlCl_3 \cdot 2CH_3OH$)
is not an alkylation catalyst.   The alkylation reactions of aluminum
chloride monomethanolate are generally promoted by hydrogen
chloride.

f. Aluminum chloride–nitroalkane complexes

Aluminum chloride dissolves in as little as its own weight of the readily available nitroalkanes (nitromethane, nitroethane, and the two nitropropanes). The resulting solutions, particularly those employing nitromethane, are catalysts for alkylation reactions, hydrogen chloride serving as promoter (288,536). In contrast, solutions of aluminum chloride in a molecular excess of ethers, ketones, and alcohols are catalytically inactive for the alkylation of hydrocarbons.

The active component of the nitroparaffin solutions of aluminum chloride has shown to be not the metal chloride *per se*, but rather the addition compound $AlCl_3 \cdot RNO_2$. In this connection it is significant to contrast the nitroparaffin solutions with the ether, ketone, and alcohol solutions. All four types of compounds are monomolecular addition complexes (usually crystalline compounds of low melting points) which are catalytically active under specific conditions (536).

In comparison, aluminum chloride monomethanolate (the crystalline, isolated complex) catalyzes the alkylation of isoparaffins. Solutions of aluminum chloride monomethanolate in excess methanol are, however, catalytically inert. Indeed even the solid product of the reaction of aluminum chloride with two molecular proportions of methanol (*i.e.*, $AlCl_3 \cdot 2CH_3OH$) is not an alkylation catalyst. Similarly aluminum chloride monetherates and monoketonates lose their catalytic activity, when dissolved in ethers and ketones respectively. In direct contrast, the aluminum chloride–nitroalkane complexes retain their activity even with a large excess of the nitroparaffin solvent. Alkylations of aromatics and isoparaffins with olefins, as well as alkylations with alkyl chlorides are examples of the effectiveness of aluminum chloride–nitromethane complex ($AlCl_3 \cdot CH_3NO_2$) as catalyst.

When a nitroparaffin solution of aluminum chloride is added to benzene a single clear solution is obtained. The nitroparaffin may be considered to be a solubilizer for aluminum chloride in aromatic hydrocarbon. It makes possible the solution of almost any desired quantity of aluminum chloride in benzene. Pure benzene will dissolve only 0.72% by weight of aluminum chloride at 80°. Benzene containing a minor amount of nitroparaffin can at room temperature dissolve aluminum chloride in an amount at least equal to the weight of the nitroparaffin. The solid complex $AlCl_3 \cdot CH_3NO_2$ remaining after the excess nitromethane is evaporated from the solution of aluminum chloride in nitromethane is soluble in benzene and the solution thus obtained contains active catalyst. The complex contains more than twice as much aluminum chloride by weight as nitromethane. The modified nitroparaffin–aluminum chloride complexes were used by Schmerling in alkylation of aromatics with olefins and alkyl chlorides (536).

g. Aluminum chloride–nitrobenzene

The $AlCl_3$–nitrobenzene complex in nitrobenzene solution is an effective and useful catalyst (656).

h. Aluminum chloride–tertiary amine complexes

Aluminum chloride forms complexes of the type $R_3N \cdot AlCl_3$ (163) with tertiary amines. Drahowzal (155,156) found that complexes of aromatic or heterocyclic tertiary amines with aluminum chloride, like $C_6H_5N(CH_3)_2 \cdot AlCl_3$ or pyridine–$AlCl_3$ are useful liquid catalysts for the alkylation of aromatics with alkyl halides.

The tertiary amine–aluminum chloride complexes are liquids with relatively low freezing points which easily dissolve excess aluminum chloride (which is needed to provide an active catalyst system) and are themselves soluble in aromatics. It seems that the amine complexes the aluminum chloride firmly, thus rendering the chloride inactive as a Friedel-Crafts catalyst. However, the amine complex is able to dissolve excess of uncomplexed aluminum chloride, which is then the effective catalyst.

i. Effect of complexing on catalytic activity

All the liquid aluminum chloride complexes discussed are basically complexes of aluminum chloride with a suitable donor solvent. Besides allowing homogeneous reaction conditions, these complexes also act as promoters of the Friedel-Crafts reactions and are more easily handled. As pointed out by Walker (664) the effective form of aluminum halides in Friedel-Crafts reactions is the monomer. (Dimer formation itself is a proof of the strong coordinating power of the electron-deficient aluminum, which is satisfied in the halogen bridged dimers through donations by the unshared halogen pairs.) The heat of dissociation of aluminum chloride dimer is 29.0 $\pm$ 1 kcal./mole (189).

$$Al_2Cl_6 \rightleftharpoons 2AlCl_3$$

If a suitable complexing solvent is used, with which aluminum chloride forms a 1 : 1 monomeric complex (as is the case with all the previously discussed soluble complexes) the bond strength of which is smaller than that of the dimer, then the monomeric aluminum chloride will be more easily available for interaction with a reagent (organic halide, anhydride, etc.) in Friedel-Crafts reaction than if the dimeric aluminum chloride itself were used. The question of the co-catalytic effect of certain solvents in Friedel-Crafts reactions will be discussed in more detail in the case of aluminum bromide (p. 264).

It should be pointed out that the heat of dissociation of the aluminum halide dimers decreases from the fluoride to the iodide as shown in Table IX. Thus the observed order of catalyst reactivity

TABLE IX.   Heat of dissociation of aluminum trihalide dimers

	kcal./mole
Al_2F_6	48.0 ± 1
Al_2Cl_6	29.0 ± 1
Al_2Br_6	26.5 ± 0.5
Al_2I_6	22.5 ± 1

of aluminum halides could be attributed to the decreasing energy requirements for obtaining the active monomeric AlX_3 Lewis acid

$$AlI_3 \geqq AlBr_3 > AlCl_3 \gg AlF_3$$

in addition to other factors (smaller back conjugation from the halogens to aluminum, better solubility, etc.).

j. Double salts with alkali metal chlorides

Aluminum chloride forms double salts of relatively low melting points (e.g., 140–300°) with alkali metal chlorides. Such addition compounds containing equimolar quantities of sodium chloride, lithium chloride, ammonium chloride, or potassium chloride are used in alkylation and other Friedel-Crafts reactions (49).

The sodium chloride–aluminum chloride double salt is commonly used for reasons of economy and its low melting point. The compound $AlCl_3 \cdot NaCl$ is fluid at 141° (313), and is convenient for some reactions (542).

Sodium aluminum chloride was used by Friedel and Crafts during their very first investigations.

Stoichiometric quantities of sodium chloride and aluminum chloride are melted together and the reaction products are then introduced into the melt at temperature between 140° and 250°.

The activity of the double salts in alkylation reactions decreases in the order NaCl > LiCl > NH$_4$Cl > KCl, potassium chloride compounds being quite inactive. However, it has been reported to give some polymers in reactions with olefins above 300°. Melts of the aluminum chloride–alkali metal chloride double salts are used when vigorous Friedel-Crafts reaction conditions are needed, since under these conditions the reaction can be carried out in homogeneous media at relatively high temperatures.

G. Aluminum Tribromide

a. Properties

Aluminum bromide forms colorless, brilliant leaflets, with a density of $d = 3.20$, melting point 97.5°, boiling point 265°. It decomposes with effervescence when brought in contact with a little water and burns with flame in oxygen. With excess water it forms the hexahydrate AlBr$_3$·6H$_2$O similarly to aluminum chloride. It easily forms complexes of the type M^{I}AlBr$_4$. As pointed out previously it is dimeric in all states and has a bridged halogen structure, similar to that of the aluminum chloride dimer.

b. Catalytic activity

Aluminum bromide is generally considered a more reactive catalyst than aluminum chloride. This is partly due to its better solubility. Whereas aluminum chloride is substantially insoluble in hydrocarbons, aluminum bromide (which even in the solid state consists of dimeric covalent molecules instead of the ionic nature of solid aluminum chloride) is quite soluble in benzene, toluene nitrobenzene, ethylene bromide, CS$_2$, and simple hydrocarbons (241).

The greatly improved solubility accounts for the fact that in physico-chemical studies of Friedel-Crafts reactions aluminum bromide is preferentially used instead of aluminum chloride. The increased catalytic activity of aluminum bromide over aluminum chloride, however, is also due to the fact that in aluminum bromide the back donation from the unshared pairs of the halogens to aluminum, decreasing the electron deficiency, is smaller than in the case of the chloride, thus aluminum bromide generally is a stronger Lewis acid (provided that steric interactions with the donor base do not interfere to a substantial degree).

Aluminum bromide forms complexes similar to aluminum chloride with a variety of donor molecules. In different solvent systems (CH$_3$Br, (CH$_3$)$_2$O, CH$_3$NO$_2$) aluminum bromide is present as a 1 : 1 complex of the monomer with the donor solvent: CH$_3$Br–AlBr$_3$, CH$_3$NO$_2$–AlBr$_3$, (CH$_3$)$_2$O–AlBr$_3$.

According to the investigations of Walker (664) the formation of monomeric aluminum halide from the dimer plays an important role in the mechanism of Friedel-Crafts reactions. The monomer itself is the catalytically active species that catalyzes certain reactions by bonding with a $C=C$, $C-Hal$, $C=O$, $C-H$ bond of the organic molecule that is reacting, to form an activated complex. This complex then reacts with the substrate to form the end products.

Thus, according to Walker, the rate at which AlX_3 is formed determines the catalyst activity and the overall reaction rate. Normally, aluminum halides are dimers (Al_2X_6) which dissociate into two monomer units. When a co-catalyst is added to an aluminum halide, however, it forms an addition molecule with a monomeric AlX_3 unit. The bond in this addition molecule is weaker than that which binds the two AlX_3 units of the dimer. Collisions with molecules of substrate break the bond and release AlX_3 monomer, which then catalyzes the reaction.

The role of the co-catalyst is thus to boost production of free aluminum halide monomer. The key factor is the bonding strength of the addition molecule formed with the aluminum halide.

TABLE X. Bonding strengths in aluminum bromide complexes

Species	Estimated bond strength (kcal./mole)
$CH_3Br-AlBr_3$	20 ± 1
$(CH_3)_2O-2AlBr_3$	20
2-Pentene-$AlBr_3$	18–19
m-Xylene-$AlBr_3$	20
Mesitylene-$AlBr_3$	22–24
Hexamethylbenzene-$AlBr_3$	25
$(CH_3)_2O-AlBr_3$	30

For example, aluminum bromide is a dimer (Al_2Br_6) in its pure solid, liquid, and gas phases (100 to 300°). Its heat of association is 26.5 kcal./g.mol. But with a co-catalyst such as methyl bromide it forms an addition molecule $(CH_3Br-AlBr_3)$ that has a bonding strength of only about 20 kcal.

If the formation of monomeric $AlBr_3$ determines the rate of reaction, then the reaction with co-catalyst base : $AlBr_3$ species with a bond strength of 20 kcal. should be $e^{(26,500-20,000)/RT}$ ~ 50,000-fold more than with pure Al_2Br_6. Experimental data support this suggestion. Use of favorable co-catalysts can boost catalysis some 50,000 times in a reaction such as paraffin isomerization.

Bonding strength of the addition compounds formed with aluminum bromide determines whether a substance will be a good co-catalyst. Estimated bonding strengths of a number of compounds bonded with $AlBr_3$ are shown in Table X (according to Walker).

Using these data Lewis base–acid systems may be grouped according to Walker into three classes according to bonding strength (Table XI).

TABLE XI.   Classification of co-catalyst bases according to Walker (664)

Class	Effective co-catalysts?	Base–AlX_3 bond strength (kcal.)	Examples of base
1	No	< 15	Paraffins, naphthenes, benzene.
2	Yes	> 15 < 27	Alkyl halides, olefins, many aromatics, and many addition compounds containing oxygen and nitrogen, such as $Me_2O–AlX_3$.
3	No	> 27	Alcohols, ethers, esters, water, and practically any molecule that contains oxygen or nitrogen.

Class 1 bases (less than 15 kcal. bonding strength) are not effective co-catalysts for aluminum halides. They bond so weakly to the halide that no catalytically significant amount of the base–AlX_3 system exists. Any catalysis in these systems is caused by the Al_2X_6 molecule.

Class 3 bases (greater than 27 kcal. bonding strength) are not good co-catalysts either, because they bond too strongly with aluminum halide. Molecular collisions with reagents fail to produce any catalytically significant amount of free monomeric AlX_3 from these stable complexes.

Class 2 bases (between 15 and 27 kcal. bonding strength) are effective co-catalysts. The bond is strong enough that a significant amount of base–AlX_3 complex exists. But, at the same time, the bond is weaker than that which binds the two AlX_3 units in Al_2X_6.

In some systems, the mole ratio of co-catalyst to $AlBr_3$ has an important effect on catalytic activity. For example, in the isomerization of n-hexane with the $(CH_3)_2O–AlBr_3$ system, there is no activity when the $(CH_3)_2O/AlBr_3$ ratio equals 1.0, or when it equals 0. But in between activity rises to a maximum at a $(CH_3)_2O/AlBr_3$

ratio of 0.5. At high concentrations of $(CH_3)_2O$ the species present is primarily $(CH_3)_2O-AlBr_3$. This has a bonding strength of greater than 30 kcal./mol. and hence shows little catalytic activity. At lower mole ratios $(CH_3)_2O/AlBr_3$, relatively more $(CH_3)_2O/2AlBr_3$ exists, with a maximum amount at a mole ratio of 0.5. This system has a lower bonding strength (20 kcal.) and is thus effective in catalyzing the reaction.

Other compounds having oxygen or nitrogen atoms, such as water, acetone, and amines, behave in the same way, but systems such as $CH_3Br-AlBr_3$ do not. Mole ratio here has a relatively minor effect. Even when the $CH_3Br/AlBr_3$ ratio is greater than or equal to nine, it still shows activity. This is in agreement with the fact that only one type of addition compound is formed.

The fact that the active aluminum halide catalyst is the monomeric, coordinatively unsaturated Lewis acid, which must be first formed by the dissociation of the dimeric aluminum halide molecule, was also demonstrated in another way (449). It was found that certain aluminum halide catalyzed Friedel-Crafts reactions, especially those catalyzed by aluminum bromide, are considerably activated by irradiation with ultra-violet light. It was shown that the ultra-violet light has little effect on activating C–halogen bonds participating in the Friedel-Crafts reactions involved. Instead light activates the dissociation of the aluminum halide dimer

$$Al_2Br_6 \rightleftharpoons 2AlBr_3.$$

As would be expected the effect is much stronger in the case of aluminum bromide than with aluminum chloride, the strongest effect being found with aluminum iodide. It was also shown that similar reactions using boron halide catalyst (which are all monomeric) are not effected at all by ultra-violet irradiation.

H. Aluminum Triiodide

Aluminum iodide has a density of 3.95, melting point 191°, boiling point 382° and is very similar to the bromide in its properties. It is dimeric in all states, with a bridged halogen structure identical to other aluminum halide dimers. Aluminum iodide is extremely sensitive to moisture and all operations with it must be carried out under controlled anhydrous conditions (668).

Complex iodides of the type M^IAlI_4 are known. Although Friedel and Crafts in one of their first papers indicated that aluminum iodide may be substituted for aluminum chloride in the Friedel-Crafts reactions, relatively little is known of the comparative activity of the two halides. In a recent comparative work of the effect of aluminum iodide and other aluminum halide Friedel-Crafts catalysts in alkylation and acylation reactions (106,317) aluminum iodide was found to be the most reactive catalyst in the former case. The reactivity sequence being aluminum iodide > aluminum bromide > aluminum chloride. In acylation and other reactions, however, direct comparison was difficult, owing to side reactions (liberation of iodine and formation of iodinated products) taking place with aluminum iodide.

Addition of IF to fluoroolefins was reported to be catalyzed by $AlI_3 + Al$ (253).

5. Group IIIB Halides

Gallium, indium, and thallium all have the ns^2np^1 valence electron configuration met with in boron and aluminum ($Ga:4s^24p^1$, $In:5s^25p^1$, $Tl:6s^26p^1$). They differ from aluminum (and boron) largely because of a decreasing tendency for coordination.

The fluorides are ionic solids of high melting point and low solubility, similar to AlF_3. The other anhydrous trihalides are largely covalent.

A. Gallium Halides

The trihalides of gallium are very similar in their structure to the aluminum halides.

TABLE XII. Melting and boiling points of gallium halides

	M.p. °C	B.p. °C
GaF_3	950 (sublimes)	—
$GaCl_3$	77.9	201.3
$GaBr_3$	121.5	279
GaI_3	212	346

The chloride, bromide, and iodide of gallium in the liquid and vapor states are covalent dimeric molecules. This is confirmed by the fact that these substances are extremely poor conductors of electricity at temperatures just above their melting points. The fluoride, on the other hand, is a highly ionic compound, and therefore has no catalytic activity. The observed conductivities of gallium chloride, bromide, and iodide at temperatures just below the melting point have been interpreted as indicating that the solids are ionic as contrasted with the low conductivities of the molten material which show them to be covalent dimers (228).

Vapor-density determinations have shown that, in the neighborhood of the boiling point, the vapor of gallium chloride consists of double molecules, Ga_2Cl_6. At 600° the vapor density corresponds to that required by the monomeric formula $GaCl_3$ and at higher temperatures the incipient ionic dissociation of the $GaCl_3$ molecules becomes evident.

Electron-diffraction measurements have also shown that both gallium chloride and bromide exist as dimeric molecules in the vapor state. The dimeric halides doubtless have the bridged structure (like aluminum(III) and gold(III) halides) with Ga having a fully shared octet.

The bond distances are: Ga–Cl 2.22 Å, Ga–Br 2.38 Å. *Gallium iodide*, GaI_3 is only slightly associated (Ga–I = 2.49 Å), and in the fused state has a higher conductivity than the chloride or bromide, pointing to the fact that it must contain a small proportion of ions.

The nature of complexes of the gallium halides with organic compounds is similar to that of the aluminum trihalides.

Gallium chloride and bromide are excellent catalysts in alkylation and acylation reactions and are extensively used in kinetic investigations necessitating homogeneous reaction conditions because of their high solubility in organic solvents (97,582,646,647,648).

B. Indium Halides

Indium trifluoride (d = 4.39) is a high-melting, ionic compound, which does not show Lewis acid properties in Friedel-Crafts catalysis.

TABLE XIII.   Melting and boiling points of indium halides

	M.p. °C	B.p. °C	Color
InF_3	1170	1200	colorless
$InCl_3$		498 subl.	
$InBr_3$		371 subl.	colorless, light brown molten
InI_3	210	decomp.	yellowish, dark brown molten

The other indium halides, as shown from comparison of their melting points in Table XIII, are, although higher melting than the corresponding gallium halides, far less ionic in nature.

Indium trichloride, $InCl_3$ ($d = 3.46$ at 25°), sublimes readily.   At 800° the vapor density corresponds to the monomeric formula. However, at lower temperatures in the vapor phase electron diffraction shows that indium trichloride, as well as *indium tribromide* ($d = 4.75$) and *indium triiodide* ($d = 4.68$) are all dimeric, In_2X_6, covalent compounds with a bridged halide structure.   Although it has not been proved conclusively it seems they are also dimeric in the liquid phase (71,606).

Their color darkens as the atomic weight of the halogen increases and is always darker in the liquid phase.   Thus in the solid state indium trichloride and tribromide are colorless and the triiodide is yellow.   In the liquid the trichloride is yellow and the other two are pale brown.

Complex salts (chloroindates) are formed with organic halides.

Indium(III) bromide and indium(III) iodide are very similar to the chloride.   There are two modifications of indium triiodide, a red and a yellow form.

The trihalides of indium have been scantily investigated as potential Friedel-Crafts catalysts.   Indium chloride has been stated to be inactive in alkylation (227,649).   It was recently shown (442) that the trichloride, tribromide, and triodide are effective catalysts in Friedel-Crafts acylations, alkylations, and halogenations although considerably weaker than the corresponding gallium halides.

C. Thallium Halides

Owing to the strong tendency of the thallic compounds to be reduced to the thallous state, the trihalides are relatively unstable and little known.   The ease of reduction of thallium(III) compounds to thallium(I) compounds makes them strong oxidizing agents.

Anhydrous *thallic fluoride,* TlF_3, is a high-melting crystalline solid (m.p. > 500°) with no catalytic activity.   Anhydrous *thallic chloride* in contrast melts at 155°.   It is therefore quite obvious that the high-melting fluoride is ionic, but the low-melting chloride must be covalent.   Thallic bromide and thallic iodide are fairly unstable and lose halogen spontaneously, even at room temperature, to form thallous halides (303).

Thallic chloride resembles BCl_3 in giving stable addition compounds and not forming a dimer.

Thallic chloride, $TlCl_3$ and similarly *thallic bromide,* $TlBr_3$, have a strong tendency to form coordination compounds (chloro- and bromo-thallates) and are catalytically active as Lewis acid type Friedel-Crafts type catalysts.

It was demonstrated (442) that $TlCl_3$ and $TlBr_3$ are effective, but mild catalysts in Friedel-Crafts acylations and alkylations with halides and in ring halogenations of aromatics.

6. Group IVB Halides

Group IV of the Periodic Table (as well as Groups V–VIII) is markedly different from the previous ones. The elements of this group are able to form valency octets directly without the help of coordination. As they can extend their valence octets, however, through available d-orbit participation, the halides are able to show strong Lewis acid catalytic activity.

The atoms of the Group IVB elements all have the ns^2np^2 valence electron configuration and form tetrahedral bonds associated with sp^3 hybridization.

The electronic structure of the *carbon* atom gives the element a remarkable diversity of chemical properties. The four valence electrons of carbon occupy the configuration $2s^22p^2$, but hybridization occurs very freely and accounts for the characteristic valency of four which is never exceeded. With no d-orbitals available carbon cannot extend its valence octet. Carbon tetrahalides therefore cannot show Lewis acid properties.

A. Silicon Halides

Silicon has a valence electron configuration of $3s^23p^2$. It differs from carbon in that $3d$ orbitals are accessible, giving a greater variety and increased number of valencies. Silicon with its vacant

TABLE XIV.   Melting and boiling points of silicon tetrahalides

	M.p. °C	B.p. °C
SiF_4	-90.2	-95.7 (subl.)
$SiCl_4$	-70.4	$+57.0$
$SiBr_4$	$+5.2$	$+154.6$
SiI_4	$+123.8$	$+290.0$

d-orbitals can extend its valence octet. It can thus exhibit a covalency greater than 4, as in the SiF_6^{2-} ion. Therefore, silicon tetrahalides can act as Lewis acids.

A comparison of the melting and boiling points of the silicon tetrahalides seems to be in accordance with a predominantly covalent structure, even for the fluoride (Table XIV). *Silicon tetrafluoride* is a colorless gas, which forms dense fumes in moist air. The molecular weight corresponds to the formula SiF_4. When strongly cooled at atmospheric pressure, silicon tetrafluoride passes directly from the gaseous to the solid state (sublimation temperature $-95°$). It can be liquefied under pressure (critical temperature $-1.5°$, critical pressure 50 atm.). The compound has a high heat of formation (360 kcal./mol.) and is therefore very stable. Silicon tetrafluoride forms addition compounds with many donor substances, especially with fluorides (of the SiF_6^{2-} type), and is a moderately active catalyst in certain Friedel-Crafts reactions (as in acylation and nitration) (441,631).

Silicon tetrachloride, tetrabromide, and *tetraiodide* all seem to be monomeric SiX_4, based on their molecular weights. All are very hygroscopic. Their application as catalysts in Friedel-Crafts reactions, with the exception of $SiCl_4$, has not yet been reported. Silicon tetrachloride was found to be a moderately active catalyst in Friedel-Crafts acylations (415,686).

B. Germanium Halides

Germanium has the electronic configuration $4s^2 4p^2$, tin $5s^2 5p^2$, and lead $6s^2 6p^2$.

Since the atoms have relatively low electronegativities, the bonds in many of their compounds are fairly ionic.

The physical properties of their tetrahalides correspond to those of the covalent halides of carbon and silicon. GeF_4 is a gas similar to SiF_4, however SnF_4 and PbF_4 are both high-melting solids, being more ionic in character.

The germanium tetrahalides, based on comparison of their boiling and melting points (Table XV) and their molecular weights, are

TABLE XV.   Melting and boiling points of germanium tetrahalides

	M.p. °C	B.p. °C
GeF_4	-15 (4 atm.)	-37.9 (subl.)
$GeCl_4$	-49.5	$+86.5$
$GeBr_4$	$+26.1$	$+186.5$
GeI_4	144	above 300 (dec.)

covalent, monomeric compounds. Electron-diffraction studies have shown the covalent, tetrahedral structure of $GeCl_4$, $GeBr_4$, and GeI_4. Although they possess coordinating ability (they form stable octavalent complexes of the MX_6^{2-} type) their catalytic activity in Friedel-Crafts reactions has not been sufficiently investigated but their catalytic activity probably is very small. As covalent halides they show good solubility in many organic halides. $GeCl_4$ and $GeBr_4$ are very weak acylation, alkylation, and halogenation catalysts (70,442,521).

C. Tin Halides

From the Group IVB of the Periodic Table *stannic halides* have shown the most predominant catalytic activity in Friedel-Crafts type reactions. *Stannic fluoride*, SnF_4, forms a colorless mass of radiating crystals, density = 4.75, which sublime at 705°. A comparison of the melting and boiling points of tin tetrahalides points to the fact that the tetrafluoride is a somewhat ionic compound whereas the chloride, bromide, and iodide, according to their physical properties, should be mostly covalent.

TABLE XVI.   Melting and boiling points of stannic halides

	M.p. °C	B.p. °C
SnF_4	—	705 (subl.)
$SnCl_4$	− 36.2	114.1
$SnBr_4$	33	203.3
SnI_4	144.5	346 (extrapol.)

The structure of stannic fluoride is still unknown. The radius ratio Sn^{4+}/F^- of 0.52, within the required limits for sixfold coordination, argues against a three-dimensional ionic lattice, since ionic tetrahalides would require coordination numbers of 8 or more. A molecular lattice is even more improbable, because the density of molecular stannic iodide is less than that of stannic fluoride. Hückel (269) has discussed this question and suggests a chain structure with octahedrally coordinated tin. Either this type of structure or a layer lattice seems probable.

Woolf found SnF_4 to be a fairly active Friedel-Crafts acylation catalyst, between the activity of $SnCl_4$ and $SnBr_4$ in acetylation of toluene (678). However, it does not polymerize isobutylene or styrene alone but only in the presence of acetyl bromide as cocatalyst (434).

The tetrahedral, covalent structures of stannic chloride, bromide, and iodide were confirmed by electron-diffraction studies (250). *Stannic chloride* and to a lesser degree stannic bromide, form coordination compounds with a variety of donor molecules of the type SnX_6^{2-} and are frequently used as Friedel-Crafts catalysts.

It is of some interest that the earliest observation on the catalytic action of a Lewis acid type metal halide was that of Deville who, in 1839, found that stannic chloride polymerized styrene (139).

The mild catalytic activity of stannic chloride has been utilized in the acylation of thiophene (which is too vigorous and leads to decomposition with $AlCl_3$) (215).

Stannic chloride is applicable in Friedel-Crafts acylation and alkylation of active aromatics which do not require vigorous reaction conditions (594) and has been used in Friedel-Crafts reactions of furans (212). It is a valuable polymerizing agent (good solubility in most solvents and easy to remove) (75,426,522). It is widely applied as a catalyst in chloromethylations.

Stannic bromide is a suitable polymerization catalyst for styrene (462).

D. Lead Halides

Lead dihalides, as other lower valency metal halides, generally show no catalytic activity in Friedel-Crafts systems, being also more ionic in character. The fluoride is a high-melting (818°), ionic solid, the chloride, bromide, and iodide based on their considerable lower melting points and increased volatility (Table XVII) are increasingly covalent.

TABLE XVII.   Melting and boiling points of lead dihalides

	M.p. °C	B.p. °C
PbF_2	818	1285
$PbCl_2$	298	954
$PbBr_2$	373	916
PbI_2	412	

Lead tetrahalides are relatively unstable.

Lead tetrachloride is a yellow, mobile liquid, freezing to a yellow crystalline solid at about $-15°$. It explodes at 105° owing to its instability. No investigation on its catalytic activity has been reported.

Lead tetrafluoride is a more stable, but more ionic compound and dissociates into $PbF_2 + F_2$, being a good fluorinating agent, but not suitable as a Friedel-Crafts catalyst.

7. Actinide Halides

Before the properties of neptunium and plutonium became known (1941) actinium, thorium, protoactinium, and uranium had been considered to be the last members of their respective A-subgroups of the Periodic Table. There was some chemical evidence for that view. Uranium forms complexes indicating variable charge and a particularly stable 6+ cation typical of molybdenum and tungsten, while thorium nearly always occurs with a 4+ charge. Thus they seemed to fit into Groups VIA and IVA. With the isolation of neptunium, plutonium, and further transuranic elements, evidence for a different classification accumulated. The similarity to the lanthanides of the elements from actinium onwards became increasingly apparent so that they are now accepted as a closely related family known as *actinides*.

From the point of view of Friedel-Crafts catalytic activity only the halides of *thorium* and *uranium* have any importance. As do all the actinides, they have the valence electronic configuration $7s^2$ and vary in occupation of the $5f$ and $6d$ shells (thorium $6d^2 7s^2$, uranium $5f^3 6d^1 7s^2$).

A. Thorium Halides

Thorium tetrafluoride in the anhydrous state is a white powder, considerably less volatile and more ionic in nature than the chloride and bromide, with no catalytic activity.

Thorium tetrachloride, $ThCl_4$, forms colorless needles, melting point 765°, boiling point 922°. The compound sublimes at about 750° in vacuum. Thorium chloride forms addition compounds similarly to titanium and zirconium chloride of the type $M_2^I ThCl_6$. *Thorium bromide* and *thorium iodide* closely resemble thorium chloride in their behavior. Thorium bromide, however, is less inclined to form addition compounds than is thorium chloride and no addition compounds of thorium iodide are known (probably for steric reasons). *Thorium chloride* has a slight catalytic activity in Friedel-Crafts alkylations which, however, is milder than that of either titanium chloride or zirconium chloride (238). It is a quite active catalyst in acylations with acetyl halides (442).

No investigation on the catalytic activity of thorium bromide or iodide has yet been reported.

B. Uranium Halides

The uranium halides are the most widely studied halides of the actinides, but little is known about their activity as Friedel-Crafts catalysts. The melting points of known uranium halides are shown in Table XVIII.

The absence of pentahalides and hexahalides with bromine and iodine is noteworthy.

Although UF_6 and UCl_6 are the most stable halides (hexavalent uranium is the most stable valency state of uranium and is also the one in which the element most clearly shows its similarity to Group VIB halides, such as molybdenum and tungsten) little is known about their catalytic properties. UF_6, a colorless volatile solid, is the most volatile uranium compound known. As a covalent metal halide it should possess considerable Lewis acidity.

TABLE XVIII.   Melting points of uranium halides

UF_6	56.5 (subl.) 69.5 (under pressure)	UCl_6	179°				
UF_5	dec. below 400°	UCl_5	327° (est.)				
UF_4	> 1100	UCl_4	590	UBr_4	519	UI_4	506
UF_3	1430 (est.)	UCl_3	843	UBr_3	752	UI_3	680 (est.)

Uranium hexafluoride sublimes at 57° which allows the use of this halide in the vapor state for the separation of isotopes of uranium by diffusion. UF_6 melts under pressure of 1134 mm. at 64°.

The density of the solid at the melting point is about 4.8 and that of the liquid 3.62, the increase in volume upon liquefaction being unusually great.   The crystals of the compound are made up of UF_6 molecules in which the six fluorine atoms are arranged around the U atom in the form of a somewhat distorted octahedron whereas in the gaseous state the molecules are quite undistorted, regular octahedra.   The dipole moment of UF_6 in the gaseous state is therefore close to zero.

As other high valency fluorides, it is a powerful fluorinating agent for organic substances, a fact which may limit its use as a catalyst in Friedel-Crafts type reactions.   Although UF_6 has been claimed as an effective catalyst in the polymerization of isobutylene (180) and may be effective in other Friedel-Crafts reactions such as nitration with NO_2F, there are no data available on the behavior of UCl_6 or UCl_5 as catalysts, but *uranium tetrachloride* was found to be active in Friedel-Crafts alkylations and acylations (304).

8. Group IVA Halides

The Group IVA transition elements, titanium, zirconium, and hafnium all have the $(n-1)d^2ns^2$ valence electron configuration (titanium: $3d^24s^2$, zirconium: $4d^25s^2$, hafnium: $(4f^{14})5d^26s^2$. They can extend their valence octets through vacant d-orbital participation and therefore their tetrahalides can act as Lewis acids.

The atomic (and ionic) radii of Hf are very close to those of Zr, in spite of the great increase in atomic number. This is a result of the lanthanide contraction which accompanies the filling of the $4f$ shell (completed at Lu, the element preceding Hf) and is largely responsible for the exceptional similarity between Zr and Hf, unmatched by a pair of elements in any other subgroup.

The tetrahalides of the Group IVA of the Periodic Table show considerable Friedel-Crafts catalytic activity.

A. Titanium Halides

Both melting and boiling points of titanium tetrahalides, as shown in Table XIX, point to a highly covalent structure of the compounds.

TABLE XIX.   Melting and boiling points of titanium tetrahalides

	M.p. °C	B.p. °C
TiF_4	> 400	284
$TiCl_4$	− 23	136.4
$TiBr_4$	38.2	230
TiI_4	150	377.2

Indeed X-ray investigations of titanium tetrabromide and iodide show that these substances exist as molecular crystals with tetrahedral structures. Titanium tetrachloride exhibits a similar structure in the vapor state. Titanium tetrafluoride has a vapor density at 444° corresponding to the simple molecular formula. In the solid state, however, it seems to be somewhat ionic. The tetrahalides of titanium are remarkably like those of tin. The melting points and boiling points of $TiCl_4$, $TiBr_4$, and TiI_4 lie within 30 degrees of those of the corresponding tin compounds. TiF_4 is a solid like SnF_4 but is considerably more volatile and therefore more covalent in nature.

Titanium tetrahalides are capable of forming relatively stable complexes, particularly with oxygen donors and halides (esters, alcohols, ketones, ethers, phenols, etc.).

Titanium tetrafluoride, especially in solutions (CH_3NO_2,HF), is a fairly active Friedel-Crafts catalyst in alkylations, acylations, polymerizations, and isomerizations (434,446). Therefore, in solution it must be highly covalent. In the solid state it resembles other fluorides, being more ionic. The chloride, bromide, and iodide of titanium (as well as zirconium), as would be expected of simple covalent substances, are soluble in a variety of organic media, including hydrocarbons. In contrast, the fluoride is considerably less soluble. However, it dissolves to a substantial degree in nitro-methane and anhydrous HF, probably forming the complex acid H_2TiF_6 or HTi_2F_9 in the latter, and in some other organic solvents.

Titanium tetrafluoride, like the other anhydrous titanium tetra-halides, is very hygroscopic, and must be carefully kept and used under anhydrous conditions. It is a somewhat less active catalyst than $TiCl_4$ but more active than $TiBr_4$.

Titanium tetrachloride, $TiCl_4$, is an effective catalyst in Friedel-Crafts acylations and alkylations (136,204,238,464,596) but is generally less active than $AlCl_3$. As an example it catalyzes the reaction of thiophene with benzyl chloride. The reaction of benzene with benzyl chloride, however, is less readily effected by this catalyst, as benzene is considerably less reactive than thiophene (597).

$TiCl_4$ is used as catalyst for the polymerization of isobutylene (180,482,483), vinylacetylene (108), and α-methylstyrene (603).

Little is known yet on the catalytic activity of titanium tetra-bromide. It is a weaker acid catalyst than the tetrachloride and tetrafluoride (even with less back conjugation from the halogen atoms to the electron-deficient titanium) probably due to more steric hindrance.

Titanium tetraiodide has only a limited effect as a Friedel-Crafts catalyst.

B. Zirconium Halides

The tetrahalides of zirconium have properties similar to those of the titanium tetrahalides.

TABLE XX.   Melting and boiling points of zirconium tetrahalides

	M.p. °C	B.p. °C
ZrF_4		subl. above 600
$ZrCl_4$	437 (25 atm.)	331 subl.
$ZrBr_4$	450 (15 atm.)	357 subl.
ZrI_4	499 (6.3 atm.)	431 subl.

Zirconium tetrafluoride, a white, monoclinic crystalline substance, is almost insoluble in organic solvents and shows properties charac-teristic of an ionic salt with practically no catalytic activity.

Zirconium tetrachloride exists as a molecular crystal lattice of the SnI_4 type. *Zirconium tetrachloride* and *tetrabromide* show fair catalytic reactivity in Friedel-Crafts type reactions.

Zirconium tetrachloride is active as a catalyst in alkylations, acylations, and many related reactions (102,238,256,280,288). It is also used in petroleum cracking and hydrocarbon isomerization, as well as in copolymerization of iso-olefins with olefins.

$ZrCl_4$ and $ZrBr_4$ form adducts similarly to titanium halides with oxygen donors and halides, but the zirconium compounds are far less stable than the titanium analogs.

Zirconium tetraiodide has low activity as a Friedel-Crafts catalyst.

C. Hafnium Halides

The close similarity between the compounds of hafnium and zirconium, which is generally much closer than observed between any other homologous elements, indicates that hafnium tetrachloride and tetrabromide may have catalytic activities similar to the corresponding zirconium salts.

Hafnium tetrachloride forms a number of addition compounds similarly to zirconium tetrachloride. The addition compounds $3ZrCl_4 \cdot 2POCl_3$ and $3HfCl_4 \cdot 2POCl_3$ may be separated from each other by distillation, this being one of the better ways of separating zirconium from hafnium (the hafnium adduct boils 5 degrees lower than the zirconium adduct).

Hafnium tetrachloride was found to be an active catalyst although less vigorous than titanium or zirconium chlorides, in alkylations, acylations and polymerizations (442).

The most stable complexes of Ti(IV) and Zr(IV) contain the hexafluoro ions, TiF_6^{2-} and ZrF_6^{2-}. Complexes of the type $Ti_2F_9^-$ are also known. In addition, a number of salts having the ZrF_7^{3-} ions are known. The latter ion is one of the very rare examples of seven coordination and has been shown to be shaped like an octahedron with an extra fluorine above one of the triangular faces. Hexachloro and hexabromo complexes are also known.

9. Group VA Halides

The atoms of the elements of Group VA of the Periodic Table (vanadium, niobium, and tantalum) have five valence electrons in the configuration V: $3d^34s^2$, Nb: $4d^45s^1$, Ta: $5d^36s^2$.

The close similarity in the atomic and ionic radii of niobium and tantalum is reflected in the properties of their compounds.

The pentahalides of the metals of the fifth subgroup of the Periodic System, although far less investigated in their catalytic activity than the corresponding halides of Group IVA, show considerable activity and potential as Lewis acid type catalysts. The pentahalides of the elements, which have been obtained pure, are listed with their respective melting and boiling points in Table XXI.

As indicated by the melting and boiling points, the pentahalides exist as simple molecular compounds. The only pentahalide of vanadium is the pentafluoride VF_5. Oxyhalide compounds, both

TABLE XXI.   Melting and boiling points of Group VA halides

	M.p. °C	B.p. °C
VF_5	200	111 (subl.)
NbF_5	80.0	234.9
TaF_5	96.8	229.2
$NbCl_5$	194	241
$TaCl_5$	210	239
$NbBr_5$	150	270
$TaBr_5$	280	349
TaI_5	496	543

simple and complex, are more readily formed by this element in the penta-positive state than are completely halogenated derivatives. The pentafluorides display considerable activity and coordinate with fluoride ions to form complexes of the type MF_6^-.

A. Niobium and Tantalum Halides

Niobium pentafluoride appears to be somewhat more reactive than the corresponding tantalum compound. The *pentafluorides*, *pentachlorides*, and *pentabromides* of both *niobium and tantalum* have proved their good catalytic activity in Friedel-Crafts alkylation and acylation reactions (112,134,176,238).

TaI_5 was found to be catalytically inactive.

Although tantalum and niobium tetrachlorides exist, they are considerably less stable than the pentahalides and no Friedel-Crafts catalytic activity is known.

$NbCl_3$, $TaCl_3$, and $TaBr_3$ exist. They are non-volatile substances and no catalytic activity has been reported.

B. Vanadium Halides

Vanadium tetrachloride, VCl_4, is a brownish-red, heavy oily liquid, b.p. 154°, f.p. 25.7°. Its vapor density corresponds to the mono-meric formula, but soon decreases as a result of dissociation. Elec-tron diffraction shows it to have a regular tetrahedral structure in the vapor phase.

Vanadium tetrachloride was found to be a very active Friedel-Crafts catalyst in alkylation, acylation, and polymerization reactions (442).

Vanadium tetrafluoride, VF_4, exists, but is not well known. It is a hygroscopic powder which begins to disproportionate into the penta- and trifluoride at 325°.

No catalytic activity has been reported.

The complete series of trihalides of vanadium is known (Table XXII). *Vanadium trihalides* have only minor importance as Friedel-Crafts catalysts, being fairly ionic solids.

TABLE XXII.   Melting points of vanadium trihalides

	M.p. °C	Properties
VF_3	> 800	highly ionic salt
VCl_3	decomp. on heating	non-volatile, ionic, violet solid
VBr_3	decomp. on heating	
VI_3	decomp. > 280	

All vanadium trihalides have a marked tendency to form complexes, preferably of the type $M_2^I VX_5$.

10. Group VB Halides

The elements of the Group VB of the Periodic Table (nitrogen family) have the $ns^2\,np^3$ valence electron configuration. (N: $2s^2 2p^3$, P: $3s^2 3p^3$, As: $4s^2 4p^3$, Sb: $5s^2 5p^3$, Bi: $6s^2 6p^3$).

All elements with five valence electrons form bonds which are almost exclusively covalent in character. The chief differences between nitrogen and phosphorus (and the other elements) arise from the availability of d-orbitals (in the shell with principal quantum number 3) in the latter. This gives phosphorus valence properties not shared by nitrogen. For example, a capacity for 6 coordination associated with octahedral hybridization.

In the ground state all elements have three singly occupied p-orbitals. With this configuration is associated a spherically symmetrical distribution of electrons. The donor properties of the lone nitrogen pair in NH_3, to form sp^3 hybridized NH_4^+ complexes by acting as a proton acceptor, are greatly reduced in NF_3 and NCl_3. NF_3 is not hydrolyzed by water or alkali, therefore it cannot be considered as a donor molecule (Lewis base).

NCl_3, in contrast, is completely decomposed by water with formation of ammonia and hypochlorous acid. Since NCl_3 is capable of acting only as a donor toward water, the reaction probably proceeds through a hydrogen bonding mechanism. NCl_3, with a central atom (N) whose octet cannot be extended, is unable to show Lewis acid properties.

The trihalides of the other members of Group VA, although they possess unshared electron pairs, contain central atoms much less electronegative than nitrogen and have little ability to react with water through a hydrogen bonding mechanism. At the same time these elements can expand their valence octet and therefore their halides act as Lewis acids. With these trihalides water apparently is the donor molecule and the reactions result in the formation of the appropriate hydrohalic acid and Group V element hydroxide or basic salt.

Trihalides of arsenic, antimony, and bismuth can also function as acceptor molecules toward halide ion and other "bases" to form coordination compounds of varying stability.

The phosphorus trihalides are apparently less apt to behave in a similar fashion. In the field of Friedel-Crafts Lewis acid activity this means that phosphorus trihalides have only little activity as catalysts. As, Sb, and Bi trihalides, however, are somewhat active.

A. Phosphorus Halides

In contrast to many metal fluorides discussed previously *phosphorus pentafluoride* is a practically completely covalent volatile gas (b.p. $-84.6°$) of great thermal stability, whereas *phosphorus pentachloride* and *pentabromide* are solid crystalline materials of ionic nature (Table XXIII).

TABLE XXIII. Melting and boiling points of phosphorus pentahalides

	M.p. °C	B.p. °C
PF_5	-93	-85
PCl_5	167 (pressure)	163 (subl.)
PBr_5	< 100 (dec.)	*ca* 106 (dec.)

Structural studies have shown that in the vapor phase the pentafluoride and pentachloride have the trigonal bipyramid configuration, expected for molecules with sp^3d bonding. However, whereas in the pentafluoride only a single P–F bond distance (1.56 Å) is observed, in the pentachloride two P–Cl lengths (2.04 Å equatorial, 2.11 Å axial) are found. X-ray examination of the solid pentachloride strongly indicates that it is a salt consisting of tetrahedral $(PCl_4)^+$ and octahedral $(PCl_6)^-$. PCl_5 accordingly is fairly good conductor of electric current in solvents, such as nitrobenzene. The solid pentabromide appears to be composed of $(PBr_4)^+$ and Br^-. *Phosphorus pentafluoride* shows catalytic activity in Friedel-Crafts alkylations, acylations, nitration, and polymerization. It forms ionic complexes with organic fluorides of the hexafluorophosphate type, PF_6^- (16,80,337,448). No similar stable complexes with phosphorus pentachloride or bromide are formed (although the existence of PCl_6^- in the solid pentachloride has been noted previously).

Phosphorus pentachloride, although ionic in the crystalline state, shows some catalytic activity in solutions. Rearrangements (618) and cyclizations (469) are catalyzed. PCl_5 is also used as a chlorination catalyst (129). It must be noted, however, that as a chlorination catalyst the action of PCl_5 is not clear, because it is known to promote side-chain chlorination of alkylbenzenes preferentially over ring chlorination, a reaction generally characteristic of free radical conditions.

The tendency of phosphorus to extend its valence octet gives phosphorus trihalides some Lewis acid properties.

The *phosphorus trihalides* generally are covalent compounds, as may be seen from their low boiling and melting points (Table XXIV).

TABLE XXIV.   Melting and boiling points of phosphorus trihalides

	M.p. °C	B.p. °C
PF_3	-151.5	-101.8
PCl_3	-93.6	76
PBr_3	-40.5	173
PI_3	61	> 200 (dec.)

PF_3 should show some Lewis acid properties. However, so far no data are available to prove Friedel-Crafts activity of phosphorus trihalides.

B. Arsenic Halides

The only known pentahalide of *arsenic* is the fluoride, AsF_5. It is a colorless gas, condensing to a yellow liquid at $-53°$. The density of the AsF_5 vapor indicates some dissociation. Arsenic pentafluoride in accordance with its physical properties is a completely covalent compound, with a high degree of coordinating ability. It is an effective catalyst in Friedel-Crafts alkylations, acylations, polymerizations, nitrations with monofluoride reagents (337,434,448, 604,605). It forms stable complexes of the AsF_6^- type (16,337,448).

Arsenic trihalides behave as acceptor molecules (similarly to tri-halides of antimony and bismuth) toward Lewis bases. They also form complexes with halides of the type $M^I AsX_4$. According to their physical properties (Table XXV) the arsenic trihalides are covalent compounds.

TABLE XXV.   Melting and boiling points of arsenic trihalides

	M.p. °C	B.p. °C
AsF_3	− 8.5	62.8
$AsCl_3$	− 13	130
$AsBr_3$	131	220
AsI_3	141	403

Raman spectra (1,681) show that the trifluoride and trichloride are pyramidal molecules like PCl_3 and the same structure was found for the solid tribromide by X-ray analysis (63). Comparatively little is known about the catalytic activity of arsenic trihalides.

Arsenic trifluoride has some catalytic activity in olefin poly-merization (604). It is also useful in extracting basic alkylbenzenes (26).

Arsenic trichloride is a moderate catalyst in alkylation, halo-alkylation, and aromatic ring chlorination (121,201,619).

Arsenic tribromide has a catalytic effect similar to arsenic tri-chloride in alkylation and halogenation reactions.

Arsenic triiodide was claimed as a catalyst for dehydrochlorination (325). Its activity, however, needs further proof.

C. Antimony Halides

Of the pentavalent *antimony* halides, the *fluoride* and the *chloride* are known in the pure state. Both are liquids and are covalent in nature (Table XXVI).

TABLE XXVI.   Melting and boiling points of antimony pentahalides

	M.p. °C	B.p. °C
SbF_5	+ 6	+ 150
$SbCl_5$	+ 4	140 dissoc. (68°/14 mm.)

The viscous SbF_5 gives a non-ionic solid on cooling. It is com-posed of covalent trigonal bipyramidal molecules in both the solid and vapor states. Vapor density measurements of SbF_5 at 150° suggest a molecular association corresponding to $(SbF_5)_3$ and that of $(SbF_5)_2$ at 250°. Measurements of parachor and surface tension are in agreement with molecular association and support the fluorine bridged, associated nature of SbF_5. Antimony pentafluoride—in accordance with indications of the fluorine bridged polymeric structure of the liquid—is one of the strongest Lewis acid halides.

Antimony pentahalides form stable coordination complexes with halides of the SbF_6^- and $SbCl_6^-$ type. Besides the halo complexes the pentahalides give stable 1:1 addition compounds with oxygen and nitrogen donors such as ethers, alcohols, aldehydes, esters, and nitriles. Both *antimony pentafluoride* (124,337,441,448,604) and *pentachloride* (50,123,604,615) are effective Friedel-Crafts catalysts in alkylations, acylations, polymerizations, and halogenations.

As the antimony pentahalides are also powerful halogenating agents, side reactions frequently interfere with their use in a Friedel-Crafts reaction. In acylations it is preferable to prepare the stable 1:1 addition compounds with acyl halides and use these as acylating agents (modified Perrier synthesis).

Antimony trihalides, as mentioned previously, have the ability to act as Lewis acids.

According to their physical properties (Table XXVII) $SbCl_3$,

TABLE XXVII.   Melting and boiling points of antimony trihalides

	M.p. °C	B.p. °C
SbF_3	ca 292	376
$SbCl_3$	73	223
$SbBr_3$	96	288
SbI_3	170	401 (dec.)

$SbBr_3$, and SbI_3 must be fairly covalent, whereas SbF_3 seems to be a somewhat ionic solid. Electron diffraction has shown (231) that $SbCl_3$, $SbBr_3$, and SbI_3 all have pyramidal molecules in the vapor phase. X-ray analysis of solid $SbBr_3$ has confirmed the same structure (63).

Antimony trifluoride, SbF_3, similarly to HgF_2 and AlF_3, is ionic in structure and shows no Friedel-Crafts catalytic activity.

Although Friedel and Crafts reported that antimony chloride did not catalyze their reactions it was later found that the *trichloride* may be used in the ketone synthesis with acyl chlorides and in alkylation of benzene with benzyl chloride. It is considerably less active and usually requires higher temperatures than does aluminum chloride (110,390,442,590). It is also active as a polymerization catalyst (604).

Antimony tribromide is somewhat less active than $SbCl_3$ in acylations (135), alkylations (442), and polymerizations (604). Antimony triiodide is inactive as a Friedel-Crafts catalyst.

D. Bismuth Halides

Bismuth is much more metallic than the other elements of the subgroup and in many ways resembles the Group III metals. The trihalides of bismuth, similarly to those of arsenic and antimony, show a tendency to act as acceptor molecules. The acceptor properties increase from arsenic to bismuth.

TABLE XXVIII. Melting and boiling points of bismuth trihalides

	M.p. °C	B.p. °C
BiF_3	725–730	salt like
$BiCl_3$	233	447
$BiBr_3$	218	441
BiI_3	408	500 (dec.)

Bismuth trifluoride, BiF_3, has all the characteristics of a salt. It is a white powder which is scarcely volatile, even at red heat. It is much the most stable of the trihalides. No catalytic activity has been reported, in accordance with its ionic nature.

Bismuth trichloride, $BiCl_3$, at 250° has the specific conductivity of a salt, but has a normal vapor density and dissolves in a number of organic solvents, like nitrobenzene (donor solvents, with which it presumably forms complexes). Electron diffraction shows that in the vapor phase $BiCl_3$ is pyramidal (576). Bismuth trichloride is a fairly effective Friedel-Crafts catalyst in alkylation, acylation, and polymerization reactions (110,134,604,636).

Bismuth tribromide, $BiBr_3$, is similar in its catalytic action to $BiCl_3$, but is a weaker catalyst. It is soluble in aromatic hydrocarbons (434). It also acts as a ring bromination catalyst (637).

The only known stable pentahalide of bismuth, *bismuth pentafluoride* BiF_5 shows no catalytic activity or complex-forming ability. This is easily explained if it is considered that BiF_5 is a high-melting solid (subl. 550°) with an obvious ionic structure.

11. Group VIB Halides

As in the preceding group of the Periodic Table there is a gradual transition to more metallic character with increasing atomic number.

The elements have six electrons in their valence shell with ns^2np^4 configuration O: $2s^2\ 2p^4$, S: $3s^2\ 3p^4$, Se: $4s^2\ 4p^4$, Te: $5s^2\ 5p^4$.

The halides of oxygen can act only as electron donors. Therefore they have no importance from the point of view of Friedel-Crafts catalytic activity.

Beyond oxygen d-hybridization commonly occurs giving a possibility of Lewis acid properties and complex formation.

Sulfur tetrafluoride, SF_4, has the ability to extend its electron octet and thus act as a Lewis acid. However, owing to its strong fluorinating ability, no investigation relating its probable catalytic activity has yet been reported.

Tetrahalides of selenium and tellurium are capable of acting as acceptor molecules toward halides with the formation of complexes of the type $M_2^ISeX_6^{2-}$ and $M_2^ITeX_6^{2-}$. Complexes of the type $M^ITeX_5^-$ are also known. These halides have some importance as Friedel-Crafts catalysts.

A. Selenium Halides

Selenium tetrafluoride and tetrachloride are, according to their physical data, covalent, monomeric compounds.

TABLE XXIX. Melting and boiling points of selenium tetrahalides

	M.p. °C.	B.p. °C.
SeF_4	-9.5	106
$SeCl_4$		196 subl.
$SeBr_4$	unstable, decomp. at	
	70–80° to $Br_2 + Se_2Br_2$	

Selenium tetrabromide, a yellow solid, loses Br_2 to form Se_2Br_2 even at room temperature. Therefore, it is too unstable for catalytic investigation.

Selenium tetrafluoride gives the stable $(NO_2^+)_2\ SeF_6^{2-}$ with NO_2F (16), which is useful in Friedel-Crafts type nitrations (337,441).

Selenium tetrachloride is active as a dehydrohalogenating catalyst (327), and is also used as an isomerization catalyst (275).

B. Tellurium Halides

The tetrahalides of tellurium are all known, but with the exception of $TeCl_4$ they are thermally unstable and easily disproportionate. They are all somewhat active Friedel-Crafts catalysts.

Tellurium tetrafluoride forms a stable nitronium salt with nitryl fluoride (16) which is capable of Friedel-Crafts nitration.

TABLE XXX. Melting and boiling points of tellurium tetrahalides

	M.p. °C	B.p. °C
TeF_4	130	dec. above 194° to TeF_6 + Te
$TeCl_4$	225	390
$TeBr_4$	380	414 dec.
TeI_4	280 (sealed tube)	dec.

Tellurium tetrachloride, a white crystalline compound, in spite of the high conductivity has the solubility of a covalent compound, being soluble in benzene and toluene. It is a fair alkylation and acylation catalyst (134,136,186).

Tellurium tetrabromide is a dark yellow to red crystalline compound, which dissociates completely in the vapor phase. However, in solution it is quite stable and catalyzes Friedel-Crafts acylations as well as ring bromination (135,637).

Tellurium tetraiodide forms black crystals, which begin to volatilize and dissociate above 100°. It is claimed as a dehydrohalogenation catalyst (327) but no further data are available on its Friedel-Crafts catalytic activity, which may be limited.

12. Group VIA Halides

Although the first two elements have the d^5s^1 valence electron configuration (Cr $3d^54s^1$, Mo $4d^55s^1$) chromium differs from molybdenum in many respects. It forms ions of lower charge and much lower radius. On the other hand the atomic and ionic radii of both molybdenum and tungsten (W. $5d^46s^2$) are very similar, which is shown in the properties of their compounds. The higher valency halides show good complexing ability and are effective Friedel-Crafts catalysts.

A. Chromium Halides

Trivalent chromium readily forms complexes, similar to trivalent cobalt and tetravalent platinum.

Among the trihalides of *chromium*, the *fluoride*, *chloride* and *bromide* are known in the anhydrous state. The iodide has been obtained pure only as the nonahydrate. The chromic halides are highly colored substances of low volatility.

Chromic fluoride is obviously an ionic salt. The *chloride and bromide* are definitely non-ionic, but consist of crystals containing two-dimensional $(CrX_3)_n$ layers, with van der Waals forces holding the layers together (671).

Anhydrous *chromic chloride* exists as red-violet leaflets and is effective in Friedel-Crafts isomerizations (500) and some catalytic

TABLE XXXI. Melting and boiling points of chromic halides

	M.p. °C	B.p. °C
CrF_3	> 1000	> 1100
$CrCl_3$		1300 subl. with decomp.
$CrBr_3$	decomp.	

effect is also observed in alkylations with benzyl chloride and ketone synthesis with acetyl bromide (442).

Chromic bromide is a somewhat more active acylation catalyst than $CrCl_3$. It is also an effective ring bromination catalyst (442).

B. Molybdenum Halides

Whereas *molybdenum* exhibits the maximum valence of five towards chlorine it forms a *hexafluoride*, MoF_6, with fluorine. MoF_6 is a very hygroscopic, white crystalline mass, which melts at 17° to a colorless liquid and boils at 35°. According to its physical properties, it is most probably a highly covalent compound, but its Lewis acid activity in Friedel-Crafts reaction has not been reported. *Molybdenum pentachloride* forms dark, green-black crystals, density 2.93, melting point 194°, boiling point 268°. Both the liquid and the vapor are dark red. It forms trigonal bipyramid crystals with all Mo–Cl distances being 2.27 Å. In the fused state, molybdenum pentachloride is practically a non-conductor. It is soluble in hydrocarbons, carbon disulfide, chloroform, carbon tetrachloride, sulfuryl chloride, nitrobenzene, ethyl bromide, and quinoline. It is also soluble in a great variety of other polar solvents, such as ethers, alcohols, aldehydes, ketones, esters, amines, acids, and acid anhydrides. In most of these cases, however, a reaction takes place between the molybdenum pentachloride and the solvent. $MoCl_5$ is a fairly covalent compound, and it shows considerable Lewis acid catalyst activity in Friedel-Crafts alkylation and acylations, polymerizations and in ring chlorination of aromatics. Ethyl benzyl ether acts as a powerful benzylating agent in reactions with aromatic hydrocarbons in the presence of molybdenic pentachloride. Benzene and other aromatics are chlorinated by chlorine in the presence of molybdenum chloride catalyst, the reaction product includes chlorobenzene and dichlorobenzenes as well as higher chlorinated products (134,136,304,343,442).

It is also a particularly effective catalyst in chlorination of phthalic anhydride (347). No pentabromide or pentaiodide of molybdenum is yet known in the pure form.

The tetrabromide of molybdenum is not well known and may easily disproportionate although it has been claimed as an acylation catalyst in Friedel-Crafts ketone syntheses (135). Until the compound itself is more definitely established this claim should be accepted with reservation.

C. Tungsten Halides

Of the anhydrous tungsten hexahalides, the fluoride, chloride, and bromide are known.

Tungsten hexafluoride resembles the hexafluorides of molybdenum, tellurium, and uranium both in volatility and ease of hydrolysis. Its boiling point (17.5°) is lower than that of MoF_6 (35°). Tungsten hexafluoride is very reactive, practically covalent in nature and should be a fairly effective Lewis acid catalyst. However, both tungsten and molybdenum fluorides are very active fluorinating

TABLE XXXII. Melting and boiling points of tungsten hexahalides

	M.p. °C	B.p. °C
WF_6	2.3 (375 mm.)	17.5
WCl_6	275	347
WBr_6	decomp.	

agents, resembling elemental fluorine in many ways and attack organic compounds violently. Their applicability therefore may be confined to reactions with highly fluorinated organic compounds.

Tungsten hexachloride, WCl_6, exists as dark blue or violet hexagonal crystals. It melts at 275° and boils at 347.0°. Its vapor pressure indicates that it is slightly dissociated at the boiling point. Its crystal structure is interpreted as a hexagonal, close-packed chlorine lattice, slightly deformed so as to group six chlorine atoms octahedrally around a tungsten atom. In the fused state WCl_6 is a very poor conductor. It is soluble in carbon disulfide, chloroform, benzene, carbon tetrachloride, and petroleum ether giving brown to reddish brown solutions. It is also soluble in oxygen-containing solvents, such as phosphorus oxychloride, methyl alcohol, ethyl alcohol, acetone, glycerol, and also in pyridine. In ether it gives a yellow solution. Based on its properties in the liquid phase it is highly covalent in nature and has fairly good catalytic properties in Friedel-Crafts type reactions as acylations and alkylations (305).

Tungsten hexabromide, WBr_6, resembles the hexachloride in its catalytic action but it is more readily decomposed. It is effective in Friedel-Crafts ketone syntheses (135).

13. Group VIIA Halides

The valence electron configurations are Mn: $3d^5 4s^2$, Re: $5d^5 6s^2$.

The halides of Group VIIA have little importance as Friedel-Crafts catalysts.

A. Manganese Halides

Manganic fluoride, MnF_3, and *manganic chloride*, $MnCl_3$, are known and both are able to form complexes of the type $M^I MnX_4$ and $M_2^I MnX_6$. Manganic fluoride is a red, crystalline, fairly ionic substance. $MnCl_3$ (although not yet well known as the pure compound) is much more covalent. According to cryoscopic measurements in liquid HCl it is monomeric. No catalytic effect of manganic halides in Friedel-Crafts reactions is known.

The manganous dihalides (MnF_2, $MnCl_2$, $MnBr_2$) are all high-melting ionic compounds and no particular Friedel-Crafts catalytic activity can be expected. They have been claimed as dehydrochlorinating catalysts (328) at higher temperatures but this should not necessarily mean that they act as Lewis acid catalysts.

B. Rhenium Halides

Although higher valency rhenium halides exist (ReF_6 m.p. 18.8°, b.p. 47.6°, $ReCl_5$, ReF_4 m.p. 124.5°) and they readily form halide complexes, from the point of view of Friedel-Crafts catalytic activity only *rhenium trichloride* and *tribromide* have been investigated.

Rhenium trichloride is a reddish-black substance which is obviously highly covalent. The molecular weight in glacial acetic acid is that of the dimer Re_2Cl_6 (679). The structure presumably is analogous to that of the bridged structure of Al_2Cl_6 or Fe_2Cl_6.

In hydrochloric acid solution rhenium trichloride is present as the complex acid $H[ReCl_4]$. Salts of the complex acid with the composition $M^I ReCl_4$ can be crystallized from hydrochloric acid solution containing alkali metal chlorides.

$ReCl_3$ was found to be an effective Friedel-Crafts catalyst in acylations, alkylations, and polymerizations (442).

Rhenium tribromide is similar in structure and catalytic behavior to the trichloride.

14. Group VIIB Halides (Interhalogens)

The halogens form a surprisingly large number of compounds with one another, all of which are well defined and covalent substances. This peculiarity of the halogens is due to their being the only series of electronegative elements which can form acylic molecules without double links and hence such combinations are not restricted to the highest members of the group. They were reviewed by A. G. Sharpe (553) in 1950, but since then IF_3 and IF have been prepared so that at present 13 compounds of the type AB_n ($n = 1,3,5$ or 7) are known (all the possible mono- and trihalides with exception of IBr_3, the pentafluorides of bromine and iodine and the heptafluoride of iodine).

TABLE XXXIII.   Melting and boiling points of interhalogens

ClF (b.p. $-100°$)	ClF_3 (b.p. $12°$)	
BrF (b.p. $20°$)	BrF_3 (b.p. $127°$)	BrF_5 (b.p. $40°$)
BrCl (b.p. $5°$)		
IF (dec. $>0°$)	IF_3 (dec. $>-35°$)	IF_5 (b.p. $97°$) IF_7 (b.p. $4°$)
ICl (solid, 2 forms)	ICl_3 (m.p. dec. $101°$)	
IBr (m.p. $36°$)		

Iodine being relatively the most electropositive of the halogens can function as a fairly strong electron acceptor. This effect is further enhanced in the interhalogen compounds of iodine. Fluorine being the most electronegative of the halogens means that in iodine fluorides the electron-accepting power of iodine must be the strongest. Indeed, *iodine pentafluoride* is a good acceptor and forms a stable nitronium salt $NO_2^+IF_6^-$ with NO_2F (16). A similar salt, $NO_2^+BrF_4^-$, is also formed with *bromine trifluoride*. Although these nitronium salts could be used in Friedel-Crafts type nitrations, the strong fluorinating and oxidizing effect of IF_5 and BrF_3, when contacted with hydrogen-containing organic substance generally prohibits their use as catalysts except with highly fluorinated substances.

From the point of view of Friedel-Crafts catalytic activity the most promising interhalogen compounds are *iodine monochloride*, ICl, and *iodine trichloride*, ICl_3.

Iodine trichloride forms lemon-yellow needles, which melt at 101° under their own (dissociation) vapor pressure at 16 atm. to a reddish-brown liquid. Both its vapor and solution dissociates easily $ICl_3 \rightarrow ICl + Cl_2$. The absorption spectrum of ICl_3 in CCl_4 solution is identical with that of $ICl + Cl_2$ (210). It is therefore difficult to tell whether it is ICl_3 or ICl which is involved in catalyzing Friedel-Crafts reactions. ICl or ICl_3 is involved as the active catalyst in the iodine catalyzed ring chlorination of aromatics (184).

Campaigne (105) found that ICl_3 effects *vicinal* dichlorination of aromatics, but later investigation proved it to be a usual *ortho–para* chlorinating agent (104).

ICl_3 is an effective catalyst in Friedel-Crafts acylations (giving unusually high ortho-substitution) and alkylation. It is also a good chlorination catalyst (434). I_2 itself is a fairly active catalyst in certain cases.

Catalysis by *iodine* is due to the possibility of iodine acting as an electron acceptor and forming polarized complexes.

Friedel-Crafts ketone syntheses with acyl halides or acid anhydrides take place readily in the presence of I_2, the activated intermediate probably being of the type

A similar effect takes place in alkylations with alkyl halides.

15. Group VIII Halides

Group VIII of the Periodic Table is made up of the following nine elements

Fe	Co	Ni
Ru	Rh	Pd
Os	Ir	Pt

Nowhere in the Periodic Table, except among the lanthanides, are the resemblances between horizontally adjacent elements so marked as in these triads. It is convenient, therefore, to divide the elements of the transition triads into two groups, one dealing with the $3d$ triad (Fe, Co, Ni) and the other with the $4d$ and $5d$ triads (platinum metals).

Halides of the 3d transition triad (iron, cobalt, and nickel). The elements of the 3d triad have the following valence electron configurations: Fe: $3d^6 4s^2$; Co: $3d^7 4s^2$; Ni: $3d^8 4s^2$

A. Iron Halides

Of the anhydrous trihalides of iron only the iodide has not been isolated in a pure state. Iron triiodide decomposes (even at ordinary temperatures) in accordance with the equilibrium

$$2FeI_3 \rightleftharpoons 2FeI_2 + I_2.$$

Ferric fluoride is a greenish, relatively non-volatile (subl. > 1000°) fairly ionic substance which is slightly soluble in water. It is practically insoluble in organic solvents and shows no Friedel-Crafts catalytic activity.

Ferric chloride is a red deliquescent solid, which melts and sublimes at 300°. When heated in vacuum to temperatures above 500° it undergoes dissociation into the dichloride and chlorine. The vapor of the trichloride at temperatures up to at least 400° is in the dimeric form, Fe_2Cl_6. The dimeric molecules in the vapor state possess a bridged structure similar to that of aluminum chloride. That is, they are made up of two $FeCl_4$ tetrahedra with two halogen atoms being held in common. The Fe–Cl bond distance is 2.17 Å, which is very close to the value expected for a single covalent bond between these atoms.

Above 750° the vapor consists almost entirely of $FeCl_3$ monomeric molecules.

In the crystalline state, the relationship between the iron and aluminum halides is not as close. However, insufficient X-ray and electron-diffraction work on the solid iron trihalides has been reported to enable their structures to be known with certainty.

Ferric chloride is soluble in a variety of organic solvents and forms complexes of similar structure to those of aluminum chloride, for example, $FeCl_3(C_2H_5)_2O$ or $FeCl_3 \cdot CH_3NO_2$. Ferric chloride is monomeric in most donor solvents similarly to aluminum chloride.

Friedel and Crafts described first the catalytic activity of ferric chloride (198).

In 1897 Nencki (408) pointed out that ferric chloride, being a milder catalyst than $AlCl_3$, was applicable to syntheses in which

$AlCl_3$ was too drastic causing decomposition and unwanted side reactions. The use of ferric chloride in alkylations and acylation, however, has been found to result in lower yields than those obtained when $AlCl_3$ was used (672), although previously it had been reported that the action of $FeCl_3$ in Friedel-Crafts ketone syntheses gives similar yields to those obtained with aluminum chloride (52).

$FeCl_3$ does not catalyze the alkylation of benzene with ethylene at moderate temperatures (203) but is an active catalyst in alkylations with alkyl halides and more reactive olefins.

Ferric chloride, bromide, and probably the unstable iodide are the catalysts intermediately formed when iron is used as a halogenation catalyst in aromatic halogenations.

Ferric bromide is very similar in its structure and catalytic properties to ferric chloride. Ferric bromide is also a dimeric halogen bridged molecule in the vapor phase, similar to aluminum bromide. It is generally a somewhat more active catalyst both in acylation and alkylation, as well as in polymerization and ring halogenation (135,442,604,637).

B. Cobalt Halides

Anhydrous *cobaltic fluoride*, CoF_3, is the only known anhydrous trihalide of cobalt. It is a pale brown crystalline powder which turns much darker in the presence even of minute amounts of water. On heating it loses fluorine to form CoF_2 (it is therefore a strong fluorinating agent). Its crystal structure is not certain, but it seems to be related to the ionic ferric and chromic fluorides. No Friedel-Crafts catalytic activity is known but the compound $CoF_3 \cdot BF_3$ is an active polymerization catalyst (358).

Cobaltous salts of all four halogens are known. Anhydrous *cobaltous fluoride*, CoF_2, is a pink, crystalline, ionic solid which has the rutile lattice and no catalytic activity is to be expected.

Anhydrous *cobaltous chloride*, $CoCl_2$, is a pale blue solid which has the same crystal lattice as magnesium chloride—a typical ionic lattice with 6 chlorine atoms at the points of a regular octahedron around each cobalt atom. It melts at 735° ($MgCl_2$ at 718°) and boils at 1049°. Its catalytic properties are also similar to those of $MgCl_2$. Modest Friedel-Crafts activity has been claimed in alkylations, acylations, and dehydrochlorinations (37,146,326).

Cobaltous bromide, $CoBr_2$, is generally similar in its properties to the chloride. The anhydrous salt is bright green and melts at 678° That in solution $CoBr_2$ can show considerable covalent character is shown by the fact that its molecular weight by the boiling point method in pyridine and in quinoline solution is that of the mono-meric $CoBr_2$ (29). This is remarkable as one would expect it to dissolve in these solvents as the ionized bromide of a solvated cation. $CoBr_2$ is a somewhat more active Friedel-Crafts catalyst than $CoCl_2$ in acylation and halogenation reactions (68,442).

C. Nickel Halides

No higher halides than the dihalides of nickel are known and they have very limited Friedel-Crafts catalytic activity.

Nickel fluoride, NiF_2, forms brownish-green crystals of the rutile lattice type. Being a highly ionic compound no catalytic activity can be expected.

Nickel chloride, $NiCl_2$ (subl. 993°), forms yellow crystals of the cadmium chloride type. It is much less soluble in organic solvents than $CoCl_2$.

Nickel bromide, $NiBr_2$, is very much like the chloride. *Nickel iodide*, NiI_2, is very much like elemental iodine in appearance. It has a cadmium chloride lattice.

A slight Friedel-Crafts catalytic activity has been claimed for anhydrous $NiCl_2$, $NiBr_2$, and NiI_2 in alkylations, acylations, dehydro-chlorinations, and halogenations (146,165,326). Owing to the highly ionic nature of these halides this claim needs further confirmation.

Halides of the 4d and 5d transition triads (platinum metals). The elements of the 4d and 5d triads, commonly known as the platinum metals, have the following valence shell configurations Ru: $4d^7 5s^1$; Rh: $4d^8 5s^1$; Pd: $4d^{10}$; Os: $5d^6 6s^2$; Ir: $5d^9$, Pt: $5d^9 6s^1$.

Few of the halides of the platinum group metals have importance as Friedel-Crafts catalysts due, perhaps, to their expensive nature. As catalysts they must be considered to be of the mixed acidic–electronic type, as metals of the platinum group have a well-known ability to act as oxidation–reduction and hydrogenation catalysts of the electronic type.

D. Ruthenium Halides

Ruthenium pentafluoride, RuF_5, is a dark green, fairly covalent compound (m.p. 101°, b.p. 250°) which readily forms complexes of the type $M^I RuF_6$. However, no Friedel-Crafts catalytic activity has been reported.

Ruthenium tetrachloride, $RuCl_4$, has not been reported as a catalyst in reactions related to the Friedel-Crafts type.

Ruthenium trichloride, $RuCl_3$, forms brown-black crystal leaflets which have a layer lattice. It was reported to be a catalyst in the addition of olefins to silicon hydride, a suggested cationic reaction catalyzed by a metal halide Lewis acid (591).

E. Rhodium and Osmium Halides

Rhodium trichloride, $RhCl_3$, can occur in several forms, some of them certainly complex. It very easily forms complexes of the type $M_2[RdCl_5]$, $Me_3[RhX_6]$, $M_4[RhX_7]$. No complex compound corresponding to $M[RhX_4]$ has been isolated, but may be intermediately formed. The diamagnetic character of $RdCl_3$ suggests an inner orbital d^2sp^3 complex in which rhodium is coordinated octahedrally to six chlorine atoms in some sort of an infinite three-dimensional complex. $RhCl_3$ was found to be an effective ring chlorination catalyst (68). No data on its effect in other Friedel-Crafts reactions are known so far.

The catalytic effect of palladium halides has not been reported so far in Friedel-Crafts type reactions. Palladium only forms dihalides, no tetrahalide being known.

Osmium hexafluoride, OsF_6 (m.p. 34.4°, b.p. 47.5°), is the highest valency osmium halide known. It was previously reported (Ruff) that OsF_8 exists but the compound was found to be identical with OsF_6 of known structure (vapor density measurements, infra-red, and Raman spectra) which is the highest fluorination product of osmium. It has an octahedral covalent structure (670), and was claimed as a Friedel-Crafts polymerization catalyst (180).

F. Iridium Halides

Iridium hexafluoride (m.p. 44°, b.p. 53°) is an extremely reactive volatile yellow compound but does not behave as a Friedel-Crafts type catalyst.

Iridium tetrachloride, $IrCl_4$, similarly to IrF_4, is not very stable and therefore insufficiently known in the anhydrous state. It was claimed to be effective as an aromatic ring chlorination catalyst (68).

Iridium trichloride, $IrCl_3$, is an olive green powder in the anhydrous state which becomes yellow and crystalline, without change in composition, when sintered at 650°. It forms hexa-coordinated complexes of the type $M_3^I[IrX_6]$. It was described, together with $IrCl_4$, as a ring chlorination catalyst and also as a catalyst in the ionic addition of olefins to silicon hydride (591).

G. Platinum Halides

Platinous chloride, $PtCl_2$, is a greenish-grey or yellow powder. It readily forms complexes of the type M_2PtCl_4. It is active as a chlorination catalyst (68).

Platinum tetrachloride, $PtCl_4$, is a red-brown slightly hygroscopic compound, which forms complexes of the type $M_2^IPtCl_6$. It has been claimed as a Lewis acid type dehydrohalogenation catalyst (326).

16. Rare-Earth or Lanthanide Halides

The lanthanides, as elements in which inner orbitals are being occupied (generally the $4f$ orbitals are being filled), have properties characteristic of transition elements. However, they differ in a number of important aspects from transition elements in which inner d-orbitals are being filled, largely because the f-orbitals lie deeper in the atom and consequently are less readily available for chemical reaction and are better screened from the effect of surrounding electronic systems. Only the trihalides of cerium and neodymium have been investigated so far as Friedel-Crafts catalysts. Lanthanum tribromide, $LaBr_3$, is inactive as a catalyst in the Fries rearrangement (132).

A. Cerium Halides

Cerium trichloride, $CeCl_3$, is fairly soluble and readily forms complexes of the type M^ICeCl_4.

It is the active catalyst in Friedel-Crafts acylation of aromatics with acyl chlorides catalyzed by cerium powder.

$CeCl_3$ is a good catalyst (in the presence of $HgCl_2$) for the addition of hydrogen chloride to acetylene and olefins (60,61,397). It also catalyzes acetal formation.

Cerium bromide is similar in its activity to the chloride.

B. Neodymium Halides

Neodymium trichloride, $NdCl_3$, was found to catalyze polymerizations (similar in activity to TiF_4) (613).

17. Survey of Acidic Halide Catalysts

To facilitate a survey of Lewis acid type acidic halides with Friedel-Crafts catalytic action Table XXXIV summarizes reported catalysts and the main reaction types in which they have been employed. Although the Table was prepared from an extensive literature survey no claim as to completeness can be made owing to the vast number of papers and patents which are concerned with the topic.

TABLE XXXIV.   Acidic halide catalysts

Catalyst	Reactions catalyzed	Ref.
$AlBr_3$	various Friedel-Crafts reactions as alkylation, acylation, isomerization, cracking, polymerization, hydrogenation, dehydrogenation, etc.	196,288
$AlBr_3$ + alkali halide	alkylation, polymerization	196,600
$AlBr_3 \cdot AlOBr$		683
$6AlBr_3 \cdot AlOBr$	alkylation, polymerization	683
$AlCl_3$	various Friedel-Crafts reactions	623
$AlCl_3$ + alkyl halide	alkylation	196,600 288
$AlCl_3$ + metal salts		133
+ $CaCl_2$ deposited on or combined with hydrous silica	alkylation, polymerization	602
+ $FeCl_3$	various Friedel-Crafts reactions	133,196,288, 600,602,623
+ KCl on hydrous silica or melt	alkylation, polymerization	133,196,288 600,602,623
+ LiCl on hydrous silica or melt	alkylation, polymerization	133,196,288 600,602,623
+ NaCl on hydrous silica or melt	alkylation, cracking, condensation, dehydration	21,198,523,598
+ NH_4Cl on hydrous alumina or melt	alkylation, polymerization	133,196,288, 600,602,623
+ $SbCl_3$ + HCl	alkylation, isomerization, various Friedel-Crafts reactions	185,518,519, 520
+ Cu_2Cl_2	alkylation	4,289
+ Hg_2Cl_2	alkylation	4,289
+ $NiCl_2$	alkylation	4,289
+ $SiCl_4$	alkylation, various Friedel-Crafts reactions	453
+ $SnCl_4$	alkylation, various Friedel-Crafts reactions	453
+ $TiCl_4$	alkylation, various Friedel-Crafts reactions	453
"poisoned $AlCl_3$"	acetylation of olefins	25
"$AlCl_3$ sludges"	alkylation	288
Aluminum chloride monomethanolate $AlCl_3 \cdot CH_3OH$	alkylation	288,534

TABLE XXXIV (*continued*)

Catalyst	Reactions catalyzed	Ref.
$AlCl_3 \cdot CH_3NO_2$	alkylation, acylation	288,535,536
$AlCl_3 \cdot C_2H_5NO_2$	alkylation	288
$AlCl_3 \cdot C_3H_7NO_2$	alkylation	288
$AlCl_3 \cdot C_6H_5NO_2$	alkylation. acylation	656
$AlCl_2 \cdot HSO_4(AlCl_3 \cdot H_2SO_4)$	alkylation, condensation, isomerization, polymerization	334,538
$AlCl_2 \cdot O_3SR$	alkylation	320
$AlCl_2 \cdot H_2PO_4(AlCl_3 \cdot H_3PO_4)$	alkylation	627
$Al-AlCl_3$	acylation	240
$AlCl_3 \cdot$ tert.amines	alkylation	155,156
AlF_3	inactive in alkylation, acylation	103
	claimed in polymerization of cyclopentadiene, isomerizations	604 302a
AlI_3	alkylation, various Friedel-Crafts reaction	198,317
$Al(SCN)_3$	thiocyanation	584
$AsBr_3$	alkylation, halogenation, chloroalkylation	121,201 619
$AsCl_3$	alkylation, halogenation chloroalkylation	121,201 619
AsF_3	polymerization	26,604
AsF_5	polymerization, alkylation, acylation, nitration	26,604 16,337,441,448
AsI_3	dehydrochlorination	325
$AuCl_3$	acylation, alkylation, polymerization	442
$AuBr_3$	acylation, alkylation, polymerization	442
$BaCl_2$	alkylation	432
BaF_2	alkylation	432
BBr_3	alkylation, acylation, polymerization	684
BCl_3	alkylation, acylation, polymerization, condensation	24,30,209,363, 438,583,675
BF_3	various Friedel-Crafts reactions	128,266,379, 438,587
$BF_3 \cdot H_2O$	alkylation	349
$BF_3 \cdot 2H_2O$	alkylation	56
$BF_3 \cdot H_2SO_4$	alkylation	56,260
$BF_3 \cdot HF$	isomerization	357,378
$BF_3 \cdot H_3PO_4$	alkylation	635
$BF_3 \cdot P_2O_5$	alkylation	260,578

Table continued

TABLE XXXIV (*continued*)

Catalyst	Reactions catalyzed	Ref.
$BF_3 \cdot SO_3$	isomerization	676
$BF_3 \cdot HSO_3F$	isomerization	601
$BF_3 \cdot 2CH_2ClCOOH$	polymerization	
$BF_3 \cdot 2CCl_3COOH$	polymerization	
$BF_3 \cdot 2CF_3COOH$	polymerization	
$BF_3 \cdot C_6H_5SO_3H$	alkylation	260
$BF_3 \cdot HPO_2F_2$	alkylation	579
$BF_3 \cdot (CH_3)_2O$	alklyation, acylation	56
$BF_3 \cdot (C_2H_5)_2O$	alkylation, acylation	56
$BF_3 \cdot 2CH_3COOH$	alkylation, acylation, poly- merization	56
$BF_3 \cdot 2C_6H_5OH$	alkylation	56
$BF_3 \cdot 2CH_3OH$	alkylation, polymerization	56
$BF_3 \cdot 2C_2H_5OH$	polymerization	56
$BF_3 \cdot C_2H_5NH_2$	polymerization	
$BF_3 \cdot CO(NH_2)_2$	polymerization	
$BF_3 \cdot N(CH_2CH_2OH)_3$	alkylation	155,156
$BF_3 \cdot piperidine$	alkylation	155,156
$BF_3 \cdot Al_2O_3$	alkylation	373
$BF_3 \cdot FeF_2$	alkylation, isomerization, polymerization	651,652,653, 659
$BF_3 \cdot FeF_3$	alkylation, polymerization	358
$BF_3 \cdot CrF_3$	alkylation, polymerization	358
BI_3	alkylation, acylation	434
$BeBr_2$	alkylation, polymerization, halogenation	455,617
$BeCl_2$	alkylation, acylation, crack- ing, polymerization	66,134,238
BeF_2	acylation, alkylation, crack- ing, polymerization	66
$BiBr_3$	halogenation	637
$BiCl_3$	alkylation, acylation, poly- merization	107,110,134, 136,604,636
$BiF_5 \cdot HF$	isomerization	376
BrF_3	nitration (with NO_2F)	16
$CaCl_2$	alkylation	432
$CaBr_2$	acylation	434
$CaBr_2 \cdot AlCl_3$	acylation	34
CaF_2 (as natural fluor- spar)	alkylation	432
$CaF_2 \cdot Cr_2O_3$	alkylation	433
$CaF_2 \cdot BaF_2$	isomerization	432
$CaF_2 \cdot SrF_2$	isomerization	432
$CaF_2 \cdot ZnF_2$	isomerization	432
$CbCl_5$	alkylation, acylation	134,238
CbF_5	acylation	434

TABLE XXXIV (*continued*)

Catalyst	Reactions catalyzed	Ref.
$CdBr_2$	halogenation, acylation	135,637
$CdCl_2$	polymerization	333
CdI_2	polymerization	294
$CeBr_3$	alkylation, acylation	60,341
$CeCl_3$	acetal formation	4
$CoBr_2$	formylation, halogenation	37,68,146,179, 326,499
$CoCl_2$	formylation, alkylation, de-hydrohalogenation	37,68,146,179, 326,499
CoI_2	formylation	37,68,146,179, 326,499
$CoF_3 \cdot BF_3$	polymerization	358
$CrCl_3$	isomerization	423,550
$CrCl_2$	dehydrochlorination	327
CrF_2	dehydrochlorination	327
$CuBr_2$	halogenation, isomerization	362,500,674
$CuCl_2$	isomerization	362,500,674
Cu_2Cl_2	alkylation, acylation, polymerization	431
Cu_2I_2	isomerization	362,500,674
Cu halides	alkylation, isomerization, polymerization	330,419
$FeBr_3$	alkylation, acylation, halogenation, polymerization, formylation	37,135,362,434, 500,604,637, 674
$FeCl_3$	various Friedel-Crafts reactions, alkylation, acylation, formylation, polymerization	6,37,52,58,198, 367,399,408, 464,672
$FeF_3 \cdot BF_3$	alkylation	358
FeI_3	halogenation, formylation	37
$GaBr_3$	alkylation, acylation	582
$GaCl_3$	alkylation, acylation	97,647,649
$GeBr_4$	alkylation, acylation	442,521
$GeCl_4$	alkylation	70,146,521
GeF_4	nitration	16
$HgBr_2$	acylation, halogenation, polymerization	16,135,146, 615,637
$HgCl_2$	alkylation, acylation, isomerization, polymerization	50,180,424, 431,547
ICl_3	acylation, alkylation, halogenation	104,439
IF_5	nitration	16
$InBr_3$	alkylation, acylation, halogenation, deuteration	442,554,649

Table continued

TABLE XXXIV (*continued*)

Catalyst	Reactions catalyzed	Ref.
$InCl_3$	alkylation, acylation, halogenation	442,649
$IrCl_3$	halogenation, addition of olefins	591
$IrCl_4$	halogenation	68
$MgBr_2$	alkylation, acylation, halogenation	34,76,391
$MgCl_2$	alkylation, isomerization of dichlorobenzenes	46,76,392,432
MgF_2	alkylation	432
$MnBr_2$	dehydrohalogenation	328
$MnCl_2$	dehydrohalogenation	328
MnF_2	dehydrohalogenation	328
$MoBr_4$	acylation	135
$MoCl_5$	alkylation, acylation, halogenation, polymerization	134,136,304, 343
$NbBr_5$	alkylation, acylation	112,176
$NbCl_5$	alkylation, acylation	176,238
NbF_5	alkylation, acylation, isomerization	176,376
$NdCl_3$	polymerization	614
$NiBr_2$	alkylation, chlorination, formylation, dehydrohalogenation	146,165,326
$NiCl_2$	alkylation, chlorination, formylation, dehydrohalogenation	146,165,326
NiI_2	alkylation, chlorination, formylation, dehydrohalogenation	146,165,326
OsF_6	polymerization	180
$PbBr_2$	polymerization	3
$PbCl_2$	polymerization	3
PBr_3	alkylation, dealkylation	452,619
PCl_3	alkylation, dealkylation	452,619
PF_3	alkylation, dealkylation	452,619
PCl_5	acylation, cyclization, halogenation rearrangement	129,469,618
PF_5	alkylation, acylation, nitration, polymerization	16,80,337,441, 678
$PtCl_2$	chlorination	
$PtCl_4$	dehydrohalogenation	326
$RhCl_3$	chlorination	68
$RuCl_3$	addition of olefins to silicon hydride	591

TABLE XXXIV (*continued*)

Catalyst	Reactions catalyzed	Ref.
SbBr$_3$	alkylation, acylation, poly-merization	135,604,619, 675
SbCl$_3$	alkylation, acylation, poly-merization	110,135,328, 390,590,604, 619,675
SbCl$_5$	acylation, halogenation, poly-merization, isomerization	50,124,125,135, 180,465,604, 615,619,675
SbF$_5$	alkylation, acylation nitra-tion, halogenation, poly-merization	16,135,337,441, 448,604,619, 675
SbF$_3$Cl$_2$	halogenation	615
SeCl$_4$	dehydrohalogenation	327
SeF$_4$	nitration	16
SiBr$_4$	acylation	434
SiCl$_4$	acylation	415,686
SiF$_4$	alkylation, acylation, nitra-tion	441,631
SnBr$_4$	alkylation, acylation, poly-merization	135,462,604
SnCl$_2$	acylation, polymerization	590,122
SnCl$_4$	alkylation, acylation, halogen-ation, polymerization	212,215,426, 522,564,595, 686
SnF$_4$	alkylation, acylation	437,678
SnI$_4$	acylation, polymerization	442
SrCl$_2$	alkylation	432
TaBr$_5$	alkylation, acylation, dehydro-halogenation	112,325
TaCl$_5$	alkylation, acylation	134,238
TaF$_5$	alkylation, acylation	176,376
TaI$_5$	no activity	112
TeBr$_4$	acylation, halogenation	135,637
TeCl$_2$	acylation	134,136
TeCl$_4$	alkylation, acylation	134,136,186
TeF$_4$	nitration	452
TeI$_4$	dehydrohalogenation	303
TiBr$_4$	alkylation, acylation, poly-merization, haloalkylation, various Friedel-Crafts reac-tions	135,164,192, 434,555
TiCl$_3$	alkylation, acylation, poly-merization	309

Table continued

TABLE XXXIV (*continued*)

Catalyst	Reactions catalyzed	Ref.
$TiCl_4$	alkylation, acylation, polymerization, nitration, various Friedel-Crafts reactions	130,136,192, 204,238,464, 596,603
TiF_4	alkylation, acylation, isomerization, polymerization	376,446,612
TiI_4	various Friedel-Crafts reactions based on complex-forming ability	164
$ThBr_4$	acylation, halogenation	434
$ThCl_4$	alkylation, acylation	238,442
$TlCl_3$	alkylation, acylation	303
UCl_4	alkylation, acylation	304
UF_6	polymerization	180
VCl_4	alkylation, acylation, polymerization	23,442
WCl_6	alkylation, acylation, various Friedel-Crafts reactions	88,304
WBr_6	acylation	135
WI_4	dehydrohalogenation	327
$ZnBr_2$	alkylation, acylation, halogenation	135,437,637
$ZnCl_2$	alkylation, acylation, haloalkylation, polymerization, isomerization, Houben-Hoesch reaction	50,134,136, 137,198,244, 247,290,336, 401,412,496, 585
ZnF_2	alkylation, acylation	103
ZnI_2	alkylation, acylation, polymerization	437,442
$ZnCl_2$ + acetic acid	alkylation, aromatic hydrogen exchange	36,533
$ZnCl_2$ + acetic acid, HCl	alkylation, aromatic hydrogen exchange	36,533
$ZnCl_2$ on Al_2O_3	alkylation	558
$ZrBr_4$	alkylation, acylation	434
$ZrCl_4$	alkylation, acylation, various Friedel-Crafts reactions	102,238,256, 280,288

As a conclusion to the survey of the acidic halide type Friedel-Crafts catalysts it must be said that although the catalytic activity of various inorganic halides has been known since the work of Friedel and Crafts in 1877, no reliable basis for predicting catalyst activity has yet emerged. In the acylation of toluene (134,136) (probably the most extensively studied reaction for which catalysts have been arranged in order of reactivity), the halides which are active catalysts have the following properties in common: all have molecular chain, or layer lattices, never three-dimensional ionic lattices; all form complex halogen anions, stable in the solid state and in solution; all complex with organic molecules containing donor atoms (N, O, S, halogens). However, not all halides with the requisite lattices and coordinating ability were found to be catalysts. (The division between active and inactive halides varies with the reaction and experimental conditions and if based only on product yields is to some extent also governed by experimental precision.)

On the other hand, halides which in the solid state are fairly ionic compounds may become fair catalysts in solution with appropriate donor reagents. As an example SnF_4, regarded at least to a certain degree an ionic fluoride (it sublimes at 705°) in contrast to the other covalent tetrahalides of Group IVA, would not be expected to possess catalytic properties but was found quite active in the acetylation of toluene (678). It is concluded that the strong acceptor properties of stannic fluoride must be the factor which decides its behavior as a Lewis acid type catalyst in spite of its physical properties. It seems that even a high lattice energy can be compensated by the strength of suitable coordination.

IV. Amount of Catalyst Required

The alkylation of hydrocarbons, the isomerization of olefins, paraffins, and cycloparaffins and the polymerization of olefins in Friedel-Crafts type reactions require only the use of small amounts of catalyst, as there is no stable addition compound formation with the reagents. However, the ability of certain halide catalysts to form in the presence of co-catalysts stable arenonium type complexes ($ArH_2^+ AlCl_4^-$) with basic hydrocarbons (red oils) results in the successive deactivation of catalysts in hydrocarbon reactions and the need for additional catalysts to maintain the reactions. As reactions of the Friedel-Crafts type are frequently heterogeneous reactions, fresh catalyst may possess an active surface, giving rise also to surface catalytic effects general in heterogeneous reactions (593).

When alcohols, ethers, or esters are used as alkylating agents complexes are formed with the catalyst. The quantity of $AlCl_3$ (or other catalyst) required is thus higher and depends upon the alkylating agent used. Approximately 0.5, 0.75, and 1.0 mole of $AlCl_3$, for example, gave the best alkylation results respectively with tertiary, secondary, and primary alcohols.

In the aromatic ketone synthesis by condensing acid chlorides with aromatics it is necessary to employ a molar excess of catalyst. The acylation data of Groggins (234) relating to the preparation of benzophenone show the relation between the quantity of catalyst used and the yield of ketone obtained (Table XXXV). Optimum yield was obtained with about 1.1 mole $AlCl_3$ per mole of benzoyl chloride. Further increase of the amount of catalyst lowers the yield (owing to secondary condensation of the benzophenone formed) and less than the equimolar amount of catalyst gives increasingly low yields owing to incomplete complexing of reagent and product alike.

TABLE XXXV.   Relation of amount of catalyst and yield of benzophenone

Moles $AlCl_3$	Yield %
1.3	92.0
1.1	94.2
1.0	90.1
0.9	80.7
0.8	72.4
0.6	57.8
0.4	38.6
0.2	18.6

When aliphatic carboxylic acid anhydrides are used as acylating agents with aromatic compounds to form alkyl aryl ketones, the use of up to 2 moles of $AlCl_3$ results in a yield of approximately 100% based on only *one* acyl group being reactive. When 3 or more moles of $AlCl_3$ per mole of anhydride are used, then yields of 75 to 85% based on *both* acyl groups are obtainable (236).

The relation between the acylating agent (acid chloride, acid anhydride, carboxylic acid) and the catalyst (aluminum chloride) may be explained in terms of complex formation. When an acyl chloride is used as acylating agent it forms a 1:1 addition compound with aluminum halide. When aliphatic or aromatic carboxylic acid anhydrides are similarly condensed with excess catalyst reactions may occur depending on the molar ratio of aluminum chloride used. The first mole of catalyst halide can accomplish the fission of the acid anhydride with the possible production of acid halide (which then complexes with another mole of catalyst) and the formation of the aluminum chloride salt of the carboxylic acid.

$$(CH_3CO)_2O + AlCl_3 \rightarrow CH_3COCl \cdot AlCl_3 + CH_3COOAlCl_2$$

When three molecular proportions of $AlCl_3$ are employed, the acid formed itself can be used as an acylating agent, probably by first transforming it to the acyl halide. Thus aluminum chloride acts as a chlorinating agent and it is clear that the maximum yield of ketone obtainable with approximately two molecular portions of $AlCl_3$ would be about 100% based on one acetyl group and only half as much in terms of both acyl groups. However, with three molecular portions of $AlCl_3$ both acetyl groups can be used. This is confirmed by the data of Groggins (234) on the preparation of acetophenone from acetic anhydride and benzene at 90° (Table XXXVI).

TABLE XXXVI.   Relation of amount of catalyst and yield of acetophenone

Moles $AlCl_3$	Yield % (based on both acetyl groups)
1.1	12
2.3	52
2.8	65
3.3	76
4.4	81

In Friedel-Crafts ketone synthesis it has always been found advisable to incorporate a slight excess of aluminum chloride to insure complete reaction. When the aluminum halide is known to contain inactive or less active constituents, a proportionate allowance must be made. Generally an excess of 5–10% catalyst over that theoretically required has been found to give the best yields. The limiting conditions will be the purity of the halide and the absence of moisture in the reaction system.

The discovery that both acyl groups of acid anhydrides could be made reactive by employing at least three molecular proportions of aluminum chloride indicated that many carboxylic acids themselves could be employed as acylating agents in the preparation of ketones (235,274).

Using boron trifluoride no halogenation by the catalyst is to be expected. In this case the use of two moles plus excess catalyst still enables use to be made of both acyl groups with certain acids. The by-product acids themselves act as acylating agents in the presence of BF_3.

The amount of catalyst used in Friedel-Crafts polymerizations can affect the molecular weight of the polymers obtained. With more catalyst the number of starting chains as well as terminations are increased and lower molecular weight polymers are obtained than with smaller catalyst concentrations.

In Friedel-Crafts isomerization, like that of dialkylbenzenes or dihalobenzenes, the amount of catalyst used frequently has a marked effect. If one of the isomers formed (generally the *meta*-isomer) is more basic than the others, it can be selectively complexed and thus removed from the reaction mixture by excess catalyst. Thus it is possible to shift the composition of the isomer mixture over the thermodynamic equilibrium of the isomers and to obtain practically exclusively the more basic isomer. The amount of disproportionation is often dependent on the amount of catalyst used.

V. Purity of Catalysts

As discussed previously (p. 219) Friedel-Crafts reactions are nearly always carried out not only in the presence of substrate, reagent, and catalyst (solvent), but generally there is also present, in small amounts, an additional substance capable of reacting with the catalyst to activate it through proton or other cation formation. These substances are called *co-catalysts* or *promoters*. Co-catalysts are always needed to effect alkylations with olefins, polymerizations, and isomerizations. Alkylations and acylations with donor reagent molecules (containing O, N, S, halogen atoms) can also take place in the absence of co-catalyst or promoters, but general reaction conditions do not exclude these. It is obvious now that with few exceptions, the large body of experimental work reported on Friedel-Crafts reactions was carried out under reaction conditions, which provided copious amounts of co-catalysts. These include water present as moisture or impurity in the reagents and a variety of other donor compounds (present as impurities in reagents) or even oxygen from air. With the progress of new experimental techniques (infra-red and nuclear magnetic resonance spectroscopy, gas chromatography) it is now possible to obtain a fair picture of the general purity of Friedel-Crafts reactions systems. Claims of reactions carried out under the catalytic effect of "anhydrous" aluminum chloride (and related Friedel-Crafts acidic halide catalysts) are considered in the 1960's differently than they were even 10–15 years ago. The same must be said about the purity of the metal halide catalysts themselves. It must be remembered that the upsurge of interest in "anhydrous" aluminum chloride was brought about by the discovery of Saint-Claire Deville (139a) in 1854, based on the previous work of Wöhler, that $AlCl_3$ can be reduced by sodium to metallic aluminum. This route indeed was the first technical process to produce metallic aluminum and the electrochemical process was discovered only in 1888. Therefore the aluminum chloride used by

Friedel and Crafts and other early investigators was necessarily far from being chemically pure. It was prepared by chlorination of aluminum-containing clays or bauxites. Iron and other metal impurities, which co-sublime with aluminum chloride, cannot be excluded unless aluminum chloride is prepared from highest purity (> 99.99%) electrolytically prepared aluminum and pure chlorine gas. Fortunately, however, all these conditions have not hindered the pioneering work of Friedel and Crafts, or generations of subsequent chemists. Moisture and other impurities are, as we now know, to a certain degree favorable co-catalysts in Friedel-Crafts reactions. Similarly, almost from the beginning of investigations on the aluminum chloride reaction it was clearly understood that other metal halide impurities in aluminum chloride enhanced its activity in many reactions. This observation led Friedel and Crafts to the findings that in their reactions $AlCl_3$ may be replaced by other metal halide catalysts ($FeCl_3$, $ZnCl_2$, etc.).

Commercial aluminum chloride containing iron is more active in the ketone syntheses than pure $AlCl_3$. Ferric chloride is the cause of increased activity. Similar observations were made in alkylations (58,367,510,511).

$AlCl_3$ is activated for the Gatterman-Koch synthesis by $TiCl_4$ and $FeCl_3$ (178) and by $SnCl_4$, $SiCl_4$, or $TiCl_4$ for the acetylation of cyclohexane to acetyl methylcyclopentane (453). These are only a few of many examples of the activating effect of other metal halides on aluminum chloride. Similar effects must be operative in the case of various other acidic halide catalysts. This should be kept in mind in considering reactions with metal or metal halide catalysts of uncertain purity. For example the effectiveness of cerium powder as catalyst in Friedel-Crafts type reactions with alkyl and acyl chlorides (341) is difficult to evaluate in view of possible iron impurities giving catalytically active $FeCl_3$. Similar considerations may arise in the case of many other examples in the literature where the purity of catalysts has not been clearly stated. Of course the same must be also valid for the effect of moisture or other impurities, which could react with otherwise catalytically inactive metal halides (above all with fluorides) giving rise to hydrogen halides, which then themselves could cause Friedel-Crafts type reactions. For example phosphorus trifluoride, PF_3, which under absolutely dry conditions is a weak catalyst, if active at all, polymerizes styrene and initiates Friedel-Crafts alkylations readily if the reaction mixture contains even traces of water. This effect is suggested to be due to HF formed by hydrolysis of the catalyst. Similar phenomena are frequently encountered with other halide catalysts (434) and must be considered before evaluating the alleged effect of a catalyst system.

VI. Effect of Solubility and Particle Size of Catalyst

1. Effect of Physical Nature of Catalyst

Most of the original work of Friedel and Crafts was carried out in heterogeneous reaction systems. Aluminum chloride and related metal halide Lewis acid catalyst generally show only limited solubility in hydrocarbon solvents. The solubility increases in solvents having polar donor groups such as oxygen, sulfur, nitrogen, and halogens.

Since the activity of a heterogeneous catalyst in general increases as its surface is increased, finely powdered aluminum chloride is more useful in many heterogeneous reactions than is a coarsely ground lump aluminum chloride. The use of the finely powdered catalyst, however, often results in too rapid reactions, which may be undesirable (such as secondary condensations effected with sensitive materials or with very active compounds). For these reactions lump aluminum chloride is more efficient than the powdered grade. In most cases both grades of aluminum chloride form a soluble complex with polar solvents, diluents or reactants with equal ease. Here, of course, the particle size of the aluminum chloride (or related halide catalyst) has no effect on the course of the reaction. Neither is it of importance in reactions effected under conditions providing close contact of the reactants, as in a ball mill, which is used in phthalic anhydride addition reactions in the absence of solvents or diluents.

In different industrial processes various grades of aluminum chloride are used. For example in ethylbenzene production the granular product is preferred. The particle size chosen is usually dependent upon the process equipment available. A granular product lends itself to a uniform continuous feeding by screw conveyors better than would a very fine powder. Also in many operations finely divided aluminum chloride cannot be conveniently handled because its greater reactivity results in rapid heat evolution, making temperature control difficult. Thus, equipment design greatly influences preferences in the physical form of aluminum chloride.

The activity of aluminum chloride in heterogeneous reactions can be further increased by depositing it on a porous carrier (13). Heating the deposited Friedel-Crafts catalyst on a porous carrier in the presence of oxygen (or other oxidizing agent) at 750–800° until multivalent compounds are oxidized to the highest oxidation state, is generally the most suitable procedure of preparing a catalyst with highest activity.

Aluminum bromide is in general considerably more soluble than aluminum chloride. The reason for this is that aluminum chloride is an ionic solid, having an ionic lattice with a heat of fusion of 17 kcal./mole. As a consequence of its high lattice energy the chloride has a low solubility in inert solvents, *e.g.*, carbon disulfide, light petroleum, benzene. Aluminum bromide, a covalent dimeric compound even in the solid state, has a molecular lattice with a heat of fusion of only 5.4 kcal./mole and is readily soluble in many hydrocarbons.

2. Solution by Complex Formation

The question of the existence of solution complexes between aromatic hydrocarbons and aluminum halides as well as other Lewis acid metal halide catalysts attracted the attention of a large number of workers. Similarly the formation of ternary complexes involving aromatic hydrocarbons, Lewis acid halides, and hydrogen halides giving rise to the so called "red oils" (which are basically δ-complexes) was widely investigated. It was shown that in a strictly binary aluminum halide–hydrocarbon system, involving neither hydrogen chloride nor moisture, only a π-complex type solvent–solute interaction takes place and the aluminum halide is present mostly as the dimer Al_2X_6 (85,161,177,403,485,655).

Tetrahedrally bonded aluminum chloride ($AlCl_4^-$) in σ-complexes is catalytically inactive. However, the ionic "red oils" are capable of dissolving substantial amounts of uncomplexed aluminum chloride.

Aliphatic hydrocarbons like petroleum ether are frequently used as diluent media for Friedel-Crafts reactions, although their actual solvent properties are limited.

In the catalytic action of aluminum chloride and other Friedel-Crafts catalysts Korshak and Lebedev (329) stressed the importance of the powerful dipole existing in solutions where the dimeric molecules are present, as for example Al_2Cl_6. They attributed the form $AlCl_2^+ AlCl_4^-$ to the dipole and called attention to the fact that the highly polar catalyst dimers can introduce strong polarization into the reactant molecules. This they believed to be an important step in the usual Friedel-Crafts substitution reactions. Similarly the action of complexes of Lewis acid with solvents, like boron trifluoride–ether, should be similar, owing to the high dipole present in the addition compound between the positive oxygen and the

negative boron atoms.   In view of our present understanding of
the action of aluminum chloride (and related catalysts) this theory is
less plausible.   The catalytically active species seems to be well
established as the monomeric Lewis acid halide.   The solvent complexes, if they have lower bonding energies than the heat of dissociation of the aluminum halide dimers, help the dissociation of the
catalyst into the monomer and make it more easily available for the
reagents.   High dipoles can be equally well explained on the basis
of polarized donor–acceptor complexes.

Aluminum chloride is readily soluble in compounds which contain
donor atoms or groups, *e.g.*, alcohols, ethers, ketones, acids, acyl
chlorides, nitriles, and aromatic and aliphatic nitro compounds.
The freezing-point method shows that in these solvents the aluminum
halide is present as the monomeric AlX_3.   Thus it is evident that
coordination complexes of the type [solvent: AlX_3] are formed.
This ability to form complexes is explained on the basis that the
aluminum halide (and related acidic halide Lewis acids) in the
monomeric form completes its group of eight electrons by attaching
to itself a lone pair belonging to another donor atom.   In many
cases these addition compounds have been isolated as stable entities
under anhydrous conditions, *e.g.*, $ROH \cdot AlX_3$; $R_2O \cdot AlX_3$; $R_2CO \cdot$
AlX_3; $RCOX \cdot AlX_3$; $RCN \cdot AlX_3$; $RNO_2 \cdot AlX_3$; $C_5H_5N \cdot AlX_3$, etc.
In other cases, however, the intermediate complexes may exist only
momentarily.   This is the basis of the catalyst action in Friedel-
Crafts reactions.   Loss of the original form (which is mostly dimeric
or polymeric) of the catalyst is inherent in all of these reactions, the
active catalyst being the monomeric Lewis acid halide.   Further,
since acidic halides such as aluminum halides are polar catalysts, the
reactions catalyzed by them are of an ionic or polar type, *e.g.*, bonds
are made and broken heterolytically.

3. Solvents

Work on the use of solvents in Friedel-Crafts reactions has indicated that efficient solvents make possible a smaller reagent: catalyst
ratio, higher purity of product, and higher yields.

Although the actual technique of the Friedel-Crafts reactions
cannot be detailed here, a word must be said concerning the solvents
most frequently used, since they have an important effect on the
reactions.   One of the most generally applied practices is to use as
the solvent an excess of the hydrocarbon or substituted hydrocarbon
that is undergoing substitution.   Thus for the substitution of alkyl

and acyl groups into benzene, the latter compound is almost invariably used as solvent. Although this practice yields ketone effectively, it is not generally desirable in alkylation where a definite product and high yield are the major interests. Other aromatics such as toluene, alkylbenzenes, fluoro-, chloro-, and bromobenzene may also be used in excess as solvents. However, there are often reasons why one of the reactants may not be used as the solvent. In this case an inert solvent may be used. It must also be remembered that, in many cases, hydrocarbons are rather poor solvents for acidic halide catalysts and serve predominantly only as diluents in heterogeneous systems.

The most frequently used inert solvent in Friedel-Crafts type reactions is carbon disulfide. Another frequently used is petroleum ether. Carbon disulfide appears to have a specific action on the Friedel-Crafts syntheses, tending to prevent migration of groups (524). It is generally an excellent solvent for a variety of alkylations (212) and acylations (425,450). In certain cases, however, carbon disulfide is known to reduce the yields considerably (398) and slow down the velocity of the reactions (511). Still it remains as one of the best general solvents for the Friedel-Crafts type reactions.

Ethylene chloride and methylene chloride are fairly good solvents although they may themselves react with the catalysts at elevated temperatures.

Nitrobenzene and nitroalkanes (nitromethane, nitroethane, nitropropanes) are also good solvents for Friedel-Crafts systems not only because of their good solvent powers, but also because they form addition complexes with aluminum chloride and other acidic metal halides and in this way ameliorate the action of the catalyst. The complex formation lessens the tarring and disporportionation action frequently caused by aluminum chloride, thereby making the introduction of alkyl and acyl groups essentially a mild reaction, applicable to sensitive compounds like phenols, amines, benzyl halides, or quinolines.

When highly reactive substrates are to be substituted and mild condensing agents are employed, solvents such as o-dichlorobenzene, acetylene tetrachloride or diethyl ether may be used. Under these conditions the alkylation or acylation of the substrate in question must be so rapid that the solvent under the same reaction conditions does not have an opportunity to react. The same is true when benzene is applied as solvent, for example in the acylation of furan by acetic anhydride in the presence of stannic chloride. The major product is acetylfuran, although benzene itself may undergo slow acetylation under the same conditions. Anthracene also can be acylated in benzene solution by phthalic anhydride and aluminum chloride.   Ether is often recommended as a solvent of Friedel-Crafts reactions (451,654) although with strong Lewis acid halides, like aluminum chloride, it reacts quite vigorously on heating and serves as an alkylating agent.

Liquid ethylene and propane are sometimes employed as solvents for low-temperature alkylations and polymerizations. Excess of halide reagents such as acetyl or benzoyl halides or ethyl halides can be used as solvents for Friedel-Crafts acylations and alkylations. Liquid sulfur dioxide is also a good solvent for aluminum chloride and many other metal halide catalysts because of its low reactivity and high solvent properties (517).   There is some evidence for the formation of a 1 : 1 aluminum chloride–SO_2 complex but it is unstable under atmospheric pressure and ordinary temperatures. Dimethyl sulfoxide and tetramethylene sulfone have excellent solvent properties and are rapidly acquiring considerable importance as solvents for homogeneous Friedel-Crafts reactions.

The solubility of aluminum chloride and bromide in a number of solvents is summarized in Table XXXVII.

TABLE XXXVII. Solubility of $AlCl_3$ and $AlBr_3$ in organic solvents at 20° (g./100 g. solvent)

	$AlCl_3$	$AlBr_3$
$CHCl_3$	0.8	
CCl_4	0.15	
Ether	20^a	
C_2S_2	0.5	59
SO_2	$8^{-8°}$	$0.02^{0°}$
CH_3NO_2	37^{00}	
C_2H_5Cl	5^{00}	
$C_2H_4Cl_2$		$C_2H_4Br_2$
		42.1
CH_3COCl	150	
$(CH_3)_2S$	70^a	
C_6H_6	0.1	46.5
$CH_3C_6H_5$	0.3	42.5
$C_6H_5NO_2$	23	45
C_6H_5COCl	19	34
C_6H_5CN		24
C_6H_{12}		38

^{a} Reacts after a period of time.

Few precise solubility data exist for acidic metal halide catalysts in the organic and non-aqueous inorganic solvents used in Friedel-Crafts systems. The same must be said for Brønsted type proton acids in similar systems. It would be of considerable aid in finding the most appropriate solvents for any Friedel-Crafts reaction if future work provided such data.

If it is necessary to obtain very powerful reaction conditions and no suitable solvent can be found, it is possible to carry out the aluminum chloride catalyzed reactions in a melt of sodium chloride–aluminum chloride. Stoichiometric quantities of sodium chloride and aluminum chloride are melted together and the reaction products are then introduced into the melt at temperatures between 110–250° (542).

Few quantitative studies exist on solubilities of boron trifluoride and other boron halides in unreactive solvents and very little qualitative information is available on general solubilities (56).

Straight-chain, unsaturated, aliphatic hydrocarbons are reported to dissolve boron trifluoride. Kerosene, high-boiling naphtha (356), benzene (395), pentane (138), propane and ethane (599), dichlorobenzene, carbon tetrachloride, chloroform (356), and tetrachloroethane (152) have been listed as inert solvents used in catalytic processes involving boron fluoride as the catalysts. However, the solubilities in apolar solvents are generally small (98). As an example the solubility of boron trifluoride in n-pentane was found to be in the order of 1–12 ml. of boron fluoride gas/g. of pentane (depending on the temperature).

The solubility of boron trichloride, tribromide, and triiodide in hydrocarbon solvents (both aliphatic and aromatic) is considerably higher than that of BF_3. This probably is at least partly due to their higher boiling points.

The great tendency of boron fluoride to form complexes depends upon its small molecular volume, high electropolarity, and the unsaturated character of the boron atom, which favors coordination. Molecules forming complexes with boron fluoride generally have donor atoms with unshared electron pairs (O, N, S, F, etc.) which can share their lone pair with the electron-deficient boron atom. As a result the electron sextet envelope of boron is made up to an octet. Between boron fluoride and the complexing substance a donor–acceptor bond is formed and in the newly formed compound the boron atom becomes negatively charged.

The $sp^2 \rightarrow sp^3$ hybridization of boron leads to a change of planar configuration of the BF_3 molecule (sp^2 hybrid state) to a tetrahedral one (sp^3 hybrid state). The three fluorine atoms and the donor atom (O, N, S, F, etc.) are at the apices of a tetrahedron, the boron atom being at the center.

With substances capable of being found in the associated state (water, alcohol, acids) boron fluoride can form coordination compounds which may contain one, two or a larger number of associated molecules per mole of BF_3. With all other unassociated (amines, amides, acid anhydrides) or weakly associated (ethers, esters, ketones) compounds, boron fluoride forms coordination compounds of mole for mole composition. Molecules having more than one donor atom can form complexes with molar amounts of BF_3 per donor atom. Thus acid anhydrides form complexes $(RCO)_2O \cdot 3BF_3$.

Boron fluoride coordination compounds are often formed as intermediate products in reactions in which BF_3 is used as a catalyst. In the formation of coordination compounds of donor molecules with boron trifluoride the increase of valency of the boron and donor atoms leads to a weakening of the remaining bonds of the donor atom and to an increase in the interatomic distances. The organic molecules in these compounds are therefore activated and exist in excited states from which they readily enter into different reactions.

Many boron fluoride coordination compounds surpass free BF_3 in their catalytic activity and are therefore used as specific independent catalysts.

Some of the most frequently used catalytic addition compounds are $BF_3 \cdot (C_2H_5)_2O$; $CH_3OH \cdot BF_3$; $CH_3COOH \cdot BF_3$; $BF_3 \cdot H_3PO_4$; $BF_3 \cdot H_2SO_4$ (see p. 230).

Sometimes addition compounds simply serve as a convenient source of BF_3 catalyst.

The coordination compounds of BF_3 have been reviewed by Booth and Martin (56) and by Topchiev (635).

Boron trichloride, boron tribromide, and boron triiodide also tend to form addition compounds. However, in these compounds secondary reactions leading to cleavage and formation of secondary products with the complexing molecule frequently occur and prohibit their use as catalytic halides.

VII. Solvent Effects

The course of Friedel-Crafts type reactions is in some instances greatly affected by the solvent used in the reaction.

The acylation of naphthalene with acetyl chloride in solvents like ethylene chloride or carbon disulfide yields chiefly 1-acetylnaphthalene, whereas in nitrobenzene or nitromesitylene the product contains substantial amounts of the 2-isomer. This method of obtaining 2-acetyl-naphthalene is important because it is one of the very few satisfactory ways of directly introducing a side chain into the β-position of naphthalene.

The α- and β-isomer ratios obtained in the acetylation of naphthalene using aluminum chloride catalyst in methylene chloride and nitrobenzene solvents are shown in Table XXXVIII (22), in the presence of added agents capable of complexing with the catalyst.

TABLE XXXVIII.   Isomer distribution in acetylation of naphthalene

Acylating agent	Solvent	Added agent	% Isomer acetyl naphthalene	
			α-	β-
$CH_3COCl \cdot AlCl_3$	CH_2Cl_2	—	98	2
$CH_3COCl \cdot AlCl_3$	CH_2Cl_2	CH_3COCl	60	40
$CH_3COCl \cdot AlCl_3$	$C_6H_5NO_2$		34	66
$CH_3COCl \cdot AlCl_3$	CH_2Cl_2	nitromesitylene	29	71
$CH_3COCl \cdot AlCl_3$	CH_2Cl_2	$C_6H_5NO_2 \cdot AlCl_3$	97.5	2.5

The formation of 2-acetylnaphthalene is also favored by the use of an excess of the acylating agents.   A possible reason for the preferential attack at the 2-position is steric hindrance, since solvents that favor this position form bulky complexes with the acylating agent (22).

The effect of the solvent on isomer ratios in substitution reactions is also observed in many other instances.   For example, when β-naphthol methyl ether is acetylated with acetyl chloride in carbon disulfide or benzene solution, the product of the reaction is 2-methoxy-1-acetylnaphthalene (200).   However, if the reaction is carried out in nitrobenzene solution, the product is mostly 2-methoxy-6-acetylnaphthalene (254).

In the acylation of perylene in carbon disulfide solution, the product is preferentially the 3,9-substituted product but in the absence of carbon disulfide and with an excess of acyl or aroyl chlorides the main product is the 3,4-diacylperylene (486).

Eitel and Fialla (160) have investigated the influence of solvents on the relative proportions of o-1-naphthoyl- and o-2-naphthoyl-benzoic acid formed by the Friedel-Crafts reaction of phthalic anhydride and naphthalene.   It was found that the solvent used has little effect on the proportion of the isomers, but it does influence the overall yield.   The mixed naphthoylbenzoic acids obtained from the synthesis contain 60–70% of the o-1-isomer.   Reaction temperatures above 20° generally decrease the amount of this isomer and lower the overall yield.   The yield at 0°–20° with various solvents are shown in Table XXXIX.

TABLE XXXIX.   Effect of solvent on yields of naphthoylbenzoic acids

Solvent	Yield %
benzene	87–91
o-dichlorobenzene	91
tetrachloroethane	43
nitrobenzene	28
carbon disulfide	15–18

The solvent can also affect the isomer ratios obtained in other Friedel-Crafts acylations and alkylations.   In acylations of phenols with acyl halides in the presence of aluminum chloride catalyst it was found that the *para* : *ortho* ratio was the lowest in carbon disulfide solution while the highest occurred in nitrobenzene solution (504).   Similar observations with other acyl halide acylations have been reported (503).

Solvent effects influencing Friedel-Crafts nitration with nitronium salts were summarized recently (440).   It is probable that most of these solvent effects are due to the fact that the active acylating or alkylating species is solvated to a certain degree giving rise to a bulkier substituting agent, with a different steric effect.

At the same time solvation of the active substituting agent (acylium, nitronium, incipient carbonium ion) may also alter its activity and influence not only the positional but also the substrate selectivity.

Pepper (460) has shown that the rate and degree of polymerization of styrene and methylstyrene by tin tetrachloride are increased by an increase in the dielectric constant of the solvent.   Traces of moisture reduce the rate in solvents of high dielectric constant, but appear to increase it in hydrocarbons.   Solutions of $SnCl_4$ in solvents of high dielectric constants show appreciable conductivity in the presence of traces of water.   The polymerization has an ionic mechanism, but in solvents of high dielectric constants, for example in nitroparaffins, it does not involve hydrated ions.

VIII. Effect of Temperature

In organic syntheses the general effect of operating at higher temperatures is to increase the rate of reactions.   In some reactions, *e.g.*, sulfonation of naphthalene or related reactions, variations in temperature also affect the orientation of the new substituent. However, in many Friedel-Crafts reactions other deep-seated changes occur when the reaction temperature is not rigidly controlled at the optimum level for a particular synthesis. Even a moderate elevation in temperature may bring about secondary condensations. With excessive heating almost all Friedel-Crafts reaction mixtures can be largely converted to oily or resinous masses of complex or uncertain constitution.

The aluminum chloride catalyzed reactions with active reactants usually proceed rapidly with evolution of heat. A rise in temperature to the point at which decomposition may occur must be prevented. This can be effected by external cooling of the reaction vessel or by slow addition of the catalyst or reactant to the reaction mixture with vigorous stirring. Active reagents which react rapidly must be handled with some or all of these precautions in order to prevent decompositions, resinifications, or charring. On the other hand, syntheses with unreactive materials generally require external heating, the temperature necessitated being governed by the reaction rate. In alkylations, acylations, and most addition reactions temperatures of over 90–100° are rarely needed, whereas dehydrating and dehydrogenating condensations are usually carried out at higher temperatures in aluminum chloride–sodium chloride melts. Catalysts of low volatility, as boron fluoride, can be generally used only at lower temperatures. However, when its addition complexes like boron fluoride–ether, boron fluoride–acetic acid, boron fluoride–alcohol are used, the boiling point of these complexes determines the temperatures applicable. Proton acids, like sulfuric acid and anhydrous HF, are generally used only at lower temperatures, partly owing to secondary decomposition which readily occurs when organic compounds are heated to elevated temperatures with sulfuric acid and partly to the volatility of anhydrous HF (b.p. 20°). When higher reaction temperatures are needed anhydrous HF may be used at superatmospheric conditions.

At elevated temperatures a number of Friedel-Crafts type reactions are reported to take place even in the absence of any catalyst.

As an example α-benzoylnaphthalene is obtained in a Friedel-Crafts type ketone synthesis with $AlCl_3$ as catalyst in carbon disulfide as solvent at 0° (99).

At the reflux temperature of the components (naphthalene, benzoyl chloride) α-benzoyl naphthalene was obtained in the absence of any catalyst, even before the Friedel-Crafts synthesis was known (239).

9-Benzoylanthracene is obtained from benzoyl chloride and anthracene in the presence of $AlCl_3$ in nitrobenzene as solvent at $-10°$ with good yield (335).

The same *meso*-ketone was obtained by Nenitzescu (413) when the components were heated in refluxing nitrobenzene solution, *e.g.*, around 200° in the absence of catalyst.

The chloride of 2-tetralylbutyric acid undergoes ring closure to 1-anthracenone during its preparation, in the absence of $AlCl_3$ (548). In the presence of $AlCl_3$ the ring closure is so violent even at lower temperatures that only low yields of the ring ketone are obtained.

Nenitzescu (413) has shown that when benzyl chloride is heated with such highly reactive aromatic compounds as biphenyl, naphthalene, anisole, and mesitylene, the expected diarylmethane derivatives are obtained in the absence of any catalyst.

Acetylations of toluene and mesitylene with acetyl chloride takes place at 220° with substantial yields in the absence of any catalyst (439).

Similarly benzoyl chloride may acylate biphenyl, naphthalene, or anthracene simply by heating the reactants in the presence of nitrobenzene. An almost quantitative yield of 9-benzoylanthracene may be obtained in this way. Since these condensations proceed in an analogous manner, but at considerably lower temperatures in the presence of aluminum chloride and related catalysts, the Friedel-Crafts reactions may not necessarily always involve the intermediate formation of a complex with the metal halide catalyst.

The fact that certain reactions can be effected either by thermal excitation or by $AlCl_3$ catalyst indicates only that $AlCl_3$, like heat, is capable of activating the reactants involved. However, it does not explain the mechanism through which the activation is brought about. The possibility that both heat activation and metal halide catalysis may follow the same path should be borne in mind (Thomas) (623).

The conclusions of Thomas require consideration, but nevertheless should be regarded with some reservation in view of the work of Simons and Hart (572), who proved that HCl can act as a catalyst in certain Friedel-Crafts type reactions. Some of the reaction investigated also occur in the absence of HCl, but with substantially lower yields. As hydrogen chloride may be formed from halide reagents at elevated temperatures and as it is also the by-product of condensation with these reagents, the autocatalytic effect of the reactions can be easily interpreted.

Recently there have been indications that thermal excitation in the "non-catalytic" acylations takes place at least partly through homolysis and not heterolysis. The isomer ratio of the metyl acetophenes obtained in non-catalytic acetylation of toluene with acetyl chloride at 240° is: 55% *ortho*, 6% *meta* and 39% *para*, compared with the isomer ratio in heterolytic Friedel-Crafts acetylation of toluene: 1.5% *ortho*, 0.5% *meta*, 98% *para* (439).

IX. Catalytic Activity of Acidic Halides in Aqueous Media

In a series of investigations it has been shown by Jenny (296,297, 298) that contrary to general belief certain metal halide Lewis acids possess considerable catalytic activity in alkylation reactions in aqueous media. Thus he found that $SnCl_2 \cdot 2H_2O$, $SnCl_3 \cdot 3H_2O$, $SnCl_4 \cdot 5H_2O$ catalyzes the benzylation of aromatics with benzyl chloride, as does also $H_2SnCl_6 \cdot 6H_2O$. Yields of the diphenyl-methanes obtained were between 23–67%. It is interesting to note that the corresponding anhydrous halide in a similar reaction has either no catalytic activity or gives considerably lower yields. $ZnCl_2 \cdot 2H_2O$ and $FeCl_3 \cdot 6H_2O$ are also active catalysts in benzylations, as are $SbCl_3 \cdot 4H_2O$, $SbCl_5 \cdot 5H_2O$, $BiCl_2 \cdot H_2O$, and $TiCl_4 \cdot 6H_2O$. The hydrates of $FeCl_3$, $SnCl_4$, and $SbCl_5$ are the most active catalysts (proven by the evolution of HCl in the alkylation reaction). $ZnCl_2$ and $BiCl_3$ hydrates are moderately reactive, whereas hydrates of $SbCl_3$, $SnCl_2$, and $TiCl_4$ are less active. $ZnBr_2$ and $FeBr_3$ are equally active catalysts in hydrochloric or hydrobromic acid solutions and compare favorably with the corresponding chlorides. $FeBr_2$ is a good catalyst in benzylation of phenol but the corresponding $FeCl_2$ is inactive.

In the investigations of Jenny the following metal chlorides have been found to be inactive in attempted benzylation reactions in aqueous media: $NaCl$, KCl, NH_4Cl, $MgCl_2$, $CaCl_2$, $BaCl_2$, $CuCl_2$, $CdCl_2$, $HgCl_2$, $MnCl_2$, $CoCl_2$, $NiCl_2$, $FeCl_2$, $CrCl_3$, $AlCl_3$. The inactivity of certain metal halides catalysts could be due either to a specific inactivity concerning catalytic power, or to low solubility in the applied aqueous hydrochloric acid solutions. The first case may be illustrated with $CaCl_2$ which is very soluble but has less specific catalytic activity. The second case may be present with aluminum chloride, which is one of the most active Lewis acid metal halide catalysts in anhydrous media, but is inactive in aqueous media, partly due to the fact of its very low solubility, particularly in aqueous hydrochloric acid solutions.

Allen, however, found that $AlCl_3 \cdot 2H_2O$, $AlCl_3 \cdot 3H_2O$, $AlCl_3 \cdot 4H_2O$, and $AlCl_3 \cdot 5H_2O$ are all capable of the catalysis of the propylation of toluene (8).

A more plausible explanation of the catalytic activity of the investigated metal halide hydrates is to be found in their Brønsted and not Lewis acidity. It is improbable that there is any Lewis acid activity of the coordinatively saturated hydrates in aqueous media. However, the hydrates of the metallic Lewis acid halides can be looked upon as conjugate acids in the hydrated form and therefore capable of certain proton acid catalytic activity. The importance of dissolved HCl in the reactions, like benzylation, is unknown. Hydrolysis of benzyl chloride itself must go through a carbonium-ion type of intermediate, capable under suitable conditions of effecting concurrent alkylation. It is, therefore, not entirely unexpected to observe Friedel-Crafts type reactions even in aqueous media. As concurrent hydrolysis represents strong competition, fast reactions, such as benzylation must be chosen. It should be remembered that acid-catalyzed esterification of acid anhydrides with alcohols can also be carried out in aqueous media, as esterification is generally faster than concurrent hydrolysis.

Reported catalytically active metal halide hydrate catalysts are summarized in Table XL.

TABLE XL.   Metal halide hydrate catalysts

Catalyst	Reactions catalyzed	Ref.
$AlCl_3 \cdot 2H_2O$	alkylation	8
$AlCl_3 \cdot 3H_2O$	alkylation	8
$AlCl_3 \cdot 4H_2O$	alkylation	8
$AlCl_3 \cdot 5H_2O$	alkylation	8
$AlCl_3 \cdot 6H_2O$	dealkylation	452
$BiCl_3 \cdot 2H_2O$	alkylation	297
$FeCl_3 \cdot H_2O$	alkylation	296
$SbCl_3 \cdot 4H_2O$	alkylation	297
$SbCl_5 \cdot 5H_2O$	alkylation	297
$SnCl_2 \cdot 2H_2O$	alkylation	298
$SnCl_2 \cdot 3H_2O$	alkylation	298
$SnCl_4 \cdot 5H_2O$	alkylation	298
$TiCl_4 \cdot 6H_2O$	alkylation	297
$ZnCl_2 \cdot H_2O$	alkylation	298

X. Acidic Oxyhalide Catalysts

If it is assumed that the central atom of an oxyhalide ($POCl_3$, $COCl_2$, $SOCl_2$, $SeOCl_2$, CH_3COCl, $VOCl_3$) is electron deficient or bears some positive charge, owing to polarization by the electron-attracting halogen atoms, then it should behave as an acceptor toward sufficiently strong donors.

However, it must be remembered that oxyhalides act as donors under general conditions and in systems containing strong Lewis acid halide acceptors (556). Their acceptor properties are observed only if the system contains strong electron donors and no other more powerful acceptors.

It has been shown from the freezing-point diagrams of pyridine and certain oxychlorides, that 2 : 1 and 1 : 1 addition compounds are formed with $SeOCl_2$, CH_3COCl, and $SOCl_2$ (556).

Zeffert has shown that $POCl_3$ and C_5H_5N form no compound (687). While it is considered that donor–acceptor bonding is responsible for bond formation in most of the above cases, the oxyhalide being an electron acceptor, it is probable that such bonding is not the only intermolecular attraction in these binary systems, particularly with the more polar $POCl_3$ and $SeOCl_2$. Induced dipole–dipole interactions offer explanations for some of the peculiarities observed in the freezing-point diagrams of the systems.

Concerning the Friedel-Crafts catalytic activity of oxyhalides phosgene, phosphorus oxychloride, and thionyl chloride are extensively used in the Vilsmeier type aldehyde and ketone syntheses.

Aluminum oxyhalides, like AlOBr, are sometimes involved in the catalysis of alkylations and polymerizations (683).

$SeOCl_2$ and $VOCl_3$ show catalytic ability in acylation of aromatics with acyl halides (442).

TABLE XLI.   Acidic oxyhalide catalysts

Catalyst	Reactions catalyzed	Ref.
AlOBr	alkylation, polymerization	
$COCl_2$	Vilsmeier acylation	57
$POCl_3$	acylation, Vilsmeier aldehyde synthesis, rearrangements (Beckmann)	28,35,181,319, 630,662
$SOCl_2$	Vilsmeier acylation, rearrangements (Beckmann)	57,630
$SeOCl_2$	acylation	
$VOCl_3$	acylation	

XI. Metal Powder Catalysts Effecting Reactions with Halide Reagents (*in situ* formation of acidic halides)

As discussed in Chapter I in connection with the history of the Friedel-Crafts reactions, it was Zincke (198,239,405,470,690) who first observed the condensation of alkyl and acyl halides with aromatic hydrocarbons in the presence of metal powders (Ag, Cu, Zn). It was, however, Friedel and Crafts who established that the active catalysts in these and related systems were indeed the metal halides formed *in situ*.

These early observations have been used successfully ever since in carrying out Friedel-Crafts type reactions with halide (alkyl, aralkyl, acyl, sulfonyl and nitryl halides, elementary halogens, etc.) reagents in the presence of metal powders, instead of preformed anhydrous metal halides.

The method has achieved widest application in the ring halogenation (especially chlorination and bromination) of aromatics. It is also used in alkylations, acylations, and other related reactions.

Radzivanovskii has done pioneering work in the use of activated aluminum metal in place of aluminum chloride catalyst. Tsukervanik and Azatyan and their co-workers extended this work to other metal powders such as copper, titanium, chromium, molybdenum, tungsten, etc.

Table XLII summarizes some of the metal catalysts reported to give Friedel-Crafts activity through *in situ* formation of catalytic acidic metal halides with halide reagents or halogens.

TABLE XLII.   Metal powder catalysts in Friedel-Crafts type reactions

Catalyst	Reactions catalyzed	Ref.
Al	alkylation (including that of phenols and aromatic amines), acylation, halogenation	17,19,197,609,610,611, 637,642,643
Al + HCl	various Friedel-Crafts reactions	147,148,149,268
Al + Cl_2, Br_2, I_2	various Friedel-Crafts reactions	492,493
Al + Cu	alkylation	
Al + Hg	alkylation, acylation	1,53,74,79,142,203,265, 507,508
Al + $HgCl_2$	alkylation	53,54
Al + Zn	alkylation, acylation	
Au	halogenation	434
Bi	halogenation	637
Ca + Cu	alkylation	48
Cd	halogenation	637
Ce	alkylation, acylation	341
Co	halogenation	434
Cr	alkylation	18,635
Cu	alkylation	
Fe	alkylation, acylation, halogenation	637,639

TABLE XLII (*continued*)

Catalyst	Reactions catalyzed	Ref.
Ga	alkylation, acylation, halo-genation	434
Hg	alkylation, halogenation	76,637
In	halogenation	434
Mg	halogenation	
I₂	alkylation, acylation, halo-genation	114,248,270,310,641
Mn	halogenation	434
Mo	alkylation	638
Nb	alkylation, acylation, halo-genation	434
Pb	alkylation, halogenation	637,638
S	halogenation	637
Sb	halogenation	434
Se	alkylation	638
Sn	alkylation	638
Ta	alkylation, acylation, halo-genation	434
Te	halogenation	637
Th	halogenation	434
Ti	alkylation	551
Tl	alkylation, acylation, halo-genation	434
U	halogenation	434
V	halogenation	434
W	alkylation	638
Zn	alkylation, acylation, halo-genation	198,239,405,470,690 637
Zr	alkylation, acylation, halo-genation	434

XII. Metal Oxide Catalysts Effecting Reactions with Halide Reagents (*in situ* formation of acidic halides)

In addition to metal powders metal oxides can also in certain cases catalyze reactions with halide reagents through intermediate

TABLE XLIII. Metal oxide catalysts active in reactions with halide reagents

Catalyst	Reactions catalyzed	Ref.
P_2O_5	alkylation, acylation	346,366
TcO_2	alkylation	186,494
ZnO_2	acylation	144
Al_2O_3 + HCl	alkylation	191
Al_2O_3	halogenation	365
Fe_2O_3	halogenation	233

formation of the catalytically active metal halides (or oxyhalides).
Table XLIII shows some reported cases of these reactions.

XIII. Metal Alkyl Lewis Acid Catalysts

Although most interest in Lewis acid type catalytic activity in
Friedel-Crafts type reactions is concerned with acidic metal halides,
metal alkyl Lewis acids are gaining increasing interest. This is
particularly the case since the Ziegler-Natta type isotactic polymerization of olefins focused interest on metal alkyls in general and
many of the previously unavailable alkyls became readily available
as commercial chemicals.

It is not within the scope of the present discussion to treat in any
detail the general chemistry and properties of metal alkyls.   Excellent summaries are available on organoboranes by H. C. Brown (78)
and on organo-aluminum compounds by K. Ziegler (689).

The Lewis acidity of alkyls of aluminum, beryllium, boron,
magnesium, etc., is due to the electron deficiency of the central metal
atoms, very much as in the case of the corresponding halides.

All aluminum or boron compounds of the type AlX_3 or BX_3
including the metal alkyls have 6 valence electrons.   Consequently
there is an electron deficiency (360,456,608) which is the cause of the
tendency to add molecules with electron donor properties.

Aluminum trialkyls are generally associated being dimeric,
bridged molecules with the structure:

$$\begin{array}{ccc} R & R & R \\ \diagdown & \vdots\,\cdots & \diagup \\ & Al \quad Al & \\ \diagup & \vdots\,\cdots & \diagdown \\ R & R & R \end{array}$$

The dimers have a distinct tendency to dissociate according to the
equilibrium

$$(AlR_3)_2 \rightleftharpoons 2AlR_3$$

This can be clearly demonstrated in the vapor phase at temperatures
up to 160°.   The heat of dissociation for $[Al(CH_3)_3]_2$ is 20.2 kcal.   In
solution, especially in benzene, the degree of dissociation is rather
low if R is an n-alkyl, but probably higher for the higher alkyl
compounds. Aluminum tri-isobutyl, tri-isopropyl, tri-neopentyl,
and their analogs generally show monomolecular weights.   Even
in the pure liquids and in highly concentrated solutions the degree
of association is small.   Thus it seems that steric hindrance
diminishes the tendency of association through bridge formation.

Dialkylaluminum halides, R_2AlX and alkyl aluminum dihalides, $RAlX_2$ are also associated, being present mostly as dimers (or trimers) with halogen bridged structures.

$$R_2Al \diagdown_{X}^{X} AlR_2$$

The structure and coordinating power of aluminum alkyls and alkylaluminum halides thus shows close similarity with aluminum trihalides. Similarities also exist with the other metal alkyls.

The Friedel-Crafts type catalytic activity of metal alkyls has been much less extensively investigated than that of the Lewis acid metal halides. Table XLIV summarizes the systems investigated so far.

TABLE XLIV. Metal alkyl Lewis acid catalysts

Catalyst	Reactions catalyzed	Ref.
Alkyl aluminum halides	alkylation	252,418,560, 561,658
$Al_2Cl_3Me_3$ + HCl	alkylation	376
$AlCl_2C_2H_5$	alkylation, polymerization	
$AlCl(C_2H_5)_2$		
$Al_2Br_3Et_3$	various Friedel-Crafts reactions, alkylation	237
$Al(C_2H_5)_2Br$	alkylation	418
$Al(C_2H_5)Br_2$	alkylation	418
$Al(C_2H_5)_3$	ploymerization, alkylation	405,406
$Al(CH_3)_3$	alkylation	439
$Al(t\text{-butyl})_3$	alkylation	439
$Al(i\text{-}C_4H_9)_3$	alkylation, polymerization	439
$AuBr_2C_2H_5$	alkylation	
$Be(C_2H_5)_2$	polymerization	405,406
BR_3	alkylation	439
MgR_2	alkylation, polymerization	46
$R_2Mg \cdot MgX_2$	alkylation, polymerization	46
$TiCl_3CH_3$	alkylation	439
$TiCl_3(C_2H_5)$	alkylation	439
$Zn(C_2H_5)_2$	alkylation	439

No reference is made in the present discussion to catalyst systems of the Ziegler-Natta type, although certain similarities with Friedel-Crafts systems are apparent. For example alkylation of aromatics with olefins in the presence of Ziegler-Natta catalyst, resulting in longer chain alkylates is obviously related to the principle of Friedel-Crafts alkylation.

As metal alkyls themselves are reactive compounds in alkylations it is generally advantageous to use the same alkyl substituted metal alkyls as the alkylating agent in order to avoid trans-alkylation of the catalyst and introduction of alkyl groups from the catalyst into the substrate.

It seems to be that the application of metal alkyls as Lewis acid type catalysts in systems related to the Friedel-Crafts type has so far been very limited. High solubility of the generally liquid catalysts may be just one of the advantages, allowing homogeneous reaction conditions. It is considered justifiable to say at this point that these catalysts may have extended applications in the future.

XIV. Metal Alkoxide Lewis Acid Catalysts

Owing to the strong conjugative effect of the oxygen atoms attached to the central metal atoms, metal alkoxides of coordinatively unsaturated elements such as aluminum, boron, titanium, generally show only weak Lewis acidity. Aryl substituents, allowing conjugation of the unshared oxygen electron pairs with the aromatic ring system, often give somewhat stronger Lewis acidity. Reactions catalyzed by metal alkoxides are summarized in Table XLV.

TABLE XLV.   Metal alkoxide Lewis acid catalysts

Catalyst	Reactions catalyzed	Ref.
$Al(OC_6H_5)_3$	alkylation of phenols	323,610,611
Aluminum alkoxides	alkylation	47
$AlCl_2OR$	various Friedel-Crafts reactions	539
$AlCl_3 \cdot Ti(OR)_4$	polymerization	685
$Ti(BuO)_4$	polymerization	555
$Ti(iso-PrO)_4$	polymerization	555

The most important application of metal alkoxides in Friedel-Crafts type reactions is that of *aluminum phenolate* as an alkylation catalyst in phenol alkylation (323). Aluminum phenolate is also the active catalyst when aluminum metal is itself used as a catalyst in phenol alkylation. Phenol is sufficiently "acidic" to react with aluminum with the formation of $(C_6H_5O)_3Al$.

Aluminum phenolate when dissolved in phenol greatly increases the acidic strength. It is believed that as with the alkoxo-acids of

Meerwein (383) an aluminum-phenoxo acid is formed which is a strong conjugate acid of the type $HAl(OC_6H_5)_4$. This acid is then the catalytically active species.

XV. Brønsted Acid Catalysts

Strong proton acids such as sulfuric (69,279,331) chlorosulfonic, fluorosulfonic, phosphoric (283,286), and polyphosphoric acid (645), hydrogen fluoride (567), hydrogen chloride, and alkane sulfonic acids (495), all possess strong Friedel-Crafts type catalytic activity, particularly in such reactions as alkylations with olefins, polymerizations, and condensations.

Of the acids most frequently used sulfuric acid has the disadvantage of causing undesirable side reactions because of its strong oxidizing and dehydrating properties and sulfonating ability. However, the cheapness of sulfuric acid and its ease of handling have led to its wide use in aromatic alkylation, in spite of its disadvantages. The easy recovery of hydrogen fluoride coupled with its excellent catalytic ability has resulted in its large-scale utilization for industrial alkylations, despite its higher price.

1. Sulfuric Acid

It has long been known that sulfuric acid catalyzes the reaction between olefins and aromatic hydrocarbons (69,331). Brochet considered the reaction to be of general use with all olefins. However, Ipatieff (286) later found that it is not applicable to the ethylation of benzene either at atmospheric or superatmospheric pressures. Phosphoric acid, in contrast, will bring about the ethylation of benzene.

Brochet alkylated benzene with hexene in the presence of sulfuric acid but was unsuccessful in a similar reaction with pentene. Brochet's lack of success was probably due to the formation of amyl sulfate, which decomposes to tar when the crude product is distilled. It is a general precaution that esters of sulfuric acid should be removed before distillation and this is readily accomplished by treatment with cold sulfuric acid.

Ipatieff and his co-workers (279) should be credited with the general application of sulfuric acid as an alkylation catalyst and for determining the influence of concentration of sulfuric acid on the alkylation reaction. Their investigations established that with proper choice of the acid concentration (and other reaction conditions) it is possible to control the competing reactions, *viz.*, polymerization of olefin and addition of H_2SO_4 to olefin to form esters (279).

In the sulfuric acid catalyzed alkylation of aromatics with olefins three competing reactions can take place: (1) addition of olefin to the aromatic; (2) polymerization of the olefins; (3) addition of olefin to sulfuric acid to form esters. The predominant reaction of propene in the presence of benzene and 96% sulfuric acid is alkylation (a small amount of ester is also formed, but no polymerization takes place). With 80% acid propene reacts almost equally in ester formation and alkylation. Threefold competition for olefins is especially noticeable with isobutylene. By suitable choice of acid concentration (and temperature) any one of the three reactions can be made to predominate. For example, in the presence of 96% acid the main reaction is alkylation, together with some ester formation. With 80% acid the chief reaction is polymerization. There is also a small amount of ester formation, but no alkylation. With 70% acid isobutylene is converted into ester without polymerization or alkylation. Similar results were obtained with butene, pentene, octene, nonene, and dodecene.

There is a vast difference in the ease with which different olefins react with aromatics in the presence of sulfuric acid catalyst. Isobutylene will alkylate aromatics with sulfuric acid of 85–90% concentration, while propene requires acid up to 96%. Ethylene requires sulfuric acid in the neighborhood of 98% concentration for alkylation. Since acid of this strength rapidly converts both benzene and the alkylated product into the sulfonic acid, the use of sulfuric acid for ethylation of aromatics is impractical (324).

The need for high octane gasolines has greatly increased the use of alkylation in the petrochemical industry. Sulfuric acid (besides anhydrous HF) is widely used in processes for the production of alkylate of high octane blending power. Sulfuric acid is also used in refining petroleum distillates by simultaneous removal of sulfur and gum-forming compounds, presumably effecting alkylation and polymerization reactions of sulfur-containing ingredients, mainly mercaptans (with simultaneous H_2S elimination).

The use of sulfuric acid as an alkylation catalyst was developed during World War II to such dimensions that about 12% of the total H_2SO_4 production in the U.S.A. was used by the petroleum industry (60% of this as alkylation catalyst, the remaining part in refining for removal of sulfur and gum-forming ingredients).

Sulfuric acid is also widely used as an isomerization catalyst. Such isomerization of saturated hydrocarbons differ from isomerizations effected by acidic halides in two important aspects: (1) No additional promoter or initiator is needed; (2) only saturated hydrocarbons having at least one tertiary hydrogen atom are isomerized or produced by isomerization (96,218,324,400,454,513,607).

The reason why no promoter is needed with sulfuric acid appears to be that sulfuric acid oxidizes saturated hydrocarbons (96,218, 454,673) and the carbonium ions formed serve through hydride abstraction as initiators.   The oxidation has been represented as (454):

$$(CH_3)_3CH + 2H_2SO_4 \rightarrow (CH_3)_3C^+OSO_3H^- + SO_2 + 2H_2O$$

and sulfur dioxide has been identified (96,218,607,673).

Moreover, an induction period has been observed (454,607).   The induction period can be eliminated and the reaction rate increased by careful addition of small amounts of olefins (218,454) or sodium alkyl sulfates (218).

The oxidation occurs most readily with hydrocarbons having at least one tertiary hydrogen atom such as 2-methylbutane and 2-methylhexane (218) and very slowly or not at all with hydrocarbons having no tertiary hydrogen atom (324) such as n-octane (218,324). Sulfuric acid is also a good catalyst for the Friedel-Crafts ketone synthesis, principally with acid anhydrides.

2. Hydrogen Fluoride

Anhydrous hydrogen fluoride is a highly mobile liquid with low boiling ($+19.5°$) and freezing points ($-83°$).   This gives hydrogen fluoride a distinct advantage over other acid catalysts, like sulfuric acid, as the catalyst is easily evaporated from reaction mixtures and thus recovered.   The low freezing point insures that the catalyst will not freeze out even when low reaction temperatures are employed.   Its low viscosity and surface tension assist in intimate mixing in heterogeneous reactions and greatly hastens settling time in larger installations.   The low viscosity also allows easy pumping. The low surface tension and viscosity, however, make leaks in the equipment more serious.   Solvent properties are also quite important in the use of hydrogen fluoride as catalyst.   The high solubility of oxygen-, nitrogen-, and sulfur-containing compounds and even to some extent of hydrocarbons in hydrogen fluoride, as well as the solubility of the acid in these substances enables the catalyst to function in the liquid phase.   This provides the advantage of speed of reaction and high specificity.

As hydrogen fluoride is a toxic material and strong acid, certain precautions must be considered with its use.   Its effects are serious

if inhaled or allowed to contact the skin. If unwashed and un-
treated, a small drop of the acid on the skin will cause a painful
wound. The effects are not immediate with the aqueous acid but
after some time a painful throbbing will be felt and in a few days an
abscess will develop, which is very slow to heal. Anhydrous
hydrogen fluoride, owing to its strong dehydrating effect immediately
causes severe burning pains. However, hydrogen fluoride if handled
with care is not more dangerous than many other chemicals used in
large volumes. Rubber gloves, face shields, and safety clothing
should always be worn when working with anhydrous HF. If skin
which has been in contact with hydrogen fluoride is quickly and
properly treated there are no after-effects. It must be remembered,
however, that hydrogen fluoride is readily absorbed in tissue. As
both hydrogen and fluoride ions are toxic, the methods of treatment
are indicated. Immediate copious washing with water removes any
acid from the skin and then some material is applied which will
precipitate the fluoride and neutralize the acid. If the exposure is
not great calcium hydroxide is very effective. Pastes made of
organic calcium salts, such as the gluconate or lactate, with pre-
cipitated magnesium oxide are good and not as corrosive as lime, but
are probably not as effective for immediate and short use. Such
pastes are commercially available. They should be kept moist on
the affected surfaces for an extended period, sometimes for as long
as three days. If there has been a considerable inhalation or large
body surface area contaminated the physician will probably give
the patient a calcium salt internally or intravenously to counteract
the precipitation of calcium ions in the organism by the fluoride ions.
Even a neglected wound caused by hydrofluoric acid will be helped
by proper treatment, but the sooner the treatment is applied the
better.

From the point of view of its catalytic activity the most important
chemical property of hydrogen fluoride is its extremely high acidity.
H_0 for anhydrous HF is -9.9 as compared with -11 for 100%
$H_2SO_.$. This is despite the fact that in aqueous solution hydrogen
fluoride is an apparently weak acid (563).

All substances appreciably soluble in liquid hydrogen fluoride behave either as bases or salts. No acids relative to solvent hydrogen
fluoride have yet been found. This in itself shows the strong acidity
of liquid hydrogen fluoride. Addition of BF_3, SbF_5, AsF_5, TaF_5,
NbF_5, TiF_4, and other acidic fluorides considerably increases the
acid strength.

Another important chemical property of anhydrous hydrogen

fluoride is its very powerful dehydrating action. No chemical drying agent has yet been found capable of extracting water from it (563). It dehydrates sulfuric acid and produces water. Reasonably dry hydrogen fluoride can be obtained by distillation, but the remaining water can only be removed by electrolysis. The importance of this property of HF is largely in dehydrating condensations. Although it would be expected to degrade rapidly all oxygen-containing organic substance and remove the elements of water, it does this to a much less extent than H_2SO_4 or other drying agents. The reason for this probably lies in its extremely high acidity. All the oxygen-containing substances, even carboxylic acids, are bases relative to hydrogen fluoride and form, through protonation by HF, positive ions in solution. These positive ions are then much more resistant to dehydration.

The catalytic activity of hydrogen fluoride in promoting condensation reactions was first demonstrated by Simons and co-workers (563). Anhydrous hydrogen fluoride is a remarkable Friedel-Crafts catalyst for bringing about a wide variety of reactions. Some of the most important reactions catalyzed by hydrogen fluoride are alkylation, acylation, sulfonation, nitration, polymerization, miscellaneous condensations, and rearrangements.

The nuclear alkylation of aromatic compounds is, in general, effectively catalyzed by HF. The alkylating agents which may be used include olefins (100,101,565,566,592), alcohols (101,566,568), alkyl halides (565,566), esters (569), ethers (101), sulfides, and mercaptans. Aromatics which have been substituted include benzene, toluene, xylenes, various alkylbenzenes, naphthalene, tetrahydronaphthalene, mono- and di-isopropylnaphthalene, α-nitronaphthalene, phenanthrene, anthracene, phenol, cresols, diphenyl oxide, β-naphthol, 2,3-dihydroxynaphthoic acid, naphthalene-β-sulfonic acid, N,N-dimethyl-p-aminophenol, p-aminophenol, anisidine, N,N-diethylamino-3-ethoxybenzene, 1-amino-2-methoxynaphthalene, and many others.

Many Friedel-Crafts catalysts will bring about reactions with aromatics like phenols and alkylbenzenes. The conversion of materials so difficult to alkylate such as benzoic acid (100,101) and aromatic nitro compounds (100,101) is, however, remarkable. This success has been ascribed to the fact that the reactants can be heated in contact with HF for extended periods of time without the formation of undue amounts of tar.

While the conditions required thus vary with the aromatic compound used, the relative activities of the various alkylating agents can be compared by the conditions required for the alkylation of benzene. Thus, tertiary alkyl halides are the most active of their class, the conversion to the corresponding tertiary alkyl benzene taking place rapidly at 0°. Secondary alkyl halides react fairly well at room temperature, but heating under pressure at approximately 100° is necessary for satisfactory alkylation by primary halides. Both primary and secondary halides yield secondary alkyl benzenes, owing to isomerization during alkylation.

Aliphatic olefins such as propene, and cyclic olefins such as cyclohexene, alkylate benzene at 0°. In this case it is necessary to add the olefin slowly to the agitated mixture of HF and benzene to minimize polymerization. It is also advisable to use a large excess of benzene if a monoalkylated product is desired.

Alcohols are similar to the alkyl halides, in that the reactivity decreases in the order tertiary > secondary > primary. However, a considerably larger amount of HF is required to catalyze the reaction as water is eliminated instead of hydrogen halide, which reduces the catalytic activity of HF by dilution. In general, ethers and esters resemble alcohols in their behavior, except that esters yield ketones as well as alkylated aromatics.

Alkyl sulfides and mercaptans behave similarly to ethers and alcohols as alkylating agents in the presence of anhydrous hydrogen fluoride. Hydrogen sulfide is the by-product of the condensations instead of water and it escapes as a gas (not being significantly soluble in liquid hydrogen fluoride). The catalyst acid is thus not diluted and considerably less is needed than in alkylations with the corresponding oxygen compounds.

Paraffins, cycloparaffins, and alkylated aromatics themselves can act as alkylating agents, using HF as catalyst. Cyclopropane, for example, gives good yields of n-propylbenzene (and di-n-propylbenzene) when it reacts with benzene. Intermolecular transalkylations are easily carried out as in the case of the reaction of t-butylbenzene with phenols.

Both diaryl ketones and alkyl aryl ketones are formed by the reaction of aromatic hydrocarbons with acids, acid anhydrides, and acyl halides in the presence of excess hydrogen fluoride. Except for its greater effectiveness with acids, hydrogen fluoride generally catalyzes acylations in the same manner as aluminum chloride and other Friedel-Crafts catalysts. It is especially preferred, however, in those cases where the use of aluminum chloride or sulfuric acid is undesirable owing to side reactions. Examples of such cases are ring closures to form ketones of the 2-hydrindone, 2-tetralone, and anthrone types. The three ketones mentioned are formed by the hydrogen fluoride catalyzed intramolecular condensations of hydrocinnamic acid, 2-phenylbutyric acid, and o-benzylbenzoic acid which take place without undesirable side reactions.

In certain cases hydrogen fluoride causes acylation to take place on a different carbon atom of the aromatic nucleus to that substituted in the presence of aluminum chloride. For example, acetic acid in the presence of HF acylates acenaphthene to give 1-acetyl acenaphthene instead of the 3-isomer formed in substitution with aluminum chloride.

Hydrogen fluoride catalyzed polymerizations of olefins usually yield products which are rather complex mixtures. When heated for four hours at 100°, cyclohexene gives a mixture containing 17% dimer, 6% trimer, 8% tetramer, and varying amounts of higher polymers.

Other unsaturated compounds such as oleic acid, linseed oil, amylene, butadiene, dipentene, and indene also yield complex mixtures and acetaldehyde, furfural, and acetone yield polycondensate mixtures.

A wide variety of other hydrogen fluoride catalyzed reactions have been noted. Among these are rearrangements of the Fries, Beckmann, and other types.

The preparation of acids from alcohols or alkyl halides by condensation with carbon monoxide in the presence of HF has recently gained substantial interest.

Carbon monoxide also adds to aromatics in a modified Gatterman-Koch reaction using $HF + BF_3$ as catalyst.

Sulfonation by sulfuric acid and fluorosulfonic acid as well as sulfonylations by sulfonyl halides are successfully catalyzed by HF. Similarly it is an effective catalyst in aromatic nitrations with nitric acid or potassium nitrate.

3. Hydrogen Chloride

Simons and Hart (572) were the first to prove that alkylation of aromatics with alkyl halides and olefins and acylations with acyl halides could be successfully accomplished with anhydrous HCl as catalyst.

To actually demonstrate the catalytic activity of hydrogen chloride t-butyl chloride and toluene were heated to 235° for 24 hours under nitrogen pressure. 84% of toluene and 55% of t-butyl chloride were recovered unchanged and only 10% t-butyltoluene was found.

Using HCl as catalyst an 88% yield of *t*-butyltoluene was obtained, thus showing a strong catalytic effect. In the "uncatalyzed" reaction it may be argued that owing to prolonged heating at elevated temperature some *t*-butyl chloride may have decomposed to give a small amount of HCl which is sufficient to catalyze a certain amount of alkylation but because of dilution with nitrogen the HCl so formed was not sufficient to give a good yield of alkyltoluene.

Schmerling (537) alkylated benzene with propylene at 300° in the presence of a small amount of hydrogen chloride.

Cationic polymerization catalyzed by HCl has been observed by Pepper (461).

Anhydrous hydrogen chloride is also a frequently used catalyst in esterifications, acetal formation, various acylations and condensations.

The use of a variety of inorganic chlorides, chlorinated hydrocarbons, and other organic chlorinated compounds of low molecular weight, as homogeneous catalysts in Friedel-Crafts type reactions has been investigated. They are relatively unstable and reactive at elevated temperature, liberating hydrogen chloride which is then responsible for their catalytic action. Their use was formerly limited to hydrogenation and cracking operations but they have now been employed in the alkylation of benzene and phenol with olefins (529,531). Chloroform and other halogenated organic compounds were used at temperatures above 300°. When phenol was condensed with amylene under these conditions in the presence of 5% chloroform a 93% yield of amylphenols was obtained.

It was also found that active aromatic hydrocarbons such as diphenyl and naphthalene could be alkylated and acylated with alkyl and acyl halides at elevated temperatures without the addition of any extraneous catalyst (413). Relatively inactive aromatic compounds, such as benzene, do not react in this manner except under special conditions (527). The acylation of benzene with benzoyl chloride was the principal reaction investigated. The opinion has been expressed that this type of seemingly non-catalytic condensation reaction is analogous to those catalyzed by chlorinated or halogenated compounds.

The chlorine-containing reactant splits off hydrogen chloride at elevated temperatures and thus acts simultaneously as catalyst. The reaction of benzoyl chloride with benzene carried out at 220–350° gave a 37–85% yield of benzophenone.

Recent work (439) on the acetylation of toluene with acetyl chloride at 240°, however, gave isomer distributions different from regular ionic substitution. Thus in the high temperature, non-catalytic substitution of aromatics with halide reagents, the contribution of homolytic reaction conditions cannot be left without consideration.

4. Hydrogen Bromide

Anhydrous hydrogen bromide is very similar in its catalytic activity to hydrogen chloride in alkylation and acylation reactions.

Pepper (462) found that anhydrous hydrogen bromide polymerizes styrene.

5. Hydrogen Iodide

Hydrogen iodide was found to be an effective catalyst in acetylation of thiophene and furan by acetic anhydride (248). Its reductive effect, however, limits its usefulness in halide reactions.

6. Phosphoric Acid

Phosphoric acid was found to be a highly effective catalyst for the alkylation of aromatic hydrocarbons with olefins (286). Liquid 85–89% orthophosphoric acid can be used as a catalyst for the direct alkylation of aromatic hydrocarbons. Even more advantageous is the use of supported solid phosphoric acid catalysts (phosphoric acid, P_2O_5, on Kieselguhr) developed by Ipatieff (278).

The ethylation of benzene, naphthalene, and tetrahydro-naphthalene in the presence of phosphoric acid at 300° and the alkylation of naphthalene and fluorene with propene at 200° have been investigated. Products obtained from the reaction consisted of the corresponding mono-, di-, and polyalkyl aromatics. Propylation of aromatics with propene has also been used on a large scale in the presence of phosphoric acid catalyst although propene undergoes polymerization in addition to effective alkylation of aromatics under certain conditions.

Phosphoric acid is also an effective catalyst in the destructive alkylation of benzene with n-hexane and 2,2,4-trimethylpentane (283) and for the polymerization of olefins (277,284).

$H_3PO_4 \cdot BF_3$ is an even more effective catalyst than phosphoric acid itself and its alkylation power as well as its application in other Friedel-Crafts type reactions has been thoroughly investigated (635).

7. Polyphosphoric Acid (PPA)

In recent years a mixture of polyphosphoric acids, generally known simply as "polyphosphoric acid" (PPA), has found increasing use as an acid catalyst in certain Friedel-Crafts type reactions.

First recognized as an exceptional reagent for cyclizations it has since found application in various acid-catalyzed acylations, rearrangements, esterifications, etc.

Prior to the recognition of PPA as a useful acid catalyst of wide scope, there were scattered reports of the use of mixtures of phosphoric acid and phosphoric pentoxide for the cyclization of aryl substituted carboxylic acid to cyclic ketones (cycliacylation) (321, 541). The active catalyst in these reactions was obviously polyphosphoric acid. The first mention of a polyphosphoric acid as catalyst for a Friedel-Crafts type reaction was by Rosen, who carried out alkylations of branched paraffins with olefins over a catalyst containing tetraphosphoric acid (515). However, the general potential value of polyphosphoric acid in reactions of the Friedel-Crafts type was recognized only in the 1950's.

Pure individual polyphosphoric acids cannot be obtained by the methods of preparation generally used. Therefore, it is to be understood that the term "polyphosphoric acid" is used to indicate mixtures of polymers of the composition

$$HO-\overset{\overset{O}{\uparrow}}{\underset{OH}{P}}-O-\left(\overset{\overset{O}{\uparrow}}{\underset{OH}{P}}-O\right)_n-\overset{\overset{O}{\uparrow}}{\underset{OH}{P}}-OH$$

where n is 1–7 or higher. Much of the work with PPA as a reagent has been carried out with a commercial material, having a phosphoric anhydride equivalent of 82–84%. The composition of mixtures approximating to this percentage is shown in Table XLVI. Polyphosphoric acid is a liquid and is a reasonably good solvent for

TABLE XLVI. Composition of two typical polyphosphoric acid mixtures (%) (271)

P_2O_5 equivalent	81.6	84.2
Orthophosphoric acid	8	4
Pyrophosphoric acid	27	11
Polyphosphoric acid:		
$\quad n=1$	22	11
$\quad n=2$	17	13
$\quad n=3$	11	12
$\quad n=4$	6	10
$\quad n=5$	4	8
$\quad n=6$	2	6
$\quad n=7$	2	5
all higher values	1	20

organic compounds of many types, a fact which gives it a great advantage over many other catalysts. It contains both acidic and anhydride groups. Water, produced during the course of reactions carried out in it, is rapidly removed by reaction with the anhydride linkages generating new acid functions. Since water is a stronger base than many of the organic compounds involved in the reactions, its removal in this manner may be of fundamental importance in preserving the effective acidity of the reaction mixture if a water-producing process occurs. As a dehydrating agent PPA is much milder than phosphoric anhydride. As a result there is seldom any difficulty with local overheating in a PPA reaction mixture. In practice most of the reactions are carried out by heating the mixtures.

PPA is often used in place of a much more strongly acidic reagent such as aluminum chloride or concentrated sulfuric acid. As a result molecules containing functions which are sensitive to the more vigorous reagent, such as ester groups, may be dealt with satisfactorily with PPA. The fact that PPA is not a strong oxidizing agent, and has little or no tendency to enter into substitution of aromatic nuclei, gives it further advantage over concentrated sulfuric acid (which tends to sulfonate aromatics). Compared to hydrogen fluoride, particularly for cyclizing aryl substituted carboxylic acids, it has advantages in that it hardly requires any special precautions so that simple equipment can be used. An excellent summary of the reactions catalyzed by PPA was recently published by Uhlig and Snyder (645).

8. Fluoroboric and Hydroxyfluoroboric Acids

Phases which have been claimed to be isolated from the BF_3–HF–H_2O system are $BF_3 \cdot H_2O$, $BF_3 \cdot 2H_2O$ (381,388), $BF_3 \cdot 3H_2O$ (374), $BF_3 \cdot HF \cdot H_2O$ (459). $BF_3 \cdot HF \cdot H_2O$ may probably be formulated as (H_3O^+) (BF_4^-) but there is some doubt as to its existence (194). Nothing is known of the structure of $BF_3 \cdot 3H_2O$. The mono- and dihydrates are formed by interaction of water and boron trifluoride. A study of the vapor pressure of a solution of boron trifluoride in ethylene dichloride in the presence of varying quantities of water (119) has shown that with excess of boron trifluoride all the water is present as the 1:1 complex. With excess of water there is no pressure of BF_3.

From conductivity, electrochemical dissociation, and viscosity measurements, Greenwood and Martin (220,222) favor hydroxy-fluoroboric structures for the ionized forms of these complexes, $H^+BF_3OH^-$ and $H_3O^+BF_3OH^-$. The lower limit for the dissociation of both complexes is put at 10% with an upper limit of 50% for the monohydrate and 100% for the dihydrate; there is extensive ion-pair formation in the liquid phase. X-ray powder photographs of solid $BF_3 \cdot 2H_2O$ is in favor of an ionized structure but nuclear magnetic resonance studies favor an unionized structure for the slowly cooled solid. N.m.r. studies are in agreement with a partly ionized structure for the rapidly cooled adduct. It has been shown that dilution of both the mono- and dihydrates with water gives solutions of hydroxytrifluoroboric acid (666).

An aqueous solution of hydroxytrifluoroboric acid, in equilibrium with its hydrolysis products, may be obtained by dissolving boron trifluoride mono- or dihydrates in excess of water; by addition of three moles of HF to one mole of $B(OH)_3$; or by passing potassium hydroxytrifluoroborate through an ion-exchange column (666). It is a strong acid, the conductivity being very similar to that of tetrafluoroboric acid.

The dihydroxydifluoroborate ion is an intermediate in the hydrolysis of hydroxytrifluoroborates. The free acid has been prepared in various ways (56), all involving the partial hydrolysis of boron trifluoride. Many physical properties of the acid have been measured but these results are not very useful as it is certain that the aqueous solutions will contain hydroxytrifluoroboric acid and the hydrolysis products of dihydroxydifluoroboric acid. It has been suggested that the predominant species in such a mixture are the tetrafluoroborate ion and a cyclic species $(B_3F_6O_3)$ (525,526). Dihydroxydifluoroboric acid has found extensive use as a catalyst (55,56,73,249).

Hydroxyfluoroboric acids are good alkylation, depolyalkylation, and isomerization catalysts (349).

Many other fluoroboric acids are mentioned in the older literature but there is no modern evidence for their existence and they will not be considered in any detail. If these acids do exist it seems very likely that they are condensed acids with chains or rings linked together by B–O–B linkages.

As pointed out previously anhydrous fluoroboric acid, HBF_4, does not exist in the free state. However, it plays an important role in the presence of proton acceptors in the $HF + BF_3$ catalyzed reactions (especially in alkylations, isomerizations, and polymerizations) (120,376,377). Aqueous fluoroboric acid was used successfully as catalyst in certain alkylations (109,302). Solutions of the anhydrous acid in nitromethane and tetramethylene sulfone (where they are stable) are also advantageously applied as catalysts.

9. Miscellaneous Inorganic Acids

A number of other Brønsted acids also find application as catalysts in Friedel-Crafts type reactions. Some of the more frequently used of these acids are *perchloric acid* (92,151,557,621), *pyrosulfuric acid* (382,385), *chlorosulfonic acid* (382,385), *fluorosulfonic acid* (382,385).

Fields of application include alkylation, polymerization and acylation.

10. Organic Acid Catalysts

The use of proton acids to catalyze Friedel-Crafts type reactions is by no means limited to inorganic Brønsted acids. A variety of strong organic acids like haloacetic acids and alkanesulfonic acids as well as arenesulfonic acids find application in the catalysis of miscellaneous reactions.

Chloroacetic acid was found to be a suitable catalyst in the ketone synthesis, as well in polymerizations (styrene, methylstyrene) (77,650). *Dichloroacetic acid* is a catalyst in the ring chlorination of aromatics (311) and *trichloroacetic acid* is effective in the polymerization of styrene.

Trifluoroacetic acid is a useful catalyst in aromatic ketone syntheses with acid anhydrides (417). It is far more effective than trichloroacetic acid. Trifluoroacetic acid also finds application as a catalyst in ring chlorination, isomerization, etc. *Trifluoroacetic anhydride* is an acylation catalyst which has found widespread application in Friedel-Crafts acylations, in cyclialkylations and esterifications (59).

Alkanesulfonic acids as well as *arenesulfonic acids* are useful catalysts for alkylation, polymerization, and isomerization reactions (273,491,495). Alkanesulfonic acids, like ethanesulfonic acid, catalyze alkylations with olefins and miscellaneous isomerizations and rearrangements. Ethanesulfonic acid also oxidizes paraffins having at least one tertiary hydrogen atom, but more slowly than sulfuric acid (218). It is therefore applicable as a catalyst in alkylations with alkanes. Ethanesulfonic acid and other sulfonic acid catalysts generally require higher reaction temperatures than sulfuric acid.

Arenesulfonic acid catalysts, as benzene sulfonic acid and *p*-toluenesulfonic acid, are used in alkylation, isomerization, and polymerization reactions.

11. Survey of Brønsted Acid Catalysts

Tables XLVII and XLVIII summarize some of the reported applications of Brønsted proton acid catalysts in Friedel-Crafts type reactions.

TABLE XLVII.   A. Inorganic acids

Catalyst	Reactions catalyzed	Ref.
HBF_4 (as solution in water, nitromethane, tetra-methylene sulfone)	alkylation, polymerization	38,109,120, 302,376, 377,489
HBr	polymerization	434,462
HCl	alkylation, polymerization	195,461,572
HF	alkylation, acylation, polymer-ization, various Friedel-Crafts reactions	562,565,566, 568,569, 570,571, 573,574, 575
$HF + AsF_5$, $HF + BF_3$, $HF + PF_5$, $HF + SbF_5$, $HF + SiF_4$, $HF + SnF_4$, $HF + TaF_5$, $HF + TiF_4$	alkylation, isomerization	120,376,377
HI	acylation	248,382,385
$HClO_4$	alkylation, acylation, polymer-ization	92,151,557, 621
$BF_3 \cdot H_2O$ [HBF_3OH]	alkylation, depolyalkylation	348,349
$BF_3 \cdot 2H_2O(H_3O[BF_3OH])$	alkylation, depolyalkylation	348,349
HIO_3	polymerization	382,385
$HClSO_3$	polymerization	382,385
$HFSO_3$	polymerization	382,385
H_2SO_4	alkylation, acylation, polymer-ization, halogenation, nitra-tion, isomerization	27,69,279,331, 393,471, 577,680
$H_2S_2O_7$	polymerization	382,385
H_3PO_4	alkylation, acylation	283,286,308
Polyphosphoric acid	acylation, cyclization, re-arrangements	487,645

TABLE XLVIII.   B. Organic acids

Catalyst	Reactions catalyzed	Ref.
$CH_2ClCOOH$	acylation, polymerization	77,650
$(CH_2ClCO)_2O$	acylation	77,650
CCl_2HCOOH	polymerization	
CCl_3COOH	ring chlorination, polymeriza-tion	312
CF_3COOH	acylation, ring chlorination, isomerization	311,417,489
$(CF_3CO)_2O$	alkylation, acylation, cycliza-tion, esterification	59,182,259, 616
alkanesulfonic acids ($C_nH_{n2+1}SO_3H$)	alkylation, polymerization	495
$C_6H_5SO_3H$	alkylation, polymerization	273
p-$CH_3C_6H_4SO_3H$	alkylation, isomerization	417,489,491

XVI. Acidic Oxide (Sulfide) Catalysts (Acidic Chalcides)

1. Types of Acidic Chalcide Catalysts

Typical Friedel-Crafts type reactions, catalyzed by acidic halide catalysts or proton acids, may also be carried out in the presence of heterogeneous catalysts of the silica–alumina or related acidic oxide (sulfide) type.

The elements of Group VIA of the Periodic Table, namely, oxygen, sulfur, selenium, and tellurium, have been called "chalcogens," a term analogous to "halogens," used for the elements of Group VIIA. Accordingly, compounds of these elements are sometimes called "chalcides" (analogous to halides). Acidic chalcide catalysts include a large variety of solid oxides and sulfides, the most widely used comprising alumina, silica, and mixtures of alumina and silica, either natural or synthetic, in which other oxides such as chromia, magnesia, molybdena, thoria, tungstic oxide, zirconia, etc., may also be present, as well as sulfides of molybdenum. An infinite number of compositions is possible and many have been used as catalysts. Examples are bauxite, floridin, Georgia clay, and other natural aluminosilicates, the composition and certain features of structure of which are still not well known.

Some synthetic catalysts, other than those of silica–alumina composition, representative of the acidic chalcide group are BeO, Cr_2O_3, P_2O_5, ThO_2, TiO_2, $Al_2(SO_4)_3$ (which may be regarded as $Al_2O_3 \cdot 3SO_3$), $Al_2O_3 \cdot Cr_2O_3$, $Al_2O_3 \cdot Fe_2O_3$, $Al_2O_3 \cdot CoO$, $Al_2O_3 \cdot MnO$, $Al_2O_3 \cdot Mo_2O_3$, $Al_2O_3 \cdot V_2O_3$, $Cr_2O_3 \cdot Fe_2O_3$, MoS_2, and MoS_3 (125).

In contrast to sulfuric acid, which may be regarded as a fully hydrated chalcide, the chalcides of this group are seldom very highly hydrated under conditions of use. However, adsorbed protons are probably essential to their activity as catalysts. They are sometimes activated for use by treatment with aqueous mineral acid and then dried at a higher temperature.

The acidic oxides possess the properties of being physically and chemically stable and catalytically active at temperatures approaching the threshold of thermal decomposition of hydrocarbons. Yet their acidity is not so great as to lead them to form stable complexes with unsaturated hydrocarbons, as do the aluminum halides, for example. For these reasons, they are frequently used at higher temperatures and are preferred for isomerization of unsaturated hydrocarbons, which are polymerized by strongly acidic catalysts at lower temperatures. Polymerization is thermodynamically relatively unfavorable at higher temperatures.

Chalcide catalysts are not very effective for isomerization of saturated hydrocarbons unless they possess electronic as well as acidic properties. Electronic properties (hydrogen activity) may be imparted by the presence of a transition metal or transition metal oxide, such as cobalt, nickel, platinum, molybdenum, tungstic oxide (WO_3), and zirconia. Less effective in conferring activity for isomerization of saturated hydrocarbons are the transition metals, copper, and iron. The use of hydrogen with these catalysts is beneficial, perhaps essential, in the isomerization of saturated hydrocarbons.

2. Nature of Acidity

It is now well established that the catalytic activity of silica–alumina and related substances is due to their acidity (230,396,624). Their stoichiometric acidity can be measured by titration with alkali (116) and the acid strength of the surface of some of these solids can be gauged by color changes in adsorbed indicators (665). As acids, they are "neutralized" by adsorption of alkali metal ions (essentially an ion exchange with hydrogen) ammonia or quinoline. Their catalytic activity has been related quantitatively to their stoichiometric acidity (396,624).

The acidity may be due to constitutional hydrogen ions or, in the case of silica, to physically adsorbed mineral acid (665). The catalytic activity of molybdenum sulfide has been attributed to the presence of a surface layer of hydrogen sulfide (245). That is, the catalytically active acidic chalcides are not to be regarded as pure oxides or sulfides, the presence of hydrogen ion, often in an amount too small to be detected by elementary analysis, being responsible for the catalytic activity.

A. Alumina

Although certain synthetic chalcide catalysts and natural clays give the acid color with a variety of indicators, alumina gives no color change. This is surprising since the activity of alumina, in reactions like isomerization of cyclohexene, classifies it as a relatively strong acid. Discussing the acidity of alumina, Pines (474) quoted several reasons for this discrepancy. If one assumes that the catalytic action of alumina is due to protonic acids on the surface, then it is conceivable that alumina is a strong acid only at the high temperatures of catalytic reactions, but not at room temperature where the color tests are made. Performing the color tests in refluxing o-xylene (b.p. 144°), however, gave no positive results. Thus it appears that the Brønsted acids on alumina, if present at all, are of very low acid strength.

The catalytic activity of alumina could also be due to the presence of Lewis acid sites. Pines and Haag (474) recently presented evidence that the change of color, which certain compounds (triphenylmethane derivatives, phenolphthalein) undergo when adsorbed on alumina is indeed due to the presence of Lewis acid sites on the surface of the solid catalysts. These studies have shown that the intrinsic acidity of alumina, which manifests itself in certain catalytic reactions, can be measured by chemical means. There is a good correlation between catalytic activity (as in the isomerization of alkenes and cycloalkenes) and color formation with certain indicator substances. Catalysts of the alumina type which are not active for isomerization gave no color.

B. Silica–Alumina

Considering the nature of the silica–alumina catalyst, the following facts must be, according to Thomas, first of all considered (624).

Silica alone is either an inactive or only slightly active catalyst.

Alumina itself is better than silica but still an inferior catalyst.

The proper combination of silica–alumina is very much more active than either of its components.

Silica–alumina catalysts are prepared from the hydrated oxides; mixtures of an anhydrous oxide with a hydrated oxide do not produce active catalysts.

The silica–alumina catalyst apparently has certain acidic properties.

The catalysts are solids.

Solid acidic oxides have structures in much the same sense as organic molecules. In addition, they bear a resemblance to the organic polymers in which a characteristic group is repeated indefinitely in two or three dimensions.

It is important to realize that there is no such simple species as $O=Si=O$ in the solid state. Rather, each silicon is attached to four oxygen atoms and these oxygen atoms are equidistant from the silicon. The centers of the oxygen atoms are located at the corners of a tetrahedron with the silicon in the center.

This arrangement for silicon and oxygen is characteristic of all known crystalline forms of silica and for all solid inorganic silicates. This unit is then a monomer or building unit from which the higher molecular weight solids are built.

Aluminum, in some compounds, shares its valences with 4 oxygen atoms equally (tetrahedrally) spaced around it. Aluminum can also share its valences with six oxygen atoms equally (octahedrally) spaced around it. These two types of aluminum–oxygen combinations may be regarded as monomers for building more complex polymers or co-polymers.

If silica and alumina are combined, which type of aluminum–oxygen combinations should be used? The silica–alumina catalyst gives no X-ray diffraction pattern that would permit a decision. Therefore, the following assumptions have been made by Thomas:

1. The aluminum in the silica–alumina catalysts that contributes to catalytic activity is tetrahedrally linked to 4 oxygen atoms.

2. A positive hydrogen ion is associated with tetrahedral aluminum in the catalyst.

3. The catalytic activity of properly prepared silica–alumina masses is due to the acidic hydrogen ion associated with it.

Using assumption 1 the building blocks for the catalysts become tetrahedral silica and tetrahedral alumina. Figure 2 illustrates in two dimensions, a three-dimensional network of silica and oxygen:

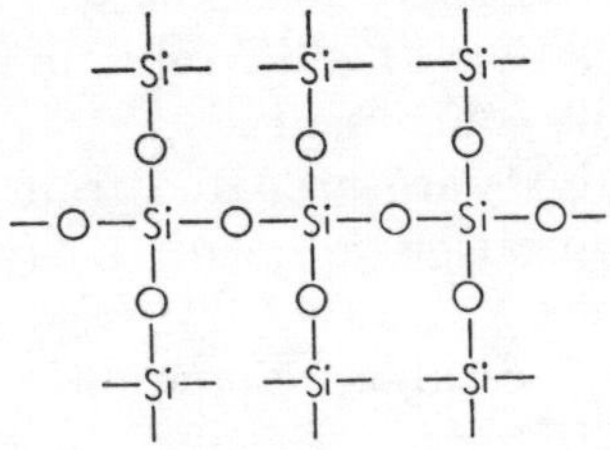

Fig. 2.   Silica.

Each line represents one valence unit. Each silicon atom has 4 such lines to satisfy its valence of four. Each oxygen atom has 2 such lines and its valence of two is satisfied. If the central silicon atom is removed and replaced with a tetrahedral aluminum atom, the result is as shown in Fig. 3.

It is known that the valence of aluminum is three. Each line attached to the aluminum atom represents three-fourths of a valence unit. With four such lines the valence of aluminum is satisfied. The oxygen atoms share one valence with a silicon cation and three-fourths of a valence with aluminum. Each oxygen atom has available to it only one and three-fourths valence units when its valence demands two. Each such oxygen atom is unsatisfied by one-fourth of a valence unit and there are four such oxygens for each aluminum. It follows that the AlO_4 part of the molecule is unsatisfied by a whole valence unit and, since oxygen is a negative element, this is a negative valence.

The hydrogen ion (assumption) has been drawn to show its relation to the tetrahedral aluminum. Owing to its association with four oxygen atoms it seems logical that this hydrogen atom should be ionized and possess acidic properties.

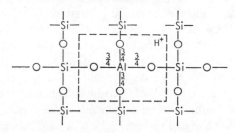

Fig. 3.   Acidic silica–alumina.

On the basis of assumption 3, the silica–alumina catalyst is an acid that is stable at high temperatures. Since this conclusion assigns a definite function to the source of activity, it is desirable to recapitulate at this point and compare the picture that has been built up with the facts outlined. It is immediately apparent that neither silica nor alumina should be active according to this picture. A definite function is assigned to the silica–alumina combination that is not present in either one.

The –Si–O–Al– bonds could be formed readily by dehydration of intimately mixed hydroxides, *i.e.*,

$$\diagdown\text{SiOH} + \text{HOAl}\diagup \rightarrow \diagdown\text{Si—O—Al}\diagup + \text{H}_2\text{O}.$$

Such bonds could not be formed readily if both or either component was anhydrous. Finally the source of the acidic nature of the catalyst is explained and a definite function is ascribed to it. The picture fits the facts reasonably well.

If the acidity is the source of catalyst activity, then maximum activity should result when maximum acidity is obtained. Maximum acidity should result from catalysts in which the atomic ratio of silica to alumina is one.

In many methods of making catalysts a silica hydrogel is first formed. It is not possible to form a silica hydrogel without forming

Si–O–Si bonds. Si–O–Si bonds reduce the number of –Si–O–Al– bonds potentially available. From this it is concluded that such methods of making silica–alumina catalysts cannot yield catalysts of maximum acidity.

If the catalyst could be prepared so that every oxygen atom present is shared by one silicon and one aluminum atom, then the composition of the catalyst would be $HAlSiO_4$. Since the complex acid exists as a solid of high molecular weight it should be written $(HAlSiO_4)_x$ just as solid NaCl should be written $(NaCl)_x$. $(HAlSiO_4)_x$ is then the catalytic part of the "catalyst." Any silica not directly associated with the acid-forming part of the catalyst does not contribute to the activity. The commercial silica–alumina catalysts ordinarily contain 10–15% Al_2O_3 and 85–90% SiO_2, i.e., they contain too much silica for all of it to be in the acid form $(HAlSiO_4)_x$. In one sense the commercial catalysts might be regarded as $HAlSiO_4$ supported on inactive silica gel. This is an incomplete picture, however, as it fails to show that the support is chemically combined with the catalyst and that one is inseparable from the other.

It seems entirely reasonable that for a hydrogen ion to be effective in a catalytic reaction the hydrogen ion and the substance to be catalyzed must come together. In catalytic reactions only the hydrogen ions on the surface of the catalyst should be effective. In other words, hydrogen ions that are buried in the solid are not available to the reagents, and therefore are not effective. The catalyst activity then should increase as the surface area of the catalyst increases and should increase as the concentration of hydrogen ions on the surface increases.

Silica plus octahedral alumina as catalysts. So far, only tetra-hedral silica and tetrahedral alumina have been considered. In octahedral alumina (Fig. 4) each Al–O line represents one half a

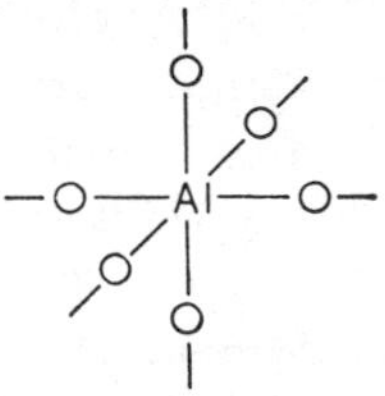

Fig. 4.   Octahedral alumina.

valence unit. Silica can react with such a system (Fig. 5) so that each oxygen shares one tetra hedral silicon atom and two octahedral aluminum atoms.

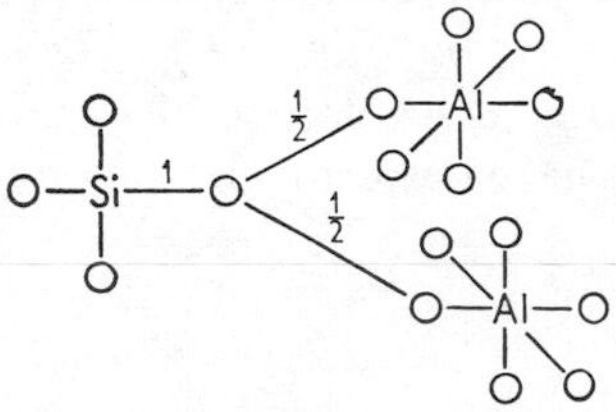

Fig. 5.   Silica and octahedral alumina (neutral aluminum silicate).

This produces a saturated aluminum silicate that should have no acidity and, therefore, no activity.   There are several aluminum silicates known in which the aluminum is octahedral.   So far, none of these has been found to be a catalyst.

This tends to substantiate assumption 1 made by Thomas.

C. Silica–Magnesia

Magnesium occurs in both tetrahedral and octahedral types with oxygen.   The tetrahedral magnesia gives rise to the following type of silica–magnesia structure (Fig . 6).   Each Si–O–Mg oxygen is

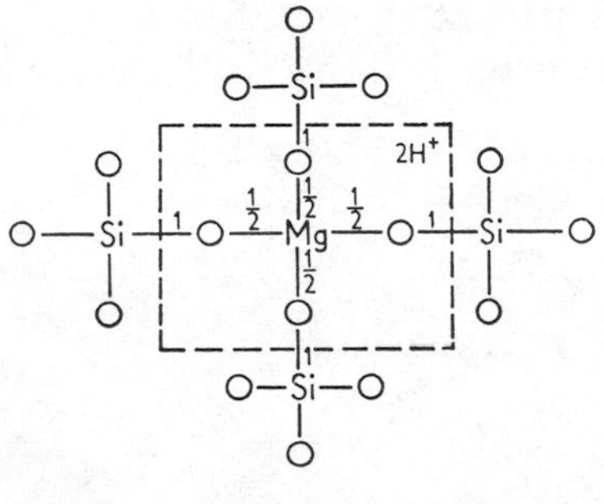

Fig. 6.   Acidic silica–magnesia.

unsatisfied by one half of a valence unit.   There are four of these oxygens so there are two full negative valences to be satisfied. Two hydrogen ions are postulated to satisfy this deficiency.   The catalytic part of the catalyst is then $(H_2MgSiO_4)_x$.   This means that catalysts of maximum activity should be obtained when the ratio of magnesia to silica is one, provided they are properly prepared.

It should be mentioned that most of the better silica–magnesia catalysts that have been prepared have a magnesia to silica ratio of about one.

Is $(H_2MgSiO_4)_x$ a stronger acid than $(HAlSiO_4)_x$? This question relates to the activity to be expected. It seems likely that the $(H_2MgSiO_4)_x$ may well be a weaker acid because it is a dibasic acid and because of the basic nature of the magnesium that forms part of the anion.

D. Silica–Zirconia

Zirconium is associated with eight oxygen atoms in zirconia (Fig. 7).

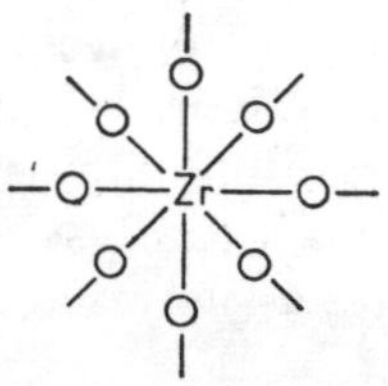

Fig. 7.   Zirconia.

Since zirconium has a valence of four, each Zr–O line represents one half a valence unit, the same value that was encountered in tetrahedral magnesia. This leads to $(H_4ZrSi_2O_8)_x$ as the formula of the active acid catalytic part of the catalyst and to the conclusion that the catalyst of maximum activity will have a Si/Zr ratio of two (Fig. 8).

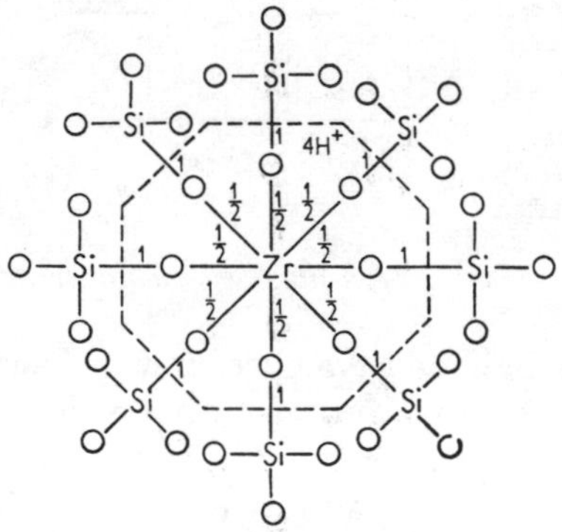

Fig. 8.   Acidic silica–zirconia.

If the catalyst actually has four hydrogen ions per zirconium ion there is a reasonable doubt that more than one, or at most two, is active enough to be catalytic.

E. Alumina–Boria

Boron usually shares three oxygen atoms $\begin{matrix} O \\ \diagdown \\ B-O- \\ \diagup \\ O \end{matrix}$. Active

alumina–boria catalysts can be made from γ-alumina. γ-Alumina is thought to contain two octahedral aluminum atoms to each tetrahedral aluminum atom (Fig. 9). In the absence of any other

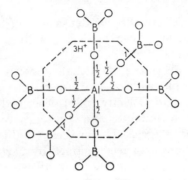

Fig. 9.   Acidic alumina–boria.

basis for selections, one would be inclined to favor the octahedral alumina as the one that gives an active acidic catalyst with boria. On this basis the catalyst is $(H_3AlB_2O_6)_x$ and the catalyst of maximum activity has a B/Al ratio of two.

F. Titania–Boria

Titanium shares six oxygen atoms in the same way that octahedral aluminum does. Except for the difference in valence, titania–boria looks very much like alumina–boria (Fig. 10).

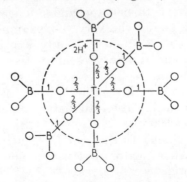

Fig. 10.   Acidic titania–boria.

From this the acidic catalyst is $(H_2TiB_2O_6)_x$ and the catalyst of maximum activity should have a B/Ti ratio of two.

3. Friedel-Crafts Type Reactions Catalyzed by Acidic Chalcide Catalysts

The alkylation of aromatic hydrocarbons with olefins, long established in Friedel-Crafts and related proton acid catalyzed syntheses, was the first application of acidic oxide catalysts to reactions of this type.

Michel (394) described the condensation of naphthalene with propene under pressure over Fuller's earth to produce tetra-isopropylnaphthalene.

Schollkopf (543) alkylated naphthalene with ethylene at 230° under 20–40 atmospheres pressure over activated hydrosilicate catalyst. Sachanen and O'Kelly (530) described the alkylation of benzene with propene, butenes, and pentenes over silica–alumina at 450° and 100 atmospheres. Under these conditions the alkylation proceeded smoothly but was accompanied by partial cracking of the paraffinic side chains. As a result toluene, ethylbenzene, and xylenes were produced in substantial quantities.

Destructive alkylation reactions catalyzed by aluminum chloride, as observed by Ipatieff and co-workers (276), were carried out over silica–alumina catalysts by Sachanen and David (528). These investigators allowed benzene to react with pentanes over an activated clay at 1050 pounds per square inch for 45 minutes at 480°. Sachanen (246) and Thomas (625) pioneered the application of silica–alumina to dealkylation reactions formerly catalyzed by Friedel-Crafts and proton acid catalysts. They reported the dealkylation of alkylaromatic hydrocarbons in the presence of silica–alumina at 450–550°. An example of these reactions is the conversion of ethylbenzene to benzene and ethylene. Although this observation may be one of the very few examples of a reversible Friedel-Crafts reaction, it is more readily explained as a hydrodealkylation, as ethane is also present in the reaction products. Sachanen also described the conversion of alkylaromatic hydrocarbons by the transfer of alkyl radicals from one aromatic nucleus to another or by a dealkylation–alkylation reaction catalyzed by silica–alumina at 450–500°.

These results are similar to those of Anschütz (9,51,257,340,546) and other previous investigators as Ipatieff and Pines (285), who used aluminum chloride as a catalyst.

Thomas was the first to investigate in detail the catalytic

cracking of alkylaromatics over silica–alumina type catalysts. He observed that these reactions bear a resemblance to those obtained with anhydrous aluminum chloride. Ipatieff and Pines have shown that aluminum chloride in the presence of a hydrogen source effects the cleavage of the side chain from the nucleus to form aromatic hydrocarbon and an alkane. Alkyl groups of three or more carbon atoms are easily removed from the benzene nucleus by aluminum chloride in the presence of either cyclohexane or decalin to give the corresponding alkane and benzene. Ethylbenzene and toluene do not react under these conditions. The contrast between the results obtained by thermal cracking and catalytic cracking of alkylaromatics is noteworthy.

It was found (143) that the liquid product from the thermal cracking of cumene was mainly styrene with some benzene and a trace of toluene. The conversion of butyl- and amylbenzenes leads to rather complex mixtures, with the reactions following no clear-cut course but being effected by the configurations of the alkyl groups. It is evident that the thermal and catalytic reactions are different, for the reaction products are different. During the cracking of ethylbenzene, cumene, butylbenzene, and technical amylbenzene in the presence of a synthetic catalyst of the silica–alumina type at 400–500° and atmospheric pressure Thomas and co-workers obtained benzene and the corresponding alkane with ease and great selectivity when the side chain contained three or more carbon atoms. The yield of benzene being at least 90% in each case. The order of decreasing susceptibility to cracking is cumene > technical amylbenzene > butylbenzene $\gg$ ethylbenzene. No toluene or styrene was found in any products of these reactions. The alkane formed is independent of side reactions to some extent although some hexanes are formed in the catalytic cracking of cumene. Presumably the initially formed propylene is dimerized prior to hydrogen transfer. The olefinic product from butylbenzene undergoes hydrogen transfer readily and butanes are formed.

Acidic oxide catalysts of the silica–alumina type are also used in isomerization of dialkylbenzenes. It has been claimed that an increased proportion of p-xylene is obtained from the catalytic isomerization of xylenes over silica–alumina catalyst (644).

Pines recently initiated a systematic study of the catalytic activity of alumina and other chalcide catalyst in isomerization, alkylation, and related reactions (472,473,474). It is not possible to give here a detailed account of this work, as the present survey was largely completed by the time of these publications, but it certainly is one of the most important investigations on the solid acidic oxide catalysts so far published.

Specific interest in acidic–electronic catalysts has arisen in connection with the hydrodealkylation process of making toluene from benzene and naphthalene from methylnaphthalenes. These catalysts are a combination of acidic oxide catalysts with metallic mono-electronic (hydrogen active) catalysts of the nickel, platinum, and charcoal type. The basic reaction

$$C_6H_5CH_3 \xrightarrow[\text{Catalyst}]{2H} C_6H_6 + CH_4$$

$$C_{10}H_7CH_3 \xrightarrow[\text{Catalyst}]{2H} C_{10}H_8 + CH_4$$

consists of a dealkylation step (catalyzed by the acid) and a hydrogenation step (catalyzed by the mono-electronic metal catalyst). Insufficient information is available so far on catalyst details.

The analogy between the catalytic effects of acidic chalcides and other acidic Friedel-Crafts catalysts is further supported by the successful application of acidic chalcide catalysts in ketone syntheses.

Benzene gives anthraquinone with phthalic anhydride over silica–alumina (527). The reaction is carried out at 370–385°.

Although the overall yield of anthraquinone formation in the aluminum chloride catalyzed reaction of phthalic anhydride with benzene is higher than that obtained over silica–alumina catalyst, this process consumes 1.5–2 moles of aluminum chloride per mole of phthalic anhydride charged. The catalyst consumption using silica-alumina is extremely low. The activity of the solid acid catalyst is not destroyed by one charge of phthalic anhydride and benzene. When the deposit of carbonaceous material on the catalyst after prolonged use has substantially reduced its activity, it is burned off as in the conventional cracking processes. In this manner the life of the silica–alumina catalyst can be extended to periods ranging from 4 months to a year. The conversion per pass over silica-alumina is small. However, the stability of the reactants and the product at the reaction temperature and the short catalyst contact time permits the recovery of most of the unchanged benzene and phthalic anhydride for recycling over the catalyst bed. The use of silica–alumina offers the possibility of eliminating the use of large quantities of Friedel-Crafts catalysts in acylation reactions and also offers a simple way of regenerating catalyst for re-use over considerable periods of time.

Table XLIX gives a survey of reported acidic chalcide catalysts in Friedel-Crafts type reactions.

TABLE XLIX. Acidic chalcide catalysts

Catalyst	Reaction catalyzed	Ref.
Alumina	alkylation, cracking	282
$Al_2O_3 \cdot CoO$	isomerization	175
$Al_2O_3 \cdot Cr_2O_3$	isomerization	175
$Al_2O_3 \cdot Fe_2O_3$	isomerization	175
$Al_2O_3 \cdot MnO$	isomerization	175
$Al_2O_3 \cdot Mo_2O_3$	isomerization	117
$Al_2(SO_4)_3 \cdot Al_2O_3 \cdot 3SO_3$	isomerization	211,255,640
$Al_2O_3 \cdot V_2O_3$	isomerization	480
Aluminosilicates (natural)	isomerization	688
Bauxite	isomerization	33
Bentonite clay	alkylation	301
BeO	isomerization	10,11
Clay, acid activated	alkylation	217
Chromia (with silica-alumina)	isomerization	354
Cr_2O_3 (synthetic)	isomerization	352
$Cr_2O_3 \cdot Fe_2O_3$	isomerization	353
Floridin	isomerization	580
Georgia clay	isomerization	255
Gumbrin clay	alkylation	205
Magnesia (with silica-alumina)	isomerization	660
Molybdena–alumina	isomerization	115,117
MoS_2	isomerization	369,420,421, 497,501,620
MoS_3	isomerization	389,468
$MoS_2 \cdot CoS$	isomerization	498
Montmorillonite clay	alkylation	12
Nickel–alumina	hydrodealkylation	559
P_2O_5	isomerization	157
Silica–alumina	alkylation, dealkylation, isomerization, cracking, acylation	246,394,472, 473,474,527, 530,625
Thoria (with silica-alumina)	isomerization	45
ThO_2 (synthetic)	isomerization	216
TiO_2	isomerization	10
WO_3 (with silica–alumina)	isomerization	115
Zirconia (with silica-alumina)	isomerization, polymerization	45

XVII. Acidic Cation Exchanger Catalysts

1. Types

Cation exchangers long used for water treatment and related ion exchange processes, are basically solid acids. Different types of cation exchangers exist, such as sulfonated coals, sulfonated phenol–formaldehyde resins, sulfonated styrene–divinylbenzene cross linked polymers and exchangers with carboxylic (RCOOH), phenolic (ArOH), or alumina–silicate groupings. Apparently the nature of the cation exchanger is of relatively minor importance as long as it contains strongly acidic groups. Both natural and synthetic ion exchanger materials are used. The synthetic organic exchangers were developed by Adams and Holmes in 1935 (2).

The sulfonic acid resins are the only type of strongly acidic resins of commercial importance. As an example Dowex 50, a widely used synthetic acidic resin, is made by the nuclear sulfonation of cross-linked polystyrene–divinylbenzene beads (153). Dowex 50 is comparable in acid strength with hydrochloric acid. Weakly acidic resins contain carboxylic groups as the functional sites. There are also acidic resins with phenolic OH groups as functional sites.

Some of the commercially available cation exchangers, according to Kunin (338), are shown in Table L.

TABLE L. Acidic cation exchangers

Name	Manufacturer	Type
Amberlite IR–100	Rohm and Haas	Sulfonated phenol–formaldehyde
Dowex 30	Dow Chemical	Sulfonated phenol–formaldehyde
Duolite C–3	Chemical Process	Sulfonated phenol–formaldehyde
Wofatit P,K,KS	former I.G. Farben	Sulfonated phenol–formaldehyde
Zeo Rex	Permutit	Sulfonated phenol–formaldehyde
Zeo-Karb	Permutit	Sulfonated coal
Amberlite IR–120	Rohm and Haas	Sulfonated styrene–divinylbenzene
Dowex 50	Dow Chemical	Sulfonated styrene–divinylbenzene
Permutit Q	Permutit	Sulfonated styrene–divinylbenzene
Alkalex	Research Products	Carboxylic
Amberlite IRC–50	Rohm and Haas	Carboxylic
Duolite CS–100	Chemical Process	Carboxylic
Permutit H	Permutit	Carboxylic

TABLE L (*continued*)

Name	Manufacturer	Type
Wofatit C	former I.G. Farben	Carboxylic
Montmorillonite		Aluminum silicate
Kaolinite		Aluminum silicate
Glauconite		Aluminum silicate
Permutit	Permutit	Aluminum silicate
Decalso	Permutit	Aluminum silicate
Zeo Dur	Permutit	Aluminum silicate
Silica gel		Acidic silica

2. Catalytic Activity

The use of acid regenerated cation exchangers as catalysts for organic reactions was started during World War II simultaneously in Germany and in the United States (141,614). The German researchers have used Wofatit (a sulfonated phenol–formaldehyde type cation exchange resin) as a catalyst for certain esterifications, ester interchanges and hydrolysis reactions.

The American work was started with Zeo-Karb H, a sulfonated coal type of cation exchanger (628), although some work was also carried out with phenol–formaldehyde types of resins containing sulfonic acid groups (402).

Cation exchangers were used in reactions such as esterification, acetal synthesis, ester alcoholysis, acetal alcoholysis, alcohol dehydration, ester hydrolysis, and sucrose inversion. The solid acid type cation exchanger catalyst permits simplified procedures in reactions involving high boiling and viscous compounds because the catalyst can be separated from the reaction products by simple filtration. Compounds which polymerize in the presence of proton acids could be esterified directly without any polymerization by the use of acidic cation exchangers.

Sulfonated styrene–divinylbenzene cross-linked polymers have been applied in many of the previously mentioned reactions and successfully used in acylation of thiophene with acetic anhydride and acetyl chloride (428).

Resins of this type (Dowex 50, Amberlite IR–112, and Permutit Q) are particularly effective catalysts in the alkylation of phenols with olefins (as propene, isobutylene, diisobutylene, 1-nonene) and also with alkyl halides and alcohols (131,359).

Isagulyants and co-workers carried out alkylations of reactive aromatic substrates, such as phenols, with olefins in the presence of cation exchanger resins of the sulfonated styrene–divinylbenzene and phenol–formaldehyde types (291,292,293).

Cation exchanger resins have also been used as catalysts in different condensation reactions. As catalysts for reactions of the Friedel-Crafts type, except for certain selectivity effects and special properties due to surface activity of resins, they behave as acid catalysts of the sulfuric acid and hydrochloric acid type. In certain cases reductions in side reactions and certain specificity of the catalyst have been noted (351).

The application of ion-exchanger resins as catalysts for organic reactions is still in the very early stages of development. It is probable that further development in the synthesis of suitable cation exchanger resins will bring about extended use of these very advantageous catalysts. However, temperature limitations associated with resin stabilities may cause certain problems. As the cation exchangers generally do not form stable complexes with either the reagents or products, they can be removed by simple filtration from the reaction mixtures and re-used repeatedly. This represents definite advantages in their use. Even if the resin surface after longer use becomes contaminated and consequently deactivated, regeneration of exchanger resins for other uses is well known and easy to carry out.

3. Survey of Acidic Cation Exchanger Catalysts

Table LI summarizes reported acidic cation exchanger catalysts in Friedel-Crafts type reactions.

TABLE LI. Acidic cation exchanger catalysts

Reaction catalyzed	Type of cation exchanger or form of synthetic resin	Ref.
Alkylation	Sulfonated coal	131,629
Esterification, inversion	Sulfonated coal	614
Esterification	RSO_3H	370
Alkylation	RSO_3H	131,291,292, 293,359,434
	Permutit Q	359
Alkylation	$RSO_3H + BF_3$	312
Inversion	RSO_3H	339
	RCOOH	
Acetal formation	RSO_3H	145
Polymerization	RSO_3H	339
Depolymerization	RSO_2H	339
Isomerization	RSO_3H	339
Acylation	RSO_3H	428
Esterification	Wofatit	141

XVIII. Metathetic Cation-forming Substances

All the catalysts found active in reactions of the Friedel-Crafts type are electron acceptors or acids in Lewis's generalized sense.

So far the Lewis acid type acidic halide, Brønsted type proton acid, acidic oxide (chalcide), and cation exchanger catalysts have been discussed. There is one last group of cation-forming substances which must be considered.

Metathetic cation-forming salts cannot really be classified as catalysts since they are used up in the course of the reactions.

Compounds of this type, however, are able to initiate reactions of the Friedel-Crafts type and it was therefore felt justifiable to discuss them here. They were first categorized as such by Pepper (463), although some members have been known for their ability for a long time. Pepper quoted $AgClO_4$ and $(C_6H_5)_3CCl$ as cation-forming substances capable of initiating cationic polymerization.

The theory of co-catalysis in Friedel-Crafts reactions shifts the emphasis from the Friedel-Crafts halide to the co-catalyst, since it is the latter (alkyl halide, acyl halide, etc.) which provides the initiating cation. The function of the catalyst is merely to facilitate the dissociation of the co-catalyst into the cation, by complexing with and stabilizing the anionic residue, i.e., it behaves as an "ansolvo" acid in Meerwein's sense (380).

In certain cases, as in the case of triphenylmethyl chloride, dissociation takes place so easily that cation formation is practically spontaneous in solvents of sufficiently high dielectric constant or in the presence of molecules capable of acting as chloride acceptors (32).

Anhydrous silver salts ($AgClO_4$, $AgBF_4$, $AgSbF_6$, $AgAsF_6$, $AgPF_6$, Ag_3PO_4, etc.) act not as acids but as metathetic cation-forming substances when reacted with halide reagents, although the electron-deficient silver itself is "acidic."

The designation of silver perchlorate as a Lewis acid (355) has given rise to the idea that silver perchlorate can act as a Friedel-Crafts catalyst.

Burton and Praill (93), however, were unable to catalyze the polymerization of styrene with the pure, dry $AgClO_4$ and concluded that the polymerization reported previously by Eley and Salamon must have been initiated by traces of perchloric acid present as impurity.

The more important cation-forming substances capable of effecting reactions of the Friedel-Crafts type are summarized in Table LII.

TABLE LII.   Metathetic cation-forming substances

Catalyst	Reaction catalyzed	Ref.
$AgAsF_6$	alkylation, acylation	434
$AgClO_4$	alkylation, acylation, polymerization	91
$AgBF_4$	alkylation, acylation, nitration, formylation, polymerization, sulfonation, sulfonylation, various Friedel-Crafts reactions	443
$AgNO_3$	halogenation	183
$AgOOCCF_3$	halogenation	251
$AgPO_4$	acylation	545
$AgPF_6$	alkylation, acylation	434
$AgSbF_6$	alkylation, acylation	434
Ag_2SO_4	halogenation	315
$AgNbF_6$	alkylation, acylation	434
$AgTaF_6$	alkylation, acylation	434
$AgTi_2F_9$	alkylation, acylation	434
$NaClO_4$	acylation	371

XIX. Stable Carbonium and Related Complexes as Catalysts for Reactions of the Friedel-Crafts Type

The availability of an increasing number of stable carbonium, oxocarbonium, oxonium, nitronium, etc., complexes has made it possible to use these positive ions as initiators in reactions related to the Friedel-Crafts type.   A detailed discussion of these reactions is not considered in the scope of the present survey, but Table LIII gives a number of representative references to this type of reaction.

TABLE LIII.   Stable carbonium-ion type catalysts

Catalyst	Reaction catalyzed	Ref.
Oxocarbonium (acylium) tetrafluoroborates, hexafluoroantimonates and related · salts	alkylation, polymerization	445
Acylium chlorozincates	polymerization	361
Acylium perchlorates	polymerization, alkylation	361,445
Alkyl carbonium tetrafluoroborate	polymerization, alkylation	361,445
Alkyl perchlorates	polymerization	361
Carboxonium salts $[RC(OR)_2]^+$	polymerization	384,385,386

TABLE LIII (*continued*)

Catalyst	Reaction catalyzed	Ref.
Trialkyl oxonium salts $[(C_2H_5)_3O]^+BF_4^-$, $SbCl_6^-$, $FeCl_4$, $AlCl_4$	polymerization	384,385,386
Nitronium salts	polymerization	445
Benzenonium salts	polymerization	445
Diazonium tetratetrahalo-borates	polymerization	445

XX. Recent Advances

1. Solid Acids

Ordinary Friedel-Crafts type reactions catalyzed by acidic halide catalysts or proton acids can also be successfully carried out in the presence of solid acid catalysts of the silica–alumina type. Acidic oxides are stable and active at temperatures approaching the thermal decomposition point of hydrocarbons. However, their acidity is relatively low and they do not form stable complexes with unsaturated hydrocarbons, as do aluminum halides. Thus they are often employed at higher temperatures and are preferred for isomerization of unsaturated hydrocarbons, compounds that are polymerized by strong acid catalysts at less stringent temperatures. The catalytically active solid acidic chalcides should not be regarded as pure oxides (or sulfides), since it appears that it is the presence of protic sites, often in trace amounts, that initiates the catalytic activity.

Rare earth-exchanged X- and Y-type faujasites are versatile catalysts for electrophilic reactions such as alkylation and related reactions (691,692), as well for condensations (693). Generally, the patterns of substrate reactivity and product distribution in zeolite-catalyzed alkylations are similar to the corresponding selectivities reported for alkylations with sulfuric acid, hydrofluoric acid, and water-promoted Lewis acids. The mechanism proposed for Friedel-Crafts alkylation of simple aromatic hydrocarbons with olefins consists of initial transfer of a proton from the catalyst to the olefin. This creates an absorbed carbenium ion like species on the surface of the zeolite. This electrophile is then attacked in the normal manner to bring about formation of the alkylaromatic. Ethylation of toluene and benzene using ethylene and a rare earth-exchanged X-type zeolite gave results indicating substantial correlation with the Brown linear free-energy selectivity relationship (694) in a heterogeneous system.

The research group of Venuto (691–693) also discussed the scope and applicability of a series of organic reactions catalyzed by crystalline aluminosilicate acids. For strongly acidic faujasites operation in the liquid phase at temperatures of 140–230° is generally necessary in order to achieve efficient alkylation of simple aromatic hydrocarbons using a variety of alkenes, alcohols, and haloalkanes. Similar temperatures are needed to effect related Friedel-Crafts type transethylations. For higher energy reactions, such as dealkylation, Friedel-Crafts alkylations using alkanes, and the disproportionation of toluene, temperatures in excess of 230° were found to be necessary. Venuto and co-workers found that when the alkylating agent was one of low molecular weight (e.g., ethylene) a high molar ratio of aromatic hydrocarbon to alkene was required in order to prevent aging of the catalyst. In typical alkylations of substituted benzenes, ortho- and para-orientation was observed, as predicted. When alkylating agents containing three or more carbons were used, the monalkylate in the product mixture generally consisted of a mixture of arylalkanes, as might be predicted. When the aromatic substrate being alkylated was heteroaromatic (thiopene), the major reaction product was the expected 2-alkylthiophene.

Under comparable reaction conditions a more highly acidic faujasite zeolite catalyzes Friedel-Crafts reactions at lower temperatures than a silica–alumina type catalyst does, but nonetheless generally at higher temperatures than those required if strong Lewis acids ($AlCl_3$ or BF_3, for example) or strong Brønsted acids (sulfuric acid, hydrofluoric acid, or phosphoric acid) are the catalysts (691).

A further study was made to determine product distributions, substrate reactivities, and related parameters in alkylation reactions catalyzed by acidic faujasites (692). The rare earth-exchanged X- and Y- and hydrogen Y-faujasites catalyze alkylations which ordinarily proceed by means of carbocation type mechanisms. Sites that are active for alkylation in these three acidic faujasites are visualized as being strongly acidic and ultimately protonic in nature.

A subsequent study addressed itself to condensation reactions catalyzed by zeolites (693). Acidic hydrogen Y-faujasite derived from the thermal decomposition of ammonium Y-faujasite shows high catalytic activity for the condensation of carbonyl compounds with simple aromatic substrates, such as phenol or m-xylene, and for the aldol self-condensation of ketones as well. The observed product distributions generally suggest the operation of carbocation type mechanisms.

$$\begin{array}{c} R_1 \\ \diagdown \\ C{=}O \\ \diagup \\ R_2 \end{array} \longrightarrow \quad /\!/\!/\!/\!/\!/ \; O^-\!\!-\!\text{Zeolite} \; /\!/\!/\!/\!/\!/$$

$$+$$

$$H^+ \; O{-}\text{Zeolite}^-$$

$$\begin{array}{c} H \\ | \\ O = C \\ _{\delta-} _{\delta+} \end{array} \begin{array}{c} R_2 \\ \diagup \\ \diagdown \\ R_1 \end{array}$$

$$\diagdown \; C_6H_6$$

$$C_6H_5{-}\overset{\displaystyle R_1}{\underset{\displaystyle R_2}{\overset{|}{\underset{|}{C}}}}{-}OH + H^+O^-\!\!-\text{zeolite}$$

Aldol type condensations proceeded smoothly in the liquid phase at temperatures between 150 and 200°.

An attempt has been made to survey the general applicability of linear free-energy relationships in heterogeneous systems using solid acid catalysts (695–699). A correlation has been reported between the reactivity of a series of alkylbenzenes in dealkylation on several solid acids and the enthalpy change in the hydride abstraction process. Isomerization of both o-xylene and of cyclohexene under Friedel-Crafts conditions was similarly found to follow linear free-energy relationships using a wide variety of solid acids as catalysts.

Mechanistic studies of carbocationic reactions of silica–alumina and related catalysts have been reported (700,701). A report by Csicsery has been published (702) on the reactions of n-butylbenzene over supported platinum catalysts. These catalysts, often employed in crude oil reforming techniques, are dual-functionality catalysts in that they exhibit acidic and dehydrogenating activities. He was able to distinguish acid- and metal-catalyzed processes from one another. In this manner he was thus able to establish the relative contribution of each reaction pathway.

Silica–alumina cracking catalysts and calcium–ammonium Y-type faujasites catalyze the isomerization and translkylation of poly-alkylbenzenes (703). Two possible mechanisms were suggested: (a) sequential intramolecular 1,2-shifts, or (b) an intermolecular reaction involving transalkylated intermediates. At temperatures below 200° intermolecular reactions involving transalkylated intermediates predominate. However, at increased temperatures (in excess of 300°), dialkylbenzenes apparently isomerize primarily by intramolecular 1,2-shifts under the influence of either silica–alumina or calcium–ammonium Y-type catalysts.

A silica–alumina catalyst has been employed in the Friedel-Crafts reaction of 3-phenylpentane (704). The reaction proceeded at a temperature of approximately 450° and followed three separate reaction pathways: (a) *dealkylation* to benzene and five-carbon olefins (90.4% of product mixture); (b) *fragmentation* to toluene, ethylbenzene, *n*-propylbenzene, and *sec*-butylbenzene (6.1% of product); and (c) *ring closure* to ethylindane (3.5% of product). These reactions are all believed to occur by means of well-defined, standard Friedel-Crafts alkylation mechanisms.

The silica–alumina-catalyzed dehydrocyclization of β-amylnaphthalene to 2-methylanthracene and anthracene has been reported (705). The reaction occurs concurrently with the expected isomerization, fragmentation, and dealkylation. The dealkylation and isomerization are determined by the formation of σ-complex type

carbenium ion centers, while fragmentation and dehydrocyclization apparently come about *via* alkylcarbenium ion formation.

Isotopic tracer studies (using C^{14}) have been performed (706) to determine relative rate constants for *n*-butene interconversion reactions over alumina, fluoridated alumina, and silica–alumina. The overall reactivities for this Friedel-Crafts isomerization reaction followed the order: 1-butene > *cis*-2-butene > *trans*-2-butene.

Fluorination enhances the catalytic properties of alumina and related catalysts in carbocation type and other Friedel-Crafts reactions. Covini, Fattore, and Giodrano (707,708) investigated the acidity of fluorinated aluminas and the extent of their catalytic activity in cumene hydrodealkylation. The highest activities were exhibited by catalysts having between 5 and 20 wt.% fluorine. It was also found that in this range the acidity sites of the catalyst were of higher acidity than 88% sulfuric acid.

The relationship between catalytic activity and the nature of the acidity of crystalline zeolites has been investigated (709). A determination of the disproportionation of toluene, and the reaction of *n*-butane and *n*-pentane catalyzed by the crystalline zeolites as a function of the extent of decomposition of the ammonium zeolites to their hydrogen (Brønsted acid) and hydrogen-free (Lewis acid) forms, was made. Removal of the water of constitution from the hydrogen forms destroys their catalytic activity. Conversely, upon readdition of water the catalytic activity is restored. These catalysts then, are seen to be water-promoted. In addition, it thus appears that the seat of the catalytic activity is the hydrogen ion (Brønsted acid) released upon decomposition of the ammonium forms of the zeolites. The zeolites mordenite and Y-type faujasite were employed in this study since the hydrogen (Brønsted acid) form could be converted to the hydrogen-free (Lewis acid) form without destruction of the crystal lattice of the zeolite.

As the sodium content of a zeolite is decreased (701), the hydroxyl content, the Brønsted acid site concentration, and the catalytic activity all increase. Under the conditions employed the Lewis acid site concentration was constant and kept very small, thus leading to the conclusion that Lewis acid sites are themselves not responsible for the catalytic activity. On the basis of this and earlier results

(79,711,712), it was concluded that the Brønsted acid sites are important centers of activity. The complexity of the true nature of the catalyst is indicated by the fact that catalytic activity reaches a maximum near 600°, a temperature at which the Brønsted acid site concentration is only about 5% of its maximum value.

The acidic sites of fluoridated alumina appear to undergo a reversible dehydration-rehydration analogous to the interconversion of Brønsted acid sites and Lewis acid sites in silica–alumina catalysts (713). It was, however, not possible to elucidate the specific nature of the roles of the two types of acid sites. In a related article (714) measurements of the Brønsted acidity and catalytic activity of silica–alumina catalysts were reported. Brønsted activity was measured by means of infra-red spectra of chemisorbed pyridines. Moisture, an ever-present problem in all catalytic work, was carefully taken into account in this solid acid study.

This investigation demonstrated that Brønsted acidity in silica–alumina type catalysts can be detected down to the equivalent of 0.25% alumina content. For catalysts containing as much as 14% alumina, there is a direct correlation between the Brønsted acidity and the catalytic activity for o-xylene isomerization. It is also possible for silica–alumina catalysts showing no detectable Brønsted acidity by means of infra-red analysis to catalyze carbocation type reactions.

Ward studied the nature of the active sites of X- and Y-type zeolites (715) and found that in general the Brønsted acid sites are more important.

Hansford and Ward (714) investigated the relative catalytic activities of crystalline and amorphous aluminosilicates and concluded that the activities of the most catalytically active faujasites are only about 40 times the activity of silica–alumina. Differences in catalytic activity of various catalysts can be rationalized in terms of the concentration of the Brønsted acid sites.

Hightower and Hall (716) investigated alkylcyclopropane isomerization over silica–alumina. For example:

$$CH_3CH_2CH=CH_2 \ (\sim 25\%)$$

Tracer experiments revealed that one hydrogen atom was transferred intermolecularly in this Friedel-Crafts isomerization.

These results led the investigators to believe that a Brønsted acid intermediate (protonated cyclopropane) was involved in these reactions rather than a Lewis acid intermediate (cyclopropyl cation type intermediate). Work by Karabatsos and Graham (717) and Skell, Starer, and Krapcho (718), however, showed that protonated cyclopropanes would not be operative and involved with 1,1-dimethylcyclopropane, and the evidence is largely negative for such involvement with respect to methylcyclopropane.

It has recently been found (719) that Y-type zeolite-catalyzed isomerization of xylenes is always accompanied by extensive transalkylations. The degree of isomerization seems to be proportional to the extent of the transalkylation. The investigators invoke a diphenylalkane type transition state for the intermediate in the xylene isomerizations.

It has been reported that sulfonated and phosphonated polyphenyls are ion exchangers of high thermal stability and can be used as heterogeneous acid type catalysts (720).

Studies on the thermal behavior of crystalline aluminosilicate catalysts (721), the nature of the active acidic sites on zeolite type solid acids (722), and the use of polyphenyl and polyaromatic aluminosilicates as catalysts of the Friedel-Crafts type (723) have recently been published.

2. Superacids

Although we customarily refer to acids such as HCl, H_2SO_4, and HNO_3 as strong mineral acids because they are fully ionized in solution in water, these aqueous solutions are by no means the most highly acidic media known. It has in fact been well known since Hammett's work (724) that in the sulfuric acid–water system the acidity, i.e., the proton-donating ability of the medium, increases continuously from very dilute solutions up to 100% H_2SO_4. Any definition of a superacid is necessarily somewhat arbitrary, but we may consider all media more highly acidic than 100% H_2SO_4 as *superacids*.

The term superacid was introduced in 1927 by Hull and Conant (725), who studied salt formation of weak organic bases, such as carbonyl compounds with perchloric acid. They were indeed able to protonate a series of weak organic bases and obtain perchlorate salts. Consequently, they called perchloric acid a superacid. Extensive study of superacid systems in recent years was carried out by Gillespie and his co-workers (726), and their chemical application was developed by Olah and co-workers (727), particularly in the study of carbocations.

The strengths of acids are customarily compared by measuring the extent of their ionization in water. However, all strong inorganic acids, *e.g.*, HCl, H_2SO_4, and $HClO_4$, are fully ionized in dilute aqueous solution; their acidities are in fact leveled to that of the common product of their ionization in water, namely, H_3O^+

$$HA + H_2O \longrightarrow H_3O^+ + A^-$$

This method is clearly inappropriate for comparing the relative acidities of the 100% anhydrous acids. The proton-donating ability of the H_2SO_4–H_2O system over the whole concentration range was determined by Hammett by measuring the extents to which suitable

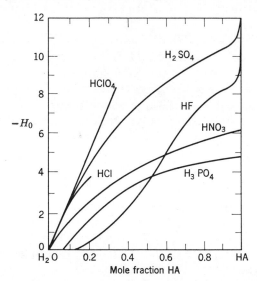

Fig. 11 H_0 values for water–acid systems

weak bases such as series of nitroanilines are protonated (724). On the basis of measurements of this kind, Hammett defined the H_0 function, which is a logarithmic measure of the acidity of a medium which becomes equal to pH in dilute aqueous solutions. 100% sulfuric acid has an H_0 value of -12.6, whereas, a 0.1 N solution of sulfuric acid in water has an H_0 (pH) value of 1.0. Thus there is an increase in the acidity over this composition range of more than 10^{13}.

The H_0 values for several acid–water system are given in Fig. 11.

The high acidity of sulfuric acid as a solvent medium has been extensively studied (728) and illustrated by the observation that it can protonate many molecules that behave as acids in aqueous solution (such as H_3PO_4, HNO_3, etc.).

The high acidity of sulfuric acid can, however, be further increased by the addition of solutes that, behave as acids of the sulfuric acid–solvent system, i.e.,

$$HA + H_2SO_4 \rightleftharpoons H_3SO^+ + A^-$$

because these solutes increase the concentration of the highly acidic $H_3SO_4^+$ cation in just the same way as the addition of an acid to water increases the concentration of H_3O^+ ion. Fuming sulfuric acid or oleum contains a series of such acids, the polysulfuric acids, the simplest of which is disulfuric acid $H_2S_2O_7$ (728). This ionizes as a moderately strong acid in sulfuric acid

$$H_2S_2O_7 + H_2SO_4 \rightleftharpoons H_3SO_4^+ + HS_2O_7^-$$

The higher polysulfuric acids $H_2S_3O_{10}$, $H_2S_4O_{13}$, etc., also behave as

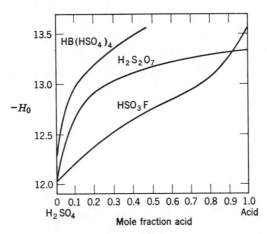

Fig. 12 H_0 values for sulfuric acid–acid systems.

acids and appear to be somewhat stronger than $H_2S_2O_7$. The value of the acidity function H_0 increases continuously up to and beyond the composition $H_2S_2O_7$ as SO_3 is added to H_2SO_4 (Fig. 12). At the composition $H_2S_2O_7$ the very weak base nitrobenzene is fully ionized, whereas in a 0.1 molal solution in H_2SO_4 it is protonated only to the extent of about 30%. Only three other simple protonic acids are known that are capable of protonating sulfuric acid, $HClO_4$, HSO_3Cl, and HSO_3F, and they are ionized only to a very small extent. Nevertheless, the acidity increases continuously in the H_2SO_4–HSO_3F and the H_2SO_4–HSO_3Cl systems, reaching a value of -13.6 for both HSO_3F and HSO_3Cl (Fig. 12). The strongest known acid of the sulfuric acid solvent system, according to Gillespie (726), appears to be the complex acid $HB(HSO_4)_4$, but this has only limited solubility in sulfuric acid. We see in Fig. 12 that H_0 increases more rapidly with increasing concentration for this acid than for the weaker acids $H_2S_2O_7$ and HSO_3F.

In the last few years fluorosulfuric acid has been used quite extensively as a highly acidic medium. It is readily available from commercial sources; it is relatively easily purified by distillation; and it does not attack glass. In addition to having a higher acidity than H_2SO_4, it has several other advantages over sulfuric acid; in particular, it has a convenient wide liquid range from -89 to $161°$ and is much less viscous. The low freezing point has proved to be a very considerable advantage, as it has greatly facilitated the study of protonation reactions (727). At ordinary temperatures proton-exchange reactions in all these acid solvents are much too fast to allow the detection of both protonated and unprotonated forms of a base, for example, by n.m.r. spectroscopy. At low temperatures near the freezing point of fluorosulfuric acid, however, many proton-exchange reactions are slowed down sufficiently so that n.m.r. measurements become useful.

The acidity of fluorosulfuric acid can be still further increased (726,729) by the addition of SbF_5 which has been shown to ionize in the following manner, increasing the concentration of the highly acidic $H_2SO_3F^+$

$$SbF_5 + 2HSO_3F \rightleftharpoons H_2SO_3F^+ + SbF_5(SO_3F^-)$$

ion. The acidity function H_0 is increased to values of the order of -18.

Studies of solutions in fluorosulfuric acid and the still more highly acidic HSO_3F–SbF_5 (Magic Acid) system (729) have further extended our knowledge of cations that are much too electrophilic (i.e., acidic in the Lewis sense) to exist as stable species in more basic solvents. Thus, for example, stable solutions of alkylcarbenium ions, such as the trimethylcarbenium ion, have been obtained for the first time by Olah and co-workers (730). The simple acetyl cation CH_3CO^+ is also stable in these media. In the realm of inorganic chemistry, Gillespie and co-workers (726) found that the reaction of oxidizing agent peroxydisulfuryl difluoride, $S_2O_6F_2$, with various nonmetallic elements in solution in HSO_3F or HSO_3F–SbF_5 led to the production of several new low-valency-state cations whose existence was previously unknown and unsuspected, e.g.,

$$2I_2 + S_2O_6F_2 \rightarrow 2I_2^+ + 2SO_3F^-$$

$$3I_2 + S_2O_6F_2 \rightarrow 2I_3^+ + 2SO_3F^-$$

There is similar evidence that bromine gives Br_2^+ and Br_3^+, although these ions appear to be stable only in the most highly acidic media, e.g., HSO_3F–SbF_5. Even Cl_2^+ was observed in SbF_5 media (731), although the question was raised whether the species was Cl_2O^+ or $ClOCl^+$ (732,733). Related HF based superacid systems, such as HF–SbF_5, have also been widely studied, but there are yet not sufficient data to compare their acidities with HSO_3F–SbF_5 even in a semiquantitative manner. Fluoroantimonic acid, HF–SbF_5 (1 : 1) is, however, based on its protonating ability, considered even more acidic, than FSO_3H–SbF_5. Recent studies by Brouwer (734) estimate H_0 for this superacid as > -20!

Even HSO_3F–SbF_5 (Magic acid® (735)) is not the most adicic fluorosulfuric acid medium that has been investigated, for it has been found by Gillespie that addition of SO_3 to this system still further increases the acidity. It is, however, aside from HF–SbF_5, the most frequently used superacid system. It is not yet known what the limit of attainable acidity in the liquid state is. What is known, however, is that there exist acids that are at least 10^7 times stronger acids than 100% sulfuric acid and against these superacids an increasingly large number of weak bases, such as methane, carbon dioxide, and sulfur dioxide, show marked reactivity, allowing substantial extension of the application of acid catalysis to new areas.

References

1. Adams, R., and K. R. Eilar, *J. Am. Chem. Soc.*, **73**, 1149 (1951).
2. Adams, B. A., and E. L. Holmes, *J. Soc. Chem. Ind.*, **54**, 1 (1935).
3. Adelson, D. E., R. P. Ruh, and H. F. Gray, Jr., U.S. Pat. 2,402, 483 (1946).
4. Adkins, H., and B. H. Nissen, *J. Am. Chem. Soc.*, **44**, 2749 (1922).

5. Ador, E., and A. Rilliet, *Ber.*, **11**, 1627 (1878).
6. Ahmad, M. S., G. Baddeley, and R. M. Topping, *Chem. and Ind.*, 1327 (1958).
7. Alden, R. C., F. E. Frey, H. J. Hepp, and L. A. McReynolds, *Oil Gas J.*, **44**, No. 40, 70 (1946).
8. Allen, R. H., and L. R. Yats, *J. Am. Chem. Soc.*, **83**, 2799 (1961).
9. Anschütz, R., *Ann.*, **235**, 177 (1886).
9a. Anschütz, R., and H. Immendorff, *Ber.*, **18**, 657 (1885).
10. Arbuzov, Yu. A., and N. D. Zelinskii, *Compt. rend. acad. sci. U.S.S.R.*, **30**, 717 (1941): through *C.A.*, **37**, 625 (1943).
11. Arbuzov, Yu. A., N. D. Zelinskii *et al.*, *Compt. rend. acad. sci. U.S.S.R.*, **46**, 150 (1945); through *C.A.*, **39**, 4837 (1945).
12. Aries, R. S., U.S. Pat. 2,930,819 (1960): U.S. Pat. 2,930,820 (1960).
13. Arnold, C., Brit. Pat. 570,157 (1945).
14. Aschan, O., *Ann.*, **324**, 10 (1902).
15. Atoji, M., and W. H. Lipscomb, *J. Chem. Phys.*, **27**, 195 (1957); **28**, 355 (1958).
16. Aynsley, E. E., G. Hetherington, and P. L. Robinson, *J. Chem. Soc.*, 1119 (1954).
17. Azatyan, D., *Proc. Acad. Sci. U.S.S.R.*, **59**, 901 (1948) and subsequent papers.
18. Azatyan, V. D., *Dokl. Akad. Nauk Arm. S.S.R.*, **28**, 7 (1959).
19. Azatyan, V. D., *Izv. Akad. Nauk Arm. S.S.R., Div. Chem. Soc.*, **13**, 189 (1960).
20. Bachman, G. B., and J. L. Dever, *J. Am. Chem. Soc.*, **80**, 5871 (1958).
21. Baddar, F. G., A. M. Fleifel, and S. Sherif, *J. Chem. Soc.*, 1009 (1959).
22. Baddeley, G., *J. Chem. Soc.*, 99 (1949).
23. Badische Anilin-Soda Fabrik Akt.-Ges., Brit. Pat. 794,785 (1958).
24. Barry, A. J., U.S. Pat. 2,499,561 (1950).
25. Baserga, E., and H. Hopff, *Helv. Chim. Acta*, **39**, 1438 (1956).
26. de Bataafsche, N. V., Petroleum Maatschappy, Germ. Pat. 421,118 (1934); French Pat. 952.262 (1949); through *C.A.*, **45**, 5397 (1951).
27. Battegay, M., *Bull. Soc. Chim. France*, **43**, 109 (1928).
28. Bayer, O., in Houben-Weyl, "Methoden der Organischen Chemie," IV. Aufl. Stuttgart (1954).   Vol. VII, 1, pp. 29.
29. Bechman, E., *Z. anorg. allgem. Chem.*, **51**, 242 (1906).
30. Bell, R. P., and B. G. Skinner, *J. Chem. Soc.*, 2955 (1952).
31. Benesi, H. A., *J. Am. Chem. Soc.*, **78**, 5490 (1956).
32. Bentley, A., A. G. Evans, and J. Halpern, *Trans. Faraday Soc.*, **47**, 711 (1951).
33. Berg, L., G. S. Sumner, Jr., and C. W. Montgomery, *Ind. Eng. Chem.*, **38**, 734 (1946).
34. Berlin, A. A., *J. Gen. Chem. U.S.S.R.*, **14**, 1096 (1944).
35. Berti, F. A., and L. M. Ziti, *Arch. Pharm.*, **285**, 372 (1952); *Arch. Pharm.*, **286**, 253 (1953).
36. Bethell, D., and V. Gold, *J. Chem. Soc.*, 1930 (1958).
37. Bhattacharyya, S. K., and D. Vir, *Advances in Catalysis*, **9**, 625 (1957).
38. Bikales, N. M., and L. Rapoport, *Text. Res. J.*, **28**, 737 (1958).
39. Biltz, W., *Z. anorg. allgem. Chem.*, **133**, 312 (1924).
40. Biltz, W., and W. Klemm, *Z. anorg. allgem. Chem.*, **152**, 267 (1926).

41. Biltz, W., and A. Voigt, *Z. anorg. allgem. Chem.*, **126**, 39 (1923).
42. Biltz, W., and W. Wien, *Z. anorg. allgem. Chem.*, **121**, 257 (1922).
43. Bjerrum, N., *Ber.*, **62**, 1091 (1929).
44. Bloch, H. S., H. Pines, and L. Schmerling, *J. Am. Chem. Soc.*, **68**, 153 (1946).
45. Bloch, H. S., and C. L. Thomas, *J. Am. Chem. Soc.*, **66**, 1589 (1944).
46. Blues, E. T., and D. Bryce-Smith, *Proc. Chem. Soc.*, 245 (1961).
47. Blues, E. T., and D. Bryce-Smith, *Chem Ind.*, 1533 (1960).
48. Bluestein, B. A., U.S. Pat. 2,887,501 (1959); U.S. Pat. 2,887,502 (1959); through *C.A.*, **53**, 18865 (1959).
49. Blunck, F. H., and D. R. Carmody, *Ind. Eng. Chem.* **32**, 328 (1940).
50. Bodendorf, K., and H. Bohme, *Ann.*, **516**, 1 (1935).
51. Boedtker, E., and O. M. Halse, *Bull. Soc. Chim. France*, **19**, 444 (1916).
52. Boeseken, J., *Rec. Trav. Chim.*, **22**, 315 (1903).
53. Boeseken, J., and A. A. Adler, *Rec. Trav. Chim.*, **48**, 474 (1929).
54. Boeseken, J., and A. A. Adler, *Rec. Trav. Chim.*, **32**, 184 (1913).
55. Bohlbro, H., U.S. Pat. 2,728,805 (1955).
56. Booth, H. S., and D. R. Martin, "Boron Trifluoride and Its Derivatives," John Wiley and Sons, New York (1949).
57. Bosshar, H. H., and H. Zollinger, *Helv. Chim. Acta*, **42**, 1659 (1959).
58. Boswell, M. C., and R. R. McLaughlin, *Can. J. Research*, **1**, 400 (1929).
59. Bourne, E. G., M. Stacey, J. C. Tatlow, and J. M. Tedder, *J. Chem. Soc.*, 718 (1951).
60. Boyd, T., U.S. Pat. 2,446,124 (1948).
61. Boyd, T., U.S. Pat. 2,446,123 (1948).
62. Brackman, D. S., and P. H. Plesch, *J. Chem. Soc.*, 1289 (1953).
63. Braekken, H., *Kgl. Norske Videnskab. Selskabs, Forh.*, **8**, No. 10, 1 (1935); through *C.A.*, **31**, 6076 (1937).
64. Braekken, H., and W. Scholten, *Z. Kryst.*, **89**, 448 (1934).
65. Braune, H., and S. Knoke, *Z. Physik. Chem.*, **B23**, 163 (1933).
66. Bredereck, H., G. Lehmann, Ch. Schonfeld, and E. Fritzsche, *Ber.*, **72b**, 1414 (1939).
67. Bredereck, H., G. Lehmann, Ch. Schonfeld, and E. Fritzsche, *Angew. Chem.*, **52**, 445 (1939).
68. Britzinger, H., and H. Orth, *Monatsh.*, **85**, 1015 (1954).
69. Brochet, A., *Compt. rend.*, **117**, 115 (1893).
70. Brockway, L. O., *J. Am. Chem. Soc.*, **57**, 958 (1935).
71. Brode, H., *Ann. Phys.*, **37**, 344 (1940).
72. Bronov, B. V., and L. A. Pershina, *J. Gen. Chem. U.S.S.R.*, **24**, 1608 (1954).
73. Brooks, J. W., *Ind. Eng. Chem.*, **41**, 1694 (1949).
74. Broome, J., B. R. Brown, and G. H. R. Summers, *J. Chem. Soc.*, 2071 (1957).
75. Bruson, H. A., and H. Staudinger, *Ind. Eng. Chem.*, **18**, 381 (1926).
76. Bryce-Smith, D., and W. J. Owen, *J. Chem. Soc.*, 3319 (1960).
77. Brown, C. P., and A. R. Mathieson, *J. Chem. Soc.*, 3612, 3620, 3631 (1957); 3445 (1958).
78. Brown, H. C., in "Organometallic Chemistry," ACS Monograph No. 147. Ed. H. Zeiss, Reinhold Publ. Corp., New York, N.Y., 1960, p. 150.
79. Brown, H. C., and W. Grayson, *J. Am. Chem. Soc.*, **75**, 6285 (1953).

80. Brown, H. C., and W. S. Higbey, U.S. Pat. 2,767,230 (1956); 2,819,328 (1958).
81. Brown, H. C., and R. R. Holmes, *J. Am. Chem. Soc.*, **78**, 2173 (1956).
82. Brown, H. C., and H. W. Pearsall, *J. Am. Chem. Soc.*, **73**, 4681 (1951).
83. Brown, H. C., and H. W. Pearsall, *J. Am. Chem. Soc.*, **74**, 191 (1952).
84. Brown, H. C., H. Pearsall, and L. P. Eddy, *J. Am. Chem. Soc.*, **72**, 5347 (1950).
85. Brown, H. C., and W. J. Wallace, *J. Am. Chem. Soc.*, **75**, 6265 (1953).
86. Brown, H. C., and W. J. Wallace, *J. Am. Chem. Soc.*, **75**, 6279 (1953).
87. Brown, H. C., and W. J. Wallace, *J. Am. Chem. Soc.*, **75**, 6268 (1953).
88. Burk, R. E., "12th Catalyst Report," National Research Council, p. 251, John Wiley and Sons, New York, 1940.
89. Burk, R. E., H. E. Thompson, A. J. Weith, and I. Williams, "Polymerisation," p. 80, Reinhold Publ. Corp., New York (1937).
90. Buroway, A., and C. S. Gibson, *J. Chem. Soc.*, 217 (1935).
91. Burton, H., and D. A. Munday, *J. Chem. Soc.*, 1718 (1957).
92. Burton, H., and P. F. G. Praill, *J. Chem. Soc.*, 1203 (1950).
93. Burton, H., and P. F. G. Praill, *J. Chem. Soc.*, 837 (1953).
93a. Burton, H., and Praill, P. F. G., *Chem. and Ind.*, 90 (1954).
94. Burwell, R. L., Jr., and L. M. Elkin, *J. Am. Chem. Soc.*, **73**, 502 (1951).
95. Burwell, R. L., Jr., L. M. Elkin, and A. D. Shields, *J. Am. Chem. Soc.*, **74**, 4567, 4570 (1952).
96. Burwell, R. L., Jr., and G. S. Gordon, III, *J. Am. Chem. Soc.*, **70**, 3128 (1948).
97. Byrne, J. J., Dissertation, Purdue University (1958).
98. Cade, G. N., R. E. Dunn, and H. J. Hepp, *J. Am. Chem. Soc.*, **68**, 2454 (1946).
99. Caille, E., *Bull. Soc. Chim. France*, (4) **3**, 916 (1908).
100. Calcott, W. S., J. M. Tinker, and V. Weinmayr, *J. Am. Chem. Soc.*, **61**, 949 (1939).
101. Calcott, W. S., J. M. Tinker, and V. Weinmayr, *J. Am. Chem. Soc.*, **61**, 1010 (1939).
102. Calloway, N. O., *Chem. Rev.*, **17**, 376 (1935).
103. Calloway, N. O., *J. Am. Chem. Soc.*, **59**, 1474 (1937).
104. Campaigne, E., and J. R. Leal, *J. Am. Chem. Soc.*, **75**, 230 (1953).
105. Campaigne, E., and W. Thompson, *J. Am. Chem. Soc.*, **72**, 629 (1950).
106. Campbell, B. N., Jr., "Certain aspects of the Friedel-Crafts reaction," Dissertation, University of Connecticut (1958).
107. Carothers, W. H., G. J. Berchet, and R. A. Jacobson, U.S. Pat. 1,963,074 (1934).
108. Carter, A. S., and F. B. Downing, U.S. Pat. 1,896,162, *C.A.*, **27**, 2591.
109. Caserio, M. C., J. D. Roberts, M. Neeman, and W. S. Johnson, *J. Am. Chem. Soc.*, **80**, 2584 (1958).
110. Cauquil, G., and H. Barrera, *Bull. Soc. Chim. France*, 84 (1951).
111. Cauquil, G., H. Barrera, and R. Barrera, *Compt. rend.*, **233**, 1117 (1951).
112. Cawley, A., F. Fairbrother, and N. Scott, *J. Chem. Soc.*, 3133 (1958).
113. Chalmers, W., *Can. J. Research*, **7**, 113, 472 (1932); *J. Am. Chem. Soc.*, **56**, 912 (1934).
113a. Chiba, T., *J. Phys. Soc. Japan*, **13**, 860 (1958).

114. Chodroff, S., and H. C. Klein, *J. Am. Chem. Soc.*, **70**, 1647 (1948).
115. Ciapetta, F. G., and J. B. Hunter, *Ind. Eng. Chem.*, **45**, 147 (1953).
116. Ciapetta, F. G., *Ind. Eng. Chem.*, **45**, 159 (1953).
117. Clark, A., M. P. Matuszak, N. C. Carter, and J. S. Cromeans, *Ind. Eng. Chem.*, **45**, 803 (1953).
118. Clark, D., "Cationic Polymerization and Related Complexes", edited by P. H. Plesch, Academic Press, New York, p. 99 (1953).
119. Clayton, J. M., and A. M. Eastham, *J. Am. Chem. Soc.*, **79**, 5368 (1957).
120. Clifford, A. F., H. C. Beachell, and W. M. Jack, *J. Inorg. Nucl. Chem.*, **5**, 57 (1957).
121. Cockerille, F. O., U.S. Pat. 2,541,408 (1951).
122. Colonge, J., and L. Bonnard, *Bull. Soc. Chim. France*, 742 (1958).
123. Comstock, W. J., *J. Am. Chem. Soc.*, **18**, 547 (1896).
124. Comstock, W. J., and G. Perrier, *Ann. Chim. Phys.*, **18**, 547 (1896).
125. Condon, F. E., in "Catalysis," Vol. VI, ed. P. H. Emmet, p. 43, Reinhold Publ. Corp., New York (1953).
126. Cook, O. W., and V. J. Chamber, *J. Am. Chem. Soc.*, **43**, 334 (1921).
127. Crabtree, J. M., C. S. Lees, and K. Little, *J. Inorg. Nucl. Chem.*, **1**, 213 (1956).
128. Croxall, W. J., F. J. Sowa, and J. A. Nieuwland, *J. Am. Chem. Soc.*, **56**, 2054 (1934).
129. Csürös, Z., M. Groszmann, and B. Zsuffa, *Periodica Polytech.*, **2**, 183 (1958); through *C.A.*, **53**, 6674 (1959).
130. Cullinane, N. M., and D. M. Leyshon, *J. Chem. Soc.*, 376, 4106 (1952); 2942 (1954).
131. d'Alelio, G. F., U.S. Pat. 2,802,884 (1957).
132. d'Ans, J., and H. Zimmer, *Ber.*, **85**, 585 (1952).
133. Dayan, J. E., and S. M. Roberts, U.S. Pat. 2,853,522 (1958).
134. Dermer, O. C., and R. A. Billmeier, *J. Am. Chem. Soc.*, **64**, 464 (1942).
135. Dermer, O. C., P. T. Mori, and S. Suguitan, *Proc. Okla. Acad. Sci.*, **29**, 74 (1948).
136. Dermer, O. C., D. M. Wilson, F. M. Johnson, and V. H. Dermer, *J. Am. Chem. Soc.*, **63**, 2881 (1941).
137. Desai, R. D., and R. M. Desai, *J. Sci. Ind. Res. (India)*, **14B**, 498 (1955).
138. DeSimo, M., and F. E. Hilmer, U.S. Pat. 2,085,524 (1937).
139. Deville, H. St. C., *Ann. Chim. (Paris)*, **75**, 66 (1839).
139a. Deville, H. St. C., *Ann. Chim. Phys.*, (3), **46**, 438 (1856).
140. Diehl, P., *J. Inorg. Nucl. Chem.*, **8**, 468 (1958); *Helv. Phys. Acta*, **31**, 685 (1958).
141. Dierichs, A., Report PB 866, Office of Technical Services, Dept. of Commerce, Washington, 1945.
142. Diuguid, L. I., *J. Am. Chem. Soc.*, **63**, 3572 (1941).
143. Dobryansky, A. F., E. K. Kanep, and S. V. Katsman, *Trans. Exp. Research Lab. "Khemgas"* (Leningrad), **3**, 1 (1936); through *C.A.*, **31**, 5334 (1937).
144. Doebner, O., and Stackman, *Ber.*, **9**, 1918 (1876).
145. Doleniek, A. A., and M. Potash, U.S. Pat. 2,479,559 (1949).
146. Dolgov, B. N., Yu. I. Khudobin, and N. P. Kharitonov, *Bull. Acad. Sci. U.S.S.R., Div. Chem. Sci.*, 1238 (1959).

147. Dolgov, B. N., and N. H. Kuchmova, *J. Gen. Chem. U.S.S.R.*, **20**, 445 (1950).
148. Dolgov, B. N., N. I. Sorokina, and A. S. Cherkasov, *J. Gen. Chem. U.S.S.R.*, **21**, 563 (1951).
149. Dolgov, B. N., N. I. Sorokina, and A. S. Cherkasov, *J. Gen. Chem. U.S.S.R.*, **21**, 509 (1951).
150. Donnel, C. K., and R. M. Kennedy, *J. Am. Chem. Soc.*, **74**, 4162 (1952).
151. Dorofeenko, G. N., *Zh. Vses. Khim. Obshchestva. im. D.I. Mendeleeva*, **5**, 354 (1960); through *C.A.*, **54**, 22563 (1960).
152. Dorris, T. B., and F. J. Sowa, *J. Am. Chem. Soc.*, **60**, 358 (1938).
153. Dow Chemical Company, "Dowex Ion Exchange," 1958.
154. Dow, D. A., *J. Chem. Phys.*, **31**, 1637 (1959).
155. Drahowzahl, F., *Monatsh.*, **88**, 842 (1957).
156. Drahowzahl, F., and H. Bildstein, Austrian Pat. 190,501 (1957).
157. Dumontet, J., *Compt. rend*, **234**, 1173 (1952).
158. Eastham, A. M., *J. Am. Chem. Soc..* **78**, 6040 (1956).
159. Egloff, G., G. Hulla, and V. I. Komarewsky: "Isomerisation of Pure Hydrocarbons," Reinhold Publ. Corp., New York, p. 28, 1942.
160. Eitel, A., and R. Fialla, *Monatsh.*, **79**, 112 (1948).
161. Eley, D. D., and P. J. King, *Trans. Faraday Soc.*, **47**, 1287 (1951).
162. Eley, D. D., and A. W. Richards, *Trans. Faraday Soc.*, **45**, 425 (1949).
163. Eley, D. D., and H. Watts, *J. Chem. Soc.*, 1914 (1952).
164. Emeleus, H. J., and G. S. Rao, *J. Chem. Soc.*, 4245 (1958).
165. Esterka, F., and M. Dolezal, *Chem. Prumysl.*, **10**, 62 (1960); through *C.A.*, **54**, 10281 (1960).
166. Evans, A. G., *J. Appl. Chem.*, **1**, 240 (1951).
167. Evans, A. G., and S. D. Hammann, *Proc. Roy. Dublin Soc.*, **25**, 139 (1950).
168. Evans, A. G., D. Holden, P. H. Plesch, M. Polanyi, H. A. Skinner, and M. A. Weinberger, *Nature*, **157**, 102 (1946).
169. Evans, A. G., and G. W. Meadows, *J. Polymer Sci.*, **4**, 359 (1949).
170. Evans, A. G., and G. W. Meadows, *Trans. Faraday Soc.*, **46**, 327 (1950).
171. Evans, A. G., G. W. Meadows, and M. Polanyi, *Nature*, **158**, 94 (1946).
172. Evans, A. G., G. W. Meadows, and M. Polanyi, *Nature*, **160**, 869 (1947).
173. Evans, A. G., and F. T. Perkins, *Research*, **2**, S394 (1949).
174. Evans, A. G., and M. Polanyi, *J. Chem. Soc.*, 252 (1947).
175. Ewell, R. H., and P. E. Hardy, *J. Am. Chem. Soc.*, **63**, 3460 (1941).
176. Fairbrother, F., and W. C. Frith, *J. Chem. Soc.*, 3051 (1951).
177. Fairbrother, F., N. Scott, and H. Prophet, *J. Chem. Soc.*, 1164 (1956).
178. I. G. Farbenindustrie, Brit. Pat. 334,0091; through *C.A.*, **25**, 710 (1931).
179. I. G. Farbenindustrie, Brit. Pat. 372,321 (1932); through *C.A.*, **27**, 3481 (1933).
180. I. G. Farbenindustrie, Brit. Pat. 421,118 (1934); through *C.A.*, **29**, 3353 (1935).
181. Felman, J. H., *Nature*, **179**, 265 (1957).
182. Ferrier, R. J., and J. M. Tedder, *J. Chem. Soc.*, 1435 (1957).
183. Fialkov, Y. A., and A. I. Gengrinovich, *Zapiski Inst. Khem. Akad. Nauk Ukr.R.S.R.*, **7**, 125 (1940); through *C.A.*, **35**, 3617 (1941).
184. Fieser, L. F., and M. A. Fieser, "Organic Chemistry", 3rd ed., Reinhold Publ. Corp., New York, 1956.
185. Fife, G., Brit. Pat. 553,976 (1943); through *C.A.*, **38**, 6060 (1944).

186. Fischer, C. H., and A. Eisner, *J. Org. Chem.*, **6**, 169 (1941).
187. Fischer, R. F., U.S. Pat. 2,915,563 (1959).
188. Fischer, W., *Z. anorg. allgem. Chem.*, **184**, 333 (1929).
189. Fischer, W., and O. Rahlfs, *Z. anorg. allgem. Chem.*, **205**, 32 (1932).
190. Fischer, W., and W. Weidemann, *Z. anorg. allgem. Chem.*, **213**, 106 (1933).
191. Flid, R. M., and N. F. Skvortsova, *Khim. Nauka i Promy*, **3**, 286 (1958).
192. Floria, V. D., U.S. Pat. 2,889,377.
193. Fontana, C. M., and R. J. Harold, *J. Am. Chem. Soc.*, **70**, 2881 (1948).
194. Ford, P. T., and R. E. Richards, *J. Chem. Soc.*, 3870 (1956).
195. Forryan, J. L., Brit. Pat. 774,716 (1957); through *C.A.*, **51**, 16554.
196. Frey, F. E., P. V. McKinney, and W. H. Wood, U.S. Pat. 2,296,511 (1943); through *C.A.*, **37**, 1594, 1943.
197. Friedel, C., and J. M. Crafts, *Compt. rend.*, **84**, 1392 (1877); *ibid.*, 1450.
198. Friedel, C., and J. M. Crafts, *Compt. rend.*, **85**, 74 (1877).
199. Friedel, C., and J. M. Crafts, *Compt. rend.*, **86**, 884 (1878).
200. Fries, K., and K. Schimmelschmidt, *Ber.*, **58**, 2835 (1925).
201. Fürst, H., J. Lang, and P. P. Rammelt, *Chem. Tech. (Berlin)*, **5**, 359 (1953).
202. Gage, D. M., and E. F. Barker, *J. Chem. Phys.*, **7**, 445 (1939).
203. Gallay, W., and G. S. Withby, *Can. J. Research*, **2**, 31 (1930).
204. Galle, R. R., *J. Gen. Chem. U.S.S.R.*, **8**, 402 (1938).
205. Gavrilov, B. G., and L. F. Andreeva, *J. Gen. Chem. U.S.S.R.*, **30**, 593 (1960).
206. Gerding, H., and H. Houtgraff, *Rec. Trav. Chim.*, **72**, 21 (1953).
207. Gerding, H., and E. Smith, *Z. Physik. Chem.*, **B50**, 171 (1941).
208. Gerding, H., and E. Smith, *Z. Physik. Chem.*, **B51**, 217 (1942).
209. Gerrard, W., M. F. Lappert, and H. B. Silver, *Proc. Chem. Soc.*, 19 (1957).
210. Gillam, A. E., and R. A. Morton, *Proc. Roy. Soc. (London)*, **124**, 606 (1929).
211. Gillet, A., *Bull. soc. chim. Belg.*, **29**, 192 (1920).
212. Gilman, H., and N. O. Calloway, *J. Am. Chem. Soc.*, **55**, 4197 (1933).
213. Glasebrook, A. L., N. E. Phillips, and W. G. Lovell, *J. Am. Chem. Soc.*, **58**, 1944 (1936).
214. Gmelin: "Handbuch der anorganischen Chemie," 8. Auflage, Band 35B, pp. 170, 183 (1933).
215. Goldfarb, I., *J. Russ. Phys. Chem. Soc.*, **62**, 1073 (1930); through *C.A.*, **25**, 2719.
216. Goldwasser, S., and H. S. Taylor, *J. Am. Chem. Soc.*, **61**, 1751 (1939).
217. Goodrich, B. F. Co., Brit. Pat. 831,828 (1960), through *C.A.*, **54**, 20345 (1960).
218. Gordon, G. S. III, and R. L. Burwell, *J. Am. Chem. Soc.*, **71**, 2355 (1949).
219. Greenwood, N. N., and R. L. Martin, *J. Chem. Soc.*, 1795 (1951).
220. Greenwood, N. N., and R. L. Martin, *J. Chem. Soc.*, 1915 (1951).
221. Greenwood, N. N., and R. L. Martin, *J. Chem. Soc.*, 751 (1953).
222. Greenwood, N. N., and R. L. Martin, *J. Chem. Soc.*, 1427 (1953).
223. Greenwood, N. N., and R. L. Martin, *Quart. Rev. (London)*, **8** 1 (1954).
224. Greenwood, N. N., R. L. Martin, and H. J. Eméleus, *J. Chem. Soc.*, 3030 (1950).

225. Greenwood, N. N., R. L. Martin, and H. J. Eméleus, *J. Chem. Soc.*, 1328, (1951).

226. Greenwood, N. N., and B. G. Perkins, *J. Chem. Soc.*, 1141 (1960).

227. Greenwood, N. H., and K. Wade, *J. Chem. Soc.*, 1533 (1956).

228. Greenwood, N. N., and Worrall, *J. Inorg. Nucl. Chem.*, **3**, 357 (1957).

229. Greensfelder, B. S., R. C. Archibald, and D. L. Fuller, *Chem. Eng. Progr.*, **43**, 561 (1947).

230. Greensfelder, B. S., H. H. Voge, and G. M. Good, *Ind. Eng. Chem.*, **41**, 2573 (1949).

231. Gregg, A. H., G. C. Hampson, G. I. Jenkins, P. L. F. Jones, and L. E. Sutton, *Trans. Faraday Soc.*, **33**, 852 (1937).

232. Grignard, V., and R. Stratford, *Compt. rend.*, **178**, 2149 (1924); *Bull. Soc. Chim. France*, (4), **35**, 931 (1924).

233. Gringraux, L., *Helv. Chim. Acta*, **12**, 921 (1929).

234. Groggins, A. H., "Unit Processes in Organic Syntheses," Fourth Edition, McGraw-Hill Book Co., New York, p. 873 (1952).

235. Groggins, P. H., and R. H. Nagel, U.S. Pat. 1,966,797 (1934).

236. Groggins, P. H., and R. H. Nagel, *Ind. Eng. Chem.*, **26**, 1313 (1934).

237. Groizebeau, L., *Compt. rend.*, **244**, 1223 (1957).

238. Grosse, A. V., and V. N. Ipatieff, *J. Org. Chem.*, **1**, 559 (1936).

239. Grucarevic, S., and V. Merz, *Ber.*, **6**, 1240 (1873).

240. Grummit, O., and E. N. Case, *J. Am. Chem. Soc.*, **64**, 878 (1942).

241. Grummit, O., E. E. Seusel, W. R. Smith, R. E. Burr, and H. P. Laurelma, *J. Am. Chem. Soc.*, **67**, 910 (1945).

242. Haendler, H. M., L. H. Towle, E. F. Bennett, and W. L. Patterson, Jr., *J. Am. Chem. Soc..*, **76**, 2178 (1954).

243. Hamman, K., *Angew. Chem.*, **63**, 231 (1951).

244. Hanning, E., *Arch. Pharm.*, **288**, 560 (1955).

245. Hansford, R. C., in "Physical Chemistry of the Hydrocarbons," A. Farkas, ed., Vol. II, p. 287, Academic Press, New York (1953).

246. Hansford, R. C., C. D. Myers, and A. N. Sachanen, *Ind. Eng. Chem.*, **37**, 671 (1945).

247. Harries, C. D., *Chem. Ztg.*, **34**, 315 (1910).

248. Hartough, H. D., and A. I. Kosak, *J. Am. Chem. Soc.*, **68**, 2639 (1946).

249. Hartough, H. D., and A. T. Kosak, U.S. Pat. 2,458,513, 2,458,520 (1949).

250. Hassel, O., and A. Sandbo, *Z. Physik. Chem.*, **B41**, 75 (1938).

251. Haszeldine, R. N., and A. G. Sharpe, *J. Chem. Soc.*, 993 (1952).

252. Haubein, A. M., L. D. Apperson, and H. Gilman, *Proc. Iowa Acad. Sci.* **46**, 219 (1939).

253. Hauptschein, M., and M. Braid, *J. Am. Chem. Soc.*, **83**, 2383 (1961).

254. Haworth, R. D., and G. Sheldrick, *J. Chem. Soc.*, 864 (1934).

255. Hay, R. G., C. W. Montgomery, and J. Coull, *Ind. Eng. Chem.*, **37**, 335 (1945).

256. Heine, H. W., D. L. Cottle, and H. L. VanMater, *J. Am. Chem. Soc.*, **68**, 524 (1946).

257. Heise, R., and A. Tohl, *Ann.*, **270**, 155 (1892).

258. Heldman, J. D., *J. Am. Chem. Soc.*, **66**, 1789 (1944).

259. Henne, A. L., and J. M. Tedder, *J. Chem. Soc.*, 3628 (1953).

260. Hennion, G. F., and L. A. Auspos, *J. Am. Chem. Soc.*, **65**, 1603 (1943).

261. Hennion, G. F., H. D. Hinton, and J. A. Nieuwland, *J. Am. Chem. Soc.*, **55**, 2857 (1937).
262. Hevesy, G. V., *Kgl. Danske. Vid. Selsk. Med.*, **3**, 13 (1921).
263. Hevesy, G. V., *Z. Physik. Chem.*, **127**, 406 (1927).
264. Hinton, H. D., and J. A. Nieuwland, *J. Am. Chem. Soc.*, **54**, 2017 (1932).
265. Hirst, H. R., and J. B. Cohen, *J. Chem. Soc.*, **67**, 826 (1895).
266. Hofmann, F., and C. Wulff, Brit. Pat. 307,902 (1928).
267. Hollaway, C., Jr., and W. S. Bonnel, *Ind. Eng. Chem.*, **38**, 1231 (1946).
268. Horny, Ch., and M. Guinet, French Pat. 63525 (1955); through *C.A.*, **53**, 7014 (1959).
269. Hückel, W., "Structural Chemistry of Inorganic Compounds," Elsevier Co., London, Vol. II, p. 474 (1951).
270. Huett, G., and W. O. Ranky, U.S. Pat. 2,900,418 (1959).
271. Huhti, A. L., and P. A. Gartaganis, *Can. J. Chem.*, **34**, 785 (1956).
272. Hyman, H. H., L. A. Quarterman, M. Kilpatrick, and J. J. Katz, *J. Phys. Chem.*, **65**, 123 (1961).
273. Iaon, Y., L. Teodorescu, S. Titeica, and C. D. Nenitzescu, *Studii si Cer. Chem.* (*Acad. R.P.R.*), **7**, 345 (1959).
274. Illari, G., *Gazz. Chim. Ital.*, **78**, 687 (1948).
275. Inoue, Y., K. Onodera, I. Karasawa, and Y. Nishisawa, *J. Agr. Chem. Soc. Japan*, **25**, 499 (1951–52); through *C.A.*, **47**, 8021 (1953).
276. Ipatieff, V. N., "Catalytic Reactions at High Pressures and Temperatures," p. 720, MacMillan Co., New York, 1936.
277. Ipatieff, V. N., *Ind. Eng. Chem.*, **27**, 1067 (1935).
278. Ipatieff, V. N., U.S. Pat. 2,067,764 (1937); *C.A.*, **31**, 1603 (1937); *Ind. Eng. Chem.*, **38**, 400 (1946).
279. Ipatieff, V. N., B. B. Corson, and H. Pines, *J. Am. Chem. Soc.*, **58**, 919 (1936).
280. Ipatieff, V. N., and A. V. Grosse, *J. Am. Chem. Soc.*, **57**, 1621 (1935).
281. Ipatieff, V. N., and A. V. Grosse, *J. Am. Chem. Soc.*, **58**, 915 (1936).
282. Ipatieff, V. N., and V. I. Komarewsky, *J. Am. Chem. Soc.*, **58**, 922 (1936).
283. Ipatieff, V. N., V. I. Komarewsky, and H. Pines, *J. Am. Chem. Soc.*, **58**, 918 (1936).
284. Ipatieff, V. N., and H. Pines, *Ind. Eng. Chem.*, **27**, 1364 (1935).
285. Ipatieff, V. N., and H. Pines, *J. Am. Chem. Soc.*, **59**, 56 (1937).
286. Ipatieff, V. N., H. Pines, and V. I. Komarewsky, *Ind. Eng. Chem.*, **28**, 222 (1936).
287. Ipatieff, V. N., and O. Routal, *Ber.*, **46**, 1748 (1913).
288. Ipatieff, V. N., and L. Schmerling, "Advances in Catalysis," Vol. I, p. 46, Academic Press, New York, 1948.
289. Ipatieff, V. N., and L. Schmerling, U.S. Pat. 2,211,207, U.S. Pat. 2,296,511, U.S. Pat. 2,342,865.
290. Ipatieff, V. N., and B. S. Sdzitovetsky, *J. Russ. Phys. Chem. Soc.*, **39**, 897 (1907).
291. Isagulyants, V. I., U.S.S.R. Pat. 109,978 (1958); *C.A.*, **52**, 15579 (1958).
292. Isagulyants, V. I., and N. A. Slavskaya, *J. Appl. Chem. U.S.S.R.*, **33**, 953 (1960).

293. Isagulyants, V. I., and V. N. Tishkova, U.S.S.R. Pat. 110,953 (1958); C.A., 52, 1468 (1958).
294. Jacobs, T. L., and P. Tuttle, Jr., J. Am. Chem. Soc., 71, 1313 (1949).
295. Jander, G., and K. Kraffczyk, Z. anorg. allgem. Chem., 282, 121 (1955).
296. Jenny, R., Compt. rend., 246, 3477 (1958).
297. Jenny, R., Compt. rend., 250, 1659 (1960).
298. Jenny, R., Compt. rend., 248, 3555 (1959).
299. Jordan, D. O., and A. R. Mathieson, J. Chem. Soc., 611, 621 (1952).
300. Jordan, D. O., and F. E. Treloar, J. Chem. Soc., 734 (1961).
301. Joris, G. G., U.S. Pat. 2,945,072 (1960).
302. Juracka, F., Collection Czech. Chem. Commun., 24, 1388 (1959).
302a. Kaiser, J. R., L. D. Moore, and R. C. Odioso, Ind. Eng. Chem., Prod. Res. Develop., 1, 127 (1962).
303. Kashtanov, L. I., J. Gen. Chem. U.S.S.R., 2, 515 (1932).
304. Kashtanov, J. Gen. Chem. U.S.S.R., 3, 229 (1933).
305. Kashtanov, L. I., J. Gen. Chem. U.S.S.R., 3, 279 (1933).
306. Kästner, D., Angew. Chem., 54, 273, 296 (1940).
307. Kästner, D., in "Newer Methods of Preparative Organic Chemistry," Interscience Publishers, New York, p. 249 (1948).
308. Kataev, E. G., and M. V. Palkina, Uch. Zap. Kazansk. Gos. Univ., 113, 115 (1953); through C.A., 50, 937 (1956).
309. Kataoka, S., Nippon Senbai Kôsha Chûô Kenkyûsho Kenkyû Hôkoku, No. 95, 79 (1956); through C.A., 53, 17561 (1959).
310. Kaye, I. A., H. C. Klein, and W. J. Buralant, J. Am. Chem. Soc., 75, 745 (1953).
311. Keefer, R. M., and L. J. Andrews, J. Am. Chem. Soc., 82, 4547 (1960).
312. Kelly, J. T., U.S. Pat. 2,834,642 (1958).
313. Kendall, J., E. D. Crittenden, and K. H. Miller, J. Am. Chem. Soc., 45, 963 (1923).
314. Ketelaar, J. A. A., C. H. MacGillavry, and P. A. Renes, Rec. Trav. Chim., 66, 501 (1947).
315. Kikindai, T., Ann. Chim., [13], 1, 273 (1956).
316. Kilpatrick, M., and A. E. Lukorsky, J. Am. Chem. Soc., 76, 5863 (1954).
317. Kline, E. R., B. N. Campbell, and E. C. Spaeth, J. Org. Chem., 24, 1781, (1959).
318. Klinkberg, L. J., and J. A. A. Ketelaar, Rec. Trav. Chim., 54, 959 (1935).
319. Klosa, J., Arch. Pharm., 288, 48 (1955).
320. Kluge, H. D., and F. W. Moore, U.S. Pat. 2,868,823 (1959).
321. Koebner, A., and R. Robinson, J. Chem. Soc., 1944 (1938).
322. Kohlrausch, K. W. F., and J. Wagner, Z. Physik. Chem., B52, 185 (1942).
323. Kolka, A. J., J. P. Napolitano, and G. G. Ecke, J. Org. Chem., 21, 712 (1956).
324. Komarewsky, V. I., and W. E. Ruther, J. Am. Chem. Soc., 72, 5501 (1950).
325. Konrad, F., U.S. Pat. 2,765,349 (1956); through C.A., 51, 7399 (1957).
326. Konrad, F., U.S. Pat. 2,765,350 (1956); through C.A., 51, 7399 (1957).
327. Konrad, F., U.S. Pat. 2,765,351 (1956); through C.A., 51, 7399 (1957).
328. Konrad, F., U.S. Pat. 2,765,352 (1956); through C.A., 51, 7399 (1957).
329. Korshak, V., and N. N. Lebedev, J. Gen. Chem. U.S.S.R., 18, 1766 (1948).
330. Kozlov, N. S., and P. N. Fedoseev, J. Gen. Chem. U.S.S.R., 6, 250 (1936); through C.A., 30, 4864 (1936).

331. Kraemer, G., and A. Spilker, *Ber.*, **23**, 3169 (1890).
332. Kränzlein, G., "Aluminium Chlorid in der organischen Chemie," 3rd ed. by P. Kränzlein, Verlag Chemie, Berlin, 1939.
333. Krauss, W., and H. Grund, *Naturwissenschaften*, **40**, 18 (1953).
334. Krentsel, B. A., A. V. Topchiev, and L. N. Andreev, *Proc. Acad. Sci. U.S.S.R.*, *Chem. Sect.*, **98**, 75 (1954); **98**, 411 (1954); **92**, 781 (1953); **107**, 265 (1956).
335. Krollpfeiffer, F., *Ber.*, **56**, 2360 (1923).
336. Kuchkarov, A. B., and I. P. Tsukervanik, *J. Gen. Chem. U.S.S.R.*, **20**, 458 (1950).
337. Kuhn, S. J., and G. A. Olah, *J. Am. Chem. Soc.*, **83**, 4564 (1961).
338. Kunin, R., in "Encyclopaedia of Chemical Technology," ed. R. E. Kirk and D. F. Othmer, Vol. 8, p. 13, Interscience Publ., New York, 1952.
339. Kunin, R., "Ion Exchange," John Wiley and Sons, New York, p. 256, 1958.
340. Lacourt, A., *Bull. soc. chim. Belg.*, **38**, 1 (1929).
341. Lal, J. B., and S. Dutt, *J. Ind. Chem.*, **9**, 565 (1932).
342. Lappert, M. F., *J. Chem. Soc.*, 784 (1955).
343. Larson, M. L., *J. Am. Chem. Soc.*, **82**, 1223 (1960).
344. Laubengayer, A. W., and G. F. Condike, *J. Am. Chem. Soc.*, **70**, 2274 (1948).
344a. Laurita, W. G., and W. S. Koski, *J. Am. Chem. Soc.*, **81**, 3179 (1959).
345. Lebeau, P., *Ann. Chim. Phys. (Paris)*, (7), **16**, 493 (1899).
346. Lecher, H., *Ber.*, **46**, 2664 (1913).
347. Lee, J. A., *Chem. Eng.*, **54**, No. 11, 118 (1947).
348. Lee, R. J., H. M. Knight, and J. T. Kelly, U.S. Pat. 2,836,634 (1959).
349. Lee, R. J., H. M. Knight, and J. T. Kelly, *Ind. Eng. Chem.*, **50**, 1001, (1958).
350. Leighton, P. A., and J. D. Heldman, *J. Am. Chem. Soc.*, **65**, 2276 (1943).
351. Leum, L. M., S. J. Macuga, and S. I. Kreps, U.S. Pat. 2,480,940 (1949).
352. Levina, R. Ya., and P. Ya. Kiryushov, *J. Gen. Chem. U.S.S.R.*, **9**, 1834 (1939); through *C.A.*, **34**, 4051 (1940).
353. Levina, R. Ya., *J. Gen. Chem. U.S.S.R.*, **9**, 2287 (1937); through *C.A.*, **34**, 4730 (1940).
354. Levina, R. Ya. *et al.*, *J. Gen. Chem. U.S.S.R.*, **16**, 817 (1940); **19**, 305 (1949); **23**, 562 (1952); *Proc. Acad. Sci. U.S.S.R.*, **71**, 1065 (1953).
355. Lewis, G. N., *J. Franklin Inst.*, **226**, 293 (1938).
356. Lieber, E., and H. T. Rice, U.S. Pat. 2,364,454 (1944).
357. Lien, A. P., and D. A. McCaulay, *J. Am. Chem. Soc.*, **75**, 2407 (1953).
358. Linn, C. B., U.S. Pat. 2,886,525 (1959).
359. Loev, B., and J. T. Massengele, *J. Org. Chem.*, **22**, 988 (1957).
360. Longuet-Higgins, H. C., *Quart. Rev. (London)*, **11**, 121 (1957).
361. Longworth, W. R., and P. H. Plesch, *Proc. Chem. Soc.*, 117 (1958).
362. Loon, M., and J. P. Wibaut, *Rec. Trav. Chim.*, **56**, 815 (1937).
363. Luvisi, J. P., and L. Schmerling, U.S. Pat. 2,910,515 (1959).
364. Mackor, E. L., A. Hofstra, and J. H. Van der Waals, *Trans. Faraday Soc.*, **54**, 186 (1958).
365. MacMullin, R. B., *Chem. Eng. Progress*, **44**, 183 (1948).
366. Malishev, B. W., *J. Am. Chem. Soc.*, **57**, 883, (1935).
367. Martin, L. F., P. Pizzolato, and L. S. McWaters, *J. Am. Chem. Soc.*, **57**, 2548 (1935).

368. Martin, R. L., unpublished observations quoted by N. N. Greenwood and R. L. Martin, *Quart. Rev. (London)*, **8**, 1 (1954).
369. Maslyanskii, G. N., *J. Gen. Chem. U.S.S.R.*, **13**, 540 (1943).
370. Mastagli, P., G. Austerweil, and E. Dubois, *Compt. rend.*, **232**, 20, 1848 (1951).
371. Mathur, K. B. L., J. N. Sharma *et al.*, *J. Am. Chem. Soc.*, **79**, 3582 (1957).
372. Mavity, J. M., U.S. Pat. 2,388,428 (1945).
373. May, P. D., and J. T. Kelly, U.S. Pat. 2,804,491 (1957).
374. McAlevy, A., U.S. Pat. 2,135,454 (1938).
375. McCaulay, D. A., personal communication.
376. McCaulay, D. A., W. S. Higley, and A. P. Lien, *J. Am. Chem. Soc.*, **78**, 3009 (1956).
377. McCaulay, D. A., and A. P. Lien, *J. Am. Chem. Soc.*, **73**, 2013 (1951).
378. McCaulay, D. A., B. H. Shoemaker, and A. P. Lien, *Ind. Eng. Chem.*, **42**, 2103 (1950).
379. McKenna, J. F., and F. J. Sowa, *J. Am. Chem. Soc.*, **59**, 470 (1937).
380. Meerwein, H., *Ann.*, **455**, 227 (1927).
381. Meerwein, H., *Ber.*, **66**, 411 (1933).
382. Meerwein, H., Germ. Pat. 766,208 (1939).
383. Meerwein, H., and T. Bersin, *Ann.*, **476**, 113 (1929).
384. Meerwein, H., K. Bodenbenner, P. Borner, F. Kunert, and K. Wunderlich, *Ann.*, **632**, 38 (1960).
385. Meerwein, H., D. Delfs, and H. Morschel, *Angew. Chem.*, **72**, 929 (1960).
386. Meerwein, H., U. Eisenmenger, and H. Matthiae, *Ann.*, **566**, 150 (1950).
387. Meerwein, H., V. Hederich, and K. Wunderlich, *Arch. Pharm.*, **291/63**, 541 (1958).
388. Meerwein, H., and W. Pannwitz, *J. Prakt. Chem.*, **141**, 123 (1934).
389. Mescheryakov, A. P., and E. P. Kaplan, *Bull. Acad. Sci. U.S.S.R., Div. Chem. Sci.*, 1055, (1938); *C.A.*, **33**, 6230 (1939).
390. Menshutkin, B. N., *J. Russ. Phys. Chem. Soc.*, **45**, 1710 (1913).
391. Menshutkin, B. N., *Z.*, 1906 II, 1720.
392. Merkel, K., Germ. Pat., 1,020,323 (1956).
393. Meyer, H., and K. Bernhauer, *Monatsh.*, **53**, 721 (1929).
394. Michel, R., U.S. Pat. 1,741,472 (1929); 1,741,473 (1929); 1,766,344 (1930); 1,767,302 (1930); 1,878,963 (1932).
395. Mikeska, L. A., and A. H. Gleason, U.S. Pat. 2,092,889 (1937).
396. Mills, G. A., E. R. Boedeker, and A. G. Oblad, *J. Am. Chem. Soc.*, **72**, 1554 (1950).
397. Monsanto Chemical Co., Brit. Pat. 600,785 (1948); through *C.A.*, **42**, 7312 (1948).
398. Montagne, P. J., *Rec. Trav. Chim.*, **40**, 247 (1921).
399. Morgan, G. T., and F. H. Burstall, *J. Chem. Soc.*, 307 (1932).
400. Morton, F., and A. R. Richards, *J. Inst. Petrol.*, **34**, 133 (1948).
401. Mustafa, A., and O. H. Hishmat, *J. Am. Chem. Soc.*, **79**, 2225 (1957).
402. Myers, R. J., J. W. Eastes, and F. J. Myers, *Ind. Eng. Chem.*, **33**, 697 (1941).
403. Nagy, F., D. Dobis, G. Litvan, and I. Telcs, *Acta. Chim. Acad. Sci. Hung.*, **21**, 397 (1959).
404. Naragon, E. A., *Ind. Eng. Chem.*, **42**, 2490 (1950).

405. Natta, G., G. Mazzanti, G. Pregaglia, M. Binaghi, and M. Braldo, *J. Am. Chem. Soc.*, **82**, 4742 (1960).

406. Natta, G., G. Mazzanti, G. F. Pregaglia, and M. Binaghi, *Makromol. Chem.*, **44-46**, 537 (1961).

407. Nelson, K. L., and H. C. Brown, "Aromatic Substitution—Theory and Mechanism," in "Chemistry of Petroleum Hydrocarbons," Vol. 3, Reinhold Publishing Co., New York, p. 514 (1955).

408. Nencki, M., *Ber.*, 1766 (1897).

409. Nenitzescu, C. D., *Bull. Sect. Chim. Sci. Acad. Ruman.*, **27**, No. 10 (1945).

410. Nenitzescu, C. D., M. Avram, and E. Sliam, *Bull. Soc. Chim. France*, 1266. (1955).

411. Nenitzescu, C. D., and E. I. P. Cantuniari, *Ber.*, **66**, 1097 (1933).

412. Nenitzescu, C. D., and D. A. Isacescu, *Ber.*, **66**, 1100 (1933).

413. Nenitzescu, C. D., D. A. Isacescu, and C. N. Ionescu, *Ann.*, **491**, 189, 210 (1931).

414. Nenitzescu, C. D., and I. Necsoiu, A. Glatz, and M. Zalman, *Ber.*, **92**, 10 (1959).

415. Nesmeyanov, A. N., and N. A. Volkenau, *Proc. Acad. Sci. U.S.S.R., Chem. Sect.*, **111**, 605 (1956).

416. Neumann, D., and H. Richter, *Z. Electrochem.*, **31**, 484 (1925).

417. Newman, M. S., *J. Am. Chem. Soc.*, **67**, 345 (1945).

418. Nicolescu, I. V., M. Iovu, and G. I. Nikishin, *Bull. Acad. Sci. U.S.S.R., Div. Chem. Sci.*, 84 (1960).

419. Nieuwland, J. A., W. S. Calcott, F. B. Downing, and A. S. Carter, *J. Am. Chem. Soc.*, **53** 4197 (1931).

420. Nikolaeva, A. F., *J. Gen. Chem. U.S.S.R.*, **16**, 1819 (1946); through *C.A.*, **41**, 5432 (1947).

421. Nikolaeva, A. F., and P. V. Puchkov, *Compt. rend. acad. sci. U.S.S.R.*, **24**, 345 (1939); *C.A.*, **34**, 977 (1940).

422. Nilson, L. F., and O. Pettersson, *Ber.*, **17**, 987 (1884).

423. Nixon, G. R., and R. E. Downer, Brit. Pat. 394,414 (1933).

424. Noack, K., Brit. Pat. 398,474 (1933).

425. Noller, C. R., and R. Adams, *J. Am. Chem. Soc.*, **46**, 1889 (1924).

426. Norrish, R. G. W., and K. E. Russell, *Nature*, **160**, 543 (1947); *Trans. Faraday Soc.*, **48**, 91 (1952).

427. Norris, G. F., and B. M. Sturgis, *J. Am. Chem. Soc.*, **61**, 1413 (1939).

428. Norton, T. R., U.S. Pat. 2,711,414 (1955).

429. Nowacki, W., *Z. Kryst.*, **82**, 360 (1932).

430. Oblad, A. G., T. H. Milliken, Jr., and G. A. Mills, "Chemical Characteristics and Structures of Cracking Catalyst," in "Advances in Catalysis," Vol. III, Academic Press, Inc., New York (1953).

431. Ogata, I., and J. Ishiguro, *Science (Japan)*, **19**, 134 (1949); through *C.A.*, **45**, 5137.

432. O'Kelly, A. A., and D. P. J. Goldsmith, U.S. Pat. 2,347,790 (1944).

433. O'Kelly, A. A., and R. H. Work, U.S. Pat. 2,315,129 (1943).

434. Olah, G. A., unpublished.

435. Olah, G. A., and S. J. Kuhn, *J. Am. Chem. Soc.*, **80**, 654 (1958).

436. Olah, G. A., and S. J. Kuhn, *J. Am. Chem. Soc.*, **80**, 6535 (1958).

437. Olah, G. A., and S. J. Kuhn, U.S. Pat. 2,996,554 (1961).

438. Olah, G. A., and S. J. Kuhn, U.S. Pat. 3,000,986 (1961).

439. Olah, G. A., and S. J. Kuhn, unpublished.
440. Olah, G. A., and S. J. Kuhn, *J. Am. Chem. Soc.*, **84**, 3684 (1962).
441. Olah, G. A., S. J. Kuhn, and S. Mlinko, *J. Chem. Soc.*, 4257 (1956).
442. Olah, G. A., and M. E. Moffatt, unpublished.
443. Olah, G. A., A. E. Pavlath, and S. J. Kuhn, *Chem. and Ind.*, 50 (1957).
444. Olah, G. A., A. E. Pavlath, and J. A. Olah, *J. Am. Chem. Soc.*, **80**, 6540 (1958).
445. Olah, G. A., H. W. Quinn, and S. J. Kuhn, *J. Am. Chem. Soc.*, **82**, 426 (1960).
446. Olah, G. A., H. W. Quinn, and S. J. Kuhn, U.S. Pat. 3,052,733 (1962).
447. Olah, G. A., and W. S. Tolgyesi, *J. Org. Chem.*, **26**, 2319 (1961).
448. Olah, G. A., S. J. Kuhn, W. S. Tolgyesi, and E. B. Baker, *J. Am. Chem. Soc.*, **84**, 2733 (1962).
449. Olah, G. A., W. S. Tolgyesi, and R. E. A. Dear, *J. Org. Chem.*, **27**, 3441 (1962).
450. Olivier, S. C. J., *Rec. Trav. Chim.*, **33**, 91 (1913).
451. Oliverio, A., and E. Lugbi, *Gazz. Chim. Ital.*, **78**, 16 (1948).
452. Orlov, N. N., and P. G. Vaisfeld, *J. Appl. Chem. U.S.S.R.*, **10**, 861 (1937); *C.A.*, **31**, 7043.
453. Ott, E., and W. Brugger, *Z. Elektrochem.*, **46**, 105 (1940).
454. Otvos, J. W., D. P. Stevenson, C. D. Wagner, and O. J. Beeck, *J. Am. Chem. Soc.*, **73**, 5741 (1951).
455. Pajeau, R., *Compt. rend.*, **204**, 1347 (1937); **202**, 1795 (1936).
456. Patat, F., and H. Sinn, *Angew. Chem.*, **70**, 496 (1958).
457. Pauling, K., J. Palmer, and N. Elliot, *J. Am. Chem. Soc.*, **60**, 1852 (1938).
458. Paushkin, Ya. M., and M. V. Kurashev, *Bull. Acad. Sci. U.S.S.R., Div. Chem. Sci.*, 133 (1954).
459. Paushkin, Ya. M., and A. V. Topchiev, *J. Gen. Chem. U.S.S.R.*, **19a**, 657 (1949).
460. Pepper, D. C., *Trans. Faraday Soc.*, **45**, 397 (1949).
461. Pepper, D. C., in "Cationic Polymerization and Related Complexes," ed. P. H. Plesch, Academic Press, New York, p. 70 (1953).
462. Pepper, D. C., and A. E. Somerfield, in "Cationic Polymerization and Related Complexes," ed. P. H. Plesch, Academic Press, New York, p. 75 (1953).
463. Pepper, D. C., *Quart. Rev. (London)*, **8**, 88 (1954).
464. Perrier, G., *Ber.*, **33**, 815 (1900).
465. Petrova, A. M., *J. Gen. Chem. U.S.S.R.*, **24**, 491 (1954).
466. Petrov, A. A., *Proc. Acad. Sci. U.S.S.R., Chem. Sect.*, **50**, 249 (1945).
467. Petrov, D., and M. A. Cheltsova, *Compt. rend. acad. sci. U.S.S.R.*, **15**, 79 (1937).
468. Petrov, A. D., A. P. Meshcheryakov, and D. N. Andreyev, *Ber.*, **68**, 1 (1935).
469. Piazzi, F., *Atti. Accad. Nazl. Lincei*, **22**, 629 (1957); *C.A.*, **52**, 7278 (1958).
470. Piccard, *Ber.*, **7**, 1785 (1874).
471. Pinck, L. Q., *J. Am. Chem. Soc.*, **49**, 2536 (1927).
472. Pines, H., and G. Benoy, *J. Am. Chem. Soc.*, **82**, 2483 (1960).
473. Pines, H., and C. T. Chen, *J. Am. Chem. Soc.*, **82**, 3563 (1960).
474. Pines, H., and W. O. Haag, *J. Am. Chem. Soc.*, **82**, 2471, 2488 (1960).

475. Pines, H., and R. C. Wackher, *J. Am. Chem. Soc.*, **68**, 595 (1946).
476. Pines, H., and R. C. Wackher, *J. Am. Chem. Soc.*, **68**, 599 (1946).
477. Pines, H., and R. C. Wackher, *J. Am. Chem. Soc.*, **68**, 1642 (1946).
478. Pines, H., and R. C. Wackher, *J. Am. Chem. Soc.*, **68**, 2518 (1946).
479. Pines, H., and R. C. Wackher, *J. Am. Chem. Soc.*, **70**, 1742 (1948).
480. Plate, A. F., *J. Gen. Chem. U.S.S.R.*, **15**, 156 (1945).
481. Plesch, P. H., "Cationic Polymerization and Related Complexes," Academic Press, New York (1953).
482. Plesch, P. H., M. Polanyi, and H. A. Skinner, *J. Chem. Soc.*, 257 (1947).
483. Plesch, P. H., *J. Chem. Soc.*, 543 (1950).
484. Plotnikoff, W. A., S. S. Fortunatoff, and W. P. Mochowetz, *Z. Electrochem.*, **37**, 85 (1936).
485. Plotnikoff, V. A., and N. N. Gratianskii, *Mem. Inst. Chem. Acad. Sci. Ukrain S.S.R.*, **5**, 213 (1938); through *C.A.*, **33**, 2432 (1939).
486. Pongratz, A., and G. Markgraf, *Monatsh.*, **66**, 176 (1935).
487. Popp., F. D., and W. E. McEwen, *Chem. Rev.*, **58**, 321 (1958).
488. Porter, R. F., and E. E. Zeller, *J. Chem. Phys.*, **33**, 858 (1960).
489. Powell, J. W., and M. C. Whiting, *Proc. Chem. Soc.*, 412 (1960).
490. Powell, P. M., and E. B. Reid, *J. Am. Chem. Soc.*, **67**, 1020 (1945).
491. Pratt, E. F., and H. J. E. Sègrade, *J. Am. Chem. Soc.*, **81**, 5369 (1959).
492. Prey, V., and B. Metzner, Austrian Pat. 166,212, *C.A.*, **47**, 235 (1953).
493. Prey, V., B. Metzner, and M. Berbalk, *Monatsh.*, **81**, 760 (1950).
494. Prill, E. J., U.S. Pat. 2,802,032 (1957).
495. Proell, W. A., C. A. Adams, and D. H. Shoemaker, *Ind. Eng. Chem.*, **40**, 1129 (1948).
496. Profft, E., *Chem. Tech. (Berlin)*, **7**, 577 (1955).
497. Prokopets, E. I., and S. M. Boguslavskaya, *J. Appl. Chem. U.S.S.R.*, **11**, 1471 (1938); *C.A.*, **33**, 5816 (1939).
498. Prokopets, E. I., and A. N. Filaratov, *J. Appl. Chem. U.S.S.R.*, **11**, 1631 (1938); *C.A.*, **33**, 5817 (1939).
499. Pungs, W., and H. Rabe, U.S. Pat. 1,923,583 (1933); *C.A.*, **27**, 5181 (1933).
500. Pudovik, A. N., *J. Gen. Chem. U.S.S.R.*, **21**, 1811 (1951).
501. Puchkov, P. V., and A. F. Nikoaeva, *J. Gen. Chem. U.S.S.R.*, **8**, 1153 (1938); *C.A.*, **33**, 3766 (1939).
502. Rahlfs, O., and W. Fischer, *Z. anorg. allgem. Chem.*, **211**, 349 (1933).
503. Ralston, A. W., A. Ingle, and M. R. McCorkle, *J. Org. Chem.*, **7**, 457 (1942).
504. Ralston, A. W., M. R. McCorkle, and S. T. Bauer, *J. Org. Chem.*, **5**, 654 (1940).
505. Ralston, O. C., *Chem. News*, **127**, 246 (1923).
506. Rasch, E., *J. Phys. Chem.*, **45**, 995 (1913).
507. Ray, J. N., *J. Chem. Soc.*, **117**, 1335 (1920).
508. Reindel, F., and F. Seigel, *Ber.*, **56**, 1550 (1923).
509. Rennes, P. A., and C. H., MacGillavry, *Rec. Trav. Chim.*, **64**, 275 (1945).
510. Riddell, W. A., and C. R. Noller, *J. Am. Chem. Soc.*, **52**, 4365 (1930).
511. Riddell, W. A., and C. R. Noller, *J. Am. Chem. Soc.*, **54**, 290 (1932).
511a. Ring, M. A. J., D. H. Donnay, and W. S. Koski, *J. Inorg. Chem.*, **1**, 109 (1962).

512. Roe, A., *Org. Reactions*, Vol. 5, p. 200, John Wiley and Sons, New York, 1949.

513. Roebuck, A. K., and B. L. Evering, *J. Am. Chem. Soc.*, **75**, 1631 (1953).

514. Rollier, M. A., and A. Riva, *Gazz. Chim. Ital.*, **77**, 361 (1947).

515. Rosen, R., U.S. Pat. 2,257,193 (1941); through *C.A.*, **36**, 94 (1942).

516. Rosenheim, A., and P. Woge, *Z. anorg. allgem. Chem.*, **15**, 310 (1897).

517. Ross, J., J. H. Perdy, R. L. Brandt, A. I. Gebhart, J. E. Mitchell, and S. Yolles, *Ind. Eng. Chem.*, **34**, 924 (1942).

518. Ross, W. E., and G. J. Carlson, U.S. Pat. 2,393,569 (1946); through *C.A.*, **40**, 1996 (1946).

519. Ross, W. E., S. H. McAllister, and J. Anderson, U.S. Pat. 2,370,195 (1949); U.S. Pat. 2,389,550 (1945).

520. Ross, W. E., S. H. McAllister, and J. Anderson, Brit. Pat. 572,793 (1945), Can. Pat. 414,940 (1940).

521. Roualt, M., *Compt. rend.*, **206**, 51 (1938).

522. Russel, K. E., in "Cationic Polymerization and Related Complexes," ed. P. H. Plesch, Academic Press, New York, p. 114 (1953).

523. Ruthruff, R. F., and W. E. Kuentzel, U.S. Pat. 2,082,520 (1937); through *C.A.*, **31**, 5576 (1937).

524. Ruzicka, L., P. Pieth, J. T. Reichstein, and L. Ehmann, *Helv. Chim. Acta*, **16**, 275 (1933).

525. Ryss, I. G., and L. P. Bogdanova, *J. Inorg. Chem. U.S.S.R.*, **1**, 2264 (1956).

526. Ryss, I. G., *Proc. Acad. Sci. U.S.S.R., Chem. Sect.*, **97**, 691 (1954).

527. Sachanen, A. N., and P. D. Caesar, *Ind. Eng. Chem.*, **38**, 43 (1946).

528. Sachanen, A. N., and S. B. Davis, U.S. Pat. 2,234,984 (1941).

529. Sachanen, A. N., and S. B. Davis, U.S. Pat. 2,361,355 (1944).

530. Sachanen, A. N., and A. A. O'Kelly, *Ind. Eng. Chem.*, **33**, 1540 (1941).

531. Sachanen, A. N., A. A. O'Kelly, and C. G. Myers, U.S. Pat. 2,410,107 (1946).

532. Salamon, G., *Rec. Trav. Chim.*, **68**, 903 (1949).

533. Satchell, D. P. N., *J. Chem. Soc.*, 1927 (1958).

534. Schmerling, L., *J. Am. Chem. Soc.*, **68**, 275 (1946).

535. Schmerling, L., Paper presented at the New York meeting of the American Chemical Society, September, 1947.

536. Schmerling, L., *Ind. Eng. Chem.*, **40**, 2072 (1948).

537. Schmerling, L., and A. W. Durinsky, U.S. Pat. 2,357,978 (1944).

538. Schmerling, L., and V. N. Ipatieff, U.S. Pat. 2,369,691 (1945); through *C.A.*, **39**, 5254 (1945).

539. Schmerling, L., and V. N. Ipatieff, U.S. Pat. 2,406,639 (1946).

540. Schneider, A., and R. M. Kennedy, *J. Am. Chem. Soc.*, **73**, 5013, 5017, 5024 (1956).

541. Scholl, R., and K. Meyer, *Ber.*, **65**, 902 (1932).

542. Scholl, R., K. Meyer, and J. Donat, *Ber.*, **70**, 2180 (1937).

543. Schollkopf, K., U.S. Pat. 2,115,884 (1938).

544. Scholten, W., and J. N. Vijvoet, *Z. Kryst.*, **103**, 415 (1941).

545. Schoot, C. J., J. J. Ponjee, and K. H. Klaassens, *Rec. Trav. Chim.*, **80** 1084 (1961).

546. Schorger, A. W., *J. Am. Chem. Soc.*, **39**, 2671 (1917).

547. Schroeder, W. D., and R. Q. Brewster, *J. Am. Chem. Soc.*, **60**, 751 (1938).
548. Schroeter, C., *Ber.*, **57**, 2017 (1924).
549. Schuit, G. C. A., H. Hoog, and J. Verheus, *Rec. Trav. Chim.*, **59**, 793 (1940).
550. Sebrell, L. B., Can. Pat. 352,830 (1935).
551. Sharma, V. N., and S. Dutt, *J. Ind. Chem. Soc.*, **12**, 774 (1935).
552. Sharp, D. W. A., in "Advances in Fluorine Chemistry," Vol. I, p. 113, Butterworth, London, 1960.
553. Sharpe, A. G., *Quart. Rev. (London)*, **4**, 115 (1950).
554. Shatenshtein, A. I., K. I. Zhdanova, and V. M. Basmanova, *J. Gen. Chem. U.S.S.R.*, **31**, 250 (1961).
555. Shearer, N. H, Jr., H. J. Hegenmeyer, and M. B. Edwards, Brit. Pat. 813,798 (1959).
556. Sheldon, J. C., and S. Y. Tyree, Jr., *J. Am. Chem. Soc.*, **81**, 2290 (1959).
557. Shokal, E. C., and P. A. Devlin, U.S. Pat. 2,778,855 (1957).
558. Shuikin, N. I., and N. A. Pozdnyak, *Bull. Acad. Sci. U.S.S.R., Div. Chem. Sci.*, 697 (1951).
559. Shuikin, N. I., N. F. Kononov, and L. K. Kashkouskaya, *J. Gen. Chem. U.S.S.R.*, **29**, 2230 (1959); **30**, 424 (1960).
560. Sidorova, N. G., I. P. Tsukervanik, and Z. K. Abidova, *Proc. Acad. Sci. Uzbek S.S.R.*, 33 (1953).
561. Sidorova, N. G., I. P. Tsukervanik, and E. Pak, *J. Gen. Chem. U.S.S.R.*, **24**, 94 (1954).
562. Simons, J. H., *Petrol. Refiner*, **22**, 83, 189 (1943).
563. Simons, J. H., "Advances in Catalysis," Vol. 2, 197 (1950).
564. Simons, J. H., and S. Archer, *J. Am. Chem. Soc.*, **60**, 986, 2952 (1938), **62**, 451, 1623 (1940).
565. Simons, J. H., and S. Archer, *J. Am. Chem. Soc.*, **60**, 2953 (1938).
566. Simons, J. H., and S. Archer, *J. Am. Chem. Soc.*, **61**, 1521 (1939).
567. Simons, J. H., S. Archer, and E. Adams, *J. Am. Chem. Soc.*, **60**, 2955 (1938).
568. Simons, J. H., S. Archer, and H. J. Passino, *J. Am. Chem. Soc.*, **60**, 2956 (1938).
569. Simons, J. H., S. Archer, and D. J. Randall, *J. Am. Chem. Soc.*, **61**, 1821 (1939).
570. Simons, J. H., S. Archer, and D. J. Randall, *J. Am. Chem. Soc.*, **62**, 485, (1940).
571. Simons, J. H., and G. C. Bassler, *J. Am. Chem. Soc.*, **63**, 880 (1941).
572. Simons, J. H., and H. Hart, *J. Am. Chem. Soc.*, **66**, 1309 (1944).
573. Simons, J. H., and A. C. Meunier, *J. Am. Chem. Soc.*, **63**, 1921 (1941).
574. Simons, J. H., H. J. Passino, and S. Archer, *J. Am. Chem. Soc.*, **63**, 608 (1941).
575. Simons, J. H., D. J. Randall, and S. Archer, *J. Am. Chem. Soc.*, **61**, 1795 (1939).
575a. Skinner, H. A., in "Cationic Polymerization and Related Complexes," ed. P. H. Plesch, Academic Press, New York, p. 28 (1953).
576. Skinner, H. A., and L. E. Sutton, *Trans. Faraday Soc.*, **36**, 681 (1940).
577. Slanina, S. J., F. J. Sowa, and J. A. Nieuwland, *J. Am. Chem. Soc.*, **57**, 1547 (1935).
578. Slaughter, J. I., and D. A. McCaulay, U.S. Pat. 2,906,788 (1959).
579. Slaughter, J. I., and D. A. McCaulay, U.S. Pat. 2,908,728 (1959).

580. Slobodin, J. M., *J. Gen. Chem. U.S.S.R.*, **4**, 778 (1934), through *C.A.*, **29**, 2145 (1935); *ibid.*, **5**, 48 (1935), *C.A.*, **29**, 4132 (1935); **6**, 1806 (1936), *C.A.*, **31**, 4264 (1937); **6**, 1892 (1936), *C.A.*, **31**, 4264 (1937); **7**, 1664 (1937), *C.A.*, **31**, 8501 (1937); **7**, 2376 (1937), *C.A.*, **32**, 2081 (1938); **8**, 1220 (1938), *C.A.*, **33**, 4209 (1939); **9**, 272 (1939), *C.A.*, **33**, 6258 (1939); **22**, 1958 (1952), *C.A.*, **47**, 6830 (1953).

581. Smits, A., and J. Meyering, *Z. Phys. Chem.*, **B41**, 98 (1938).

582. Smoot, C. R., and H. C. Brown, *J. Am. Chem. Soc.*, **78**, 6245 (1956); **78**, 6249 (1956).

583. Soday, F. J., U.S. Pat. 2,413,893 (1947).

584. Soderbach, E., *Acta Chem. Scand.*, **8**, 1851 (1954).

585. Sontag, D., *Ann. Chim.*, **1**, 359 (1934).

586. Sowa, F. J., H. D. Hinton, and J. A. Nieuwland, *J. Am. Chem. Soc.*, **54**, 2019 (1932).

587. Sowa, F. J., H. D. Hinton, and J. A. Nieuwland, *J. Am. Chem. Soc.*, **54**, 3694 (1932).

588. Sowa, F. J., H. D. Hinton, and J. A. Nieuwland, *J. Am. Chem. Soc.*, **55**, 3402 (1933).

589. Sowa, F. J., and J. A. Nieuwland, *J. Am. Chem. Soc.*, **55**, 5052 (1933).

590. Spanagel, E. W., and W. H. Carothers, *J. Am. Chem. Soc.*, **57**. 929 (1935).

591. Speier, J. L., J. A. Webster, and J. H. Barnes, *J. Am. Chem. Soc.*, **79**, 974 (1957).

592. Spiegler, L., and J. M. Tinker, *J. Am. Chem. Soc.*, **61**, 1002 (1939).

593. Spryskov, A. A., and Yu. G. Erykalov, *J. Gen. Chem. U.S.S.R.*, **28**, 1688 (1958).

594. Stadnikoff, G., and A. Banyschewa, *Ber.*, **61**, 1996 (1928).

595. Stadnikov, G. L., and I. Goldfarb, *Ber.*, **61**, 2341 (1928).

596. Stadnikov, G. L., and L. I. Kashtanov, *Ber.*, **61**, 1389 (1928).

597. Stadnikov, G. L., and L. I. Kashtanov, *J. Russ. Phys. Chem. Soc.*, **60**, 1117 (1928); through *C.A.*, **23**, 2170.

598. Stahly, E. E., and E. M. Hattox, U.S. Pat. 2,180,374 (1939).

599. Standard Oil Development Co., Brit. Pat. 483,453 (1938); French Pat. 826,933 (1938); through *C.A.*, **32**, 6666 (1938).

600. Standard Oil Development Co., Brit. Pat. 524,252 (1940); through *C.A.*, **35**, 6778 (1941).

601. Standard Oil Development Co., Brit. Pat. 551,961 (1943); through *C.A.*, 38, 2965 (1944).

602. Stanley, L. N., and S. M. Roberts, U.S. Pat. 2,861,105 (1958).

603. Staudinger, H., and F. Breusch, *Ber.*, **62B**, 442 (1929).

604. Staudinger, H., and H. A. Bruson, *Ann.*, **447**, 110 (1926).

605. Staudinger, H., and H. A. Bruson, Germ. Pat. 421,115 (1934).

606. Stevenson, D. P., and V. Schomaker, *J. Am. Chem. Soc.*, **62**, 2514 (1942).

607. Stevenson, D. P., C. D. Wagner, O. Beeck, and J. W. Otvos, *J. Am. Chem. Soc.*, **74**, 3269 (1952).

608. Stone, F. G. A., *Quart. Rev. (London)*, **9**, 174 (1955).

609. Stroh, R., K. Heyns, and M. Paulsen, *Angew. Chem.*, **69**, 124 (1957).

610. Stroh, R., and R. Seydel, Germ. Pat. 944,014 (1956); through *C.A.*, **53**, 321 (1959).

611. Stroh, R., R. Seydel, and W. Hahn, *Angew. Chem.*, **69**, 699 (1957).
612. Sullivan, J. K., U.S. Pat. 2,729,620 (1956); through *C.A.*, **50**, 6100 (1956).
613. Sullivan, J. K., U.S. Pat. 2,729,619 (1956); through *C.A.*, **50**, 6099 (1956).
614. Sussman, S., *Ind. Eng. Chem.*, **38**, 1288 (1946).
615. Swarts, F., *Bull. Acad. Sci. Belg.*, **22**, 784 (1936).
616. Szmant, H. H., and D. A. Irving, *J. Am. Chem. Soc.*, **78**, 4386 (1956).
617. Taboury, M. F., and R. Pajaeau, *Compt. rend.*, **202**, 328 (1936).
618. Tada, R., and N. Tokura, *Bull. Chem. Soc. Japan*, **31**, 387 (1958).
619. Tadema, H. J., Dutch Pat. 65,640 (1950); through *C.A.*, **44**, 7348 (1950).
620. Tanaka, K., S. Yabuki, and M. Sato, *Bull. Inst. Phys. Chem. Research*, **21**, 190 (1942); through *C.A.*, **43**, 7912 (1949).
621. Tauber, S. J., and A. M. Eastham, *J. Am. Chem. Soc.*, **82**, 4888 (1960).
622. Taylor, M. D., and L. R. Grant, *J. Am. Chem. Soc.*, **77**, 1507 (1955).
623. Thomas, C. A., "Anhydrous Aluminum Chloride in Organic Chemistry," Reinhold Publishing Co., New York (1941).
624. Thomas, C. L., *Ind. Eng. Chem.*, **41**, 2564 (1949).
625. Thomas, C. L., J. Hoekstra, and J. T. Pinkston, *J. Am. Chem. Soc.*, **66**, 1694 (1944).
626. Thompson, R. B., and J. A. Chenicek, *Ind. Eng. Chem.*, **40**, 1265 (1948)
627. Thompson, K. M., U.S. Pat. 2,875,257 (1959).
628. Tiger, H. L., *Trans. Am. Soc. Mech. Eng.*, **60**, 315 (1938).
629. Toffe, I. I., and V. Kombulova, U.S.S.R. Pat. 121,133 (1959).
630. Tokure, N., R. Asami, and R. Toda, *Sci. Rept. Res. Inst. Tohoku Univ. Serv. A.*, **8**, 149 (1956); through *C.A.*, **51**, 4944 (1957).
631. Topchiev, A. V., and N. F. Bogomolova, *Proc. Acad. Sci. U.S.S.R. Chem. Sect.*, **88**, 691 (1953).
632. Topchiev, A. V., and Ya. M. Paushkin, *Neft. Khoz.*, **25**, No. 6, 54 (1947); through *C.A.*, **42**, 1182 (1948).
633. Topchiev, A. V., Ya. M. Paushkin, T. P. Vishnyakova, and M. V. Kurashov, *Proc. Acad. Sci. U.S.S.R. Chem. Sect.*, **80**, 381 (1951).
634. Topchiev, A. V., Ya. M. Paushkin, T. P. Vishnyakova, and M. V. Kurashov, *Proc. Acad. Sci. U.S.S.R. Chem. Sect.*, **80**, 611 (1951).
635. Topchiev, A. V., S. V. Zavgorodnii, and Ya. M. Paushkin, "Boron Fluoride and Its Compounds as Catalysts in Organic Chemistry" (English translation), Pergamon Press, New York (1959).
636. Tronov, B. V., and A. M. Petrova, *J. Gen. Chem. U.S.S.R.*, **23**, 1019 (1953).
637. Tronov, B. V., and L. A. Pershina, *J. Gen. Chem. U.S.S.R.*, **24**, 1608 (1954).
638. Tsukervanik, I. P., and N. K. Rozhova, *Proc. Acad. Sci. Uzbek S.S.R.*, **23** (1958).
639. Tsukervanik, I. P., U.S.S.R. Pat. 121,787 (1959).
640. Turkevich, J., and R. K. Smith, *J. Chem. Phys.*, **16**, 466 (1948).
641. Turner, D. L., *J. Am. Chem. Soc.*, **71**, 612 (1949).
642. Turova-Polyak, M. B., and N. P. Bobreselkaya, *J. Gen. Chem. U.S.S.R.*, **29**, 1072 (1959).

643. Turova-Polyak, M. B., and M. A. Maslova, *J. Gen. Chem. U.S.S.R.*, **27**, 897 (1957), **26**, 3019 (1956), and related papers.
644. Twasaki, T., and R. Hatta, *J. Chem. Soc. Japan, Ind. Chem. Sect.*, **63**, 1975 (1960).
645. Uhlig, F., and H. R. Snyder, "Advances in Organic Chemistry," Vol. I, pp. 35, Interscience Publ., New York (1960).
646. Ulich, H., *Angew. Chem.*, **55**, 37 (1942).
647. Ulich, H., and G. Heyne, *Z. Elektrochem.*, **41**, 509 (1935).
648. Ulich, H., and G. Heyne, *Z. Physik. Chem.*, **B49**, 284 (1941).
649. Ulich, H., A. Keutmann, and A. Geierhaas, *Z. Elektrochem.*, **49**, 292 (1943).
650. Unger, F., *Ann.*, **504**, 267 (1933).
651. Universal Oil Products Co., Brit. Pat. 795,574 (1958); through *C.A.*, **53**, 4300 (1959).
652. Universal Oil Products Co., Brit. Pat. 794,153 (1958); through *C.A.*, **53**, 4204 (1959).
653. Universal Oil Products Co., Brit. Pat. 795,713 (1958); through *C.A.*, **53**, 2596 (1959).
654. Urech, P., and R. Sulzberger, *Helv. Chim. Acta*, **27**, 1328 (1944).
655. Van Dyke, R. E., *J. Am. Chem. Soc.*, **72**, 3619 (1950).
656. Van Dyke, R. E., and H. E. Crawford, *J. Am. Chem. Soc.*, **73**, 2018 (1951).
657. Van der Meulen, P. A., and H. A. Heller, *J. Am. Chem. Soc.*, **54**, 4404 (1932).
658. Vdovtsova, E. A., I. P. Tsukervanik, *Sbornik Statei Khim.*, **2**, 1027 (1953); *Proc. Acad. Sci. U.S.S.R., Chem. Sect.*, **80**, 61 (1951).
659. Vesley, J. A., and C. B. Linn, U.S. Pat. 2,891,966 (1959); *C.A.*, **53**, 15006 (1959).
660. Voge, H. H., and N. C. May, *J. Am. Chem. Soc.*, **68**, 550 (1946).
661. Voigt, A., and W. Biltz, *Z. anorg. allgem. Chem.*, **133**, 280 (1924).
662. Vilsmeier, A., and A. Haack, *Ber.*, **60**, 119 (1927).
663. Waddington, T. C., and J. A. White, *Proc. Chem. Soc.*, 315 (1960).
664. Walker, D. G., *J. Phys. Chem.*, **64**, 939 (1960); *ibid.*, **65**, 1367 (1961); Abstr. 140th ACS Meeting, Chicago, Ill., September, 1961, p. 9T.
665. Walling, C., *J. Am. Chem. Soc.*, **72**, 1164 (1950).
666. Wamser, C. A., *J. Am. Chem. Soc.*, **73**, 409 (1951).
667. Warren, B. E., and C. F. Hill, *Z. Kryst.*, **89**, 481 (1934).
668. Watt, G. W., and G. L. Hall, "Inorganic Syntheses," Vol. IV, p. 117, McGraw Hill Book Comp. Inc., New York, N.Y., 1953.
669. Wehrly, M., *Helv. Phys. Acta*, **11**, 339 (1938).
670. Weinstock, B., and G. J. Malm, *Proc. U.N. 2nd Intern. Conf. on Peaceful Uses of Atomic Energy, Geneva*, 1958, **28**, 125.
671. Wells, "Structural Inorganic Chemistry," 2nd ed., Oxford University Press, Oxford, pp. 116, 278, 279 (1950).
672. Wertyporoch, E., I. Kowalski, and A. Roeske, *Ber.*, **66**, 1232 (1933).
673. Whitmore, F. C., and H. H. Johnson, Jr., *J. Am. Chem. Soc.*, **63**, 1481 (1941).
674. Wibaut, J. P., L. M. F. Lande, and G. Wallagh, *Rec. Trav. Chim.*, **52**, 794 (1933).

675. Williams, G., *J. Chem. Soc.*, 775 (1940).

676. Wood, J. E., and C. S. Lynch, U.S. Pat. 2,354,565 (1944).

677. Woolf, A. A., *J. Chem. Soc.*, 252 (1954).

678. Woolf, A. A., *J. Inorg. Nucl. Chem.*, **3**, 285 (1956).

679. Wrigge, F. W., and W. Biltz, *Z. anorg. allgem. Chem.*, **227**, 372 (1936).

680. Wunderly, H. L., F. J. Sowa, and J. A. Nieuwland, *J. Am. Chem. Soc.*, **58**, 1007 (1936).

681. Yost, D. M., and J. E. Sherborne, *J. Chem. Phys.*, **2**, 125 (1934).

682. Yost, D. M., D. DeVoult, T. F. Anderson, and E. N. Lassettre, *J. Chem. Phys.*, **6**, 424 (1938).

683. Young, D. W., U.S. Pat. 2,542,610 (1951); through *C. A.*, **45**,7134 (1951).

684. Young, D. W., and S. G. Gallo, U.S. Pat. 2,589,317 (1952).

685. Young, D. W., and H. S. Kellog, U.S. Pat. 2,440,498 (1948); through *C.A.*, **42**, 5628 (1948).

686. Yurev, Y. K., G. B. Elyarov, and Z. V. Belyarova, *J. Gen. Chem. U.S.S.R.*, **26**, 2353 (1956).

687. Zeffert, B. M., P. B. Coulter, and R. Macy, *J. Am. Chem. Soc.*, **75**, 752 (1953).

688. Zharkova, V., and V. Moldavskii, *J. Gen. Chem. U.S.S.R.*, **17**, 1268 (1947); through *C.A.*, **42**, 1869 (1948).

689. Ziegler, K., in "Organometallic Chemistry," ed. H. Zeiss, Reinhold Publ. Corp., New York, p. 194 (1960).

690. Zincke, A., *Ber.*, **2**, 737 (1869); *Ann.*, **159**, 367 (1871); *Ber.*, **4**, 298 (1871); *Ann.*, **161**, 93 (1872).

691. Venuto, P. B., L. A. Hamilton, P. S. Landis, and J. J. Wise, *J. Catalysis*, **5**, 81 (1966).

692. Venuto, P. B., L. A. Hamilton, and P. S. Landis, *J. Catalysis*, **5**, 884 (1966).

693. Venuto, P. B., and P. S. Landis, *J. Catalysis*, **6**, 237 (1966).

694. Brown, H. C., and K. L. Nelson, *J. Am. Chem. Soc.*, **75**, 6292 (1953).

695. Mochida, I., and Y. Yoneda, *J. Catalysis*, **7**, 386 (1967).

696. Mochida, I., and Y. Yoneda, *J. Catalysis*, **7**, 393 (1967).

697. Mochida, I., and Y. Yoneda, *J. Catalysis*, **8**, 223 (1967).

698. Mochida, I., and Y. Yoneda, *J. Catalysis*, **9**, 51 (1967).

699. Mochida, I., and Y. Yoneda, *J. Catalysis*, **9**, 57 (1967).

700. Gerberich, H. R., J. W. Hightower, and W. K. Hall, *J. Catalysis*, **8**, 391 (1967).

701. Wu, C. Y., and W. K. Hall, *J. Catalysis*, **8**, 394 (1967).

702. Csicsery, S. M., *J. Catalysis*, **9**, 336 (1967).

703. Csicsery, S. M., *J. Org. Chem.*, **34**, 3338 (1969).

704. Dimitrov, C., and P. Ignatiev, *J. Catalysis*, **7**, 103 (1967).

705. Dimitrov, C., and Z. Popova, *J. Catalysis*, **9**, 1 (1967).

706. Hightower, J. W., H. R. Gerberich, and W. K. Hall, *J. Catalysis*, **7**, 57 (1967).

707. Covini, R., V. Fattore, and N. Giordano, *J. Catalysis*, **7**, 126 (1967).

708. Covini, R., V. Fattore, and N. Girodano, *J. Catalysis*, **9**, 315 (1967).

709. Benesi, H. A., *J. Catalysis*, **8**, 368 (1967).

710. Ward, J. W., and R. C. Hansford, *J. Catalysis*, **13**, 364 (1969).

711. Ward, J. W., *J. Catalysis*, **10**, 34 (1968).

712. Turkevich, J., and S. Ciborowski, *J. Phys. Chem.*, **71**, 3208 (1967).
713. Hughes, T. R., H. M. White, and R. J. White, *J. Catalysis*, **13**, 58 (1969).
714. Ward, J. W. and R. C. Hansford, *J. Catalysis* **13**, 154 (1969).
715. Ward, J. W., *J. Catalysis*, **13**, 321, (1969); **14**, 365 (1969).
716. Hightower, J. W., and W. K. Hall, *J. Am. Chem. Soc.*, **90**, 851 (1968).
717. Karabatsos, G. J., and J. D. Graham, *J. Am. Chem. Soc.*, **82**, 5250 (1960).
718. Skell, P. S., I. Starer, and A. P. Krapcho, *J. Amer. Chem. Soc.*, **82**, 5257 (1960).
719. Lanewala, M. A., and A. P. Bolton, *J. Org. Chem.*, **34**, 3107 (1969).
720. Manhassen, J., and Sh. Khalif, *J. Catalysis*, **7**, 110 (1967).
721. Hickson, S. A., and S. M., Csicsery, *J. Catalysis*, **10**, 27 (1968).
722. Ward, J., *J. Catalysis*, **10**, 34 (1968).
723. Danforth, J. D., and J. H. Roberts, *J. Catalysis*, **10**, 252 (1968).
724. Hammett, L. P. "Physical Organic Chemistry," McGraw-Hill, New York, 1940, and references therein.
725. Hull, N. F., and J. B., Conant, *J. Am. Chem. Soc.*, **49**, 3047 (1927).
726a. Gillespie, R. J., *Accounts Chem. Res.*, **1**, 202 (1968), and references therein.
726b. Gillespie, R. J., *Can. Chem. Educ.*, **4**, 9, April 1969.
727. Olah, G. A., *Chem. Eng. News*, **45**, March 27, 1967, pp 76–88, *Science*, **168**, 1298 (1970), and references therein.
728. Gillespie, R. J., and E. A. Robinson, in "Non-Aqueous Solvent Systems," T. C. Waddington, ed., Academic Press, New York, 1965, p. 117.
729. Commeyras, A., and G. A. Olah, *J. Am. Chem. Soc.*, **90**, 2929 (1968).
730. Olah, G. A., E. B. Baker, J. C. Evans, W. S. Tolgyesi, J. S. McIntyre, and I. J. Bastieu, *J. Am. Chem. Soc.*, **86**, 1360 (1964).
731. Olah, G. A., and M. B. Comisarow, *J. Am. Chem. Soc.*, **90**, 5033 (1968).
732. Eachus, R. S., T. P. Sleight, and M. C. R. Symons, *Nature*, **222**, 769 (1969).
733. Gillespie, R. J. and M. J. Morton, *Inorg. Chem.*, **11**, 591 (1972).
734. D. M. Brouwer, and J. A. van Doorn, *Rec. Trav. Chim.*, **91**, 903 (1972).
735. Registered trade name of **Calionics Inc.** 653 Alpha Drive, Cleveland, Ohio 44143.

CHAPTER V

Reactivity and Selectivity

I. Comparison of Activity of Lewis Acid Catalysts

In discussing the effects influencing the Friedel-Crafts catalytic activity of acidic halide catalysts, the following factors having a bearing on the generalized acid–base interaction of the acid catalysts with reagent bases have to be considered:

1. Electronegativity of halogen ligands of Lewis acid halides ($F > Cl > Br > I$).
2. Reorganization energies involved in hybridizations of the central atoms of metal halide catalysts during complex formation with reagent bases.
3. Dissociation energies of dimeric metal halides and the availability of the Lewis acid catalysts as monomers or dimers.
4. Electronegativity of donor reagent molecules ($F > N > O > Cl > Br$).
5. Bond lengths and strengths in donor–acceptor complexes of catalyst:reagent interactions.
6. Lattice energies involved with solid catalysts or solid (insoluble) catalyst complexes.
7. Solvation energies involved with liquid catalysts and soluble catalyst complexes.
8. Steric effects.

In heterogeneous Friedel-Crafts systems still further factors must be considered, such as particle size, solubility, absorption, diffusion, etc. Many of these factors are of course of general importance in heterogeneous catalysis involving insoluble catalysts or gas-phase reaction on solid surfaces and their discussion is outside the scope of present considerations.

1. Empirical Sequences of Catalyst Activity

In the course of the investigation of Friedel-Crafts reactions, it was from the beginning of specific interest to compare the activity of different catalysts. As a result of investigations intended to establish the reactivity sequence of catalysts a considerable number of empirical "orders of activity" of acidic halide and related Friedel-Crafts acid catalysts was published. Data, from which the relative effectiveness of Friedel-Crafts catalysts can be judged, are widely scattered in the literature and are largely based on product yields.

A. Racemization of α-Phenylethyl Chloride

One of the earliest comprehensive studies of catalyst activity concerned the racemization of optically active α-phenylethyl chloride (33,30).

Although the results were complicated by concentration and solvent effects (racemization can be carried out in different solvents, including sulfur dioxide (25)), the following activity sequence was indicated (in benzene):

$$SbCl_5 > SnCl_4 > BF_3 > HgCl_2 > TiCl_4.$$

It should be mentioned that only "mild" catalysts can be used in this isomerization because stronger Lewis acids or related Friedel-Crafts acid catalysts, such as aluminum chloride, ferric chloride, zirconium chloride, HF, H_2SO_4, etc., tend to dehydrochlorinate and subsequently polymerize or polycondense α-phenylethyl chloride.

B. Aromatic Ketone Synthesis

In 1935 Calloway (61) summarized the available data in regard to the acylation of benzene and concluded that the relative effectiveness of metal halides was

$$AlCl_3 > FeCl_3 > ZnCl_2 > SnCl_4 > TiCl_4, \sim ZrCl_4.$$

A more extensive reactivity series of catalytic ability in the acetylation of toluene to p-methylacetophenone was presented by Dermer (95) (based on relative yields of products)

$$AlCl_3 > SbCl_5 > FeCl_3 > TeCl_2 > SnCl_4 > TiCl_4 > TeCl_4 > BiCl_3 > ZnCl_2.$$

A similar investigation using metal bromide catalysts gave the reactivity sequence:

$$AlBr_3 > FeBr_3 > SbBr_3 > ZnBr_2 > TiBr_4 > TeBr_4 > MoBr_4 > WBr_5 > HgBr_2 > SnBr_4.$$

As can be seen the positions of the halides of zinc, tin, and antimony are different in the two series. Antimony trichloride has been reported as a non-catalyst in the chloride series but has been tested only in boiling toluene where any small amount of acetophenone produced would have been destroyed.

C. Gatterman-Koch Reaction

The relative catalyst reactivity in the Gatterman-Koch reaction was found by Dilke and Eley (100), based on product yields as

$$AlBr_3 > AlI_3 > AlCl_3 > FeCl_3 > SbCl_5 > TiCl_4 > SnCl_4.$$

A difficulty in any evaluation of the catalytic activity of acidic halide catalysts, based on relative yields of products (e.g., acetophenone, benzaldehyde), is that such catalysts also show a tendency to cause secondary reactions of the product (e.g., aldol type condensations). Reliable information is therefore difficult to obtain.

D. Alkylation of Aromatics

Gross and Ipatieff (134) have considered the catalytic ability of various acidic halides in the reaction between benzene and ethylene in the presence of hydrogen chloride (based on yields of ethylbenzene obtained). They found catalytic ability to decrease in the order,

$$AlCl_3 > TaCl_5 \sim ZrCl_4 > CbCl_4 > BeCl_2 > TiCl_4.$$

E. Isomerization and Disproportionation

Russell (267) investigated the cyclohexane–methylcyclopentane isomerization and found the catalytic activity sequence to be:

$$AlBr_3 > GaBr_3 > GaCl_3 > FeCl_3 > SbCl_5 > ZrCl_4 > BF_3, BCl_3, SnCl_4, SbCl_3.$$

A similar sequence of catalytic activity was observed by Russell in the alkylation of benzene with alkyl halides and in the polymerization of styrene. Aluminum bromide was found to be a more effective isomerization catalyst than gallium bromide when isopropyl bromide was used as a co-catalyst, while in the presence of water aluminum bromide was a more effective catalyst than gallium chloride. Alkylations are much faster when catalyzed by aluminum bromide than when gallium bromide is used (295).

In the disproportionation of ethyltrimethylsilane and other substituted silanes, the catalyst sequence observed was (266):

$$AlBr_3 > AlCl_3 > AlI_3 > GaBr_3 > BCl_3, GaCl_3, ZrCl_4, SnCl_4, SbCl_3.$$

Based on his experimental results Russell suggested the following general sequence of Lewis acidity and accordingly catalytic activity:

$$AlBr_3 > AlCl_3 > AlI_3 > GaBr_3 > GaCl_3 > FeCl_3 > SbCl_5 > ZrCl_4 > BCl_3, BF_3, SbCl_3.$$

F. Polymerization

The reactivity of acidic halide Lewis acid catalysts was compared in Friedel-Crafts polymerization of isobutylene at $-78°$ by Fairbrother and Seymour (282) and found to be in the order (as judged by the yield of polymer formed in a given period of time):

$$BF_3 > AlBr_3 > TiCl_4 > TiBr_4 > SnCl_4 > SnCl_4 > BCl_3 > BBr_3.$$

The polymer molecular weight decreased in the same order, with the exception of the stannic chloride polymer, which had the lowest molecular weight. This behavior was in accordance with the often quoted "rule of thumb" in polymer chemistry (246): that the most active catalyst also gives the highest molecular weight. If, however, the activity of a catalyst is judged by the rate of polymerization rather than by the yield, then this rule shows many exceptions. Thus with styrene in carbon tetrachloride solution at $25°$ the order is (325):

$$SbCl_5 \gg SnCl_4 > BCl_3.$$

but the polymer molecular weight shows only small variation and in the reverse order. With styrene in ethylene dichloride solution (247) the rate with stannic bromide is some 400 times less than with stannic chloride, but the molecular weight is twice as great.

Fairbrother and Seymour's order of catalyst activity follows that found by Bodendorf and Böhme (30) for the racemization of α-phenylethyl chloride and that of Dermer and Billmeier (93) for the acetylation of toluene. It differs from the order quoted by Calloway (61) in the relative positions of $TiCl_4$ and $SnCl_4$.

According to the results of Topchiev and Paushkin (244,309) the catalysts employed for the polymerization of unsaturated hydrocarbons at temperatures ranging from $2-100°$ can be arranged in the following order of decreasing activity:

$$BF_3 \cdot H_3PO_4 > H_4P_2O \cdot 2BF_3 > BF_3 \cdot HPO_3 > BF_3 \cdot H_2O > AlCl_3 > HF > BF_3 \cdot 2\tfrac{1}{2}H_2SO_4$$
$$> BF_3 \cdot ROH > BF_3 \cdot RCOOH > BF_3 > BF_3 \cdot O(C_2H_5)_2 > BF_3 \cdot RCOOR > H_3PO_4.$$

Comparison of relative catalyst strengths of different Lewis acid catalysts in Friedel-Crafts copolymerizations gave the sequence (118)

$$AlCl_3 > AlBr_3 > SnCl_4 > TiCl_4 > BF_3 > ZnCl_2.$$

G. Color Change of Hammett Indicators

A qualitative comparison of relative acid strengths of Friedel-Crafts catalysts was carried out by Steigman (145) comparing color changes of unchanged basic Hammett indicators in benzene and chlorobenzene solutions with different acids. The following apparent relative strengths of acids were obtained (in decreasing order) from indicator reactions:

$AlBr_3 > AlCl_3 > FeCl_3 > H_2SO_4$ (fuming)
H_2SO_4 (96%) $> HClO_4$ (70%) $> SnCl_4 > SbCl_5 > SbCl_3$
$BiCl_3 > PCl_5 > PCl_3$
$HCl > AsCl_3 > ZnCl_2$
$CdCl_2 > CuCl_2 > HgCl_2 > AgNO_3.$

No particular relative order between the groups is implied. Data correlate fairly well with other qualitative relative reactivity (acid strength) sequences.

H. Ionization of Triarylmethyl Halides

The ionization of triarylmethyl chlorides (diphenyl-p-tolylmethyl chloride, triphenylmethyl chloride) by various Friedel-Crafts catalysts has been studied spectrophotometrically in acetic acid solution by Cotter and Evans (81). The ionization leads to the reversible formation of ion pairs $R^+ MCl_{x+1}^-$. The ionizing power of the metal chlorides decreases along the sequence

$$SbCl_5 > FeCl_3 > SnCl_4 > BiCl_3 > HgCl_2 > SbCl_3.$$

I. Infra-red Stretching Frequency of Carbonyl Complexes

Relative acid strengths of acidic Lewis acid halide catalysts can also be established from the shifts of the characteristic infra-red $C{=}O$ stretching frequencies of their ketone complexes. Susz (303) found the following $C{=}O$ shifts (in cm.$^{-1}$ relative to the $C{=}O$ stretching frequencies of uncomplexed ketones) as summarized in Table I. Accordingly the decreasing order of Lewis acid strength is

$$AlBr_3 \ggg FeCl_3 > AlCl_3 > SnCl_4 > TiCl_4 > BF_3 > ZnCl_2 > HgCl_2.$$

TABLE I. Shift of infra-red carbonyl stretching frequency of ketone: catalyst complexes (cm.$^{-1}$)

Acceptor Lewis acid	Benzophenone (1657, cryst.)	Acetophenone (1675, cryst.)	Acetone (1710, liq.)	Dipropyl ketone (1705, liq.)
$HgCl_2$		18		
$ZnCl_2$		47		
BF_3	112	107	70	70
$TiCl_4$	144	118		
$SnCl_4$			75	
$AlCl_3$	122	120		
$FeCl_3$	145	130		
$AlBr_3$	142	130	85	

The infra-red spectroscopic shifts in carbonyl stretching frequencies of complexes of ethyl acetate with Group III and IV halides have been recorded by Lappert (178a) and interpreted in terms of relative acceptor strengths of the halides, within related series and with respect to ethyl acetate (which has very low steric requirements) as the reference base.   This provided the order Br > Cl for boron and aluminum halides, but Cl > Br for indium and the Group IV halides.

Group trends found by Lappert were

$$B > Ga > Al > In \qquad Sn > Ge > Si \qquad Ti > Zr.$$

It should, however, be emphasized, as pointed out by Drago (103), that shifts of infra-red carbonyl stretching frequencies give no real thermodynamic sequence of complex stabilities.   These can be obtained, according to Drago, from correlating enthalpies of Lewis acid–base interactions.

J. Ring Deuteration of Aromatics

Shatenshtein (283) compared the catalytic activity of certain metal bromide catalysts in the ring deuteration of aromatics with DBr and found the following relative order of catalyst reactivity (based on rate constants determined, the values of which are shown in parentheses).

$AlBr_3 (5 \times 10^5) > GaBr_3 (10^5) > FeBr_3 (10^4) > BBr_3 (3 \times 10^1) > SbBr_3 (6) > TiBr_4 (1) > SnBr_4.$

Even if the absolute rate determinations had certain experimental errors, Shatenshtein's work must be considered one of the first attempts to compare in a quantitative way catalyst reactivities of acidic Lewis acid halide catalyst based on exact physical measurements.

K. Suggested General Empirical Sequence

Burk (57), in a compilation of available data, suggested the following general sequence of catalyst reactivity:

$AlBr_3 > AlCl_3 > FeCl_3 > ZrCl_4 > TaCl_5 > BF_3 > VCl_4 > TiCl_3 > WCl_6 > CbCl_5 > ZnCl_2 > SnCl_4 > TiCl_4 > BeCl_2 > SbCl_5 > HgCl_2 > CuCl_2 > BiCl_3 > AgF_3.$

Although the sequence of catalytic activity given by Burk was based on the most complete survey reported in the literature it nevertheless suffers from the shortcomings of any effort to try to establish a general sequence of Friedel-Crafts catalyst reactivity.

2. Factors Influencing Catalyst Reactivities

A. *Limitations of Unique Sequence of Reactivity*

Pepper (246) pointed out, in discussing ionic polymerizations with Friedel-Crafts catalysts, that "no unique order of catalyst activity in all systems could be expected, as specific interactions with monomer and solvent are likely to play an important part." To take an extreme example, the order of catalyst activity in Friedel-Crafts depolymerization of paraldehyde in ether solution (23) is markedly different from that obtained in the polymerization of isobutylene or styrene. Moreover, it has been demonstrated in many systems that in addition to the catalyst another substance—the co-catalyst—is necessary for polymerization. We must therefore expect activity to depend not on the catalyst alone but on the catalyst–co-catalyst combination.

In dealing with any empirical series of relative catalyst activity (or Lewis acid strength) we must always keep in mind that any reactivity (acidity) series is necessarily dependent on the nature of the reference base, *e.g.*, reactant in Friedel-Crafts reactions and on a number of other conditions (solubility, solvent, etc). Therefore it should be emphasized that *in all probability in Friedel-Crafts type reactions no simple monotonic series of catalytic activity of Lewis acid and related proton acid catalysts is possible*. In different systems the acid catalysts interact with different acceptor bases. Consequently to compare acidity, or what is equivalent to it, catalyst activity, we must always apply the principles of the generalized acid–base theory. We can state only relative catalyst strengths relating to a given specific acceptor base in a specific system. Catalyst activities change considerably depending on the nature of the reagents used (bases). Concerning the effect of the reference base (Friedel-Crafts reagent) the situation is quite similar to the extensively investigated question of the relative strength of organic bases (H. C. Brown and co-workers (39)) in which it was convincingly demonstrated that many factors (called steric "strains" by Brown) can influence the relative acid strength and change it from base to base.

For example, boron trifluoride shows little tendency to unite with chloride bromides and therefore to catalyze alkylations with alkyls or chlorides or acylations with acyl chlorides. Consequently boron trifluoride does not catalyze the reaction of cyclohexyl bromide or acetyl chloride with benzene (58,133). At the same time it possesses a great affinity for fluorides. It is thus a very active catalyst in alkylations with alkyl fluorides and acylations with acyl fluorides (223,224,225,242). Similarly, it is an active catalyst in alkylations with alcohols and olefins. If the relative sequence of catalyst activity is established in any of these reactions, boron trifluoride must be placed amongst the most reactive catalysts. On the other hand, if the catalyst activity sequence is investigated in alkylations with alkyl chlorides or bromides, boron trifluoride must be placed amongst the least reactive catalysts.

In comparison of the catalytic activity of certain homologous Friedel-Crafts metal catalysts in alkylations with alkyl fluorides, Olah and Kuhn found the following relative orders of activity:

$$BI_3 > BBr_3 > BCl_3 > BF_3$$
$$AlI_3 > AlBr_3 > AlCl_3$$
$$FeBr_3 > FeCl_3.$$

Steric hindrance must be smallest with fluorine substituents which at the same time also show the highest electronegativity. The observed catalyst reactivity sequences show, however, a reverse order, decreasing from iodides toward fluorides. This can be explained by a correlation with the decrease of electron deficiency at the central atom of the catalytic halides, owing to back conjugation caused by the unshared halogen electron pairs, which increases with less shielding as we go from iodine to fluorine.

In explaining the inability of boron trihalides to catalyze reactions of organic chlorides, Brown (213) suggested in 1953 that this may be due to steric difficulties, as the ion BCl_4^- is very unstable compared with the stable ions BF_4^- or $AlCl_4^-$. Although tetrachloroborate (BCl_4^-) and tetrabromoborate (BBr_4^-) complexes are still considerably less known than the corresponding tetrafluoroborates, in the last few years a substantial number of these stable salts have been isolated. Alkali salts (Muetterties (212)), pyridinium salts (Lappert (178)), amine salts (Gerrard (127)), and diazonium salts (Olah and Tolgyesi (237)) are a few of the representative compounds isolated.

Steric hindrance, although obviously playing an important part, therefore cannot be considered as the main reason for the catalytic inactivity of boron fluoride in reactions with organic chlorides or bromides. Disregarding steric factors the higher degree of back conjugation from the fluorine atoms to the electron-deficient boron, as compared with similar but smaller effects in the case of other halogens, would suggest that BF_3 indeed is a weaker acid than BCl_3 or BBr_3. It therefore needs a stronger base to achieve effective donor–acceptor interaction and covalent C–Cl or C–Br bonds seem to be insufficiently strong nucleophiles for this purpose. Ionized halides, however (as present in metal halides, pyridinium, or diazonium halides), being sufficiently strong nucleophiles coordinate readily.

A comparative investigation of the relative catalyst strengths of the corresponding metal halides of certain catalytically active elements in the acetylation of toluene with acetyl bromide has been carried out keeping reaction conditions constant and determining activity by the speed of the formation and yields obtained of methyl acetophenones (222). Some of the relative reactivity sequences obtained were

$$FeBr_3 > FeCl_3$$
$$SbCl_3 > SbBr_3 > SbI_3$$
$$CrBr_3 > CrCl_3$$
$$HgBr_2 > HgCl_2$$
$$ZnI_2 > ZnBr_2 > ZnCl_2$$
$$SnCl_4 > SnBr_4 > SnCl_4$$
$$TiCl_4 > TiF_4 > TiBr_4 > TiI_4.$$

As is seen, no general order of catalytic activity concerning the halogen ligands could be established. In the case of ferric, chromic, zinc, and mercury halides the bromides were catalytically more active than the chlorides. However, in the case of antimony, stannic, and titanium halides a reverse sequence of reactivity was observed. It is obvious that other factors besides the electron deficiency of the central atom of the Lewis acid halide and the donor properties of the reagent base (in this case CH_3COBr) must be operative in determining the catalyst activity. Steric factors must become quite important with the bulkier metal bromides and iodides and other contributing conditions (to be discussed later) may also cause reversal of the catalytic reactivity sequence.

B. Acid Strength of Acidic Halide Catalysts

From the point of view of their catalytic activity, the acid strength of different Lewis acid type metal halide catalysts is of great importance. It is therefore of some use to survey the available data on factors influencing the electron deficiency (acid strength) of these catalysts and serve as a means of gaining comparative data useful in selecting the most promising catalysts.

a. Boron trihalides

In the boron halide series boron trifluoride was long considered the only catalytically active boron halide in Friedel-Crafts type reactions.

Boron trichloride and tribromide were frequently reported to have practically no catalytic activity in reactions with alkyl or acyl chlorides (bromides) (133,195). It has even been stated that boron trifluoride is the strongest acceptor molecule known (34). This conclusion has no doubt arisen through the characterization of far more adducts and complex compounds of boron fluoride than with the other boron halides. Recent observations on the high catalytic activities of boron chloride and boron bromide in alkylation and acylation reactions with alkyl and acyl fluorides, far surpassing that of boron trifluoride, were, however, not entirely unexpected. Laubengayer and Sears (180) measured the heat of formation of the acetonitrile–boron halide complexes and found it to be: boron chloride > boron fluoride. The dipole moments of the complexes also show values in the order BCl_3 > BF_3 (313).

The pyridine–boron trifluoride system has been studied in the gas phase by van der Meulen and Heller (315). The results of these workers led to the value of 50.6 kcal. mole^{-1} for the heat of dissociation of $C_5H_5N \cdot BF_3$ in the gas phase. As pointed out by Brown and Horowitz (45) this value is seriously in error. The true enthalpy change with a small correction to obtain the gas-phase value must be near 25.0 kcal. mole^{-1}, the value obtained for the formation of the adduct with all components in nitrobenzene.

Brown and Holmes (43) carried out exact measurements of the heats of formation of the boron trihalide–pyridine and boron trihalide–nitrobenzene complexes and found them in the order BBr_3 > BCl_3 > BF_3 (Table II).

A similar but independent investigation on the heat of formation of the boron trihalide complexes of pyridine and piperidine was carried out by Greenwood (131) and the results were in accordance with BBr_3 > BCl_3.

An order of Lewis acidity in the boron halides contrary to electronegativity predictions is also probably true when arsine is the donor. Thus $H_3As \cdot BCl_3$ and $H_3As \cdot BBr_3$ are known (297,298) whereas $H_3As \cdot BF_3$ does not form even below $-100°$ (196). Similarly, the adduct $Cl_3P \cdot BBr_3$ exists but $Cl_3P \cdot BCl_3$ and $Cl_3P \cdot BF_3$ do not (44), contrary to a previous report (20). Furthermore although $CH_3CN \cdot BF_3$ and $CH_3CN \cdot BCl_3$ are completely dissociated in the gas phase, their relative heats of sublimation suggest that $CH_3CN \cdot BCl_3$ is the more stable adduct (180).

The boron–nitrogen bond strength in the compound $(CH_3)_3N \cdot BF_3$ is not known with certainty. The adduct is not dissociated to a sufficient extent in a convenient temperature range to permit study in the gas phase. Indeed it was reported to be undissociated at a temperature as high as 230°. However, the result suggested that the complex was actually associated to some extent (55). These observations were later shown to be incorrect when it was found that

TABLE II. Heats of reaction of boron halides with pyridine and nitrobenzene (based on the data of H. C. Brown and R. R. Holmes (43), after F. G. A. Stone (301a))

Chemical reaction	Heat of reaction $-\Delta H$ kcal. mole^{-1}
$C_5H_5N(\text{solution}) + BF_3(\text{solution}) = C_5H_5N \cdot BF_3(\text{solution})$	25.0^a
$C_5H_5N(\text{solution}) + BCl_3(\text{solution}) = C_5H_5N \cdot BCl_3(\text{solution})$	30.8^a
$C_5H_5N(\text{solution}) + BBr_3(\text{solution}) = C_6H_5N \cdot BBr_3(\text{solution})$	32.0^a
$C_6H_5NO_2(l) + BF_3(g) = C_6H_5NO_2 \cdot BF_3(\text{solution})$	9.2
$C_6H_5NO_2(l) + BF_3(l) = C_6H_5NO_2 \cdot BF_3(\text{solution})$	6.7^b
$C_6H_5NO_2(l) + BCl_3(l) = C_6H_5NO_2 \cdot BCl_3(\text{solution})$	8.7
$C_6H_5NO_2(l) + Br_3(l) = C_6H_5NO_2 \cdot BBr_3(\text{solution})$	12.5
$C_5H_5N(\text{solution}) + BF_3(l) = C_5H_5N \cdot BF_3(\text{solution})$	31.7
$C_5H_5N(\text{solution}) + BCl_3(l) = C_5H_5N \cdot BCl_3(\text{solution})$	39.5
$C_5H_5N(\text{solution}) + BBr_3(l) = C_5H_5N \cdot BBr_3(\text{solution})$	44.4

^{a} The high heats of solution of BX_3 in nitrobenzene show that the heat evolved in the reaction of pyridine with BX_3 represents the heat of the displacement reaction

$$C_5H_5N + C_6H_5NO_2 \cdot BX_3 = C_5H_5N \cdot BX_3 + C_6H_5NO_2.$$

^{b} Calculated using a value of 2.5 kcal. mole for the heat of vaporization of BF_3 after H. C. Brown and R. R. Holmes.

$(CH_3)_3N \cdot BF_3$ was monomeric and undissociated at 177° (56). The original result arose from some uncertainty in the correction to be applied for the vapor pressure of mercury. It has now been calculated that the dissociation of $(CH_3)_3N \cdot BF_3$ in the gas phase should be readily observed above 200°, but only with difficulty detected below 175° (19).

Several estimates of the dative bond strength in the compound $(CH_3)_3N \cdot BF_3$ have been made, since the value was required for comparison with the dative bond strength in other adducts. Thus, from the calorimetric studies of Brown and Lawton (47), using nitro-benzene as solvent, it is possible to calculate the enthalpy change in the dissociation

$$(CH_3)_3N \cdot BF_3 \text{ (solution)} = (CH_3)_3N \text{ (solution)} + BF_3 \text{ (solution)}$$

as 30.9 kcal. mole^{-1} (129).

Since heats of formation of molecular addition compounds apparently do not differ very greatly in nitrobenzene solution and in the gas phase it may be presumed that ΔH for the gas-phase dissociation is also close to 30.9 kcal. mole^{-1}. Another estimate ($\Delta H = 26.6$ kcal. mole^{-1}) has been made by Bauer and McCoy (19) from heat capacity data for solid $(CH_3)_3N \cdot BF_3$.

An explanation for the relative acidity of the boron halides can be found in terms of π-bonding between X and B in BX_3 molecules. This would increase along the series BBr_3, BCl_3, BF_3, thereby reducing the acceptor power of boron by successive, but not necessarily equal, amounts (43).

Reduction of acid strength through resonance between lone-pair electrons of fluorine and the boron p-orbital may be responsible for the irregularities observed.

A mesomeric effect also probably accounts for the weak Lewis acidity of alkyl borates, like $(CH_3O)_3B$ (42,293).

Although alkoxy groups are electronegative and because of this should enhance the acceptor power of boron, there is a possibility of π-bonding between the oxygen lone-pair electrons and the boron $p\pi$-orbital so that readjustment energy to free this orbital for chemical bonding to a base is high. Unlike the alkyl compounds, aryl borates form adducts with ammonia and amines (71).

In the aryl compounds boron–oxygen $p\pi$–$p\pi$ bonding would be less because of the electronegativity of the phenyl group and so a stronger Lewis acidity should be observed.

The relative dative bond strength in the trimethylamine addition compound of Group III elements is in the order aluminum > gallium > indium > boron > thallium. This sequence, except for the position of boron, reflects to some degree a decreasing desire on the part of metals to accept partial negative charge with increasing electropositive character. Undoubtedly the position of boron in the order of acceptor power is at least partly due to steric factors. Steric interference between groups on the donor with those on the acceptor would be at a maximum when the metal is the smallest element of the group, the N–metal^{III} bond length being at a minimum. However, steric factors, as explained in the fundamental work of H. C. Brown on the relative base strengths of amines in connection with different "strains," are also operative in the Lewis acid metal halide systems and must be considered.

Ligands of Group VI atoms form less stable adducts than do analogous ligands of Group V atoms with Group III acceptor molecules. This apparently unbreakable rule was first observed by Davidson and Brown (85) in their studies on trimethylaluminum addition compounds, but there are many examples of the rule in boron chemistry. Thus the order of coordination $(CH_3)_3N > (CH_3)_2O > CH_3F$ and $(CH_3)_3B > (CH_3)_2S > CH_3Cl$ is observed towards boron trifluoride, as well as borane and trimethylborane. The increase in nuclear charge in passing across the rows in the periodic classification may be responsible for the observed decrease in donor power.

The etherates of boron trifluoride provide an interesting demonstration of the importance of steric effects (38). On the basis of inductive effects diethyl ether should be about as good a donor as tetrahydrofuran. The addition compound tetrahydrofuran–boron trifluoride is, however, considerably more stable than $(C_2H_5)_2O \cdot BF_3$. From data available (262) it is possible to state that an order of coordination $(CH_2)_4O > (CH_3)_2O > (C_2H_5)_2O$ is also observed when borane is the acceptor. In the cyclic ether the base strength is greatly increased through reduction of the steric requirements. Similar behavior occurs with nitrogen as shown by Brown (52) in a demonstration of the much greater stability of quinuclidine trimethyl borane over trimethylamine trimethylborane.

Comparison of the strength of the dative bond in the etherates of boron trichloride with the strength of those in the etherates of boron trifluoride is difficult. Boron trichloride compounds often decompose irreversibly so that gas-phase dissociation studies are meaningless in terms of the evaluation of dative bond strength. The higher melting points of the boron trichloride compounds (e.g., $(CH_3)_2O \cdot BCl_3$ m.p. 76°; $(CH_3)_2O \cdot BF_3$ m.p. $-13°$) merely show that the lattice energies are larger. Dipole moments of the pyridine–boron complexes also give the sequence $BBr_3 > BCl_3 > BF_3$ (21). In addition simple molecular orbital calculation of the π-bond energies conforms to the view that the energy of reorganization from the planar (sp^2 hybridized) to the tetrahedral (sp^3 hybridized) configuration of the boron trihalides decreases in the sequence $BF_3 > BCl_3 > BBr_3$ (48.3 kcal. mole^{-1}, 30.3 kcal. mole^{-1}, 26.2 kcal. mole^{-1}) (82).

The difference between the estimated values of the reorganization energy of BCl_3 and BBr_3 is close to the difference between the gas-phase heats of formation of the pyridine complexes.

The superior acceptor strength of boron tribromide as compared with trichloride and trifluoride is therefore now well established both for reactions in solution and in the gas phase and also for the formation of the crystalline complexes themselves. An obvious explanation for this order of acceptor power must be (a) decreasing tendency to form π-bonds by back donation from the halogen atoms into the vacant boron orbitals going from fluorine to chlorine to bromine, (b) a decrease in the electrostatic repulsion between the B–X bond electrons and the lone-pair electrons of the ligand.

Boron triiodide in its catalytic activity even surpasses in certain cases boron tribromide. Instability of the iodine complexes besides other factors (e.g., steric hindrance) may account for the lack of further information and thermal data.

b. Aluminum trihalides

Similarly to the reactivity sequence of the boron trihalides, the relative acceptor strength (catalyst activity) of the aluminum halides is now well established from different data.

Aluminum iodide was found to be the most reactive catalyst in the isopropylation of benzene with isopropyl chloride, the relative sequence of activity being $AlI_3 > AlBr_3 > AlCl_3$ (168). The generally higher reactivity of aluminum bromide over aluminum chloride has been known for a long time but was mostly attributed to the higher solubility of aluminum bromide compared with that of aluminum chloride and not to the greater acceptor strength of the bromide over the chloride. That this latter is the case can be shown by comparison of the heats of formation of addition compounds and also by an investigation of the reorganization energies of monomeric $AlCl_3$ (31.6 kcal. mole^{-1}) and $AlBr_3$ (27.9 kcal. mole^{-1}) from the planar (sp^2 hybrid) to the tetrahedral (sp^3 hybrid) configuration which are very similar to the values for the corresponding boron halides. The acceptor strength of the aluminum trihalides, however, is also influenced by the heats of dissociation of their dimers (see Table III) (82).

Complications also arise when the heats of formation of complexes from crystalline aluminum trichloride and tribromide are compared, for the chloride is an ionic solid (160) whereas the bromide forms a lattice of dimeric molecules (260). Nevertheless in the absence of data for the gas-phase reactions, a comparison of the heats of formation of the aluminum trihalide complexes in condensed phase reactions is of interest (132).

The heat of formation of the pyridine complexes of aluminum trichloride, tribromide, and triiodide are 31.1, 31.9, and 31.5 kcal. mole^{-1} respectively (86,106). The corresponding heats of mixing in chlorobenzene solution are 42.6, 60.1, and 70.1 kcal. mole^{-1}, the larger values being ascribed to the formation of higher complexes in solution. The heats of formation of crystalline complexes of the aluminum halides with trimethylamine show a similar increase with increasing atomic number of halogen: 30.8, 33.0, and 33.1 kcal. mole^{-1} respectively. From the vapor pressures above the 1:2 complexes of aluminum trichloride and tribromide with trimethylamine (316) the energies of dissociation of these complexes into the 1:1 complexes and gaseous trimethylamine are 10.4 and 15.2 kcal. mole^{-1}, implying that the heats of formation of the crystalline 1:2 complexes from the parent donor and acceptors are 41.2 and 48.2 kcal. mole^{-1}, the value for the bromo complex again being the larger.

The heats of reaction of aluminum chloride (101) and bromide (182) with acetophenone (19.2 and 22.4 kcal. mole^{-1}) and with benzophenone (15.4 and 19.5 kcal. mole^{-1}) and their heats of solution in nitrobenzene (252) (12.2 and 24.5 kcal. mole^{-1}) show the same order. Likewise the heat of reaction with acetaldehyde in chlorobenzene solution increases from aluminum chloride (24.7 kcal. mole^{-1}) (101) through aluminum bromide (36.0 kcal. mole^{-1}) to the triiodide (42.4 kcal. mole^{-1}) (100).

The generality of the phenomenon is shown by heats of formations (in kcal. mole^{-1}) of the following pairs of ionic complexes (253)

$AlCl_3 \cdot NaCl$ 6.65	$AlCl_3 \cdot NaBr$ 3.32	$AlCl_3 \cdot KCl$ 12.42
$AlBr_3 \cdot NaCl$ 9.93	$AlBr_3 \cdot NaBr$ 10.1	$AlBr_3 \cdot KCl$ 14.6.

From these data it is evident that aluminum bromide is a stronger Lewis acid than aluminum chloride.

It is curious, however, that the reverse order is observed for the heats of formation of the 1:1 complexes of ammonia with both the aluminum halides (166,167), and the gallium trihalides (167). The heats of dimerization of the aluminum trihalides (115) and gallium trihalides (114) in the gas phase also decrease in the sequence Cl > Br > I, but here one is changing both the donor and acceptor simultaneously and the results are now comparable with those previously discussed in which the acceptor strength is measured relative to the same ligand in each series.

c. Gallium trihalides

The heat of formation of the gallium bromide–pyridine addition complex is larger than that of the pyridine–gallium chloride complex (132). Thus the relative acceptor strength of the gallium halides is Br > Cl. These data are in accordance with the previously discussed observation relating to boron and aluminum trichlorides and the values for heat of dimerizations.

d. Titanium and zirconium tetrahalides

Vapor-pressure measurements of complexes of the titanium halides with pyridine and acetonitrile by Eméleus (107) give the following sequence of stability of complexes I > Br > Cl.

The heat of formation of the corresponding pyridine complexes gave the corresponding sequence I > Br > Cl > F. In related investigations similar results were obtained for $ZrBr_4$ > $ZrCl_4$ > ZrF_4.

It may be said, in summary, that with few exceptions the heat of complex formation of Lewis acid metal halides and stability of the complexes increases in the order F < Cl < Br < I ·and this order therefore also gives an indication of the Lewis acid strength and catalyst activity of the halides.

C. Reagent Reactivity (Base Strength)

As the Friedel-Crafts systems must always be considered as generalized acid–base systems it is obvious that besides the electron deficiency of the catalyst (acid strength) of equal importance is the donor ability of the reagent (base strength). Calloway was the first, in 1937, to investigate the relative activities of different acyl and alkyl halides (62). He found that in the acylation of benzene catalyzed by aluminum chloride, the relative reactivity of the four acetyl halides is RCOI > RCOBr > RCOCl > RCOF.

Yamase (329,330) has recently shown, however, that this sequence (based not on kinetic data, but on relative yields obtained, with a variety of substrates) holds true only for mesitoyl halides (with anisole as substrate) and acetyl halides (with substrates as benzene, mesitylene, and anisole). In other cases the maximum reactivity may be found with the acyl bromides, or the chlorides, or equally with the bromides and chlorides. Insufficient data are available with acyl iodides and fluorides. However, acyl fluorides are quite reactive with boron halides catalysts and in this case the relative order of reactivity (based on yields of products in the acetylation of toluene) is RCOF > RCOBr > RCOCl (227).

In the aluminum chloride catalyzed alkylations of benzene with n-butyl halides the sequence of reactivity observed by Calloway was RF > RCl > RBr > RI. Although during Friedel-Crafts reactions involving organic fluorides and aluminum halides a metathetic exchange reaction takes place with the formation of aluminum fluoride, this could have little to do with any of the observed activities, because aluminum fluoride itself, under the reaction conditions used, was found to be completely inactive in either acylation or alkylation reactions with any of the halides, including the fluorides.

Using cyclohexyl halides as alkylating agents and anhydrous hydrogen fluoride as catalyst Simons (290) observed the reactivity sequence of the alkyl halides to be RF > RCl > RBr.

The fact that in alkylation reactions the alkyl fluorides are the most and the alkyl iodides are the least reactive was considered in view of the carbon–halogen bond energies to be unexpected, although it is in accordance with the relative electronegativities of the halogens. In most chemical reactivities of organic halides the opposite sequence is observed. However, there were a few previous observations, such as the reactivity of benzoyl halides with Grignard reagents (108), in which the observed reactivity was ArCOF > ArCOCl > ArCOBr > ArCOI. In the reactivity of halobenzenes to form the Wittig type benzyne intermediate, the observed sequence of halogens was also F > Br > Cl > I (157,278,327).

The observed reactivity of acyl halides in aluminum chloride catalyzed acylations is similar to that observed by Satchell (269) in proton acid catalyzed acetylations showing the reactivity:

acetyl bromide > acetyl chloride > acetyl fluoride

Comparing different types of acyl chlorides the observed reactivity according to Satchell is:

acetyl chloride > propionyl chloride > benzoyl chloride.

Thus alkanoyl halides are more reactive than aroyl halides and acetyl halides show higher reactivity than propionyl halides.

Man and Hausen (193) compared reactivities of acyl halides and acid anhydrides as acylating agents and found the sequence:

acetyl chloride > benzoyl chloride > 2-ethylbutyryl chloride

in competitive experiments and similarly:

acetic anhydride > benzoic anhydride.

In the alkyl halide series the long established sequence of reactivity is: tertiary halides > secondary halides > primary halides.

This is in accordance with the stability but not with the electron deficiency of the corresponding carbonium-ion intermediates formed in interaction with the Lewis acid catalysts, which are responsible for the Friedel-Crafts substitution reactions.

Dative bond strength is affected by the nature of the groups on the donor atom just as it is by the nature of groups on the acceptor atom. The effect is usually in the direction expected from consideration of electronegativities. Hence the ability of a ligand atom to donate a lone pair is often reduced by the attachment of electron-attracting groups. Thus the compound $H_3P \cdot BF_3$ is known whereas the adducts $F_3P \cdot BF_3$ and $Cl_3P \cdot BF_3$ are unknown (117,324). Alkyl and acyl fluorides give stable complexes with boron trifluoride (221) whereas chlorides and bromides fail to do so. However, the stability of an addition compound is not always that to be expected from predictions based on the electronegativity values even when donor and acceptor atoms are from the first row of the Periodic System, where complications arising through bonding above the $2\ sp^3$ octet are impossible. This was illustrated by the previously discussed relative acceptor powers of the boron (aluminum, etc.) halides.

D. Bond Lengths and Strengths in Donor–Acceptor Complexes

The bond lengths in the transition complexes formed between Lewis acid catalysts and donor reagent molecules contribute to the relative activity of different systems. The shorter the donor atom–central acceptor bond is, generally, the stronger it is. The central atom–halogen ligand bonds at the same time lengthen to a certain degree. Back donation from the unshared halogen pairs to the electron-deficient central atom also affects the bond length. This effect of course opposes the lengthening of the bond, but the two may not balance each other. The longer the bond, the less the π-overlaps involved influence the electron deficiency of the central atom of the Lewis acid halide.

In the interaction of halide reagents (like alkyl halides) with acidic metal halide catalysts (like aluminum halides) the carbon–halogen bond is loosened and ultimately cleaved. At the same time a new metal–halogen bond is formed (Al–halogen in the case mentioned). The heat of formation of the metal–halogen bond at least partially helps to balance the dissociation energy needed to cleave the carbon–halogen bond. This effect could be particularly important in reactions with fluoride reagents where the cleavage of the strong but polar C–F bond must be facilitated by the simultaneous formation of a strong metal–fluorine bond.

E. Steric Hindrance

Another important factor to be considered in the discussion of relative acid strength of catalysts and reactivity of catalyst:reagent systems is the question of steric hindrance. Bulky reagent groups and increasing size of halogen atoms on the Lewis acid could seriously affect the reactivity of Friedel-Crafts systems. Iodide type metal halide catalysts generally are unsuitable not only due to side reactions (iodinations) but also because they give rise to considerable steric hindrance. Although recently it was demonstrated that even bulky anions such as BI_4^- can be prepared under proper conditions (321) steric factors still interfere to a large degree in many cases with the donor–acceptor type of interaction of Lewis acid iodides with reagent donor molecules. In accordance with the expected sequence of steric requirements the use of Lewis acid iodides (BI_3, AlI_3, etc.) was found to be best suited in reactions involving alkyl or acyl fluorides as donor molecules. The small steric requirements of the fluorine atom balances (to a certain degree) the greater steric need of the Lewis acid.

According to the generalized acid–base interaction, the steric requirements of an acid can be compared only with a specific base (in this case reagent halide, anhydride, etc.).

F. Physical State of Catalyst

a. Lattice and solvation energy

In discussing the relative strength and reactivity of Friedel-Crafts catalyst systems the physical nature of the catalysts (and accordingly of the reagents) must also be considered. In systems where solid catalysts are used with limited solubility in the reaction media, and generally insoluble, often crystalline intermediate complexes are formed, the question of *lattice energies* must play an important role. In systems where soluble catalysts and complexes are present this is replaced by the role of *solvation energies*. Both lattice and solvation energies play an important role in the relative reactivity of Friedel-Crafts systems and should influence the expected selectivity of these systems to a considerable degree.

b. Heat of dissociation of dimeric and associated catalysts

Closely related to the question of lattice and solvation energies (the latter of course playing an important role relating to the effect of solvents on the course and selectivity of the reactions) is another factor frequently encountered in Friedel-Crafts systems, namely that of the dissociation energy of dimeric or associated Lewis acid catalysts to the catalytically active monomers. This has specific importance in the case of such most frequently used catalysts as aluminum, ferric, or gallium halides. As discussed previously aluminum chloride, bromide, and iodide are dimeric in the liquid and vapor phase and with the exception of the chloride, also in the crystalline solid state (aluminum chloride being ionic in the solid state). Aluminum fluoride is highly ionic both in the solid and liquid phase, although it was found recently to be somewhat dimerized in the vapor phase, but to a much lesser extent than the other halides (254).

There is no real evidence concerning the catalytic activity of the dimeric aluminum halides (similarly to the gallium halides). Although aluminum chloride forms complexes with aromatics of the composition $ArH \cdot Al_2Cl_6$ (53) and methyl chloride was reported to form 1:2 complexes with gallium chloride $CH_3Cl \cdot Ga_2Cl_6$ (41) these are probably only weak complexes and the active alkylation complex still may be $CH_3Cl \cdot GaCl_3$.

Walker (322) discussed this question in considerable detail in the case of aluminum bromide catalyzed reactions and provided interesting data to substantiate the fact that dimeric Al_2Br_6 is catalytically inactive. The active catalyst species is the monomeric Lewis acid, $AlBr_3$, and therefore the dissociation energy of the dimer plays an important role in the overall activation energy of the Friedel-Crafts reaction with dimeric metal halides. The heat of dissociation of the aluminum halides as determined by Fischer and Rahlfs (116) and Porter and Zeller (254) shows a substantial decrease from the fluoride to the iodide (Table III).

TABLE III.   Heat of dissociation of dimeric aluminum halides (kcal. $mole^{-1}$)

$Al_2F_6 \rightleftharpoons 2AlF_3$	48.0 ± 1
$Al_2Cl_6 \rightleftharpoons 2AlCl_3$	29.0 ± 1
$Al_2Br_6 \rightleftharpoons 2AlBr_3$	26.5 ± 0.5
$Al_2I_6 \rightleftharpoons 2AlI_3$	22.5 ± 1

Thus the increasing catalytic activity of aluminum halides from the fluoride to the iodide is in accordance with the decreasing energy of dissociation needed to obtain the catalytically active monomeric

Lewis acid. The decreasing values of heat of dissociation of dimeric halides are in accordance with the generally observed catalyst activity sequence of aluminum halides

$$AlI_3 > AlBr_3 > AlCl_3 > AlF_3.$$

Without elaborating further, Boeseken (31) hinted at the possibility of photosensitizing Friedel-Crafts reactions (277). In his interpretation the fundamental process of the reaction is not the formation of an intermediate compound, but simply displacement of the field within the molecule, a so-called "dislocation." The dislocation must take place at the instant of collision with the catalyst or at the instant of separation, leaving the substrate in a reactive, "opened" condition.

Recent observations on the very effective promoter effect of ultraviolet irradiation on certain Friedel-Crafts reactions, using aluminum bromide and iodide and gallium bromide as catalyst, put this question in a new light (238). It is suggested that the effect of light on Friedel-Crafts reactions in the absence of oxygen is primarily to promote the dissociation of the acidic halide dimers to the catalytically active monomers and not to activate the carbon–halogen bonds, which absorb very weakly in the ultraviolet and visible region. A proof of this suggestion was obtained indirectly when it was found that ultraviolet light has no effect at all on reactions involving boron tribromide or triiodide (which are monomeric).

The effect of light on the oxidation of aluminum halides and its consequent co-catalytic effect is discussed in Chapter XXVIII.

In considering the Friedel-Crafts systems as generalized acid–base systems and attempting to discuss the relative catalyst activity (acid strength) of various metal halide catalysts, it has been repeatedly pointed out that this is possible only in relation to defined reagent bases under specific reaction conditions and no general order of activity can be established. Changing the nature of the reagent base may entirely change the relative catalyst activity sequence, as found for example in the case of boron halide catalysts. The question is very similar to the relative base strengths of organic amines discussed in detail by H. C. Brown (36). However, the number of additional factors (as compared with Brown's "strains") is considerably larger, owing to the more complex nature of Friedel-Crafts systems.

3. Effect of Co-catalysts on Catalytic Activity

So far in our discussion we have considered only the Lewis acid strengths of various acidic halides and have deliberately omitted one of the most important factors involving all Friedel-Crafts catalytic systems.

The rates of Friedel-Crafts reactions are notoriously irreproducible and inevitably the variability has been traced to the presence of varying amounts of very small concentrations of promoting substances or co-catalysts, without which the Friedel-Crafts halides are frequently inactive. The present theory of the function of the co-catalyst is that it unites with the Lewis acid halide to form an ionic complex which provides the active catalyst.

A. Nature of Co-catalysts

a. Proton-releasing substances, especially hydroxy compounds and hydrogen acids

It has been rigorously demonstrated by Evans, Meadows, and Polanyi (111) that the polymerization of isobutylene by boron trifluoride does not take place in the absence of traces of water (or other added substances). In other systems it has been difficult to place unambiguous evidence that such traces of co-catalysts are necessary because of the exceedingly low concentrations of water which will support reaction (concentrations far below those reached by normal methods of severe drying). However, in a number of cases it was demonstrated that rates of Friedel-Crafts reactions can be markedly decreased by drying the systems and accelerated by adding traces of water.

Co-catalysts of this type have been found necessary in many simple Friedel-Crafts reactions, as was discussed previously (Chapter IV).

b. Carbonium-ion forming substances

Co-catalysis by alkyl halides accounts for the apparent absence of water co-catalysis in the stannic chloride polymerization of styrene in solution in ethylene chloride (245), the carbonium ion taking the place of the proton in initiating the polymerization chain. It is also well known that alkyl and acyl halides frequently are promoters in reactions catalyzed by aluminum halides, e.g., isomerization or alkylations with olefins.

B. Activity of Co-catalysts

It must be realized that the presence of co-catalysts shifts the emphasis from the Friedel-Crafts halide to the co-catalyst, since it is the latter which provides the initiating cation. The function of the catalyst becomes merely to facilitate the dissociation of the cation by complexing with and stabilizing the anionic residue, *i.e.*, it behaves as an "ansolvo" acid in Meerwein's sense (206). The effectiveness of a given catalyst–co-catalyst pair in providing a proton will depend on the electrophilic character of the catalyst and on the donor character of the co-catalyst and its acid strength. These influences have been considered qualitatively by Plesch (250) for the catalysts boron trifluoride, titanic chloride, stannic chloride with the co-catalysts water, *t*-butanol, and the chloroacetic acids. Russell (268) measured the effect of different co-catalysts on the rate of polymerization of isobutylene by stannic chloride in ethyl chloride at $-78°$ and found

$$CCl_3CO_2H > ClCH_2CO_2H > CH_3CO_2H > C_2H_5NO_2 > CH_3NO_2 > C_6H_5OH > H_2O$$

a sequence which follows the acid dissociation constants where known. However, other factors must also play an important part, for picric acid which has the same dissociation constant as trichloroacetic acid is found not to co-catalyze the system titanium tetrachloride–isobutylene in hexane. Topchiev (310) measured the electrical conductivity of a large number of complexes of boron trifluoride with hydroxyl compounds but found no correlation with catalytic activity.

Satchell (270) made use of the aromatic hydrogen exchange reaction to compare the co-catalytic power of a series of halogen substituted acetic acids. In the light of experimental evidence he made the tentative generalization that the co-catalytic ability ultimately depends on the electron availability on those atoms of the anion, provided by the Brønsted component which coordinates with the Lewis acid. Thus with stannic chloride he found the order of co-catalytic activity

$$HCl > H_2O > CH_3CO_2H > CH_2ClCO_2H > CHCl_2CO_2H > CF_3CO_2H.$$

Therefore it is obvious that co-catalytic power is not simply related to conventional acid strength.

This was further demonstrated (272) by the stannic chloride catalyzed tritium exchange of aromatics, using additional organic acids, both aliphatic and aromatic, as co-catalyst. The relative order of co-catalyst activity found in a series of structurally related acids was

o-toluic acid > benzoic acid > *m*-nitrobenzoic acid
pivalic (trimethylacetic) acid > phenylacetic acid > acetic acid.

The co-catalytic activity accordingly seems to be related not to simple acidity, but to the electron availability of the carboxyl oxygen atoms of the anions formed by coordination with the Lewis acid catalyst.

Benzoic acid is a better co-catalyst than acetic acid, being nearly as active as pivalic acid. These experimental data might indicate that the benzoate ion has the greater electron density oxygen than the acetate ion (perhaps through some extended conjugation which is not possible for the acetate ion). However, it is obvious that it is very difficult to predict the relative electron availabilities and thus co-catalytic effects of Brønsted type co-catalysts.

The effectiveness of the chlorinated and other polar solvents is not determined only by their co-catalytic activity. Here the effect is composite, and the general polar character of the solvent—its capacity to solvate the ions formed—plays a part, as well as the donor character of its molecules.

The relationship of the above postulated catalyst–co-catalyst complexes to the known stable complexes of Friedel-Crafts halides is not yet clearly established. Where strong complexes can be formed, e.g., of different metal chlorides with ether, alcohol, or acetone, Friedel-Crafts reactions are completely suppressed (98). However, even stable complexes of this type may have catalytic activity in other solvents; e.g., $SnCl_4 \cdot 2Et_2O$ is active in ethylene dichloride solution (98), and $BF_3 \cdot 2Et_2O$ is commonly used as a milder catalyst than boron trifluoride itself. It is not known whether in these instances the ether acts as a co-catalyst, initiating reactions by Et^+ (110), or whether the activity is due to partial hydrolysis.

II. Catalyst Activity of Proton Acid Catalysts

In previously discussing the relative strength of Lewis acid type metal halide catalysts it was pointed out that in the presence of co-catalysts the emphasis shifts from the acidic halide Lewis acids to the co-catalysts (generally a proton-releasing substance or carbonium-ion forming agent), since it is the latter which provides the required initiating cation for the reactions. From recent work of Satchell (271) it seems to be established that the co-catalyst strength of Brønsted acids is not simply dependent on their acid strengths, but more on the electron availability on those atoms of the anions formed by dissociation of the Brønsted acid, which coordinate with the Lewis acid in the formation of the anion of the conjugate acid. However, ultimately this contributes to the acid strength of the newly formed conjugate acid. This leads to the question of the relative strengths of proton acid catalysts used in Friedel-Crafts type reactions. Very little is known so far about any accurate method capable of comparing relative catalyst activity of Brønsted acid catalysts in Friedel-Crafts systems, other than comparisons based on the Hammett H_0 function.

1. Acid Strength of Proton Acids Based on Hammett's Acidity Function

Hammett (139) and his co-workers have shown that it is possible to define an acidity function which accurately expresses the tendency of a given solution to transfer a proton to a neutral base under conditions where pH or hydrogen-ion concentration lose their significance.

A. Definition of H_0

The H_0 acidity function is a quantitative measure of acidity conceived originally by Hammett and Deyrup (140). It is derived from ionization equilibria of a particular class of indicators, those behaving in the Brønsted-Lowry sense as uncharged bases

$$B + H^+ \rightleftharpoons BH^+$$

and is defined by the equation (138)

$$H_0 = pK_{BH^+} - \log \frac{C_{BH^+}}{C_B}$$

where C_{BH^+}/C_B is the directly observable concentration ratio of the indicator in its two differently colored forms and K_{BH^+} is the thermodynamic ionization constant of its conjugate acid in terms of molar concentrations, referred to ideal dilute solution in water.

The above definition is formally equivalent to

$$H_0 = - \log \left(\frac{a_{H^+} f_B}{f_{BH}} \right)$$

where a_{H^+} is the total activity of the hydrogen ion in the solution and f_B/f_{BH^+} is the ratio of the activity coefficients of a neutral base B and its conjugate acid BH^+. There is much evidence to indicate that this ratio has the same value for all bases in a given solvent (142). Hence the acidity function H_0 is independent of the base used. Values for H_0 for mixtures of various acids in both water and certain non-aqueous solvents were summarized by Paul and Long (242).

B. H_0 *Values in Aqueous and Organic Solutions*

An ever increasing body of H_0 data can be found in the literature for different solvent systems. Frequent attempts have been made to correlate the acidity function H_0 with reaction rates (141,243).

In a certain specific solvent the H_0 values give information regarding the relative acid strengths at identical concentrations.

The acidity function H_0 is not the same as the pH, but it does approach it in dilute aqueous solutions where the activity coefficient ratio approaches unity. Some values of H_0 at selected molarities of aqueous solutions at 25° show the usual acidity sequence of simple inorganic acids (242) (Table IV).

TABLE IV. Values of H_0 for aqueous solutions of acids

Acid concentration (Molality)	HNO_3	HBr	HCl	$HClO_4$	H_2SO_4	H_3PO_4	HF
1.0	−0.18	−0.20	−0.20	−0.22	−0.26	+0.63	+1.20
1.5	−0.45		−0.47	−0.53	−0.56	+0.41	+1.04
2.0	−0.67	−0.71	−0.69	−0.78	−0.84	+0.24	+0.91
3.0	−1.02	−1.11	−1.05	−1.23	−1.38	−0.08	+0.60
4.0	−1.32	−1.50	−1.40	−1.72	−1.85	−0.37	+0.40
5.0	−1.57	−1.93	−1.76	−2.23	−2.28	−0.69	+0.28
7.0	−1.99	−2.85	−2.56	−3.61	−3.32	−1.45	+0.02
8.0		−3.34	−3.0	−4.33	−3.87	−1.85	−0.11
10.0		−4.44	−3.68	−5.79	−4.89	−2.59	−0.36

The H_0 values for some of the anhydrous acids are:

HSO_4	HF	H_3PO_4
−11	−10.2	−5

Relative acidities of the anhydrous acids and their solutions in non-aqueous solvents differ very sharply from those observed in aqueous solutions. As an example anhydrous HF is a very strong acid compared with the weak aqueous HF.

Smith and Hammett (294) investigated solutions of sulfuric, hydrochloric, and methanesulfonic acids in nitromethane. The H_0 value for 0.01M solution of sulfuric acid was about −3.2, indicating a much higher acidity than that at a similar concentration in water. The H_0 value for HCl of equal concentration was, however, about 4.1 units lower than that of sulfuric acid. The reason for this discrepancy is probably to be found in the solvent. Nitromethane has a rather high dielectric constant ($\epsilon = 35$ at 25°) but has apparently very little capacity for ion solvation. Thus it differs from such oxygen base solvents as water. One is therefore dealing with solutions in which ion association is prevalent, not as a consequence of uncommonly strong interionic attraction but rather as a consequence of low solvation energy.

Another high dielectric constant solvent of particular interest is tetramethylene sulfone. According to Langford (176) a 0.01M H_2SO_4 solution in tetramethylene sulfone was found to be much more acidic than the corresponding aqueous solution. Acids generally show very high H_0 values in this solvent. HBF_4 was suggested to be stable and to have very high H_0 values in tetramethylene sulfone (255).

C. Acidity of Conjugate Friedel-Crafts Acids

The acidity of a typical Friedel-Crafts conjugate acid, HBF_4, as a 7% (weight) solution of BF_3 in anhydrous HF (in the presence of aromatics) can be derived from recent data of Mackor (191,199).

Mackor and co-workers measured the ratio of the protonated to unprotonated aromatic species in both HF and $HF \cdot BF_3$ and showed that

$$\log K^+/K = 6.6$$

where K^+ is the equilibrium constant in $HF-BF_3$ and K is the constant in HF.

The ratio of protonated to unprotonated aromatic in $HF-BF_3$ is

$$\left[\frac{C_{AH}}{C_A}\right]_{HF-BF_3} = K^+ \frac{C_{BF_4^-}}{C_{BF_3}}$$

In HF, the same ratio is:

$$\left[\frac{C_{AH}}{C_A}\right]_{HF} = KC_{F^-}$$

If we choose $C_{BF_3} = 1$ (68 g. BF_3 per 1000 g. HF, i.e., 7% BF_3) and assume $C_{BF_4^-} = C_{F^-}$, then

$$\left[\frac{C_{AH}}{C_A}\right]_{HF-BF_3} \bigg/ \left[\frac{C_{AH}}{C_A}\right]_{HF} = \frac{K^+}{K}$$

but, from the definition of H_0,

$$[H_0]_{HF-BF_3} - [H_0]_{HF} = -\log\left(\frac{C_{AH}}{C_A}\right)_{HF-BF_3} + \log\left(\frac{C_{AH}}{C_A}\right)_{HF} = -6.6$$

The H_0 of anhydrous HF is about -10.2.   Therefore,

$$[H_0]_{HF + 7\%BF_3} = -10.2 - 6.6 = -16.8.$$

H_0 is found to be -16.8 compared with about -11 for 100% H_2SO_4 and -10 for 100% HF.   Thus this Friedel-Crafts conjugate acid, which presumably has an acidity comparable to such acids as $AlCl_3 \cdot HCl$ and $AlBr_3 \cdot HBr$, is more acidic than 100% sulfuric acid by a factor of about 10^5.   This very significant difference in acidity of certain Friedel-Crafts conjugate acids could easily account for observed differences in catalytic action (for example isomerizations, alkylations, acylations).

Kilpatrick, Katz, and co-workers (152) recently investigated the hydrogen fluoride–antimony pentafluoride system.

Antimony pentafluoride and hydrogen fluoride as liquids are miscible in all proportions.   Observations on the reactivity of mixtures have been interpreted in terms of fluoride-ion acceptor behavior to form an octahedrally symmetrical SbF_6^-.   Conductivity measurements and spectroscopic investigations indicate the equilibrium $2HF + SbF_5 \rightleftharpoons H_2F^+ SbF_6^-$.   In addition to the electrical conductivity data Hammet acidity function values were determined both for SbF_5 and the related NbF_5 in anhydrous HF.

	H_0
HF + 0.02 M NbF$_5$ (0.4% Weight)	—12.5
HF + 0.36 M NbF$_5$ (6.7% Weight)	—13.5
HF + 0.36 M SbF$_5$ (6% Weight)	—14.3
HF + 3.0 M SbF$_5$ (60% Weight)	—15.2.

Thus these conjugate acids, showing very strong Friedel-Crafts catalytic activity are, similarly to HF + BF_3, very strong acids, considerably stronger than 100% H_2SO_4 or anhydrous HF itself.

That metal fluorides (BF_3, TiF_4, NbF_5, TaF_5, etc.) are acids in hydrogen fluoride, and that these co-acids considerably increase the overall acidity of the systems, was shown by McCaulay and Lien (200,202) on the basis of extraction experiments of basic aromatic hydrocarbons.   Electric conductivity investigations of the HF solutions by Kilpatrick (161,162,163) entirely supported this view.

Thus the fact has been well established that certain solutes in hydrogen fluoride such as boron trifluoride and antimony pentafluoride behaved as if they were fluoride-ion acceptors and therefore acids or co-acids.

The term co-acid was introduced by Kilpatrick for compounds which increase the acidity of a protonic solvent, without contributing a proton.   While it is obvious that the familiar term "Lewis acid" may be applied to these compounds the terms are used by Kilpatrick in a slightly different context.   The term co-acid is restricted to compounds which increase the acidity of a proton acid, while Lewis acids may react directly with the base.

Clifford (65,66) has discussed the chemical behavior of a number of possible fluoride ion acceptors.   He ranked acids as to relative strength depending on the ability of HF solutions of the potential co-acid to solvate and dissolve certain fluorides (KF, CoF_3) or complex fluorides ($AgBF_4$).   From the extensive solvolysis of BF_4^- salts he concluded that BF_3 is not a strong acid in HF, whereas PF_5 and SbF_5 are.   Owing to all other data discussed previously concerning the strong acid nature of the HF + BF_3 system this consideration cannot be accepted at the present time.

Our knowledge concerning acidity in non-aqueous solutions and in general about electrochemistry in these solutions is still in the early stage of development.   It is, however, quite obvious that no simple extrapolation of facts observed in aqueous solutions is possible.   This has particular bearing on the interpretation of relative acid strength and catalyst activities of proton acid Friedel-Crafts catalysts and so-called conjugate Friedel-Crafts acids which in organic solvents or nonaqueous inorganic solvents (like anhydrous HF, SO_2) can be far more effective acid catalysts than sulfuric or related "strong" acids.

III. Selectivity in Friedel-Crafts Systems

1. Functional Selectivity of Reagents

Although the Friedel-Crafts reactions have long been considered, together with the Grignard reactions, as one of the most versatile tools of organic chemistry, there is a serious limitation in their use. This is encountered in reactions where more than one functional reactive group or atom is present, which can interact with the catalyst and start chemical reactions.

This lack of "selectivity" is well demonstrated in examples such as the attempted alkylation with vinyl chloride or ethylene dichloride in the presence of aluminum chloride or related catalysts.   No simple chloroethyl or vinyl product is formed, but owing to the presence of two functional reagent groups (olefinic double bond and chlorine atom in one case and two chlorine atoms in the other) the reactions proceed fast over the primary interaction involving the other reactive center, leading to a mixture of secondary products.

Most of the strong Friedel-Crafts catalysts generally used, such as aluminum halides, ferric halides, etc., do not show sufficient selectivity to allow reaction of one functional group in a polyfunctional system without interfering with other reactive groups present. However, an increasing number of selective reactions with "milder" catalysts is now known. The extension of the use of these selective methods to a wider scope will add considerably in the future to the usefulness of Friedel-Crafts methods in many fields of organic chemistry.

No general survey is attempted to cover all reported "selective" Friedel-Crafts reactions, many of which will be treated in individual chapters dealing with specific reactions. Instead a number of typical examples were selected to demonstrate some of the main possibilities of carrying out selective Friedel-Crafts reactions in polyfunctional systems and also to point out certain general trends involved in finding selective conditions for specific systems.

As discussed in the first part of this chapter, the relative reactivity of Friedel-Crafts catalysts depends on a variety of factors, but is basically a generalized acid–base interaction. Consequently we cannot discuss relative "acid" strengths without considering simultaneously the appropriate reference bases. A better understanding of the nature of the acid–base interactions involved in any Friedel-Crafts system points to the fact that these systems must be treated always as an entity of the catalyst, reagent, substrate, and solvent involved, together with other factors (co-catalyst, temperature, relative concentrations of reagents and catalyst, reaction time, etc.) influencing the reaction. Weak donor "base" reagents can be very active in certain reactions when used with strong catalyst acids, and fairly strong donor reagents can remain unreactive in systems involving too weak catalyst acids. Therefore, it is possible to find selective catalysts and reaction conditions enabling the reaction of only one or certain functional reagent groups or atoms, without interfering with others simultaneously present.

A. Di- and Polyhalides

a. Dihaloalkanes

Friedel-Crafts reaction of aromatic hydrocarbons with dihaloalkanes proceeds not only with the substitution of both chlorine atoms, but also by subsequent reaction of the initial product, so that polynuclear compounds are also formed in the reaction. With methylene chloride, for example, the reaction leads intermediately to 9,10-dihydroanthracene

$$2 \; \bigcirc\!\bigcirc \; + \; CH_2Cl_2 \; \xrightarrow{\; AlCl_3 \;} \; \bigcirc\!\!\!\overset{CH_2}{\diagdown}\!\!\!\bigcirc \; + \; 2HCl$$

$$\bigcirc\!\!\!\overset{CH_2}{\diagdown}\!\!\!\bigcirc \; + \; CH_2Cl_2 \; \xrightarrow{\; AlCl_3 \;} \; \bigcirc\!\!\!\underset{CH_2}{\overset{CH_2}{\diagdown\!\!\!\diagup}}\!\!\!\bigcirc \; + \; 2HCl$$

The dihydroanthracene, however, is dehydrogenated to anthracene and the methyl chloride formed simultaneously reacts with the benzene present. The final reaction products are, therefore, diphenylmethane, anthracene, and toluene (124). The relative quantities of diphenylmethane and anthracene which are obtained depend on the proportions of dichloromethane and benzene used.

No simple chloromethylation of aromatics with methylene chloride according to the reaction

$$\bigcirc\!\bigcirc \; + \; CH_2Cl_2 \; \longrightarrow \; \overset{CH_2Cl}{\underset{|}{\bigcirc\!\bigcirc}} \; + \; HCl$$

can be achieved. Under the reaction conditions used the intermediately formed benzyl chloride reacts immediately with a second molecule of benzene to form diphenylmethane. Weak Friedel-Crafts catalysts like $ZnCl_2$, $SnCl_4$, and BCl_3 (which generally are useful in chloromethylations) do not effect alkylations with methylene halides.

In the action of anhydrous aluminum chloride with benzene and ethylene dichloride a small amount of ethylbenzene and triphenylethane is obtained (286). Diphenylethane is, however, the main product (288). With ethylene dibromide and toluene in the presence of aluminum chloride ditolylethane has been secured (123). Similarly, dimesitylethane is obtained from mesitylene and ethylene dibromide (323).

No simple chloroethylation with ethylene chloride according to the reaction

$$C_6H_6 + ClCH_2CH_2Cl \; \xrightarrow[- \; HCl]{\; AlCl_3 \;} \; C_6H_5CH_2CH_2Cl$$

is possible, as the intermediately formed β-chloroethylbenzene immediately reacts with a second molecule of benzene to form 1,2-diphenylethane.

The reaction of ethylidene chloride with benzene and aluminum chloride leads to *as*-diphenylethane (287,288).

$$CH_3CHCl_2 + 2C_6H_6 \xrightarrow{\ AlCl_3\ } CH_3CH(C_6H_5)_2 + 2\ HCl$$

Besides this liquid product a solid, 9,10-dihydro-9,10-dimethyl-anthracene has been obtained (6). Experimental proof that this compound was the dihydride, and not dimethylanthracene, has been offered (9,12).

$$2CH_3CHCl_2 + 2C_6H_6 \xrightarrow{\ AlCl_3\ } \quad C_6H_4 \begin{array}{c} CH_3 \\ | \\ CH \\ \diagup \quad \diagdown \\ \quad \quad C_6H_4 \\ \diagdown \quad \diagup \\ CH \\ | \\ CH_3 \end{array} + 4HCl$$

Higher molecular weight condensation products are also obtained in the aluminum chloride catalyzed reaction of ethylene dichloride with benzene (284,292). Propylene dichloride (1,2-dichloropropane) and benzene in the presence of aluminum chloride give 1,2-diphenyl-propane (288,289).

1,4-Dihaloalkanes are used to effect preferential cyclialkylations involving intermediate formation of a haloalkylated product which then in an intramolecular alkylation step yields the cyclocondensation products (54).

There is a lack of detailed information on the cyclialkylation of aromatics with 1,3-dihalides. Shishido and Nozaki (285) reported that 1,3-dichlorobutane failed to cyclialkylate benzene in the presence of aluminum chloride. Reppe and co-workers (261) reported a successful cyclialkylation with these reagents to yield 3-methyl-5-(3′-chlorobutyl)indane and 3-methylindane. Teyssié and Smets (307) made the interesting observation that polyvinyl chloride cyclialkylated benzene or the methylbenzenes in the presence of aluminum chloride.

When the two halogens of the dihalide are positioned 1,2 or 1,5 relative to each other, cyclialkylations can still take place but with concomitant isomerizations.

There is generally insufficient selectivity in all alkylations so far discussed involving dihaloalkanes to enable haloalkylations to be carried out without immediately effecting secondary alkylations (either intra- or intermolecular) of the intermediately formed haloalkyl compounds.

Making use of the considerably stronger donor properties of fluorides over chlorides (bromides or iodides), Olah and Kuhn found a simple direct haloalkylation method by the reaction of fluorohaloalkanes with aromatics in the presence of boron trihalide catalysts. Boron trihalides (in the absence of co-catalysts) generally do not catalyze alkylation with organic chlorides, bromides, or iodides, but are effective catalysts in alkylations with alkyl fluorides. Thus a very selective alkylation effecting only C–F bonds can be achieved (232,236).

Fluorochloro- and fluorobromoalkanes were found to be very effective chloro- or bromoalkylating agents in direct Friedel-Crafts type reactions in the presence of boron halide catalysts.

$$ArH + F(CH_2)_nCl \xrightarrow{-HF} Ar(CH_2)_n\,Cl$$
$$ArH + F(CH_2)_nBr \xrightarrow{-HF} Ar(CH_2)_n\,Br \qquad n \geqslant 2$$

b. Haloacyl halides

Halogen-substituted aralkyl ketones are usually obtained by the reaction of haloacyl chlorides with benzene or other aromatics and aluminum chloride. However, there are cases in which all the halogen has been substituted. Thus, the reaction of β-chloropropionyl chloride and benzene with aluminum chloride results in the formation of either ω-phenylpropiophenone or β-chloropropiophenone (198). The same reaction has been observed with toluene (197).

Dichloroacetyl chloride reacts in the same way. A mixture of dichloroacetyl chloride, benzene, and aluminum chloride heated for ten hours on a water-bath yielded diphenylacetophenone, each of the chlorine atoms being substituted by a phenyl residue (73). However, a thorough investigation of the behavior of a number of other halogenated acid chlorides when treated with benzene in the presence of aluminum chloride has shown that no matter how the experimental conditions may be varied, only the chlorine atom attached to the carbonyl group enters into the reaction (74). It seems that the chlorinated acyl chlorides permitting polysubstitution are exceptions to the rule.

Chloroacetyl chloride or bromide and benzene heated with aluminum chloride at modest temperatures, or at the boiling point, give only chloroacetophenone regardless of the amount of excess of benzene used.

The reaction of trichloroacetyl chloride with benzene in the presence of anhydrous aluminum chloride has been shown to take an unusual course. Delacre (88) reported the formation of a single substance which he assumed to be diphenylacetophenone, $(C_6H_5)_2CH \cdot CO \cdot C_6H_5$. Later, Anschütz and Forster (11) showed that the product was in reality triphenylvinyl alcohol. Blitz (29) reported that trichloroacetyl chloride and benzene with anhydrous aluminum chloride give triphenylvinyl alcohol, but that if moist aluminum chloride is used the reaction product is only trichloroacetophenone, thus the acyl halide halogen reacts selectively.

A 90% yield of β-chloropropiophenone has been reported by heating under reflux a mixture of β-chloropropionyl chloride, benzene, and 1 mole of aluminum chloride (135).

Under suitable conditions, however, 1-indanones can be synthesized directly in a one-step process involving acylation of the aromatic nucleus followed by a cyclizing alkylation. The intermediate β-haloalkyl aromatic ketone is cyclized by adding sulfuric acid in the same reaction (144).

A large variety of other ω-halogenated acyl halides have been allowed to react with aromatics in the presence of different Friedel-Crafts catalysts and it has always been observed that the acyl halide halogen reacted selectively to give the corresponding ω-halogenated ketones. Alkylations involving the alkyl halide halogens take place only under very forceful conditions and generally do not interfere with the haloacylations.

Table V summarizes some of the reported systems in which exclusive haloacylations have taken place.

TABLE V.   Haloacylations with haloacyl halides

Haloacyl halide	Catalyst	Ref.
FCH_2COCl	$AlCl_3$	164
FCH_2COCl	$POCl_3$	112
$ClCH_2COCl$	$AlCl_3$	104,147
$ClCH_2COCl$	$AlCl_3$	216
$ClCH_2COF$	BF_3	226
$BrCH_2COCl$	$AlCl_3$	156,259
$BrCH_2COBr$	$AlCl_3$	194,248
CF_3COCl	$AlCl_3$	159
$ClCH_2CH_2COCl$	$AlCl_3$	83
		208,279
$BrCH_2CH_2COCl$	$AlCl_3$	218
$Br(CH_2)_3COCl$	$AlCl_3$	16
$Cl(CH_2)_3COCl$	$AlCl_3$	67
$Cl(CH_2)_nCOCl$	$AlCl_3$	83
$CH_3CHClCOCl$	$AlCl_3$	104
$CH_3CHBrCOBr$	$AlCl_3$	104
$CH_3CHBr(CH_2)_2COCl$	$AlCl_3$	166

It is of particular interest that even secondary halogens in the α-position to the acyl halide group are less reactive than the acyl halide halogens. The reactions take place with the formation of halogenated ketones.

It generally requires excess of aromatic and forceful conditions to carry the reaction further and unite the haloketones with a second molecule of aromatic substrate or to achieve cyclialkylation of the same molecule.

The use of milder catalysts such as $ZnCl_2$, $SnCl_4$, BF_3, $POCl_3$, etc., may be preferred in cases where the halogen in the chain is somewhat more active.

In evaluation of the acyl halide–alkyl halide selectivity in reactions of haloacyl halides attention must be drawn to the fact that practically all examples discussed were reported prior to the introduction of more sensitive analytical methods, especially gas chromatography, in the investigation of organic reactions. A reinvestigation of the problem using more advanced analytical methods to obtain a complete and quantitative picture of all the products produced may easily show that in many cases some alkylation takes place, even if in a small degree, such that it could not be detected in previous work.

In the reaction of halogen-substituted aroyl chlorides with aromatic hydrocarbons only the acyl halogen is generally replaced. Benzene reacts with a number of halogenated aroyl chlorides and aluminum chloride with the formation of the corresponding ketones (63,87,175,209).

The higher reactivity of the acyl halide halogen over that in the alkyl or aryl group is sufficient to provide almost exclusive selective acylation, even with a strong acid catalyst such as $AlCl_3$, if equimolar amounts of the haloacyl halide and aromatics are reacted. Using excess of aromatics and forceful conditions the reaction can be carried further, effecting alkylation (cyclialkylation) of the haloketones with aromatics.

c. Dibasic acid halides

Succinyl chloride condenses with 1 mole of benzene to form β-benzoylpropionyl chloride and with 2 moles of benzene to form α,β-dibenzoylethane (190)

$$CH_2COCl$$
$$| \qquad + C_6H_6 \rightarrow C_6H_5COCH_2CH_2COCl + HCl$$
$$CH_2COCl$$

$$CH_2COCl$$
$$| \qquad + 2C_6H_6 \rightarrow C_6H_5COCH_2CH_2COC_6H_5 + 2HCl$$
$$CH_2COCl$$

Succinyl chloride may also react as a lactone in which case γ,γ-diphenylbutyrolactone is obtained with benzene and aluminum chloride. The reaction has been studied by Auger (13), who added aluminum chloride to a well-cooled benzene solution of succinyl chloride and allowed the mixture to stand for five hours. Lutz (185) reversed the procedure. He added the acyl chloride to well-stirred benzene and aluminum chloride. In this way he was able to use higher temperatures and decrease the reaction time to fifteen minutes.

Lutz (185) also effected the reaction of halogenated succinyl chlorides with benzene. Here a solution of the acyl chloride in benzene was added to a stirred mixture of aluminum chloride and benzene. The reactions proceeded normally and only the acyl halide halogens reacted

$$CHXCOCl$$
$$| \qquad + 2C_6H_6 \xrightarrow{AlCl_3} C_6H_5COCHXCHXCOC_6H_5 + 2HCl$$
$$CHXCOCl$$

Various dihalogenated *meso-* and *dl*-succinyl chlorides react in this fashion. The configuration of each dihalide of dibenzoylethylene corresponds to that of the acyl chlorides from which it was prepared.

In no case was any significant amount of isomerized or inversed product isolated.

Adipyl chloride gives 1,4-dibenzoylbutane with 2 moles of benzene (109,126).

$$\begin{array}{c}\text{CH}_2\text{CH}_2\text{COCl} \\ | \\ \text{CH}_2\text{CH}_2\text{COCl}\end{array} + 2\text{C}_6\text{H}_6 \xrightarrow{\text{AlCl}_3} \begin{array}{c}\text{CH}_2\text{CH}_2\text{COC}_6\text{H}_5 \\ | \\ \text{CH}_2\text{CH}_2\text{COC}_6\text{H}_5\end{array} + 2\text{HCl}$$

Malonyl chloride reacts with aromatic hydrocarbons in the presence of aluminum chloride to give a diketone of the type $ArCOCH_2COR$, a ketone of the type $ArCOCH_3$, and a hydrocarbon (22).

The treatment of diethylmalonyl chloride with benzene (120) in the presence of aluminum chloride results in the formation of indandiones as the chief product.

Equimolecular quantities of benzene and the chloride yield a very small amount of diethyldibenzoylmethane, $(C_6H_5CO)_2C(C_2H_5)_2$, but the chief product is 2,2-diethylindan-1,3-dione as shown in the equation above.

The condensation of benzene with dimethylmalonyl chloride does not result in a similar formation of indandiones (121). Instead, with equimolecular proportions of benzene and the acyl chloride, there is formed chiefly isopropyl phenyl ketone, $(CH_3)_2CHCOC_6H_5$.

If an excess of benzene is used in the reaction with dimethyl-malonyl chloride dimethylindandione, is also obtained,

In the reaction of phthalyl chloride with benzene and aluminum chloride Baeyer (17) reported the production of phthalophenone. Here the chloride reacts as the lactone:

According to Haller and Guyot (137), a small amount of 10,10-di-phenyl-9-anthrone is also obtained. If an insufficient amount of aluminum chloride is used, some 2-benzoylbenzoic acid is formed on hydrolysis (136).

Similarly, one mole of m-xylene reacts with terephthalyl chloride to give 4-(2,4-dimethylbenzoyl)benzoic acid (64).

Normal condensation occurs with isophthalyl chloride and benzene, with formation of isophthalophenone (2). The reaction takes place in two stages:

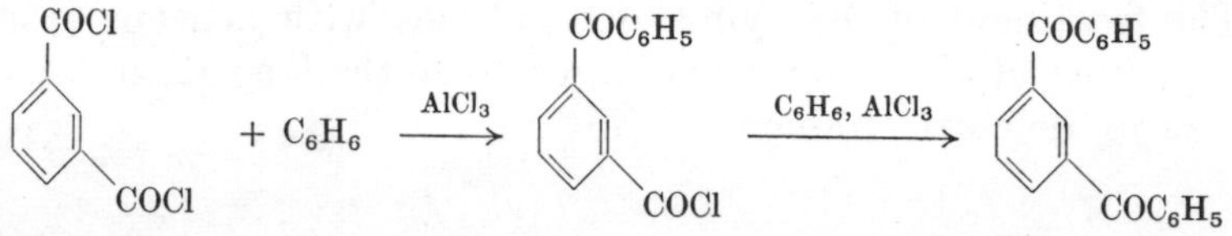

In the reactions of dibasic acyl halides it is possible to carry out the reaction selectively and stepwise, thus allowing only one of the acyl halide groups to react. This is mostly due to the deactivating effect of the already introduced keto group, which by complexing with a molar amount of Lewis acid catalyst acts as a highly electron-withdrawing group and thus deactivates the second acyl halide group. However, in the presence of 2 mole equivalents of aromatic and generally under more rigorous conditions both acyl halide groups react and diketones are obtained.

d. Halosulfonylacyl halides

Halides containing both sulfonyl and acyl halide functions such as chlorosulfonylacetyl chloride, $ClSO_2CH_2COCl$ or o-chlorosulfonyl-benzoyl chloride, o-$ClSO_2C_6H_4COCl$, show reaction of both functional groups under extreme conditions.

o-Chlorosulfonyl chloride reacts as a lactone in the presence of aluminum chloride with benzene (68,122,183).

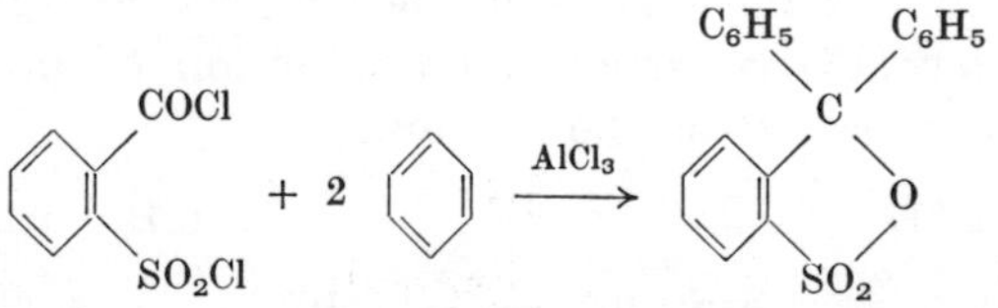

Chlorosulfonylacetyl chloride reacts with toluene or benzene in the presence of acetyl chloride at temperatures below 50° to give small amounts of chlorosulfonyl acetophenone (222).

$$ClO_2SCH_2COCl + \underset{}{\bigcirc} \xrightarrow[-HCl]{AlCl_3} \underset{}{\bigcirc}-\overset{\displaystyle O}{\underset{\displaystyle \|}{C}}-CH_2-SO_2Cl$$

With excess benzene some sulfone is formed. However, only low yields of the sulfone are obtainable, because ω-chlorosulfonylacetophenone tends to lose SO_2 preferentially (312).

B. Unsaturated Halides

a. Alkenyl halides

Böeseken and Bastet (32) found that vinyl chloride and benzene condensed in the presence of a catalyst (32) made from aluminum and mercuric chloride and produced chiefly unsymmetrical diphenylethane and a small quantity of 9,10-dimethyldihydroanthracene.

$$CH_2{=}CHCl + 2C_6H_6 \xrightarrow{AlCl_3} CH_3CH(C_6H_5)_2 + HCl$$

Vinyl bromide and benzene react in the presence of aluminum chloride (Angelbis and Anschütz (6) and Anschütz (8)) to yield the identical products described by Böeseken and Bastet. Hanriot and Guilbert (143) obtained products believed to have been β-bromoethylbenzene and p-di-(β-bromoethyl)benzene in the same reaction; however, their results were never verified.

From the general nature of the Friedel-Crafts reactions styrene or α-chloroethylbenzene would be expected as an intermediate in the reaction

$$C_6H_6 + ClCH{=}CH_2 \rightarrow C_6H_5CH{=}CH_2$$
$$C_6H_6 + ClCH{=}CH_2 \rightarrow C_6H_5CHClCH_3$$

Schramm (275) obtained unsymmetrical diphenylethane from the reaction between styrene, benzene, and aluminum chloride, but Böeseken and Bastet described preferential polymerization of styrene under these conditions. The investigation of Davidson and Lowy (84) also showed that styrene did preferentially polymerize in the presence of benzene and aluminum chloride. Unsymmetrical diphenylethane and an anthracene type resin were the other main products isolated and under special conditions 9,10-dimethyldihydroanthracene was also obtained. These data point to the fact that α-chloroethylbenzene and not styrene is formed in the primary alkylation, thus according to expectations the olefinic double bond and not the vinyl halogen atom reacts primarily.

$$C_6H_6 + CH_2{=}CHCl \xrightarrow{AlCl_3} C_6H_5CHClCH_3 \xrightarrow[C_6H_6]{AlCl_3} (C_6H_5)_2CHCH_3$$

$$2C_6H_5CHClCH_3 \xrightarrow{AlCl_3}$$

The reactions when carried out at 0–5° tended to produce only traces of 9,10-dimethyldihydroanthracene and relatively large amounts of unsymmetrical diphenylethane and resinous substances. Both of the latter substances were found to decrease if the temperature was maintained at 60–70°, and an appreciable amount of 9,10-dimethyldihydroanthracene was formed.

According to the investigations of Malinovskii the condensation of vinyl chloride with benzene in the presence of AlCl₃ (192) in the cold gave chiefly 1,1-diphenylethane, some ethylbenzene and smaller amounts of styrene and dimethyldihydroanthracene, formed by addition of HCl to styrene and further condensation with AlCl₃. A yellow oil and a dark tar are also formed. Heating of the mixture increases the amount of the tar.

Owing to the very high reactivity of α-chloroethylbenzene no direct α-chloroethylation method with vinyl chloride is so far known. ZnCl₂, SnCl₄, and BCl₃ may be the most suitable catalyst to carry out such a reaction.

The action of allyl chloride on benzene in the presence of aluminum chloride results in the formation of 1,2-diphenylpropane (174,288). An intermediate formation of α-chloropropylbenzene and subsequent reaction with the excess of benzene to form the diphenylpropane is assumed. By varying conditions, a fraction is obtained which consists principally of n-propylbenzene (326)

$$C_6H_5CH_2CHClCH_3 \xrightarrow{2H} C_6H_5CH_2CH_2CH_3 + HCl.$$

ZnCl₂ effects a similar reaction with allyl chloride (276).

According to the investigations of Nenitzescu (214,215) aluminum chloride catalyst "poisoned" with water produces a considerable amount of n-propylbenzene in the reaction of benzene with allyl chloride. The main product of the reaction is 1,2-diphenylpropane.

Some 2-chloropropylbenzene and 9,10-diethylanthracene, together with isopropylbenzene, were also isolated.

Nenitzescu established the mechanism of the formation of the isolated products by suggesting that hydride abstraction from the simultaneously formed dihydrodiethylanthracene is responsible for the reduction of 2-chloropropylbenzene to give n-propylbenzene and the corresponding diethylanthracene.

$$C_6H_6 + ClCH_2CH{=}CH_2 \xrightarrow[-HCl]{AlCl_3} [C_6H_5CH_2CH{=}CH_2] \xrightarrow{HCl} C_6H_5CH_2CHClCH_3$$

$$C_6H_5CH_2CHClCH_3 + C_6H_6 \rightarrow C_6H_5CH_2CH(C_6H_5)CH_3 + HCl$$

Phenol is propenylated by allyl chloride or bromide with reasonable yields when zinc chloride is present as catalyst (59).

Using a proton acid catalyst, such as H_2SO_4 (1) or anhydrous HF, $BF_3 \cdot H_2O$, or $HF + TiF_4$ (233), allyl chloride and bromide were found to be selective halopropylating agents.

$$ArH + CH_2{=}CHCH_2Cl(Br) \xrightarrow{H^+} ArCH(CH_3)CH_2Cl \text{ (Br)}$$

Crotyl chloride reacts with benzene in the presence of sulfuric acid to yield $C_6H_5CH(C_2H_5)CH_2Cl$ (331).

Friedel-Crafts alkylations with allyl halides are initiated by Lewis acid metal halides preferentially involving the halogen atom and primary formation of propenylated products. With proton acid catalyst, however, the olefinic double bond reacts preferentially to give chloropropylated compounds. Depending on the activity of the catalyst involved secondary reactions can follow.

Condensations of unsaturated polychlorinated aliphatic hydrocarbons with aromatic hydrocarbons have not been thoroughly studied.

Demole (92) found that a solution of dibromoethylene reacted with benzene and aluminum chloride to give asymmetrical diphenylethylene, $CH_2=C(C_6H_5)_2$. He therefore ascribed the asymmetrical structure, $CH_2=CBr_2$, to his dibromoethylene. The formation of asymmetrical diphenylethylene from vinylidene bromide and benzene was later confirmed by Anschütz (7).

Reaction of tribromoethylene with benzene and aluminum chloride likewise gives asymmetrical diphenylethylene as the main product. However, some triphenylmethane is also obtained. The production of the former was explained by the cleavage of bromine from tribromoethylene, subsequent addition of hydrogen bromide and reaction of the resulting vinylidene bromide with benzene:

$$\begin{array}{ccccccc} CHBr & & CH & & CH_2 & & CH_2 \\ \| & \rightarrow & \| \| & \rightarrow & \| & \rightarrow & \| \\ CBr_2 & & CBr & & CBr_2 & & C(C_6H_5)_2 \end{array}$$

The formation of triphenylmethane was not explained (10).

The reaction of vinylidene chloride with aromatic hydrocarbons or with their halogenated derivatives and aluminum chloride has been claimed to result mainly in the formation of resinous products (72). Although some of these appear to be diarylethylene compounds, others, from their molecular weight, are assumed to be polymers of such compounds. The reaction of trichloroethylene with an aromatic hydrocarbon containing nuclearly substituted halogen, in the presence of aluminum chloride, leads to the formation of higher molecular weight condensation products (24).

When aluminum chloride is gradually added to a solution of tetrachloroethylene, $CCl_2=CCl_2$, in benzene, and heated at $70°$, anthracene is the sole product of the reaction (210).

Schmerling (274) found that aromatic hydrocarbons react with dichloro-olefins in the presence of a Friedel-Crafts catalyst such as aluminum chloride, aluminum bromide, ferric chloride, zirconium chloride, boron fluoride, and the like, complexed by one molecular proportion of an oxygen-containing organic compound such as an alcohol, an alkyl ether, an alkyl ketone, or a nitroparaffin. The aluminum chloride–nitroparaffin complexes promote the addition reaction of 1,3-dichloro-2-methylpropene according to whereas unmodified $AlCl_3$ catalyst causes the reaction to proceed according to

$$\text{[cyclohexene]} + ClCH_2\!-\!\underset{\underset{CH_3}{|}}{C}\!\!=\!\!CHCl \longrightarrow \text{[cyclohexene]}\!-\!CH_2\!-\!\underset{\underset{CH_3}{|}}{C}\!\!=\!\!CHCl + HCl.$$

Alkylation of paraffinic hydrocarbons with allyl chloride was investigated by Schmerling in the case of isobutane (273).

Isobutane is chloropropylated when the reaction is carried out below $-10°$. The major product is 1-chloro-3,4-dimethylpentane; 1,2-dichloro-4,4-dimethylpentane is produced as by-product. At higher temperature further alkylation takes place by the interaction of chloroheptanes with isobutane.

b. Alkynyl halides

Friedel-Crafts alkylations with haloalkynes have been investigated only to a small extent. Colonge and Infarnet (76) found that in the aluminum chloride catalyzed reaction of benzene with propargyl chloride, 1-chloro-5-hexyne, or 1-ethyl-4-chloro-5-pentyne, only in the case of propargyl chloride was a definite product obtained. This was found to be 1,1,2-triphenylpropane in a yield of 10%.

It was suggested that the formation of 1,1,2-triphenylpropane was due to isomerization of propargyl chloride to 1-chloropropadiene and subsequent alkylations with the latter:

$$CH_2\!\!=\!\!C\!\!=\!\!CHCl + 2C_6H_6 \longrightarrow CH_3\!-\!CH(C_6H_5)\!-\!CH(C_6H_5)Cl \xrightarrow[-\,HCl]{C_6H_6}$$
$$\longrightarrow CH_3\!-\!CH(C_6H_5)\!-\!CH(C_6H_5)_2.$$

No selective alkylations with haloalkynes have been reported. However, when propargyl bromide reacts in the presence of sulfuric acid + mercury sulfate catalyst with benzene a small yield of 2,3-diphenyl-3-bromopropane is obtained, thus the triple bond adds to benzene with the C–Br bond being unreactive (222).

c. Alkenoyl halides

Alkenoyl chlorides react with aromatic hydrocarbons in the presence of strong Friedel-Crafts catalysts like aluminum chloride generally to give hydrindones instead of unsaturated ketones. Thus the intermediately formed unsaturated ketone undergoes cyclialkylation.

Acrylyl chloride with benzene in the presence of aluminum chloride was reported by Moureau (211) to give phenyl vinyl ketone.

$$C_6H_6 + ClOCCH\!\!=\!\!CH_2 \rightarrow C_6H_5COCH\!\!=\!\!CH_2 + HCl$$

Upon reinvestigation of the work of Moureu, Kohler (171) found that the reaction product was not the phenyl vinyl ketone claimed but 1-hydrindone. This is doubtless formed from phenyl vinyl ketone since the latter is easily changed into 1-hydrindone upon treatment with aluminum chloride. However, Kohler obtained phenyl propenyl ketone by reacting crotonyl chloride with benzene in the presence of aluminum chloride. von Auwers (319) also obtained unsaturated ketones by the reaction of crotonyl chloride with aromatics.

Milder Friedel-Crafts catalysts such as zinc and stannic chloride were used to advantage in unsaturated ketone synthesis with acrylyl chloride and crotonyl chloride. Hydrindone formation is suppressed with these catalysts (222).

Crotonyl chloride ($CH_3CH=CHCOCl$) reacts with alkylbenzenes to give crotophenones with fair yields (207).

$$ArH + ClOCCH=CHCH_3 \xrightarrow{AlCl_3} ArCOCH=CHCH_3$$

α-Chlorocrotonyl chloride gives α-chlorocrotonophenone, thus the acyl halide halogen selectively reacts over the vinylic halogen and the olefinic double bond.

$$ArH + ClOCCCl=CHCH_3 \rightarrow Ar\overset{\overset{\displaystyle O}{\|}}{C}CCl=CHCH_3$$

With the usual strong Friedel-Crafts catalysts, as $AlCl_3$, α,β-unsaturated acyl halides will react with benzene and derivatives to yield 1-indanones (60). Granger and co-workers (130) found that the cyclization of benzene with 2-ethylcrotonyl chloride gave 3-methyl-2-ethylindanone, which had the less hindered *trans*-configuration.

d. Aralkenoyl halides

Aralkenoyl chlorides generally react as bifunctional compounds not giving simple acylation products in reactions with benzene and aluminum chloride. Kohler and co-workers (172) obtained a hydrindone and β,β-diphenylpropiophenone from cinnamoyl chloride and benzene. They were not able to secure styryl phenyl ketone. These results have been substantiated by von Auwers and Risse (319).

With bromobenzene and cinnamoyl chloride, however, some normal substitution does occur. Benzal-p-bromoacetophenone, together with 6-bromo-1-phenyl-3-hydrindone, is obtained (172). Mesitylene with cinnamoyl chloride likewise gives cinnamoyl mesitylene (170).

Normal acylation products, affecting only the acyl halide and not the olefinic double bond, were reported from cinnamoyl halides using $AlCl_3$ and $ZnCl_2$ as catalysts (96,173,280).

Formation of chalcone or hydrindone depends upon the presence of substituent groups on the benzene ring, so placed that they inhibit ring closure of the chalcone primarily formed. Ring closure of chalcone to hydrindone is prevented, or at least is made difficult, if there is a methyl group in the *meta*-position to the hydrogen atom in the benzene nucleus which is to be displaced upon ring closure. This is brought out in the case of p-xylene:

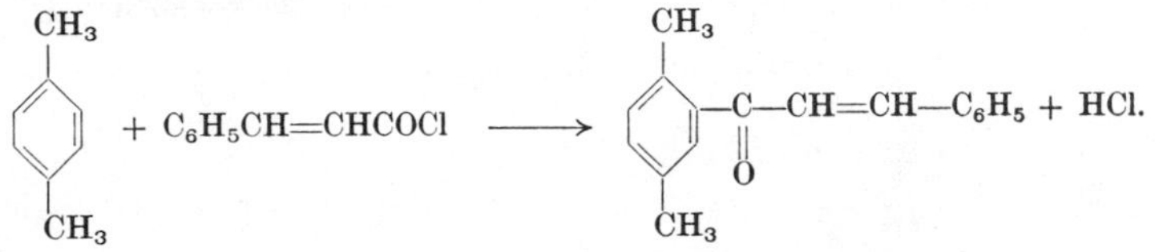

α-Methylcinnamoyl chloride and α-ethylcinnamoyl chloride were found to give mainly the normal acylation products (125,304)

$$ArH + ClOCC(R){=}CHC_6H_5 \rightarrow Ar{-}COCR{=}CHC_6H_5, \quad R{=}CH_3, C_2H_5.$$

e. Unsaturated dicarboxylic acid chlorides

The reaction of unsaturated dicarboxylic acid chlorides or their monoesters with aromatic compounds has received much attention with particular reference to possible *cis–trans* isomerization during the Friedel-Crafts reaction.

Fumaryl chloride reacts with benzene (or its substituted derivatives) in the presence of aluminum chloride to give *trans*-dibenzoylethylenes.

$$ClCOCH{=}CHCOCl + 2C_6H_6 \xrightarrow{AlCl_3} C_6H_5COCH{-}CHCOC_6H_5 + 2HCl$$

Thus both acyl halide groups react in the usual way without interference from the double bond (77,187,219).

The Friedel-Crafts reaction with the acyl chlorides of citraconic (methylmaleic) acid and mesaconic (methylfurmaric) acid or their monoesters has been investigated (186,188,189,190) in order to ascertain the influence of the methyl group on the course of the reaction. The reaction is found to proceed more readily with the *trans*-isomer than with the *cis*-compound. In the case of mesaconyl chloride, or its methyl or ethyl ester, the reaction largely involves the carbonyl chloride which is furthest removed from the methyl group.

$$
\begin{array}{ccc}
\mathrm{CH_3-C-COCl} & \xrightarrow[\mathrm{C_6H_6}]{\mathrm{AlCl_3}} & \mathrm{CH_3-C-COOH} \\
\|\| & & \|\| \\
\mathrm{ClCO-C-H} & & \mathrm{C_6H_5CO-C-H}
\end{array}
$$

With citraconyl chloride, the carbonyl chloride adjacent to the methyl group is mainly involved.

$$
\begin{array}{ccc}
\mathrm{H-C-COCl} & \xrightarrow[\mathrm{C_6H_6}]{\mathrm{AlCl_3}} & \mathrm{H-C-COOH} \\
\|\| & & \|\| \\
\mathrm{CH_3C-COCl} & & \mathrm{CH_3-C-COC_6H_5}
\end{array}
$$

Since the yield in this case is very small the tendency to react in the *cis*-form is not great.

The governing factor in inversion during the reaction has been shown to be the position of the methyl group.

f. Unsaturated sulfonyl halides

Unsaturated sulfonyl halides show a resemblance in their reactivity to unsaturated acyl halides and it is possible to obtain in a normal sulfonylation reaction the corresponding unsaturated sulfones (311).

$$
\mathrm{C_6H_6 + C_6H_5CH=CHSO_2Cl} \xrightarrow{\mathrm{AlCl_3}} \mathrm{C_6H_5CH=CHSO_2C_6H_5}
$$

$$
\mathrm{CH_3C_6H_5 + C_6H_5CH=CHSO_2Cl} \xrightarrow{\mathrm{AlCl_3}} \mathrm{C_6H_5CH=CHSO_2C_6H_4CH_3}
$$

C. Functional Carboxylic Acids

a. Unsaturated carboxylic acids

Unsaturated carboxylic acids react with aromatic hydrocarbons in the presence of aluminum chloride or related Friedel-Crafts catalysts generally by simply adding the aromatic to the double bond and saturated aralkyl carboxylic acids are formed. Thus the reactivity of the olefinic double bond is considerably higher than that of the carboxylic group and alkylation almost exclusive of acylation is achieved.

Early investigation of addition of aromatic compounds to unsaturated carboxylic acids in the presence of acid catalysts was mostly concerned with long-chain aliphatic compounds. In a review of this work Thomas (308) observes that except in the reactions with oleic acid the reports are confused and often contradictory. In some cases identification of products was not satisfactory and furthermore experimental conditions were often not recorded. Eijkman (105) found that addition of the aryl groups always takes place in the case of α,β-unsaturated acids at the β-carbon atom, except in the case of α-phenylacrylic acids, where α-addition was observed. Dippy (102) investigated in some detail the reaction of aromatic compounds with substituted acrylic acids. He substantiated the β-addition pattern of the aromatic in the case of a number of different substituted acrylic acids and also observed that hydrogen transfer from the solvent could lead to hydrogenated saturated acids. No acylation reactions were observed in any of the cases investigated. Dippy also investigated the addition of benzene, toluene, and a number of other substituted benzenes to cinnamic acid in the presence of aluminum chloride catalyst (102). Again mostly β-addition of the aromatics was observed leading to substantial yields of β,β-diphenylpropionic acids, with the substituent mostly in the *para*-position. No competing acylation reactions were reported.

The use of sulfuric acid as catalyst results in similar β-addition of aromatics to unsaturated acids, which, however, could be followed by cycliacylations leading to indanones.

Using hydrogen fluoride as catalyst the acylation of aromatics with α,β-unsaturated acids results in a one-step synthesis of 1-indanones (113,144). An excess (three equivalents) of aluminum chloride has also been used for this reaction (169).

Acylation of reactive aromatic hydrocarbons (such as thiophene, phenol, and phenol ethers) with unsaturated acids such as cinnamic, acrylic, or methacrylic acids to the corresponding unsaturated ketones can be achieved by the use of polyphosphoric acid as catalyst. Stronger acid catalyst or less reactive aromatics, necessitating higher reaction temperatures, generally cause secondary reactions involving cyclialkylation with the olefinic double bond (97,296).

$$ArH + CH_3CH{=}CHCO_2H \xrightarrow{\ PPA\ } ArCOCH{=}CHCH_3$$

$$C_6H_5OCH_3 + C_6H_5CH{=}CHCO_2H \xrightarrow{\ PPA\ } CH_3OC_6H_4{-}\underset{\underset{O}{\|}}{C}{-}CH{=}CHC_6H_5$$

b. Hydroxy acids

Little is known on the hydroxyl–carboxyl selectivity in Friedel-Crafts type reaction of hydroxycarboxylic acids. In the case of a tertiary hydroxyl group it was reported that exclusive alkylation takes place. Benzylic acid reacts in the presence of stannic chloride with benzene to give a substantial yield of triphenylacetic acid (28).

$$(C_6H_5)_2C\text{—}COOH + C_6H_6 \xrightarrow[-H_2O]{SnCl_4} (C_6H_5)_3CCOOH$$
$$\underset{OH}{|}$$

Glycolic acid condenses in the presence of sulfuric acid with aromatics giving a condensation product with tanning action, the structure of which, however, has not been described (35).

c. Halogenated acids

Little is known about Friedel-Crafts reactions with halogenated carboxylic acids. However, the few available data point to the fact that alkylation practically exclusive of acylation takes place. Diphenylchloroacetic acid and benzene in the presence of aluminum chloride give triphenylacetic acid (29).

$$(C_6H_5)_2C\text{—}CO_2H + C_6H_6 \xrightarrow{AlCl_3} (C_6H_5)_3CCO_2H + HCl$$
$$\underset{Cl}{|}$$

In this case, of course, the halogen is tertiary and therefore very reactive. In high-temperature reactions it has been reported that mono- and dichloroacetic acid as well as their esters react with reactive aromatic hydrocarbons (at temperatures of 100–275°) to give the corresponding arylacetic acids in the absence of any added catalyst (328).

Naphthalene, acenaphthene, anthracene, and fluorene are converted with chloroacetic acid to the corresponding arylacetic acids, while dichloroacetic acid under identical conditions yields dinaphthylacetic acid. Although the reactions are carried out in the absence of catalysts, as pointed out (Chapter IV) at elevated temperatures, non-catalytic reactions of halide reagents may be considered as catalyzed by hydrochloric acid, or in the present case by the rather strong haloacetic acids themselves.

D. Functional Alcohols

a. Halohydrins

Reactivity differences in the hydroxyl group and halogen atoms are generally sufficiently large to allow selective reaction of the hydroxyl group without simultaneous reaction of the halogen atoms. Aluminum halides are too strong catalysts and cause reaction of both the hydroxyl group and halogens.   Zinc chloride, boron trifluoride, stannic chloride, and other related catalysts are preferred to carry out selective reactions.

In the course of the chloromethylation reaction with CH_2O and HCl chloromethanol ($ClCH_2OH$) was suggested as an intermediate. However, no direct proof for the existence of the compound has been obtained and it has never been isolated.

Fluoromethanol (FCH_2OH) was obtained at low temperatures by Olah and Pavlath (235) and was found to react in the presence of zinc chloride as catalyst with benzene and fluorobenzene to give the corresponding fluoromethyl derivatives.

$$ArH + HOCH_2F \xrightarrow[-H_2O]{ZnCl_2} ArCH_2F$$

Diphenylmethanes, formed by the secondary interaction of fluoromethylbenzenes with excess aromatics are, however, the major products.

Fluoromethylation using formaldehyde and anhydrous HF generally yields only diphenylmethanes and higher molecular weight polycondensed products owing to the strong catalytic effect of HF on secondary alkylation with the very reactive benzyl fluorides.   Nuclear magnetic resonance investigation of the formaldehyde–hydrogen fluoride system gave evidence that monomeric FCH_2OH is not an intermediate when anhydrous HF is the solvent but a polymeric fluorohydrin, $HO-(CH_2O)_n-CH_2F$, is present which reacts similarly to fluoromethanol.   It seems probable that a similar chlorohydrin is indeed the intermediate in chloromethylations using paraformaldehyde and hydrogen chloride.   A number of highly fluorinated α-fluoroalcohols were recently prepared by Andreades (4,5), but their use in haloalkylations has not yet been reported.

Attempts to use glycol chlorohydrin in chloroethylation reactions,

$$ArH + HOCH_2CH_2Cl \rightarrow ArCH_2CH_2Cl,$$

have been unsuccessful so far.

The reaction of β-chloroethyl alcohol (glycol chlorohydrin) with benzene and aluminum chloride at 100° gave a 40% yield of bibenzyl (sym-diphenylethane)

$$2C_6H_6 + ClCH_2CH_2OH \xrightarrow{AlCl_3} C_6H_5CH_2CH_2C_6H_5 + HCl + H_2O.$$

Except for resinous by-product, bibenzyl was the only product isolated.   No β-chloroethylbenzene was obtained nor was there any phenylethylalcohol.   Both β-chloroethylbenzene and β-chloroethyl-alcohol are known, however, to give bibenzyl when they react with benzene in the presence of aluminum chloride.   Thus the reaction obviously takes place in two steps (155).

Bachman and Hellman (14) carried out halopropylation of benzene, toluene, isopropylbenzene, and their derivatives with propylene chlorohydrin $CH_3CH(OH)CH_2Cl$.   The reaction proceeds exclusively at the expense of the secondary alcohol group with the formation of haloalkylbenzene.   In addition, monoalkylbenzenes react more easily than benzene, and form chiefly p-haloalkyl sub-·stituted derivatives.   Free BF_3 appears to be sufficiently active as a catalyst, but the addition of dehydrating agents (H_2SO_4 or P_2O_5) increases the yields

$$C_6H_6 + CH_3\underset{\underset{OH}{|}}{CH}CH_2Cl \xrightarrow[-H_2O]{BF_3} C_6H_5CH(CH_3)CH_2Cl.$$

By dehydrochlorinating the chloropropylated compounds obtained α-methylstyrenes were prepared

$$C_6H_5CH(CH_3)CH_2Cl \xrightarrow{-HCl} C_6H_5\underset{\underset{}{}}{\overset{\overset{CH_3}{|}}{C}}{=}CH_2.$$

b. Unsaturated alcohols

The olefinic double bond activates the hydroxyl group of unsaturated primary aliphatic alcohols, like that of allyl alcohol, to a certain degree.   Huston and Sager (151) found that allylbenzene was formed in 16% yield in the aluminum chloride catalyzed reaction of allyl alcohol and benzene.

The yield of allylbenzene is low owing to side reactions involving the double bond, but the reaction takes place more readily than that of saturated aliphatic primary alcohols under similar conditions.

$$C_6H_6 + CH_2{=}CHCH_2OH \rightarrow C_6H_5CH_2CH{=}CH_2$$

If proton acid catalysts (H_2SO_4 or H_3PO_4) are used, the olefinic double bond of allyl alcohols reacts preferentially, but the intermediate aralkyl alcohols tend to lose water with the strong dehydrating acids and isopropenyl derivatives are formed.   These easily undergo secondary reactions, such as polymerizations or cyclialkylations. The reaction was investigated in the case of phenol, where in addition to the primary alkylation and dehydration

$$HO-\langle\bigcirc\rangle + CH_2=CH-CH_2-OH \xrightarrow[H_3PO_4]{H_2SO_4} \left[HO-\langle\bigcirc\rangle-\underset{\underset{CH_3}{|}}{CH}-CH_2-OH \right]$$

$$\xrightarrow{-H_2O} HO-\langle\bigcirc\rangle-C(CH_3)=CH_2$$

the product contained 2-methylcoumarone, chroman and a large portion of polymeric resins (184).

c. Diols

Alkylation of aromatics with diols or their derivatives generally leads only to cyclialkylated products, both functional groups taking part in the reaction. No selective reactions leading to hydroxy-alkylated products have been reported.

Ethylene glycol gives some 1,2-diphenylethane with benzene

$$2C_6H_6 + HOCH_2CH_2OH \xrightarrow{AlCl_3} C_6H_5CH_2CH_2C_6H_5$$

in addition to dioxane and polyglycols (self-condensation products).

E. Functional Substituted Aromatics

a. Aralkenes

Friedel-Crafts substitutions of aralkenes in various alkylation and acylation reactions involve the two-fold reactivity of the substrate: aromatic ring and olefinic side chain.

If the olefinic double bond is not conjugated with the aromatic ring, as in allylbenzene or aralkenes with longer side chains and double bonds isolated from the ring, the reactivity of the aromatic ring is predominant and Friedel-Crafts substitutions (alkylation, acylation, nitration, etc.) can be effected on the ring without too much interference from the olefinic side chain.

If the olefinic double bond is conjugated with the aromatic ring, as in the case of styrene, its reactivity is considerably enhanced and usual Friedel-Crafts reaction conditions result in predominant polymerization of the substrate.

Friedel-Crafts substitution of styrene deserves specific attention. The conjugated vinylic double bond shows considerably higher reactivity than the aromatic ring and consequently even in the case when reactions exclusive of the polymerization of styrene could be carried out, these are predominantly side-chain reactions and not ring substitutions.

In the presence of aluminum chloride in toluene solution styrene reacts with phosgene to give β-chlorophenylpropionyl chloride (240).

$$C_6H_5CH{=}CH_2 + COCl_2 \xrightarrow{AlCl_3} C_6H_5CHClCH_2COCl$$

Acetyl chloride and malonyl chloride in the presence of stannic chloride as catalyst add to the olefinic double bond of styrene and give chloroketones which on heating with diethylaniline can be dehydrochlorinated to the corresponding unsaturated ketones (177, 318).

$$C_6H_5CH{=}CH + CH_3COCl \xrightarrow{SnCl_4} C_6H_5CHClCH_2COCH_3$$

$$\xrightarrow{C_6H_5N(C_2H_5)_2} C_6H_5CH{=}CHCOCH_3$$

$$2C_6H_5CH{=}CH_2 + CH_2(COCl)_2 \xrightarrow[CS_2]{SnCl_4} C_6H_5CHClCH_2COCH_2COCH_2CHClC_6H_6$$

$$\xrightarrow{C_6H_5N(C_2H_5)_2} C_6H_5CH{=}CHCOCH_2COCH{=}CHC_6H_5$$

With such longer chain acyl chlorides as stearoyl, lauroyl, oleyl, linoleoyl, and linolenoyl in the presence of aluminum, however, the nucleus is acylated without effecting the vinyl group (257,258).

b. Haloaralkanes

Friedel-Crafts substitution of haloaralkanes can be considered as an example of selectivity affecting the aromatic nucleus and the aliphatic side-chain halogen. If the halogen is primary or located at carbon atoms not adjacent to the ring usual Friedel-Crafts ring substitutions can generally be effected without much interference by side reactions caused by the halogen, if mild catalysts and conditions are applied. Even benzyl chloride can be ring-substituted with fair yields. Side reactions become more predominant with stronger catalysts allowing alkylation or cyclialkylation.

When the halogen of aralkanes is secondary or tertiary and located in the α-position to the aromatic ring, Friedel-Crafts type reactions tend to take an unusual course.

Attempted alkylation of α-haloaralkanes with olefins in the presence of usually strong Friedel-Crafts catalysts, such as $AlCl_3$, $AlBr_3$, $FeCl_3$, $ZrCl_4$, BF_3, H_2SO_4, HF, etc., is unsuccessful. Owing to the high reactivity of the α-halogen atoms polycondensations and dehydrohalogenation, followed by fast polymerization, result in complex reaction products of high molecular weight. When milder catalysts, as $ZnCl_2$ or BCl_3 are used, alkylation exclusive of unwanted side reactions can be achieved. However, under these reaction conditions ring alkylation does not take place, but the α-haloaralkane adds to the olefinic double bond.

$$C_6H_5CHClCH_3 + (CH_3)_2C=CH_2 \xrightarrow{ZnCl_2} C_6H_5CHCH_2CCl(CH_3)_2 \longrightarrow$$
$$\underset{CH_3}{|}$$

$$\xrightarrow{-HCl} C_6H_5CHCH=C(CH_3)_2$$
$$\underset{CH_3}{|}$$

Dehydrochlorination of the intermediately formed haloaralkane (either thermally or by treatment with bases like pyridine or dialkyl-anilines) gives the corresponding α-aralkylated olefin (226). The reaction is of general use for a variety of α-haloaralkanes and olefins, starting with propene.

In the aromatic ring–side-chain halogen reactivity α-halogens, if secondary or tertiary, are thus sufficiently reactive for alkylation of the olefin by the haloaralkane to take place in preference to ring alkylation by the olefin. If the halogen is primary, even in the α-position (as in benzyl halides) or located in the non-conjugated position with the aromatic ring, Friedel-Crafts reactions generally take place by usual ring substitution.

Not only alkylation but also acylation reactions proceed with haloaralkanes in the manner previously discussed.

α-Chloroethylbenzene for example reacts with acetic anhydride in the presence of zinc chloride as catalyst by adding to the acid an-hydride to form α-phenylethyl acetate and no ring acylation according to the Friedel-Crafts ketone synthesis is observed (222)

$$C_6H_5CHClCH_3 + (CH_3CO)_2O \xrightarrow{ZnCl_2} C_6H_5CH(CH_3)OOCCH_3 + CH_3COCl.$$

β-Chloroethylbenzene, however, is ring acylated in the Friedel-Crafts ketone synthesis in the usual manner.

2. Substrate and Positional Selectivity of Aromatics

So far in our discussion we have treated the question of selectivity of different functional reagent groups in Friedel-Crafts reaction systems. There is, however, the important question of the kind of selectivity obtained on the substrate to be substituted. In the case of substituted aromatic compounds besides the substrate selectivity the question of positional selectivity regarding the entering electro-philic substituent is of considerable importance.

Substrate and positional selectivities are of greatest interest in the case of substitution of aromatic hydrocarbons and therefore we will confine our discussion to these systems.

In the development of the theory of aromatic substitution, attention has been directed primarily toward the aromatic component. Thus, orientation and relative rates have been interpreted qualitatively in terms of various electrical and steric effects of the aromatic substrate (153).

In general the effect of the activity of the substituting agent upon isomer distribution has been largely ignored. In a few cases where steric and polar explanations fail, as in diazonium coupling (179), such activity has been invoked occasionally to account for *ortho*: *para* ratios (89).

A. Correlation of Isomer Distribution and Substrate Reactivity

A detailed discussion of the importance of the activity of the attacking species in controlling the isomer distribution has been presented by H. C. Brown (48,50). Thus, it has been shown that in numerous reactions of aromatic nuclei, including chlorination, chloromethylation, nitration, mercuration, methylation, ethylation, and isopropylation of toluene, the orientation can be correlated with the "activity" or "selectivity" of the attacking species.

Generally the greater the reactivity of an electrophilic substituting agent, the smaller its selectivity (99). This means low selectivity with different aromatics and also a simultaneous change of the isomer distribution toward the statistical value (40% *ortho*, 40% *meta*, and 20% *para*, representing two *ortho*-positions, two *meta*-positions, and one *para*-position in a monosubstituted benzene). This generally is demonstrated by an increase in the concentration of the *meta*-isomer. The obvious explanation is the decreasing role of small activation energy differences of different individual positions compared to the overall activation energy of the reaction.

Orientations in Friedel-Crafts alkylations are frequently considered to be anomalous. Much of the early literature is confusing and misleading (119,256). However, it is now established that alkylation of aromatics with aluminum chloride and alkyl halides leads to the formation of considerable quantities of the *m*-dialkyl isomer (80,281, 291). Moreover, the exclusive formation of the *meta*-isomer in the presence of molar quantities of the catalyst has been reported (217). The possibility of forming larger yields of the *meta*-isomer than those given by the thermodynamic equilibrium value has been questioned (249), but has been amply verified (203). It was proposed (203) that the large yields of the *meta*-isomer arise from isomerization (15) of the *ortho*- and *para*-derivatives to the more stable (201) σ-complexes (40) formed by the *m*-dialkylbenzenes.

This explanation nicely accounts for the high yields of *meta*-isomer obtained under isomerizing conditions. It does not account for the formation of large amounts of the *meta*-isomer under conditions where it was claimed that isomerization is not an important factor.

B. Linear Free Energy Selectivity Relationship and its Application

H. C. Brown proposed that the high yield of the *meta*-isomer in the isopropylation of toluene and similar monoalkylbenzenes is the result of the high reactivity and resultant low selectivity of the dimethyl carbonium ion. Available data support the position that the relative yields of the *meta*-isomer in substitutions of toluene vary in a regular and predictable manner with the "activity" of the attacking agent. In a sense, then, the high yield of the *meta*-isomer is not an anomalous property which is characteristic only of Friedel-Crafts substitutions.

It has been demonstrated (80,299) in the case of many examples investigated that the reactions obey the relationship

$$\log p_\mathrm{f} = c \log (p_\mathrm{f}/m_\mathrm{f})$$

where p_f and m_f are the partial rate factors for substitution in the *para*- and *meta*-positions of toluene and of other monosubstituted benzenes. This expression was developed and tested as an empirical relationship, which became widely known as Brown's *Selectivity Relationship*.

a. Relationship between activity and meta-substitution

Examination by Brown of available data on isomer distribution in the substitutions of toluene showed that there is no sharp division between reactions which give a high proportion of *meta*-substitution and those which give little or no *meta*-substitution. Chlorination, chloromethylation, and acylation are reported to give negligible amounts of the *meta*-isomer, whereas mercuration and isopropylation result in the formation of relatively large amounts, with nitration and sulfonation giving intermediate values (Table VI).

Based on data obtained it was proposed by Brown that the importance of *meta*-substitution is related to the "activity" of the attacking species. As a convenient measure of this "activity" the relative reactivity of toluene and benzene toward the substitution reaction under consideration was utilized. According to this pro-

TABLE VI. Amount of *meta*-substitution in electrotrophilic substitutions of toluene

Reaction	Conditions	% *meta*-Isomer	Ref.
Acetylation	$CH_3COCl + AlCl_3$	0	241
Bromination	$Br_2 + $ cat.	Small	70,148,314
Chlorination	Cl_2 in HOAc at $24°$	0.5	78
Chloromethylation	$(CH_2O) + HCl$, $ZnCl_2$	Small	146
Nitration	HNO_3 in H_2SO_4 at $30°$	4.4	150,158
	$AcONO_2$ in Ac_2O at $0°$	3.7	154
	$AcONO_2$ in Ac_2O at $30°$	4.4	154
Sulfonation	100% H_2SO_4 at $35°$	6.2	149
	SO_3(gas) at 40–$55°$	9	181
Methanesulfonylation	$MeSO_2Cl + AlCl_3$ at $100°$	15	320
Mercuration	$Hg(ClO_4)_2$ in aq. $HClO_4$ at $25°$	7	165
	$Hg(ClO_4)_2$ in aq. $HClO_4$ at $85°$	13	165
	$Hg(OAc)_2$ at $110°$	13	18,69
	$Hg(OAc)_2$ at $110°$	21	165
Isopropylation	$C_3H_6 + AlCl_3$–$MeNO_2$ or BF_3–Et_2O at $5°$	29.8	80
	$C_3H_6 + AlCl_3$–$MeNO_2$ or BF_3–Et_2O at $65°$	27.5	80

posal, bromination involves a reaction of low activity. It is both highly selective between toluene and benzene (reactivity ratio of toluene/benzene = 467) and between the *meta*- and *para*-positions of toluene (no *meta* reported). Nitration involves an intermediate of moderate activity, the nitronium ion, NO_2^+, and it is only moderately selective between toluene and benzene (toluene/benzene = 23) and between the *meta*- and *para*-positions of toluene (4.4% *meta*). Finally, isopropylation involves a highly active intermediate, presumably the dimethyl carbonium ion, $(CH_3)_2CH^+$, which gives only slight selectivity between the two aromatics (toluene/benzene = 2.1) and only slight selectivity between the *meta*- and *para*-positions (29% *meta*). The available experimental data, according to Brown, are summarized in Table VII.

The parallel between the amount of *meta*-substitution and the relative reactivity of toluene and benzene suggested a definite correlation. A plot of the log of the relative toluene/benzene reactivities versus the log of the toluene *para/meta* ratios indicated a simple linear relationship between these quantities (50). In order to eliminate possible complications from steric factors, the partial rate factors for substitution were calculated and the log of the partial

TABLE VII.   Relative rate of substitution of benzene and toluene

Reaction	Conditions	Reactivity ratio $(k_{\text{toluene}}/k_{\text{benzene}})$	Ref.
Bromination	Br_2 in HOAc with I_2 at 25°	467	26
	Br_2 in HOAc with I_2 at 45°	272	26
Chlorination	Cl_2 in HOAc at 24°	353	91
Nitration	$AcONO_2$ in Ac_2O at 0°	27	301
	$AcONO_2$ in Ac_2O at 30°	23	301
Acetylation	AcCl with $AlCl_3$ in $C_2H_4Cl_2$ at 25°	128	48a
Sulfonation	H_2SO_4 in $C_6H_5NO_2$ at 40°	5.1	302
Mercuration	$Hg(OAc)_2$ in HOAc, $HClO_4$, at 50°	5.0	48
Chloromethylation	$CH_2O + HCl + ZnCl_2$ in HOAc at 60°	112	50
Isopropylation	C_3H_6 with $AlCl_3$ in CH_3NO_2 at 40°	2.1	79

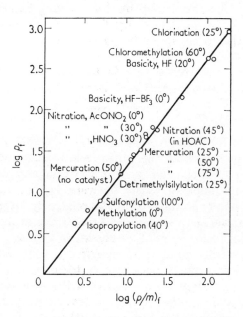

Fig. 1.   Linear relationship between the electrophilic reagents and the observed selectivity in their reactions with toluene (after Nelson (212a)).

rate factor for the *para*-position of toluene was compared with the log of the ratio of the partial rate factors for the *para*- and *meta*-positions of toluene (Fig. 1).

Data for some forty-seven electrophilic substitution reactions of toluene were compiled by Brown. These data were correlated by the Selectivity Relationship with quite satisfactory precision (299) (Table VIII).

In subsequent work H. C. Brown and co-workers extended the linear free energy correlation (Selectivity Relationship) to substitution reactions of t-butylbenzene (37), anisole (300), and to those of biphenyl and fluorene (301).

Discrepancies observed in the case of substitution of biphenyl were explained as being due to steric reasons and the aplanarity of the substrate molecule as compared with the planar fluorene molecule which obeys the relationship.

b. Selectivity Factor

It has been pointed out that a linear relationship exists between the "activity" of a substituting reaction, as measured by log p_f, and its "selectivity" as measured by log p_f/m_f. Brown found it is convenient to define the Selectivity Factor, S_f, as equal to the log of the ratio of the *para* to *meta* partial rate factors for toluene $S_f = \log (p_f/m_f)$.

In cases where the partial rate factors are not known, S_f can be calculated from the *meta*- and *para*-isomer distribution (51)

$$S_f = \log (2 \times \% \; para/\% \; meta)$$

c. Calculation of partial rate factors from the Selectivity Factor

The alkylation data fit the linear relationship which has been shown to exist between log p_f and the quantity S_f (Fig. 2). A similar linear relationship exists between log m_f and S_f (Fig. 3). The scatter of points is well within the relatively large experimental uncertainties in the analyses for small quantities of the *meta*-isomers.

A similar plot of log o_f vs. S_f is shown in Fig. 4. It is apparent that a considerable number of reactions, such as chlorination, basicity with hydrogen fluoride–boron trifluoride, nitration, detrimethylsilylation, and methylation, follow the proposed relationship. However, a number of other reactions deviate from the line. These deviations are attributed to steric and other effects.

From a consideration of Figs. 2 and 4, it is apparent that the slopes of the lines are quite similar. Presumably this arises from the similarity in resonance effects in the *ortho*- and *para*-positions. In the absence of steric effects, the following relationship holds.

$$\log p_f = 1.08 \log o_f$$

It is apparent from this relationship that resonance effects are slightly greater in the *para*-position than in the *ortho*-position. Thus the equation is merely a quantitative expression of a phenomenon which has been noted frequently.

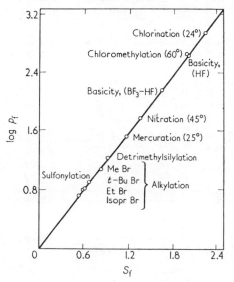

Fig. 2. Relationship between log p_f and S_f, the Selectivity Factor, according to Brown and Smoot (51).

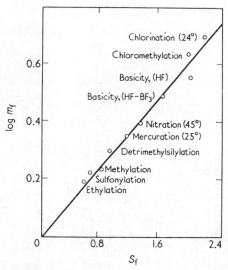

Fig. 3. Relationship between log m_f and S_f, the Selectivity Factor, according to Brown and Smoot (51).

TABLE VIII. Electrophilic substitution reactions of toluene, according to Stock and Brown (299)

Reaction	o_f	m_f	p_f	S_f	$\dfrac{\log p_f}{\log m_f}$	$-\rho$	$\Delta\rho$
Deuterium exchange, DBr, 25°	1000	6.0	6000	3.000	4.86	12.39	1.10
Bromination, Br$_2$, 85% HOAc, 25°	600	5.5	2420	2.644	4.57	11.40	0.68
Chlorination, Cl$_2$, HOAc, 24°	600	5.0	870	2.241	4.21	10.32	0.18
Chlorination, Cl$_2$, HOAc, 25°	617	4.95	820	2.219	4.19	10.24	0.17
Bromination, Br$_2$, HOAc, 25°		4.72	534	2.053	4.05	9.66	0.08
Benzoylation, C$_6$H$_5$COCl, AlCl$_3$, C$_6$H$_5$NO$_2$, 25°	32.6	5.0	831	2.221	4.18	10.28	0.15
Benzoylation, C$_6$H$_5$COCl, AlCl$_3$, 25°	30.7	4.8	589	2.090	4.07	9.88	0.01
Benzoylation, C$_6$H$_5$COCl, AlCl$_3$, C$_2$H$_4$Cl$_2$, 25°	32.6	4.9	626	2.107	4.05	10.00	0.01
Acetylation, CH$_3$COCl, AlCl$_3$, C$_2$H$_4$Cl$_2$, 25°	4.5	4.8	749	2.192	4.22	10.07	0.19
Basicity, HF, 20°	145	3.6	414	2.061	4.71	8.71	0.64
Deuteration, D$_2$O, CF$_3$CO$_2$H, 70°	253	3.8	421	2.044	4.52	8.89	0.48
Deuteration, CF$_3$CO$_2$H, 25°	263	3.57	350	1.992	4.61	8.54	0.54
Chloromethylation, CH$_2$O, HCl, ZnCl$_2$, HOAc, 60°	117	4.37	430	1.993	4.11	9.35	0.06
Basicity, HF, BF$_3$, n-C$_7$H$_{16}$, 20°	103	3.1	145	1.670	4.40	7.42	0.30
Iodination, ICl, ZnCl$_2$, HOAc, 25°	144	6.58	230	1.544	2.89	10.14	1.71
Deuteration, 68%, H$_2$SO$_4$, 25°	83	1.9	83	1.640	6.88	5.45	1.40
Tritiation, 80.8%, H$_2$SO$_4$, 25°	70	2.2	63	1.457	5.26	5.69	0.73
Bromination, HOBr, HClO$_4$, 50% dioxane, 25°	76	2.5	59	1.373	4.45	6.05	0.27
Chlorination, HOCl, HClO$_4$, H$_2$O, 25°	134	4.0	82	1.311	3.18	7.77	0.94
Brominolysis, ArB(OH)$_2$, Br$_2$, 20% HOAc, 25°		3.3	78.5	1.373	3.63	7.17	0.40
Nitration, HNO$_3$, 90% HOAc, 45°	42	2.5	58	1.366	4.43	6.04	0.26
Nitration, AcONO$_2$, Ac$_2$O, 0°	47.3	3.56	60.3	1.229	3.23	7.05	0.95
Nitration, AcONO$_2$, Ac$_2$O, 30°	40.4	3.04	51.2	1.226	3.54	6.47	0.53
Nitration, HNO$_3$, CH$_3$NO$_2$, 30°	36.6	2.33	46.1	1.297	4.53	5.63	0.31
Sulfonylation, C$_6$H$_5$SO$_2$Cl, AlCl$_3$, 25°	6.8	2.09	30.2	1.160	4.63	4.97	0.32
Detrimethylsilylation, ArSiMe$_3$, Br$_2$, 98.5% HOAc, 25°		3.2	49	1.185	3.35	6.68	0.65
Mercuration, Hg(OAc)$_2$, HClO$_4$, HOAc, 25°	4.98	2.25	32.9	1.165	4.31	5.26	0.16
Mercuration, Hg(OAc)$_2$, HClO$_4$, HOAc, 50°	4.20	2.41	28.8	1.077	3.82	5.38	0.17

Reaction							
Mercuration, Hg(OAc)$_2$, HClO$_4$, HOAc, 70°	3.24	2.23	24.5	1.041	3.99	5.00	0.04
Detrimethylsilylation, ArSiMe$_3$, HClO$_4$, 50% MeOH, 51.2°		2.3	21.2	0.964	3.66	5.00	0.26
Detrimethylsilylation, ArSiMe$_3$, HCl, HOAc, 25°		2.14	20.1	0.973	3.95	4.72	0.07
Mercuration, Hg(OAc)$_2$, HOAc, 25°	5.71	2.23	23.0	1.014	3.91	4.95	0.09
Mercuration, Hg(OAc)$_2$, HOAc, 50°	4.60	1.98	16.8	0.928	4.12	4.17	0.13
Mercuration, Hg(OAc)$_2$, HOAc, 70°	4.03	1.83	13.5	0.867	4.30	3.78	0.03
Mercuration, Hg(OAc)$_2$, HOAc, 90°	3.51	1.70	11.2	0.819	4.56	3.84	0.51
Detrimethylsilylation, ArSiMe$_3$, p-C$_7$H$_7$SO$_3$H, HOAc, 25°	17.5	2.0	16.5	0.917	4.05	4.10	0.27
Detrimethylsilylation, ArSiMe$_3$, p-C$_7$H$_7$SO$_2$H, H$_2$O, HOAc, 25°							
Detrimethylsilylation, ArSiMe$_3$, 0.02 M Hg(OAc)$_2$, HOAc, 25°	15.7	2.19	14.3	0.815	3.40	4.53	0.41
Detrimethylsilylation, ArSiMe$_3$, 0.4 M Hg(OAc)$_2$, HOAc, 25°	10.8	1.99	11.5	0.762	3.55	4.06	0.27
Methylation, CH$_3$Br, GaBr$_3$, ArH, 25°	11.3	2.15	12.3	0.758	3.29	4.35	0.46
Ethylation, C$_2$H$_5$Br, GaBr$_3$, ArH, 25°	9.51	1.7	11.8	0.842	4.66	3.06	0.27
Isopropylation, i-C$_3$H$_7$Br, GaBr$_3$, ArH, 25°	2.84	1.56	6.02	0.587	4.04	2.66	0.15
Methylation, CH$_3$Br, AlBr$_3$, ArH, 0–5°	1.52	1.41	5.05	0.554	4.72	2.54	0.38
Methylation, CH$_3$Br, AlBr$_3$, ArH, 27–32°	6.07	1.96	6.56	0.525	2.80	3.39	0.87
Methylation, CH$_3$I, AlBr$_3$, ArH, −1 to 2°	4.41	1.85	5.19	0.448	2.68	3.21	0.66
Methylation, CH$_3$Br, AlBr$_2$, 1,2,4-C$_6$H$_3$Cl$_3$, 25°	7.02	1.73	11.0	0.807	4.39	3.59	0.14
Methylation, CH$_3$Br, GaBr$_3$, ArH, 25°	8.60	1.81	9.80	0.733	3.84	3.64	0.10
Benzylation, C$_6$H$_5$CH$_2$Br, GaBr$_3$, ArH, 25°	4.91	2.32	9.43	0.609	2.66	4.39	0.91

A considerable number of electrophilic aromatic substitution reactions have been studied by Brown and have been shown to follow the log p_f–S_f relationship. It was therefore proposed that this relationship is general and that all electrophilic aromatic substitutions follow this relationship. With this assumption it

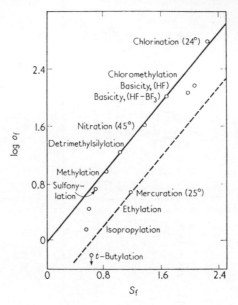

Fig. 4.   Relationship between log o_f and S_f, the Selectivity Factor, according to Brown and Smoot (51).

becomes possible to calculate the partial rate factors for any substitution reaction of toluene solely from the isomer distribution.

The following equations define the linear relationships shown in Figs. 2–4:

$$\log p_f = 1.310\, S_f$$

$$\log m_f = 0.309\, S_f$$

$$\log o_f = 1.215\, S_f$$

The correlation of o_f should be applied only to reactions which are not susceptible to steric effects, i.e., to reactions which do not deviate from the relationship. A more general expression for the *ortho* partial rate factor, applicable to all reactions, may be derived

$$\log o_f = 1.310\, S_f + \log \frac{\%\ ortho}{2 \times \%\ para}$$

Since S_f is equal to log $(2 \times \% \, para/\% \, meta)$, all of the partial rate factors are defined in terms of the isomer distribution in the reaction of toluene.

In a similar manner, the relative rate of reaction of benzene and toluene may be defined in terms of the isomer distribution: log $(k_T/k_B) = 1.310 \, S_f - \log (6 \times \% \, para/100)$.

From the definition of partial rate factors, $p_f = \% \, para \times$ relative rate $\times 6/100$, and $o_f = \% \, ortho \times$ relative rate $\times 3/100$. It is now possible to obtain o_f as a function of p_f and the isomer distribution

$$o_f = \tfrac{1}{2} \, (\% \, ortho/\% \, para) \, p_f$$

$$\log o_f = \log \% \, ortho/(2 \times \% \, para) + \log p_f.$$

d. Correlation of selectivity relationship with Hammett equation

The linear free energy expression for aromatic substitution is of the same type as that developed by Hammett for side-chain aromatic derivatives (139)

$$\log (k/k_o) = \rho\sigma$$

The relationship between the empirical expression and linear free energy expressions of the Hammett type may be demonstrated simply. Writing the Hammett equation to apply to the partial rate factors for *para*- and *meta*-substitution, we have

$$\log p_f = \rho\sigma_p$$
$$\log m_f = \rho\sigma_m$$

From these we obtain

$$\log (p_f/m_f) = \rho(\sigma_p - \sigma_m) \text{ and}$$

$$\log p_f = \left(\frac{\sigma_p}{\sigma_p - \sigma_m}\right) \log \left(\frac{p_f}{m_f}\right)$$

$$= \frac{\sigma_p}{(\sigma_p - \sigma_m)} \, S_f$$

If we consider that $c = \sigma_p(\sigma_p - \sigma_m)$ it appears reasonable, therefore, that a quantitative treatment of aromatic substitution should be possible, as suggested by Brown (p. 447). Indeed, Hammett has already pointed out that the relative reactivities in the nitration of monosubstituted benzene derivatives may be qualitatively correlated with the side-chain σ-values (140). However, attempts to develop a quantitative treatment based on these σ-values failed.

de la Mare pointed out that the Hammett σ-constants, based on side-chain reactions of benzene derivatives, did not provide any sensible agreement with experimental facts for either nitration or chlorination (90).

It was proposed by Brown that a new set of electrophilic substitution constants (σ^+) might be successful in achieving the desired correlation (50). The problem was to devise a method for testing the utility of a linear free energy relationship for these reactions without prior knowledge of either the electrophilic substituent constant, σ^+, or the reaction constant, ρ.

This difficulty was overcome by algebraic manipulation of the Hammett equation (205).

$$\log k/k_o = \rho\sigma^+$$

Writing the modified Hammett equation using the new electrophilic substituent constants σ_p^+ and σ_m^+ in accordance with previous considerations we arrive at the expression

$$\log p_f = \frac{\sigma_p^+}{\sigma_p^+ - \sigma_m^+} S_f$$

The derived expression of the Selectivity Relationship can also be written in the form

$$\log (k_R/k_H) = \frac{\sigma_p^+}{\sigma_p^+ - \sigma_m^+} S_f$$

As $\sigma_p^+/(\sigma_p^+ - \sigma_m^+) = c$, according to Brown a diagram of $\log (k_R/k_H)$ against S_f provides the means for a test of the adherence of experimental data to a linear relationship without the prior assignment of either substituent or reaction constants.

This approach differs to a certain degree from the familiar Hammett treatment in which the invariance of the substituent constants is assumed. To test a given reaction for conformity to the Hammett treatment, it is customary to plot $\log (k_R/k_H)$ versus the substituent constants. In other words, $\log (k_R/k_H)$ is plotted against σ, maintaining ρ constant. A reasonable linear correlation is generally considered to indicate adherence of the reaction to the Hammett equation.

On the other hand the basic assumption, the constancy of the

σ-values, is not necessarily valid for electrophilic substitution reactions. The most direct approach would be to assume the inductive influences operative at the *meta*-position are identical in Hammett side-chain and in electrophilic substitution reactions. Thus, the reaction constants could be established from the *meta* reactivity of the substituents under study. A plot of log (k_R/k_H) versus ρ for the various reactions, maintaining the substituent constant, would test whether the electronic contributions of that particular substituent could be represented satisfactorily by a constant, σ^+. Again this attack is not feasible because of the experimental difficulties and uncertainties involved in the evaluation of the partial rate factors for *meta*-substitution.

In the absence of adequate data to establish the reaction constant ρ with the requisite precision, another method was devised. The Selectivity Factor, S_f, is a quantity proportional to the reaction constant ρ. Likewise the quantity log p_f is proportional to ρ. Accordingly, the Selectivity Relationship provides an alternative method for the evaluation of the adherence of experimental data for a given substituent to a linear free energy relationship. Important is the fact that prior assignment of either substituent or reaction constants is unnecessary.

The new σ_p^+- and σ_m^+-values developed permitted the quantitative treatment of directive effects in aromatic substitutions.

For comparison σ^+-values versus σ-values are plotted in Fig. 5. The values for the *meta*-constants follow the relationship $\sigma_m^+ = 0.902\sigma_m$.

In view of the large uncertainty in the experimental data from which the σ_m^+-constants were derived, it appeared reasonable to assume that the σ_m^+-values are identical with the corresponding Hammett σ_m-values.

$$\sigma_m^+ = \sigma_m$$

On the other hand, there is a significant difference between the σ_p^+- and the corresponding σ_p-values.

A number of reactions were, however, not well correlated by the σ^+-values and other sets of values have therefore been devised. Knowles, Norman, and Radda (168a) proposed a modification of the Hammett equation to electrophilic aromatic substitutions, taking into account the different extents of resonance interaction between the substituents and the aromatic system in the transition state, in reactions where the reagent has different electronic requirements. It was suggested that no unique set of σ^+-values is possible and extended use of the modified Hammett equation can be expected if further adjustable parameters are being considered (as a function of solvent, reagent, etc.).

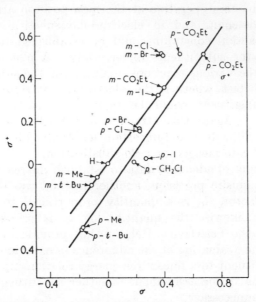

Fig. 5.   Comparison of σ^+- and σ-constants, according to McGary, Okamoto and Brown (205).

C. Limitations of the Linear Correlation of Reagent Activity and Substrate Selectivity

a. General considerations

As attractive as it seems to be, the quantitative treatment of electrophilic aromatic substitutions, including the basic Friedel-Crafts type substitution reactions, by the linear free energy correlation, which was shown to be directly correlated with Hammett's electrophilic side-chain reactivities, has limitations.

It was seriously questioned whether a relation of this type would apply to aromatic substitution reactions at all.   These arguments have been based on the concept that resonance contributions to the stabilization of the transition state would vary greatly with the reactivity of the reagent (90,128,263).

As frequently pointed out by Brown, the reliability in establishing any empirical correlation between relative rates and the corresponding isomer distributions in electrophilic substitution of aromatics is greatly hindered by the lack of sufficiently reliable data.   Any unreliable data or the use of rate data obtained under different conditions than the isomer distributions used to calculate partial rate factors may lead to serious errors. Many discrepancies were dissolved by better experimental data or by proving obvious errors in reported works.

With the advent of more advanced analytical methods and particularly of gas chromatography in the last few years, a number of quantitative investigations have been carried out which cannot be correlated with Brown's simple empirical correlations and need consideration.

b. Necessity of kinetically controlled, non-isomerizing conditions for evaluation of isomer distributions

It became more and more apparent that simple electrophilic aromatic substitutions of the Friedel-Crafts type can be carried out under considerably varied conditions. Thus reactions such as alkylation, acylation, haloalkylation, nitration, etc., can be carried out both with high or low selectivity reagent systems. No oversimplification of generalized conditions is therefore possible and the selection of either high or low selectivity data to fit the best correlation must be exercised with care. Another fact of major importance is that in the investigation of any correlation based on isomer distribution and relative rates of substitution of aromatic systems, only those data can be compared which were obtained under kinetically controlled conditions. If secondary or concurrent isomerizations (influencing considerably isomer ratios, generally through sharp increase of the *meta*-isomers) are operative, as is the case in many Friedel-Crafts type reactions, especially alkylations (3), then correlation between relative reactivity and isomer distributors becomes meaningless. Therefore any correlation of relative substrate reactivities (which are kinetically controlled, but only if no disproportionation takes place) with isomer distributions (influenced greatly by thermodynamically controlled isomerizations) is possible only when non-isomerizing conditions are proved.

c. Low substrate, high positional selectivity reactions

Investigation of a number of Friedel-Crafts type electrophilic aromatic substitutions of toluene and benzene involving strongly electrophilic reagents gave low substrate selectivity values, without, however, showing low positional selectivity (Table IX).

Calculation of partial rate factors cannot be carried out from these data of relative reactivity and corresponding isomer distributions. Similarly the Selectivity Relationship is not applicable.

TABLE IX.   Competitive substitution of toluene and benzene at 25°

Substitution (substituting agent)	Solvent	Catalyst	$k_T : k_B$	% Isomer distribution			Ref.
				Ortho	Meta	Para	
Nitration NO$_2$BF$_4$	Tetramethylene sulfone		1.67	65.4	2.8	31.8	228
Nitration HNO$_3$	Tetramethylene sulfone	H$_2$SO$_4$	1.60	56.3	2.6	41.0	231
Nitration HNO$_3$	Acetic acid	H$_2$SO$_4$	2.13	58.1	1.9	40.0	231
Benzylation C$_6$H$_5$CH$_2$Cl	CH$_3$NO$_2$	AlCl$_3$	3.20	43.5	4.5	52.0	229
Benzylation C$_6$H$_5$CH$_2$Cl	CH$_3$NO$_2$	FeCl$_3$	3.24	44.5	3.7	51.8	229
Benzylation C$_6$H$_5$CH$_2$Cl	CH$_3$NO$_2$	AgBF$_4$	2.95	39.5	3.1	57.4	229
Isopropylation i-C$_3$H$_7$Br	CH$_3$NO$_2$	AlCl$_3$	2.03	46.7	14.7	38.6	230
Isopropylation C$_3$H$_6$	CH$_3$NO$_2$	AlCl$_3$	1.96	46.1	15.1	38.8	230
t-Butylation t-C$_4$H$_9$Br	CH$_3$NO$_2$	SnCl$_4$	15.6	—	7.0	93.0	230
t-Butylation t-C$_4$H$_9$Br	CH$_3$NO$_2$	AgClO$_4$	15.3	—	5.7	94.3	230
Bromination Br$_2$	CH$_3$NO$_2$	FeCl$_3$	3.55	63.7	1.8	34.5	222
Chlorination Cl$_2$	CH$_3$NO$_2$	FeCl$_3$	13.5	67.8	2.3	29.9	234

When an attempt was made to calculate partial rate factors, the values for m_f were found to be as low as 0.1, which is impossible for toluene if we assume that individual positions are competing because there is no way of explaining how one position (such as a *meta*-position) in toluene could be deactivated ten times relative to a benzene position.

d. *Considerations of σ- versus π-complex nature of transition states*

The above discrepancy can be explained, and is even to be expected, if π- and not σ-complex formation is the main rate-determining step in the reactions. Partial rate factors and empirical selectivity factors can be used only if, in a certain reaction, competition takes place between individual positions in the molecules involved and *not* between π-electron systems, as entities, of reacting molecules. In the competition between molecules involving their π-electron systems as entities, the calculation of partial rate factors and selectivity factors is meaningless.

The stabilities of the π-complexes do not change greatly with increasing number and position of alkyl substituents. It is obvious that the absolute value of the electron density of the π-electron donor system could not change drastically by inductive and hyperconjugative effects of alkyl groups. Thus it can be explained that the relative reactivities observed show only small differences of the same order of magnitude for benzene and alkylbenzenes, or for di- and trialkylbenzenes such as mesitylene, etc. This order of reactivity of the aromatic substrates is entirely different from that in reactions where relative stabilities of intermediate σ-complexes in individual positions are involved. These, of course, frequently show relative reactivities in positions *ortho* and *para* to alkyl groups (which are highly stabilized by conjugation) a thousand times greater than that of an individual benzene position.

The next question to be answered is how the low substrate selectivity, corresponding to relative π-complex stabilities, compares with the observed isomer distributions.

The greater the reactivity of an electrophilic substituting agent, the smaller its selectivity. This means low selectivity with different aromatics and, it was suggested, a simultaneous change of the isomer distribution toward the statistical value (40% *ortho*, 40% *meta*, and 20% *para*, representing two *ortho*-positions, two *meta*-positions, and one *para*-position in a monosubstituted benzene). This is generally demonstrated by an increase in the amount of the *meta*-isomer. The obvious explanation is the decreasing role of small activation energy differences of different individual positions compared to the overall activation energy of the reaction. This, again, is valid only if the reaction is dependent on the relative stability of individual σ-complexes but *not* if competition of individual molecules is involved (according to relative π-complex stability). In this case, a low selectivity with different aromatic substrates (alkylbenzenes) is possible without necessarily involving a change of the isomer distribution in the direction of statistical distribution (demonstrated by a sharp increase of the *meta*-isomer).

By definition the transition states represent maxima on the energy diagram of the reactions and are therefore not isolable. The complexes, however, represent energy minima, and, as shown in the case of stable benzenium complexes, can be isolated in certain cases.

In considering a π-complex type transition state in the rate-determining step with subsequent "stabilization" of the entering substituent in individual positions (thus positional discrimination), it is necessary for T_1 to be *higher* than T_2, thus the formation of the π-complex is *irreversible*.

The potential energy diagram of a typical aromatic substitution reaction involving a strongly electrophilic substituting agent can be represented in the following way (Fig. 6). T_1 represents the transition state corresponding in nature to a π-complex and T_2 corresponds to the transition state of a σ-complex nature. T_2 indeed must be composed of separate transition states corresponding to the *meta-*, *para-*, and *ortho-*positions involved, from which the one leading to

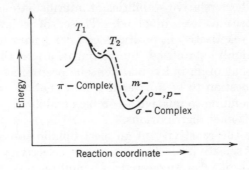

Fig. 6.   Energy diagram of typical aromatic substitution involving irreversible π-complex formation.

the *meta*-position represents the highest energy barrier. The proton elimination side of the reaction coordinate is substantially symmetrical with that of the reagent attack side, the relative heights of the actîvated states being determined by the probable existence of a kinetic isotope effect. To account for a low substrate, but high positional selectivity substitution of aromatics, it must be suggested that activated states corresponding both to π- and σ-complex nature play an important role, the first being, however, dominant in the rate-determining step, whereas the latter effects the isomer distribution.

Where does the energy resisting the formation of a π-complex of this type come from and what is the nature of such a π-complex? First of all it is necessary to point out that it is customary to call π-complexes both weak, reversible interactions of aromatics with neutral or slightly polarized reagents (like molecular halogens) and with ions (like Ag$^+$) incapable of effecting irreversible changes (substitution). Strong electrophiles, capable of effecting substitutions (like NO$_2^+$, R$^+$, Br$^+$), however, form irreversible π-complexes, as the once complexed cation is passing over the *lower* energy barrier (T_2) leading to the σ-complex and consequently to the substituted product, instead of reversing over the *higher* barrier (T_1). If the opposite is the case, T_2 being higher than T_1, then the formation of the π-complexes becomes reversible and *both* substrate and positional selectivity are determined by the activation energy needed in the formation of the σ-complex (Fig. 7). In this case the

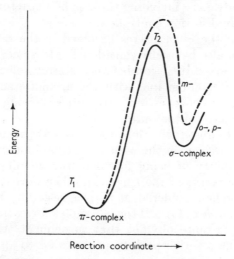

Fig. 7. Energy diagram of typical aromatic substitution involving reversible π-complex formation.

good correlation of the substrate and positional selectivities is readily understandable (Brown's Selectivity Relation). It is thus also understandable that when π-complex formation is the rate-determining step, the substrate and positional selectivities become independent of each other and low substrate selectivity is possible with simultaneous high positional selectivity. In the latter case the two are determined by separate transition states, whereas in the first case by the identical one. It is suggested that the energy resisting the formation of an irreversible π-complex of aromatics with strong electrophiles comes from the fact that no free, separated ions (like NO$_2^+$, R$^+$, Br$^+$, etc.) are present in any of the organic reaction media of the Friedel-Crafts reactions. Even in the case of isolated, stable ion salts (like $NO_2^+BF_4^-$) it was proved that in the solvent systems used they are present predominantly as solvated ion pairs (or higher conglomerates) and *not* as separated ions. This is also believed to be the case in alkylation, halogenation and like systems, where no free separated alkyl carbonium or halonium ions can be detected. Consequently to pull away the cation from the strong coulombic interaction of its ion-pair or ion-conglomerate interaction, which in turn is further strengthened by solvation, and bring it onto the uncharged aromatic molecule, involves substantial energy.

Although the Friedel-Crafts reactions of aromatic hydrocarbons with strong electrophiles (nitration, alkylations, halogenations) are fast, based on rate data, including those in flow reactor systems, they are not generally diffusion controlled.

The nature of the π-complexes involved in the rate-determining interaction must also be briefly considered. It is necessary to realize that in the primary interaction between a strong electrophile (NO_2^+, Br^+, incipient carbonium ion) with the aromatic substrate, in the case of substituted benzenes (*e.g.*, aromatics showing a non-uniform π-electron distribution) the outer complex must be an oriented one. The electrophilic substituent thus does not move around unhindered as suggested by Dewar on the π-electron orbits but has statistically a greater probability of being found in the vicinity of the higher electron density points of the ring. In other words, the π-complex formation should be considered, as first suggested by H. C. Brown, as an oriented π-complex (46). If the π-complex formation is considered in this way, it is quite obvious that subsequent transition of the reagent along the reaction path from the outer to an inner complex intermediate is facilitated in the *ortho*: *para* position over that involving the *meta* because of lower energy transition barriers.

The *ortho*:*para* ratios seem to be a highly important characteristic of electrophilic aromatic substitutions taking place under non-isomerizing (kinetically controlled) conditions. In the case of isomerization not only does the *meta* amount increase, but the *ortho* decreases sharply, thus indicating well the change of isomer ratio. Steric effects, although obviously important, have been frequently overestimated and used as the sole excuse to explain facts otherwise difficult to account for.

In a number of investigated cases it was found that low substrate selectivity does not necessarily mean low positional selectivity. These facts make a general simple empirical linear correlation covering all electrophilic aromatic substitutions in cases where π-complex formation is kinetically important questionable and also

emphasize that previous data showing this correlation were sometimes obtained from data not representing kinetically controlled but
isomerized isomer distributions.

The sometimes significant difference in isomer distributions of
alkylation systems could be accounted for by the fact that isomer
distributions are easily effected by thermodynamically controlled
isomerizations. If an alkylation reaction goes through an activated
state, corresponding in nature to the σ-complex, in the rate-determining step, it is questionable that alkylation exclusive of isomerization
is possible because the latter must have an identical intermediate
state (3). However, if the alkylation goes through an oriented π-
complex type activated state in the main rate-determining step then
alkylation is possible without concurrent isomerization.

This question is of theoretical interest when considering whether
the high *meta*-dialkylbenzene isomer ratios observed in many
aromatic alkylations were really due to kinetically controlled direct
alkylation, or at least partially to thermodynamically controlled
concurrent or consecutive isomerization.

Consecutive isomerization, caused either by the catalyst itself or
by the conjugate acid formed by the proton elimination and interaction with the catalyst, could seriously affect isomer ratios in all
alkylation reactions. Therefore, it is of predominant importance to
establish whether an alkylation system is free of conditions causing
consecutive isomerization. It is believed that a donor solvent
used (*e.g.*, nitromethane) complexes with the catalyst and the
conjugate acid formed to a degree sufficient to inactivate it for ring
protonation (σ-complex formation) and subsequent isomerization.
This suggestion is supported by the fact that Friedel-Crafts isomerization of alkylbenzenes in such solvents is difficult or non-
existent. In the absence of a suitable proton acceptor solvent (as is
the case using only excess of the aromatics as solvent) secondary
isomerizations take place easily.

Considerable differences in substrate selectivity are a consequence
of the electrophilic strength of the reagents. Strong electrophiles
(NO_2^+, Br^+, incipient carbonium ions) interact with reactive aromatic substrates in a rate-determining step corresponding in nature
to an oriented π-complex type activated state. Consequently only
slight differences in substrate selectivities are observed ($k_T : k_B \sim$
1.5–4). Weaker electrophilic reagents give increasingly higher
substrate selectivity. Solutions of elementary halogens and dilute
solutions of nitric acid in organic solvents do not contain Br^+ or NO_2^+
but only their weaker electrophile precursors, and therefore react as

considerably weaker and more selective reagents. This may explain why some of these reagents gave good linear correlations in the Selectivity Relationship reacting according to σ-complex type rate-determining activated states. It is reasonable to correlate the "activity" of the attacking agent with its electrophilic properties. The electrophilic character of bromonium ion, nitronium ion, and dimethyl carbonium ion would be expected to increase in the order listed. This is also the order found for the reactivity of these reagents: $Br^+ < NO_2^+ < (CH_3)_2CH^+$. We may also predict that the electrophilic character or "activity" of the following carbonium ions will increase in the order: $RCO^+ < (CH_3)_3C^+ < (CH_3)_2CH^+ < CH_3CH_2^+ < CH_3^+$. There is at the present time, however, no evidence available to prove the formation and reaction of ethyl or methyl carbonium ions in Friedel-Crafts type reactions.

Owing to the weak electrophilic nature of acylium ions in Friedel-Crafts acylations we would expect high selectivity. Moreover, methylation and ethylation should be less selective than isopropylation, which itself should be less selective than t-butylation. Experimental data are indeed in agreement with this sequence.

In aromatic bromination with molecular bromine, it appears that the transition state involves the aromatic molecule, bromine molecule, and some additional electrophilic agent, which may be an additional bromine or iodine molecule (264,265).

If the bromination catalyst is changed from the relatively mild electrophilic agent iodine to the more strongly acidic $FeCl_3$ or $AlCl_3$, the toluene:benzene ratio sharply decreases, because Br^+ or the incipient bromonium ion are considerably stronger electrophiles.

A reagent such as molecular bromine, which is a stable entity and only weakly electrophilic, will require a high contribution of electrons from the aromatic ring in the transition state. The more electrophilic nitronium ion will require a substantially lesser contribution, while the highly electrophilic dimethyl carbonium ion will require the least contribution.

The electrophilic nature of the reagent is also affected by the solvent and other reaction conditions.

In conclusion it could be said that although linear correlations of reagent activities and observed substrate and positional selectivities are very useful in furthering our knowledge of the selectivity of Friedel-Crafts type aromatic substitution, they also have limitations. It must be emphasized that oversimplification and generalization may frequently be misleading. There is obviously a large variety of reactions of quite different selectivity type in alkylation, haloalkylation, acylation, nitration, etc. Each reaction, under the specific conditions used, must be considered separately. It seems to be particularly questionable to use empirical correlations in cases where insufficient experimental data exist. Calculation of partial rate factors directly from isomer distributions by the use of empirical correlations, without any rate measurements, should be omitted. As isomer ratios frequently are influenced by isomerization, there is no assurance, even if a linear free energy relationship should be valid for a particular reaction, that these calculated rate factors have real meaning. The relationship cannot be applied to systems where π-complexes are of kinetic significance.

References

1. Adams, R., and T. D. Garber, *J. Am. Chem. Soc.*, **71**, 522 (1949).
2. Ador, E., *Bull. Soc. Chim. France*, (2) **33**, 56 (1880).
3. Allen, R. H., and L. D. Yats, *J. Am. Chem. Soc.*, **83**, 2799 (1961).
4. Andreades, S., and D. C. England, Abstract of Papers 14M, 140th Meeting, ACS, Chicago, Ill., September, 1961.
5. Andreades, S., and D. C. England, *J. Am. Chem. Soc.*, **83**, 4670 (1961).
6. Angelbis, A., and R. Anschütz, *Ber.*, **17**, 165, 167 (1884).
7. Anschütz, R., *Ann.*, **235**, 158 (1886).
8. Anschütz, R., *Ann.*, **235**, 299 (1886).
9. Anschütz, R., *Ann.*, **235**, 302 (1886).
10. Anschütz, R., *Ann.*, **235**, 336 (1886).
11. Anschütz, R., and P. Forster, *Ann.*, **368**, 89 (1909).
12. Anschütz, R., and E. Romig, *Ber.*, **18**, 662 (1885).
13. Auger, V., *Ann. Chim. Phys.*, (6), **22**, 310 (1891); *Bull. Soc. Chim. France*, (2) **49**, 345 (1888).
14. Bachman, G. B., and H. M. Hellman, *J. Am. Chem. Soc.*, **70**, 1772 (1948).
15. Baddeley, G., G. Holt, and D. Voss, *J. Chem. Soc.*, 100 (1952).
16. Baddeley, G., and R. Williamson, *J. Chem. Soc.*, 4647 (1956).
17. Baeyer, A., *Ann.*, **202**, 50 (1880).
18. Barduhn, A. J., and K. A. Kohe, *Ind. Eng. Chem.*, **38**, 247 (1946).
19. Bauer, S. H., and R. E. McCoy, *J. Phys. Chem.*, **60**, 1529 (1956).
20. Baumgarten, B., and W. Bruns, *Ber.*, **80**, 517 (1947).
21. Bax, C. M., A. R. Katritzky, and L. Sutton, *J. Chem. Soc.*, 1258 (1958).
22. Bebal, A., and V. Auger, *Bull. Soc. Chim. France*, (3) **9**, 696 (1894).
23. Bell, R. P., and B. G. Skinner, *J. Chem. Soc.*, 2955 (1952).
24. Bennett, N., U.S. Pat. 2,136,270 (1938).
25. Bergmann, E., and M. Polanyi, *Naturwissenschaften*, **21**, 378 (1933).
26. Berliner, E., and F. J. Bondhus, *J. Am. Chem. Soc.*, **68**, 2355 (1946); **70**, 854 (1948).
27. Bistrzycki, A., and C. Herbst, *Ber.*, **36**, 146 (1903).
28. Bistrzycki, A., and L. Mauron, *Ber.*, **40**, 4063 (1907).
29. Blitz, H., *J. Prakt. Chem.*, (2), **142**, 193 (1935).
30. Bodendorf, K., and H. Böhme, *Ann.*, **516**, 1 (1935).

31. Böeseken, J., *Verslag. Akad. Wetenschap. Amsterdam*, **31**, 226 (1922).

32. Böeseken, J., and M. C. Bastet, *Rec. Trav. Chim.*, **32**, 184 (1913).

33. Böhme, H., and O. Siering, *Ber.*, **71**, 2372 (1938).

34. Booth, H. S., and D. R. Martin, in "Fluorine Chemistry" (edited by J. H. Simons), Vol. I, p. 208, Academic Press, New York, 1950.

35. Bosshard, H., and D. Strauss, Germ. Pat. 386,012 (1922); 386,930 (1923).

36. Brown, H. C., *J. Am. Chem. Soc.*, **67**, 374, 378, 503, 1452 (1945).

37. Brown, H. C., *et al.*, *J. Am. Chem. Soc.*, **81**, 5608–5626 (1959).

38. Brown, H. C., and R. N. Adams, *J. Am. Chem. Soc.*, **65**, 2557 (1943).

39. Brown, H. C., H. Bartholomay, Jr., and M. D. Taylor, *J. Am. Chem. Soc.*, **66**, 435 (1944).

40. Brown, H. C., and J. D. Brady, *J. Am. Chem. Soc.*, **74**, 3570 (1952).

41. Brown, H. C., L. P. Eddy, and R. Wong, *J. Am. Chem. Soc.*, **75**, 2675 (1953).

42. Brown, H. C., and E. A. Fletcher, *J. Am. Chem. Soc.*, **73**, 2808 (1951).

43. Brown, H. C., and R. R. Holmes, *J. Am. Chem. Soc.*, **78**, 2173 (1956).

44. Brown, H. C., and R. R. Holmes, Abstract of Papers presented at the 129th Meeting of the American Chemical Society held in Cleveland, Ohio, p. 28, May 1956.

45. Brown, H. C., and R. H. Horowitz, *J. Am. Chem. Soc.*, **77**, 1730 (1955).

46. Brown, H. C., and H. Jungk, *J. Am. Chem. Soc.*, **77**, 5579 (1955); **78**, 2182 (1956).

47. Brown, H. C., and E. A. Lawton, unpublished work. See E. A. Lawton, Thesis, Purdue University, 1952.

48. Brown, H. C., and C. W. McGary, *J. Am. Chem. Soc.*, **77**, 2306, 2310 (1955).

48a. Brown, H. C., G. Marino, and L. M. Stock, *J. Am. Chem. Soc.*, **81**, 3310 (1959).

49. Brown, H. C., and C. W. McGary, *J. Am. Chem. Soc.*, **77**, 2300 (1955).

50. Brown, H. C., and K. L. Nelson, *J. Am. Chem. Soc.*, **75**, 6292 (1953).

51. Brown, H. C., and C. R. Smoot, *J. Am. Chem. Soc.*, **78**, 6255 (1956).

52. Brown, H. C., and S. Sujishi, *J. Am. Chem. Soc.*, **70**, 2878 (1948).

53. Brown, H. C., and W. J. Wallace, *J. Am. Chem. Soc.*, **75**, 6265 (1953).

54. Bruson, H. A., and J. W. Kroeger, *J. Am. Chem. Soc.*, **62**, 36 (1940).

55. Burg, A. B., and A. A. Green, *J. Am. Chem. Soc.*, **65**, 1838 (1943).

56. Burg, A. B., and C. L. Randolph, Third Annual Technical Report to the Office of Naval Research.   Project NR 052–050.

57. Burk, R. E., in "XIIth Report of the Committee on Catalysis, National Research Council," John Wiley and Sons, New York, p. 251, 1940.

58. Burwell, R. L., Jr., and S. Archer, *J. Am. Chem. Soc.*, **64**, 1032 (1942).

59. Buu-Hoï, N. P., B. Eckert, and P. Demerseman, *J. Org. Chem.*, **19**, 1032 (1942).

60. Buu-Hoï, N. P., and R. Roger, *Bull. Soc. Chim. France*, 812 (1947).

61. Calloway, N. O., *Chem. Rev.*, **17**, 327 (1935).

62. Calloway, N. O., *J. Am. Chem. Soc.*, **59**, 1474 (1937).

63. Cathcart, W. R., and V. Meyer, *Ber.*, **25**, 1498 (1892).

64. Clar, E., F. John, and R. Avenarius, *Ber.*, **72**, 2139 (1939).

65. Clifford, A. F., H. C. Beachell, and W. M. Jack, *J. Inorg. Nucl. Chem.*, **5**, 57 (1957).

66. Clifford, A. F., and S. Kongpricha, *J. Inorg. Nucl. Chem.*, **20**, 147 (1961).

67. Close, W. J., *J. Am. Chem. Soc.*, **79**, 1455 (1957).
68. Cobb, P. H., *Am. Chem. J.*, **35**, 486 (1906).
69. Coffey, S., *J. Chem. Soc.*, 1029 (1925).
70. Cohen, J. B., and P. K. Dutt, *J. Chem. Soc.*, **105**, 501 (1914); *Proc. Chem. Soc.*, **30**, 15, 271 (1914),
71. Colclough, T., W. Gerrard, and M. F. Lappert, *J. Chem. Soc.*, 907 (1955); 3006 (1956).
72. Coleman, G. H., G. V. Moore, and G. B. Stratton, U.S. Pat. 2,135,122 (1938).
73. Collet, A., *Bull. Soc. Chim. France*, (3) **15**, 22 (1896).
74. Collet, A., *Bull. Soc. Chim. France*, (3) **17**, 66 (1897).
75. Collet, A., *Bull. Soc. Chim. France*, (3) **17**, 506 (1897); *Compt. rend.*, **125**, 717 (1897).
76. Colonge, J., and Y. Infarnet, *Bull. Soc. Chim. France*, 1916 (1960).
77. Conant, J. B., and R. E. Lutz, *J. Am. Chem. Soc.*, **45**, 1303 (1923).
78. Condon, F. E., *J. Am. Chem. Soc.*, **70**, 1963 (1948).
79. Condon, F. E., *J. Am. Chem. Soc.*, **70**, 2265 (1948).
80. Condon, F. E., *J. Am. Chem. Soc.*, **71**, 3544 (1949).
81. Cotter, J. L., and A. G. Evans, *J. Chem. Soc.*, 2988 (1959).
82. Cotton, F. A., and J. R. Leto, *J. Chem. Phys.*, **30**, 993 (1959).
83. Dannenberg, H., and A. U. Rahman, *Ber.*, **88**, 1405 (1955).
84. Davidson, J. M., and A. Lowy, *J. Am. Chem. Soc.*, **51**, 2978 (1929).
85. Davidson, N., and H. C. Brown, *J. Am. Chem. Soc.*, **64**, 316 (1942).
86. Dear, P. J. A., and D. D. Eley, *J. Chem. Soc.*, 4684 (1954).
87. de Diesbach, H., and P. Dobbelmann, *Helv. Chim. Acta*, **14**, 369 (1931).
88. Delacre, M., *Bull. Soc. Chim. France*, (3) **13**, 857 (1895).
89. de la Mare, P. B. D., *J. Chem. Soc.*, 2871 (1949).
90. de la Mare, P. B. D., *J. Chem. Soc.*, 4450 (1954).
91. de la Mare, P. B. D., and P. W. Robertson, *J. Chem. Soc.*, 279 (1943).
92. Demole, E., *Ber.*, **12**, 2245 (1879).
93. Dermer, O. C., and R. A. Billmeier, *J. Am. Chem. Soc.*, **64**, 464 (1942).
94. Dermer, O. C., P. T. Mori, and S. Suguitan, *Proc. Oklahoma Acad. Sci.*, **29**, 74 (1948).
95. Dermer, O. C., D. M. Wilson, F. N. Johnson, and V. H. Dermer, *J. Am. Chem. Soc.*, **63**, 2881 (1941).
96. Desai, R. D., and R. M. Desai, *J. Sci. Ind. Res.* (*India*), **14B**, 498, 505 (1955).
97. Dev, S., *J. Ind. Chem. Soc.*, **33**, 703 (1956).
98. Devlin, T. R. E., and D. C. Pepper, in "Cationic Polymerization," ed. P. H. Plesch, Academic Press, New York, 1953.
99. Dewar, M. J. S., *Ann. Rept.*, **53**, 132 (1956).
100. Dilke, M. H., and D. D. Eley, *J. Chem. Soc.*, 2601, 2613 (1949).
101. Dilke, M. H., D. D. Eley, and M. G. Sheppard, *Trans. Faraday Soc.*, **46**, 261 (1950).
102. Dippy, J. F. J., and J. T. Young, *J. Chem. Soc.*, 1817 (1952); **77**, 3919 (1955).
103. Drago, R. S. "Physical Methods in Inorganic Chemistry," Van Nostrand Reinhold, New York, 1965; R. S. Drago and B. B. Wayland, *J. Am. Chem. Soc.*, **87**, 3571 (1965).
104. Drefahl, G., and F. Fischer, *Ann.*, **598**, 159 (1956).
105. Eijkman, J. F., *Chem. Weekblad*, **4**, 727 (1907); **5**, 655 (1908).
106. Eley, D. D., and H. Watts, *J. Chem. Soc.*, 1319 (1954).
107. Emeléus, H. J., and G. S. Rao, *J. Chem. Soc.*, 4245 (1958).

108. Entermann, C. E., Jr., and J. R. Johnson, *J. Am. Chem. Soc.*, **55**, 2900 (1933).
109. Etaix, L., *Ann. Chim. Phys.*, (7) **9**, 372 (1896).
110. Evans, A. G., and G. W. Meadows, *J. Polymer Sci.*, **4**, 359 (1949).
111. Evans, A. G., G. W. Meadows, and M. Polanyi, *Nature*, **158**, 94 (1946).
112. Fallmann, J. H., *Nature*, **179**, 265 (1957).
113. Fieser, L. F., and E. B. Hershberg, *J. Am. Chem. Soc.*, **61**, 1272 (1939).
114. Fischer, W., and O. Jubermann, *Z. anorg. allgem Chem.*, **227**, 227 (1936).
115. Fischer, W., and C. Rahlfs, *Z. anorg. allgem. Chem.*, **205**, 1 (1932).
116. Fischer, W., and C. Rahlfs, *Z. anorg. allgem. Chem.*, **205**, 32 (1932).
117. Fletcher, E. A., Thesis, Purdue University, 1952.
118. Florin, R. E., *J. Am. Chem. Soc.*, **73**, 4468 (1951).
119. Francis, A. W., *Chem. Rev.*, **43**, 257 (1948).
120. Freund, M., and K. Fleischer, *Ann.*, **373**, 291 (1910).
121. Freund, M., and K. Fleischer, *Ann.*, **399**, 182 (1913).
122. Freund, M., and K. Fleischer, *Ann.*, **411**, 14 (1916).
123. Friedel, C., and M. Balsohn, *Bull. Soc. Chim. France*, **35**, 52 (1881).
124. Friedel, C., and J. M. Crafts, *Ann. Chim. Phys.*, (6) **11**, 263 (1887).
125. Fujise, S., M. Suzuki, and K. Shinoda, *Sci. Rept. Tohoku Univ.*, **39**, No. 3, 185 (1956); through *C.A.*, **51**, 8084 (1957).
126. Fuson, R. C., and J. T. Walker, *Org. Syn.*, **13**, 32 (1933).
127. Gerard, W., and E. F. Mooney, *Chem. and Ind.*, 1259 (1958).
128. Gold, V., and D. P. N. Satchell, *J. Chem. Soc.*, 2743 (1956).
129. Graham, W. A. G., and F. G. A. Stone, *J. Inorg. Nucl. Chem.*, **3**, 164 (1956).
130. Granger, R., M. Corbier, J. Vinas, and P. Nau, *Compt. rend.*, **244**, 1048 (1957).
131. Greenwood, N. N., and B. G. Perkins, *J. Chem. Soc.*, 1141 (1960).
132. Greenwood, N. N., and B. G. Perkins, *J. Chem. Soc.*, 1145 (1960).
133. Greenwood, N. N., and K. Wade, *J. Chem. Soc.*, 1527 (1956).
134. Grosse, A. V., and V. N. Ipatieff, *J. Org. Chem.*, **1**, 599 (1936).
135. Hale, W. J., and E. C. Britton, *J. Am. Chem. Soc.*, **41**, 841 (1919).
136. Haller, A., and A. Guyot, *Compt. rend.*, **119**, 140 (1894).
137. Haller, A., and A. Guyot, *Bull. Soc. Chim. France*, (3) **17**, 873 (1897).
138. Hammett, L. P., *Chem. Rev.*, **16**, 67 (1935).
139. Hammett, L. P., "Physical Organic Chemistry," McGraw-Hill Book Company, New York, p. 267 ff., 1940.
140. Hammett, L. P., and A. J. Deyrup, *J. Am. Chem. Soc.*, **54**, 2721 (1932).
141. Hammett, L. P., and M. A. Paul, *J. Am. Chem. Soc.*, **56**, 830 (1934).
142. Hammett, L. P., and M. A. Paul, *J. Am. Chem. Soc.*, **56**, 827 (1934).
143. Hanriot, M., and F. Guilbert, *Compt. rend.*, **98**, 525 (1884).
144. Hart, R. T., and R. F. Teble, *J. Am. Chem. Soc.*, **72**, 3286 (1950).
145. Hawke, D. L., and J. Steigman, *Anal. Chem.*, **26**, 1989 (1954).
146. Hill, P., and W. F. Short, *J. Chem. Soc.*, 1123 (1935).
147. Hoehn, W. M., U.S. Pat. 2,781,394 (1957).
148. Holleman, A. F., *Rec. Trav. Chim.*, **33**, 183 (1914).
149. Holleman, A. F., and P. Caland, *Ber.*, **44**, 2504 (1911).
150. Holleman, A. F., J. Vermeulen, and W. J. DeMooy, *Rec. Trav. Chim.*, **33**, 1 (1914).
151. Huston, R. C., and D. D. Sager, *J. Am. Chem. Soc.*, **48**, 1955 (1926).

152. Hyman, H. H., L. A. Quarterman, M. Kilpatrick, and J. J. Katz, *J. Phys. Chem.*, **65**, 123 (1961).

153. Ingold, C. K., "Structure and Mechanism in Organic Chemistry," Cornell University Press, Ithaca, N.Y., Chapter 6, 1953.

154. Ingold, C. K., A. Lapworth, E. Rothstein, and D. Ward, *J. Chem. Soc.*, 1959 (1931).

155. Ishikawa, S., and G. Maeda, *Sci. Rept. Tokyo Kyoiku Daigaku*, **A3**, 157 (1937); through *C.A.*, **31**, 7860 (1937).

156. Iwao, J., and K. Tomio, *J. Pharm. Soc. Japan*, **76**, 808 (1956).

157. Jenny, E. F., and J. D. Roberts, *Helv. Chim. Acta*, **38**, 1248 (1955).

158. Jones, W. W., and M. Russell, *J. Chem. Soc.*, 921 (1947).

159. Kaluszyner, A., S. Reuter, and E. D. Bergmann, *J. Am. Chem. Soc.*, **77**, 4164 (1955).

160. Ketelaar, J. A. A., C. H. MacGillavry, and P. A. Renes, *Rec. Trav. Chim.*, **66**, 501 (1947).

161. Kilpatrick, M., *J. Chem. Educ.*, **37**, 403 (1960).

162. Kilpatrick, M., and T. J. Lewis, *J. Am. Chem. Soc.*, **78**, 5186 (1956).

163. Kilpatrick, M., and F. E. Luborsky, *J. Am. Chem. Soc.*, **76**, 5863 (1954).

164. Kitano, H., K. Fukui, *et al.*, *J. Chem. Soc. Japan*, **58**, 54 (1955).

165. Klapproth, W. J., and F. H. Westheimer, *J. Am. Chem. Soc.*, **72**, 4461 (1950).

166. Klemm, W., E. Clausen, and H. Jacobi, *Z. anorg. allgem. Chem.*, **200**, 367 (1931).

167. Klemm, W., W. Tilk, and H. Jacobi, *Z. anorg. allgem. Chem.*, **207**, 187 (1932).

168. Kline, E. R., B. N. Campbell, and E. C. Spaith, *J. Org. Chem.*, **24**, 1781 (1959).

168a. Knowles, J. R., R. O. C. Norman, and G. K. Radda, *J. Chem. Soc.*, 4885 (1960).

169. Koelsch, C. F., H. Hochmann, and C. D. LeClaire, *J. Am. Chem. Soc.*, **65**, 59 (1943).

170. Kohler, E. P., *Am. Chem. J.*, **38**, 511 (1907).

171. Kohler, E. P., *Am. Chem. J.*, **42**, 375 (1909).

172. Kohler, E. P., G. L. Heritage, and M. C. Burnley, *Am. Chem. J.*, **44**, 60 (1910).

173. Koike, E., and M. Okawa, *Rept. Govt. Chem. Ind. Res. Inst. Tokyo*, **50**, 1 (1955).

174. Konowaloff, M., and S. Dobrowolski, *J. Russ. Phys. Chem.*, **37**, 548 (1905).

175. Koopal, S. A., *Rec. Trav. Chim.*, **34**, 115 (1915).

176. Langford, C. H., *J. Am. Chem. Soc.*, **82**, 1503 (1960).

177. Langlois, G., *Compt. rend.*, **168**, 1052 (1919).

178. Lappert, M. F., *Proc. Chem. Soc.*, 121 (1957).

178a. Lappert, M. F., *J. Chem. Soc.*, 542 (1962).

179. Lapworth, A., and R. Robinson, *Mem. Proc. Manchester Lit. Phil. Soc.*, **72**, 243 (1928).

180. Laubengayer, A. W., and D. S. Sears, *J. Am. Chem. Soc.*, **67**, 164 (1945).

181. Lauer, K., and R. Oda, *J. Prakt. Chem.*, **143**, 139 (1935).

182. Lebedev, N. N., *J. Gen. Chem. U.S.S.R.*, **21**, 1788 (1951).

183. List, R., and M. Stein, *Ber.*, **31**, 1648 (1898).

184. Losev, I. P., O. V. Smirnova, and L. P. Ryadneva, *Sb. Statei Obshch. Khim. Akad. Nauk*, U.S.S.R., **1**, 548 (1953); through *C.A.*, **49**, 832 (1955).

185. Lutz, R. E., *J. Am. Chem. Soc.*, **49**, 1106 (1927).

186. Lutz, R. E., *J. Am. Chem. Soc.*, **56**, 1378 (1934).

187. Lutz, R. E., *Org. Syn.*, **20**, 29 (1940).

188. Lutz, R. E., D. T. Merritt, and M. Couper, *J. Org. Chem.*, **4**, 95 (1939).

189. Lutz, R. E., and R. J. Taylor, *J. Am. Chem. Soc.*, **55**, 1168 (1933).

190. Lutz, R. E., and A. W. Winne, *J. Am. Chem. Soc.*, **56**, 445 (1934).

191. Mackor, E. L., A. Hofstra, and J. H. Van der Waals, *Trans. Faraday Soc.*, **54**, 66, 186 (1958).

192. Malinovskii, M. S., *J. Gen. Chem.* U.S.S.R., **17**, 2235 (1947); *C.A.*, **43**, 591 (1949).

193. Man, E. H., and C. R. Hauser, *J. Org. Chem.*, **17**, 397 (1952).

194. Marchetti, F. A., and M. L. Stein, *Farmaco (Pavia) Ed. Sci.*, **10**, 243 (1955)

195. Martin, D. R., *Chem. Rev.*, **34**, 461 (1944); **42**, 581 (1948).

196. Martin, D. R., and R. Dial, *J. Am. Chem. Soc.*, **72**, 852 (1950).

197. Mayer, F., and W. Fischbach, *Ber.*, **58**, 1251 (1925).

198. Mayer, F., and L. van Zutphen, Dissertation, Frankfurt, 1923.

199. McCaulay, D. A., personal communication.

200. McCaulay, D. A., W. S. Higley, and A. P. Lien, *J. Am. Chem. Soc.*, **78**, 3009 (1956).

201. McCaulay, D. A., B. H. Shoemaker, and A. P. Lien, *Ind. Eng. Chem.*, **42**, 2103 (1950).

202. McCaulay, D. A., and A. P. Lien, *J. Am. Chem. Soc.*, **73**, 2013 (1951).

203. McCaulay, D. A., and A. P. Lien, *J. Am. Chem. Soc.*, **74**, 6246 (1952).

204. McDuffie, H. F., Jr., and G. Dougherty, *J. Am. Chem. Soc.*, **64**, 297 (1942).

205. McGary, C. W., Jr., Y. Okamoto, and H. C. Brown, *J. Am. Chem. Soc.*, **77**, 3037 (1955).

206. Meerwein, H., *Ann.*, **455**, 227 (1927).

207. Melamed, S., U.S. Pat. 2,776,921 (1957).

208. Mohrbacher, R. J., and N. H. Cromwell, *J. Am. Chem. Soc.*, **79**, 401 (1957).

209. Montagne, P. J., *Rec. Trav. Chim.*, **27**, 336 (1908).

210. Mouneyrat, A., *Bull. Soc. Chim. France*, (3) **19**, 557 (1898).

211. Mourue, C., *Bull. Soc. Chim. France*, (3) **9**, 568 (1893).

212. Muetterties, E. L., *J. Am. Chem. Soc.*, **79**, 6563 (1957).

212a. Nelson, K. L., *J. Org. Chem.*, **21**, 145 (1956).

213. Nelson, K. L., and H. C. Brown, in "The Chemistry of Petroleum Hydrocarbons," Vol. 3, Reinhold Publishing Corp., New York, p. 512, 1953.

214. Nenitzescu, C. D., and I. P. Cantuniari, *Ber.*, **65**, 1449 (1932).

215. Nenitzescu, C. D., and D. A. Isacescu, *Ber.*, **66**, 1100 (1933).

216. Newman, M. S., and E. Easterbrook, *J. Am. Chem. Soc.*, **77**, 3763 (1955).

217. Norris, J. F., and D. Rubinstein, *J. Am. Chem. Soc.*, **61**, 1163 (1939).

218. Oae, S., *J. Am. Chem. Soc.*, **78**, 4030 (1956).

219. Oddy, H. G., *J. Am. Chem. Soc.*, **45**, 2156 (1923).

220. Ogata, Y., and R. Oda, *Bull. Inst. Phys. Chem. Research (Tokyo)*, **21**, 728 (1942).

221. Olah, G., "Einführung in die theoretische organische Chemie," Akademie Verlag, Berlin, p. 191, 1960.

222. Olah, G. A., unpublished.
223. Olah, G. A., and S. J. Kuhn, *J. Am. Chem. Soc.*, **82**, 2380 (1960).
224. Olah, G. A., and S. J. Kuhn, *J. Am. Chem. Soc.*, **80**, 6541 (1958).
225. Olah, G., and S. Kuhn, *Ber.*, **89**, 866 (1956).
226. Olah, G. A., and S. J. Kuhn, U.S. Pat. 3,000,986 (1961).
227. Olah, G. A., and S. J. Kuhn, unpublished.
228. Olah, G. A., S. J. Kuhn, and S. H. Flood, *J. Am. Chem. Soc.*, **83**, 4571 (1961).
229. Olah, G. A., S. J. Kuhn, and S. H. Flood, *J. Am. Chem. Soc.*, **84**, 1688 (1962).
230. Olah, G. A., S. J. Kuhn, and S. H. Flood, *J. Am. Chem. Soc.*, in press.
231. Olah, G. A., S. J. Kuhn, S. H. Flood, and J. C. Evans, *J. Am. Chem. Soc.*, **84**, 3687 (1962).
232. Olah, G. A., S. J. Kuhn, and J. A. Olah, *J. Chem. Soc.*, 2174 (1957).
233. Olah, G. A., S. J. Kuhn, and H. W. Quinn, U.S. Pat. 3,052, 723 (1962).
234. Olah, G. A., and S. J. Kuhn, *J. Am. Chem. Soc.*, in press.
235. Olah, G., and A. Pavlath, *Acta Chim. Acad. Sci. Hung.*, **3**, 425 (1953).
236. Olah, G. A., A. Pavlath, S. Kuhn, and G. Varsanyi, in "Elektronentheorie der homopolaren Bindung," p. 79, Akademie Verlag, Berlin, 1955.
237. Olah, G. A., and W. S. Tolgyesi, *J. Org. Chem.*, **26**, 2319 (1961).
238. Olah, G. A., W. S. Tolgyesi, and R. E. A. Dear, *J. Org. Chem.*, **27**, 3441 (1962).
239. Olivier, S. C. J., *Rec. Trav. Chim.*, **33**, 163 (1914).
240. Pace, E., *Gazz. Chim. Ital.*, **59**, 573 (1929).
241. Pajeau, R., *Bull. Soc. Chim. France*, 544 (1946).
242. Paul, M. A., and F. A. Long, *Chem. Rev.*, **57**, 1 (1957).
243. Paul, M. A., *J. Am. Chem. Soc.*, **74**, 141 (1952).
244. Paushkin, V. M., and A. V. Topchiev, *Bull. Acad. Sci. U.S.S.R., Div. Chem. Sci.*, **7**, 813 (1947).
245. Pepper, D. C., *Trans. Faraday Soc.*, **54**, 404 (1949).
246. Pepper, D. C., *Quart. Rev. (London)*, **8**, 88 (1954).
247. Pepper, D. C., and A. E. Somerfield, in "Cationic Polymerization," ed. P. H. Plesch, p. 75, Academic Press, New York, 1953.
248. Picha, G. M., U.S. Pat. 2,774,784 (1956).
249. Pitzer, K. S., and D. W. Scott, *J. Am. Chem. Soc.*, **65**, 805 (1943).
250. Plesch, P. H., "Cationic Polymerization," p. 85, Academic Press, New York, 1953.
251. Plesch, P. H., *J. Chem. Soc.*, 5431 (1950); 1653, 1659, 1662 (1953).
252. Plotnikov, V. A., and R. G. Vaisberg, *Zap. Inst. Khim. Akad. Nauk U.S.S.R.*, **7**, No. 1, 71 (1940); through *C.A.*, **35**, 2405 (1941).
253. Plotnikov, V. A., and S. I. Yakubson, *J. Phys. Chem. U.S.S.R.*, **12**, 113 (1938).
254. Porter, R. F., and E. E. Zeller, *J. Chem. Phys.*, **33**, 858 (1960).
255. Powell, J. W., and M. C. Whitey, *Proc. Chem. Soc.*, 412 (1960).
256. Price, C. C., "Organic Reactions," Vol. 3, John Wiley and Sons, New York, N.Y., Chapter 1, 1946.
257. Ralston, A. W., and R. J. Vander Wal, U.S. Pat. 2,197,709; *C.A.*, **34**, 5466 (1940).
258. Ralston, A. W., and R. J. Vander Wal, Brit. Pat. 528,312; *C.A.*, **35**, 7418 (1941).

259. Rebstock, M. C., and C. D. Stratton, *J. Am. Chem. Soc.*, **77**, 4054 (1955).
260. Renes, P. A., and C. H. MacGillavry, *Rec. Trav. Chim.*, **64**, 275 (1945).
261. Reppe, W., *et al.*, *Ann.*, **596**, 7 (1955).
262. Rice, B., J. A. Libacy, and G. W. Schaeffer, *J. Am. Chem. Soc.*, **77**, 2750 (1955).
263. Roberts, J. D., J. K. Sanford, F. L. J. Sixma, H. Cerfontain, and R. Zagt, *J. Am. Chem. Soc.*, **76**, 4525 (1954).
264. Robertson, P. W., J. E. Allan, K. N. Haldane, and M. G. Simmers, *J. Chem. Soc.*, 933 (1949).
265. Robertson, P. W., P. B. de la Mare, and W. T. Johnston, *J. Chem. Soc.*, 276 (1943).
266. Russell, G. A., *J. Am. Chem. Soc.*, **81**, 4815, 4825, 4831 (1959).
267. Russell, G. A., *J. Am. Chem. Soc.*, **81**, 4834 (1959).
268. Russell, K. E., in "Cationic Polymerization," ed. P. H. Plesch, p. 114, Academic Press, New York, 1953.
269. Satchell, D. P. N., *J. Chem. Soc.*, 1752 (1960).
270. Satchell, D. P. N., *J. Chem. Soc.*, 1453 (1961).
271. Satchell, D. P. N., *Proc. Chem. Soc.*, 355 (1960).
272. Satchell, D. P. N., *J. Chem. Soc.*, 3822 (1961).
273. Schmerling, L., *J. Am. Chem. Soc.*, **67**, 1778 (1945).
274. Schmerling, L., U.S. Pat. 2,485,017 (1949).
275. Schramm, J., *Ber.*, **26**, 1709 (1893).
276. Schukowsky, S., *Bull. Soc. Chim. France*, (3) **16**, 126 (1896).
277. Schwab, G. M., H. S. Taylor, and R. Spence, "Catalysis," Van Nostrand Corp., New York, p. 70, 1937.
278. Scott, F. L., and R. E. Oesterling, *J. Am. Chem. Soc.*, **82**, 5247 (1960).
279. Searles, S., K. A. Dallart, and E. F. Lutz, *J. Am. Chem. Soc.*, **79**, 948 (1957).
280. Sehgal, J. M., T. R. Seshadri, and K. L. Vadehra, *Chem. and Ind.*, 252 (1956).
281. Serijan, K. T., H. F. Hipsher, and L. C. Gibbons, *J. Am. Chem. Soc.*, **71**, 873 (1949).
282. Seymour, E. L., and F. Fairbrother, Thesis (Seymour), Manchester University, 1943.
283. Shatenshtein, A. I., K. I. Zhdanova, and V. M. Basmanova, *J. Gen. Chem. U.S.S.R.*, **31**, 250 (1961).
284. Shinkle, S. H., U.S. Pat. 2,016,026 (1935).
285. Shishido, K., and H. Nozaki, *J. Am. Chem. Soc.*, **69**, 961 (1947).
286. Silva, R. D., *Bull. Soc. Chim. France*, (2), **36**, 24 (1881).
287. Silva, R. D., *Bull. Soc. Chim. France*, (2), **41**, 448 (1884).
288. Silva, R. D., *Compt. rend.*, **89**, 606 (1879).
289. Silva, R. D., *Jahresbericht über die Fortschritte der Chemie*, 379 (1879).
290. Simons, J. H., and G. C. Bassler, *J. Am. Chem. Soc.*, **63**, 880 (1941).
291. Simons, J. H., and H. Hart, *J. Am. Chem. Soc.*, **69**, 979 (1947).
292. Sisido, K., and S. Kato, *J. Soc. Chem. Ind. Japan*, **43**, 232 (1940); through *C.A.*, **35**, 1026 (1941).
293. Skinner, H. A., and N. B. Smith, *J. Chem. Soc.*, 2324 (1954).
294. Smith, L. C., and L. P. Hammett, *J. Am. Chem. Soc.*, **67**, 23 (1945).
295. Smoot, C. R., and H. C. Brown, *J. Am. Chem. Soc.*, **78**, 6245, 6249 (1956).

296. Snyder, H. R., and C. T. Elston, *J. Am. Chem. Soc.*, **77**, 364 (1955).
297. Stieber, A., *Compt. rend.*, **195**, 610 (1932).
298. Stock, A., *Ber.*, **34**, 949 (1901).
299. Stock, L. M., and H. C. Brown, *J. Am. Chem. Soc.*, **81**, 3323 (1959).
300. Stock, L. M., and H. C. Brown, *J. Am. Chem. Soc.*, **82**, 1942 (1961).
301. Stock, L. M., and H. C. Brown, *J. Am. Chem. Soc.*, **84**, 1242 (1962).
301a. Stone, F. G. A., *Chem. Rev.*, **58**, 110 (1958).
302. Stubbs, F. J., C. D. Williams, and C. N. Hinshelwood, *J. Chem. Soc.*, 1065 (1948).
303. Susz, B. P., and P. Chalandon, *Helv. Chim. Acta*, **41**, 1332 (1958).
304. Suzuki, M., *Sci. Rept. Tohoku Univ.*, **39**, 182 (1956).
305. Szmant, H. H., and J. Dudek, *J. Am. Chem. Soc.*, **71**, 3763 (1949).
306. Terenin, A., W. Filimonow, and D. Bystrow, *Z. Electrochem.*, **62**, 181 (1958).
307. Teyssié, P., and G. Smets, *J. Polymer Sci.*, **20**, 351 (1956).
308. Thomas, C. A., "Anhydrous Aluminum Chloride in Organic Chemistry," Reinhold Publ Comp., p. 468, 1941.
309. Topchiev, A. V., *Proc. Gubkin Petrol. Inst. Moscow*, **6**, 35 (1947).
310. Topchiev, A. V., *et al.*, *Proc. Acad. Sci. U.S.S.R.*, *Chem. Sect.*, **80**, 381 (1951).
311. Truce, W. E., J. A. Simmons, and H. E. Hill, *J. Am. Chem. Soc.*, **75**, 5411 (1953).
312. Truce, W. E., and C. W. Vriesen, *J. Am. Chem. Soc.*, **75**, 2525 (1953).
313. Ulich, H., and W. Nespital, *Z. anorg. allgem. Chem.*, **44**, 752 (1931).
314. Van der Laan, F. H., *Rec. Trav. Chim.*, **26**, 1 (1907).
315. Van der Meulen, B. A., and H. A. Heller, *J. Am. Chem. Soc.*, **54**, 4404 (1931).
316. Van Dyke, R. E., and H. E. Crawford, *J. Am. Chem. Soc.*, **72**, 2829 (1950).
317. Vavon, G., J. Bolle, and J. Calin, *Bull. Soc. Chim. France*, (5) **6**, 1025 (1939).
318. Venkataraman, K., *J. Indian Chem. Soc.*, **7**, 157 (1930).
319. Van Auwers, K., and E. Risse, *Ann.*, **502**, 282 (1933).
320. Vrieson, C. W., Ph.D. Thesis, Purdue University, 1952.
321. Waddington, T. C., and J. A. White, *Proc. Chem. Soc.*, 315 (1960).
322. Walker, D. G., Abstract of Papers p. 9T, 140th Meeting, American Chemical Society, Chicago, Ill., Sept. 1961.
323. Wenzel, F., and R. Kugel, *Monatsh.*, **35**, 953 (1914).
324. Wiberg, E., and U. Heubaum, *Z. anorg. allgem Chem.*, **225**, 270 (1935).
325. Williams, G., *J. Chem. Soc.*, 775 (1940).
326. Wispek, P., and R. Zuber, *Ann.*, **218**, 374 (1883).
327. Wittig, G., and H. Harle, *Ann.*, **623**, 17 (1959).
328. Wolfram, A., L. Schornig, and E. Hausdorfer, Brit. Pat. 330,916 (1930); through *C.A.*, **24**, 6031 (1930).
329. Yamase, Y., *Bull. Chem. Soc. Japan*, **34**, 480, 484 (1961).
330. Yamase, Y., and R. Goto, *J. Chem. Soc. Japan, Pure Chem. Sect.*, **81**, 1906 (1960).
331. Van Zoeren, G. J., U.S. Pat. 2,349,779 (1944).

CHAPTER VI

Mechanistic Aspects

I. Introduction

Since the preceding chapter, dealing with some questions of the mechanism (particularly relating to selectivity) of Friedel-Crafts reactions, was written, substantial progress has been made in the understanding of the more fundamental mechanistic aspects of these reactions. Therefore it seems warranted to review these more recent studies, emphasizing what is considered an emerging general picture of mechanism.

Concerning the general mechanistic aspects of the Friedel-Crafts reactions, they require an electrophilic reagent of sufficient reactivity to be able to interact with either π- or σ-donor type substrates. Generally, the polarity of a C–halogen (or other carbon–ligand bond) is not sufficiently high to allow the halide (or related reagent) to react directly with hydrocarbon substrates. Such reactions however, generally take place with heteroatom n-donor bases, frequently without catalysts, such as in Schotten-Bauman acylations, akylations, etc. Hence the need for catalysts which, by coordinating with donor sites of the reactant or protonating π- or σ-bonds, are capable of producing highly polarized or cationic reagents of sufficient electrophilicity to permit the reactions to take place.

The reaction of the formed electrophilic reagent with the substrate to be substituted (or added to) is the subsequent step in the reaction.

It must be emphasized that Friedel-Crafts reactions are of a general nature, equally adaptable to aromatic and aliphatic systems. Thus instead of artificially trying to divide them from a mechanistic point of view into aromatic and aliphatic reactions, although such a division may be still justified from a practical, preparative point of view, it is suggested that differentiation be made according to the type (electronic) of substrates involved in the reactions.

I have recently suggested (1) that electrophilic reactions in general can be divided into three major types according to the nature of the electron donor substrates involved *i.e.*, π-, σ- and n-donors.

Accordingly, it is suggested that the mechanism of Friedel-Crafts reactions can be best discussed by differentiating the reactions of π-donor systems from those of σ-donors (n-donor systems generally react in a straightforward manner with electrophiles by interaction with a lone pair of electrons and will not be discussed).

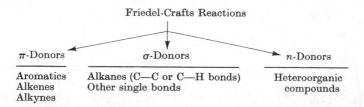

The reaction steps of Friedel-Crafts reactions can be defined as: (a) The reactant–catalyst interaction, (b) the subsequent reaction of the substrate with the reactant–catalyst complex, and (c) the final elimination or addition step of the intermediate carbocation formed in step b.

Complexing of the reactant by the acid catalyst introduces into it sufficient polarity and thus renders it active as an electrophilic substituting agent. Thus the formation of polarized and, in the limiting case, ionic reactant-catalyst complexes is an essential part of every Friedel-Crafts system.

Reaction of the electrophilic reactant with the substrate generally results in the formation of an intermediate carbocation, which then leads to final products. Knowledge of the nature of the intermediate complexes provides, together with rate and product distribution data, a better understanding of the mechanism of the reactions.

II. Reactant-Catalyst Complexes

The nature of the electrophilic reactants involved in Friedel-Crafts reactions was the focus of major research interest for many years (1). It is not considered within the scope of the present chapter to attempt to give a detailed survey of this large and complex field. From the point of view of the general mechanistic concept of the reactions, it suffices to say that both alkyl and acyl halides (as well as oxygen, sulfur, and other heteroatom containing reactants are capable of coordinating, by utilizing the non-bonded electron pairs of the heteroatoms, with acid catalysts. It is not considered essential to try to determine whether the acid catalysts are of Lewis or Brønsted nature. It should be clear, however, that Lewis acid catalysts, such as the most commonly used aluminum, boron, etc., halides can readily coordinate to non-bonded, but only weakly interact with bonded electron pairs. In the latter case (such as in the reactions of π- and σ-donor hydrocarbons), much stronger protic acid catalysts are generally needed to produce the cationic (or highly polarized) reactive intermediates. Thus reactions in these cases are preferentially catalyzed by strong conjugate Friedel-Crafts Brønsted acid catalysts.

$$R{-}X + MX_{3,5} \; \rightleftarrows \; \overset{\delta+}{R}{-}X \to \overset{\delta-}{M}X_{3,5} \; \rightleftarrows \; [R^+]MX_{4,6}^-$$

$$RCOX + MX_{3,5} \; \rightleftarrows \; \underset{\underset{X}{|}}{\overset{\delta+}{R}C}{=}O \to \overset{\delta-}{M}X_{3,5} \; \rightleftarrows \; [RCO^+]MX_{4,6}^-$$

$$RCH{=}CH_2 + HMX_{4,6} \; \rightleftarrows \; R\overset{+}{C}HCH_3 \; MX_{4,6}^-$$

$$RC{\equiv}CH + HMX_{4,6} \; \rightleftarrows \; R\overset{+}{C}H{=}CH_2 \; MX_{4,6}^-$$

$$R_2CH{-}CH_3 + HMX_{4,6} \; \underset{}{\overset{-H_2}{\rightleftarrows}} \; R_2\overset{+}{C}H \; MX_{4,6}^- + CH_4$$

Whether *de facto* free carbocations (2) (or related reactant ions such as NO_2^+, RSO_2^+, halogen$^+$, etc.) are involved in the reactions, or only highly polarized donor—acceptor complexes (or the corresponding tight ion pairs), is from the general mechanistic point of view (at least in the first approximation) of no decisive significance.

As Friedel-Crafts reactions are rarely, if ever, carried out under strictly anhydrous conditions, and the substitution reactions usually produce an equivalent amount of acid (from the eliminated protons in the reactions), there is generally more than the needed "catalytic amount" of protic acid present in the system. This is clearly beneficial, or even essential, for alkylation reactions with hydrocarbons (alkenes, alkynes, alkanes). *n*-Donor reagents, however, in principle can react with Lewis acid catalysts under strict exclusion of even traces of moisture or any other proton source.

The last decade brought about rapid progress in the investigation of cationic reactive intermediates (3), particularly carbocations (4, 5). Similarly, the nature of many polarized donor—acceptor complexes was studied (for summaries see ref. 1). Thus, indeed, we now know (4) not only through kinetic and sterochemical studies but also from direct observation, under stable ion conditions (in media of low nucleophilicity, thus generally in highly acidic solvent systems, using methods such as n.m.r., infra-red, Raman, and e.s.r. spectroscopy and X-ray crystallography) the structural aspects of many carbocationic intermediates.

Subsequent discussion first deals with structural data relating the intermediate complexes of alkylation, acylation, sulfonylation, nitration, etc., reactions (as representatives of substitutions involving typical electrophilic reagents) and their role in elucidating the mechanism of the reactions. Limitations due to the wide scope of the topic do not allow a detailed survey to be given, and the reader is referred to the original literature for details.

1. Alkylating Agents

A. Alkyl Halide–Lewis Acid Halide Systems

The transitory existence of alkylcarbenium ions in alkyl halide–Lewis acid halide systems has been inferred from a variety of early observations, such as vapor pressure depressions of CH_3Cl and C_2H_5Cl in the presence of gallium chloride (6), the electric conductivities of aluminum chloride in ethyl chloride (7) and of alkyl fluorides in boron trifluoride (8), and the effect of ethyl bromide on the dipole moment of aluminum bromide (9). However, in none of these cases has the existence of a well-defined stable ionic salt or complex been established—even at low temperatures.

Byrne (10) reported that methyl chloride forms 1:1 compounds with stannic chloride (CH_3Cl–$SnCl_4$, dissociation pressure 40.3 mm. at $-64°$, calculated m.p. $-50°$, heat of formation -4.69 kcal./mole) and antimony pentachloride (CH_3Cl–$SbCl_5$, dissociation pressure 6.5 mm. at $-50°$, calculated m.p. $90°$, heat of formation -8.92 kcal./mole.) Nelson subsequently (11) investigated the infra-red spectrum of methyl chloride in stannic chloride solution (at 30 and $-40°$) and in antimony pentachloride solution (at 28 and $-12°$). No evidence of compound formation was found in the CH_3Cl–$SnCl_4$ solutions. A temperature-dependent infra-red band at 688 cm.$^{-1}$ in the CH_3Cl–$SbCl_5$ solutions was assigned to the C–Cl stretching motion in the CH_3Cl–$SbCl_5$ addition compound. The spectra of the CH_3Cl–$SbCl_5$ solutions are consistent with a linear C--Cl-Sb bond in CH_3Cl–$SbCl_5$, but the evidence is insufficient to rule out an angular bond which seems more probable from other considerations. The results suggest that the CH_3Cl–$SnCl_4$ and CH_3Cl–$SbCl_5$ addition compounds are more accurately described as slightly polarized donor–acceptor complexes, than as ion pairs.

Alkyl fluoride–boron trifluoride complexes were first investigated by Olah, Kuhn, and Olah (8) 1:1 Addition complexes were observed at low temperatures, and their specific conductivities were measured. The specific conductivities of the propyl fluoride– and butyl floride–boron trifluoride systems were found to be three orders of magnitude larger than those of the methyl fluoride– and ethyl fluoride–boron trifluoride systems, respectively. On recovery the latter systems dissociated into their starting components, whereas the former gave polymer mixtures (12).

Nakane, Kurihara, and Natsubori (13) established the equilibrium constants of boron isotope exchange between boron trifluoride gas on one side and boron trifluoride–methyl fluoride, –methyl chloride, –isopropyl chloride, and –t-butyl chloride complexes. The value of the equilibrium constants, which represents the thermodynamic isotope effect, was related to the polarity, stability, and catalytic activity of the complexes.

The results indicated that all the investigated complexes are only polarized covalent halide complexes, and the polarity of these complexes decreases in the following order

$$\overset{\delta -}{BF_3} \leftarrow \overset{\delta +}{ClC(CH_3)_3} > \overset{\delta -}{BF_3} \leftarrow \overset{\delta +}{FCH_3} > \overset{\delta -}{BF_3} \leftarrow \overset{\delta +}{ClCH(CH_3)_2} > \overset{\delta -}{BF_3} \leftarrow \overset{\delta +}{ClCH_3}$$

In subsequent work, Nakane, Kurihara, and Natsubori (13) extended their investigation to boron trifluoride–ethyl fluoride and boron trifluoride–isopropyl fluoride systems. They found these systems also to be donor–acceptor complexes with the order of polarity

$$\overset{\delta -}{BF_3} \leftarrow \overset{\delta +}{FCH(CH_3)_2} > \overset{\delta -}{BF_3} \leftarrow \overset{\delta +}{FCH_2CH_3}$$

The boron trifluoride–alkyl fluoride (–alkyl chloride) complexes gave no evidence of alkylcarbenium ion formation. It must be emphasized, however, that (a) physical investigation of the binary systems was carried out at such low temperatures (generally below $-100°$) that ionization of the halides could hardly be expected (with the exception of highly reactive tertiary halides) and (b) the methods used could not be relied on to detect a small ionization equilibrium, even if it existed.

The fact that the boron trifluoride–t-butyl fluoride (and –t-butyl chloride) system and the boron trifluoride–isopropyl fluoride system yielded polymer mixtures on attempted recovery may indicate an equilibrium. On measuring the absorption spectra of boron trifluoride complexes—with methyl, ethyl, and isopropyl fluoride and isopropyl and t-butyl chloride—

$$\overset{\delta-}{BF_3} \overset{\delta+}{FC(CH_3)_3} \rightleftharpoons BF_4^- \overset{+}{C}(CH_3)_3$$

$$\rightleftharpoons HBF_4 + CH_2 = C(CH_3)_2$$

no absorption bands were reported above 220 nm. Finally Nakane found that the polarity of alkyl halide–boron trifluoride complexes (hence the elcetron deficiency of polar cations in the complex) decreases in the following order.

$$(CH_3)_2\overset{\delta+}{CHF}—\overset{\delta-}{BF_3} > CH_3CH_2\overset{\delta+}{F}—\overset{\delta-}{BF_3} > (CH_3)_3\overset{\delta+}{CCl}—\overset{\delta-}{BF_3} >$$

$$\overset{\delta+}{CH_3F}—\overset{\delta-}{BF_3} > (CH_3)_2\overset{\delta+}{CHCl}—\overset{\delta-}{BF_3} > \overset{\delta+}{CH_3Cl}—\overset{\delta-}{BF_3}$$

These results agree with the conductivity work of Olah.

Attempts to obtain alkylcarbenium ions by dissolving alkyl chlorides (bromides) in liquid Lewis acid halides [stannic chloride, titanium(IV) chloride, antimony pentachloride, etc.] were unsuccessful. Although stable solutions could be obtained at low temperature with, for example, t-butyl chloride, the observed n.m.r. chemical shifts generally were not more deshielded than 0.5 p.p.m. compared to the parent halides and thus could be attributed only to weak donor–acceptor complexes but not to carbenium ions. The results of these investigations seem to indicate that either the Lewis acids used were too weak to cause sufficient ionization of the C–X bond, or that the systems were not suitable to stabilize sufficiently the carbenium ion salts formed (an equilibrium containing a small concentration of the carbenium ion cannot be excluded in any of these systems but, if such is the case, rapid equilibration through the corresponding dialkylhalonium ions would take preferentially place.) It is possible that the donor strength of the alkyl halide also plays an important role. This could explain why alkyl fluorides were found to ionize more readily than chlorides or bromides. In the case of ionization with Lewis acid halide catalysts, the bond energy of the C–F bond being cleaved is compensated for by the bond energy of the metal–halogen bond being formed. It is rather unfortunate that, owing to halogen exchange, alkyl fluorides generally cannot be used in investigations involving other halogen-containing Lewis acid halides.

In 1962, Olah and co-workers, first reported (12) the formation of stable alkylcarbenium ion salts when *t*-butyl fluoride was dissolved in excess antimony pentafluoride (serving as both the Lewis acid and the solvent).

It was found that when *t*-butyl fluoride is passed into purified liquid antimony pentafluoride (with the exclusion of moisture), a stable complex layer is formed on top of excess antimony pentafluoride. When this layer was separated and its proton and carbon magnetic resonance spectra, as well as its infra-red and later Raman spectra were studied, the spectra were found to be those of the trimethylcarbenium ion.

$$(CH_3)_3CF + (SbF_5)_2 \rightleftharpoons (CH_3)_3C^+Sb_2F_{11}^-$$

The method of obtaining stable alkylcarbenium ion salts by the interaction of alkyl fluorides with antimony pentafluoride (neat or in subsequent work in sulfur dioxide, sulfuryl fluoride, or sulfuryl chloride fluoride solution) was evaluated in detail, and the investigations extended to all isomeric C_3-, C_4-, C_5-, C_6-, C_7-, and C_8-alkyl fluorides (as well as chlorides) and related precursors.

Whereas under stable ion conditions alkyl fluorides (as well as other alkyl halides) in excess SbF_5 (neat or diluted with SO_2 or SO_2ClF) generally form stable alkylcarbenium ions, methyl and ethyl fluoride give only rapidly exchanging donor–acceptor complexes (14) although, as shown in the case of the ethyl complex by isotope scrambling experiments, exchange processes involve ion pairs $(R^+SbF_6^-$ or $R^+Sb_2F_{11}^-)$.

In Lewis acid halide-catalyzed Friedel-Crafts alkylations with alkyl halides, the nature of the reactant–catalyst complexes is not necessarily that of alkylcarbenium ion salts, not only in the case of primary but also secondary or even tertiary systems. As a matter of fact, carbenium ion complexes can be observed only in very low nucleophilicity (generally highly acidic) solvent systems and at low temperatures, thus under conditions rarely encountered in carrying out Friedel-Crafts reactions.   The difference between an ion-pair complex and the corresponding donor–acceptor complex (probably both involved in rapid equilibration processes, as in the case of the methyl and ethyl fluoride–antimony pentafluoride complexes mentioned) is of lesser importance when considering their roles as electrophilic reactants.   A further aspect of alkyl halide complexes must be considered when alkyl halides other than fluorides are involved.   The non-bonded electron pairs of halogens in these halides can act as electron donor sites and consequently can be alkylated by incipient alkylcarbenium ions. For example, in the case of the methyl bromide–aluminum bromide system, dimethylbromonium ion formation takes place to the extent that the easily isolable 1 : 1 complex is indeed an equilibrating system in which the dimethylbromonium ion is a major contributor (15).

$$2\overset{\delta+}{C}H_3\overset{\delta-}{Br} \rightarrow Al\overset{}{Br}_3 \rightleftharpoons CH_3\overset{+}{Br}CH_3\ Al_2Br_7^- \rightleftharpoons CH_3Br + \overset{\delta+}{C}H_3\overset{\delta-}{Br} \rightarrow Al_2\overset{}{Br}_6$$

Methyl chloride–gallium trichloride (as well as methyl bromide–gallium trichloride) systems show similar behavior (16).

A series of dialkylhalonium ions (generally as the stable fluoro-antimonates) were prepared by Olah and DeMember (17) in pure form, and their structure and reactivity studied. Because of the high electronegativity of fluorine, alkyl fluorides do not form dialkyl-fluoronium ions and are alkylated instead on carbon.

The degree to which dialkylhalonium ions themselves are involved as alkylating agents (through nucleophilic displacement reactions by substrate aromatics or olefins) in alkylation with alkyl halides cannot at this time be determined, but in some reactions, such as alkylations by alkyl iodides, they may be of importance.

B. Alcohol- and Alkene–Brønsted Acid Systems

Electronic absorption spectra of alcohols and olefins in sulfuric acid were obtained by Rosenbaum and Symons (18). They observed absorption maxima in the 290-nm region for several simple aliphatic alcohols and ascribed this absorption to the corresponding simple alkylcarbenium ions. A reinvestigation of the problem by Finch and Symons (19) revealed that acetic acid, used as a solvent to introduce the alcohols and olefins into the sulfuric acid, participated in a condensation reaction, giving protonated mesityl oxide, which is responsible for the 290-nm absorption.

Olah and co-workers (20) carried out a detailed ultraviolet investigation of 10^{-1}–10^{-2} M solutions of alkyl cations in FSO_3H–SbF_5 solution at $-60°$ and showed that they have no ultraviolet absorption above 210 nm. Solutions of t-butyl, t-amyl, and t-heptyl cations in FSO_3H–SbF_5 are stable for hours at $-60°$, and n.m.r. studies have demonstrated that the ions are quantitatively formed from their corresponding tertiary alcohols in these systems (21).

It can now be concluded (22) that alkylcarbenium ions show no ultraviolet absorptions above 210 nm and sulfuric acid is not a suitable solvent system in which to observe these ions.

Attempts to generate simple alkylcarbenium ions in sulfuric acid or oleum solution generally result in the formation of complex mixtures in which the stable carbenium ions present are, as shown by Deno (23), of the methylated cyclopentenyl cation type (stabilized allylic cations).

Sulfuric acid and oleum as solvents have the serious disadvantage of being quite viscous and possessing relatively high freezing points. This generally results in the need to carry out investigations at or above $+10°$. At these temperatures the rate of secondary reactions leading to cyclized allylic type ions is so rapid that no simple alkyl(cycloalkyl)carbenium ions corresponding to the alcoholic precursors can be observed.

Alkyl cations thus are not directly observed in sulfuric acid systems because they are transient intermediates present in low concentrations and react with the olefins present in equilibrium. From measurement of solvolysis rates for allylic halides (24), the direct observation of allylic cation equilibria, and the equilibrium constant for the t-butyl alcohol–2-methylpropene system (25), the ratio of t-butyl cation to 2-methylpropene in 96% H_2SO_4 has been calculated to be $10^{-3.5}$. Thus it is evident that sulfuric acid is not a suitable system for the observation of stable alkyl cations. In other acid systems, such as BF_3–$2CH_3COOH$–ethylene dichloride, olefins such as butene are alkylated and undergo hydride transfer, producing hydrocarbons and alkylated alkenyl cations as end products (26). This behavior is expected to be quite general in conventional "strong" acids.

Fluorosulfuric acid is one of the strongest proton acids that has yet been studied (27). H_0 for the neat acid is about -13.5 (compared with -11 for 100% sulfuric acid and -10 for anhydrous hydrogen fluoride). At the same time, fluorosulfuric acid has a low freezing point ($-87.3°$) and can be readily purified. As expected, very few co-acids are capable of enhancing the acidity of fluorosulfuric acid. Woolf (28), as well as Thompson, Barr, and Gillespie, (29), however, observed that antimony pentafluoride acts as an acid in fluorosulfuric acid solution by enhancing its ionization

Olah and co-workers (21) were able to obtain carbenium ions from alcohols in FSO_3H by increasing the acidity of the system by adding SbF_5 as co-acid and by lowering the temperature of observation

$$(CH_3)_3COH \xrightarrow{\text{SbF}_5\text{–FSO}_3\text{H}} (CH_3)_3C^+SbF_5FSO_3^- + H_3O^+$$

It was also found that solutions of tertiary and secondary alcohols in SbF_5–SO_2 or SbF_5–SO_2ClF show formation of the corresponding carbenium ions, indicating ionization in the strong Lewis acid itself.

$$R_3COH \xrightarrow{SbF_5} R_3C^+SbF_5OH^-$$

Brouwer and Mackor (30) found that concentrated, stable solutions of a series of tertiary alkyl cations can be prepared in $HF-SbF_5$ solution, and magnetic resonance spectra were recorded. The t-butyl, t-pentyl, and t-hexyl cations were observed in this solvent system. The spectra were identical with those obtained previously by Olah in SbF_5 and in FSO_3H-SbF_5 solvent systems.

Hogeveen reported the preparation of trimethylcarbonium tetrafluoroborate through ionization of t-butyl alcohol in $HF-BF_3$ solution (31).

The generation of stable alkylcarbenium ions from alcohols clearly indicates the great advantage of increasing the acidity and using acid systems with a low freezing point.

Primary and less reactive secondary alcohols are protonated in FSO_3H-SbF_5 solution at low temperatures $(-60°)$ and show very slow, if any, exchange with the solvent (32,33).

$$CH_3CH_2CH_2OH \underset{}{\overset{H^+}{\rightleftharpoons}} CH_3CH_2CH_2OH_2^+$$

$$\underset{OH}{CH_3CHCH_2} \overset{H^+}{\rightleftharpoons} \underset{^+OH_2}{CH_3CHCH_3}$$

Temperature-dependence studies of the n.m.r. spectra of protonated alcohols allowed to follow the first-order kinetics of dehydration (33) and to obtain the activation energies.

Stable alkylcarbenium ions can be formed from olefins in superacids such as $HF-SbF_5$ or FSO_3H-SbF_5 (30,34).

$$RCH{=}CH_2 \xrightarrow[HF-SbF_5]{FSO_3H-SbF_5} R\overset{+}{C}HCH_3$$

The nmr. resonance spectra of alkylcarbenium ions obtained through the protonation of olefins are frequently poorer than spectra obtained from halides, or from alcohols as precursors. Side reactions due to the great reactivity of carbenium ions with any excess of olefin may be responsible for the more complex systems.

Other related alkylating agents, including ethers (35), thiols (36), and sulfides (36), also give the corresponding alkylcarbenium ions when ionizing under stable ion conditions.

These studies gave direct experimental proof of the formation of alkylcarbenium ions from a variety of precursors and acid catalysts.

It must be emphasized again, however, that stable ions are formed in observable concentrations only under specific, low nucleophilic conditions. Under usual conditions of preparative Friedel-Crafts reactions, if these ions are formed at all, they will react fast with substrates present to be alkylated, and no observable concentrations of carbocations are generally present. Furthermore, polarized precursor complexes themselves can act as reactants similar to those of the ions themselves, when considered as ion pairs. Alkylations in these cases can be considered as displacement by the nucleophilic substrates (aromatics or olefins) and not as electrophilic attack by the carbenium ions.

2. Acylating Agents

Acylation intermediates generally show much higher stability than those of alkylation systems and therefore were investigated not only by spectroscopic studies of their solutions (infra-red, n.m.r.) but were also isolated in some instances as crystalline salts allowing X-ray structural determinations.

Acyl halides and acid anhydrides have been known for long to form complexes with acid catalysts, but it was only in 1943 that Seel (37) reported isolation of the first stable, well-identified acyl salt. Acetyl fluoride, when reacted at low temperature with boron trifluoride, gave a solid, crystalline 1 : 1 addition compound which decomposed quantitatively into its components without melting at $+20°$, the boiling point of acetyl fluoride. Based on analytical and electrical conductivity data, the compound was characterized as the ionic acetyl tetrafluoroborate ($CH_3CO^+BF_4^-$).

The acetyl cation structure of $CH_3COF \cdot BF_3$ was further proved by infra-red spectroscopic investigations of Susz and Wuhrmann (38). The presence of two functional groups in an acyl halide, namely, the carbonyl donor group and the ionizable halogen atom, suggests that two types of intermediates may be possible in the interaction of acyl halides with Lewis acid type Friedel-Crafts catalysts. For the acetyl chloride–aluminum chloride system, this was first suggested by Pfeiffer (39) and has been more recently studied on the basis of infra-red spectroscopic data by Susz and Wuhrmann (40) and by Cook (41). The ion salt $CH_3CO^+AlCl_4^-$ and the polarized donor–acceptor complex $CH_3\overset{\delta+}{C}O \ldots \overset{\delta-}{AlCl_3}$ (with Cl below) were clearly differentiated by means of their infra-red C=O absorptions. According to Susz's data the $CH_3COF \cdot BF_3$ complex prepared and kept at low temperatures is indeed predominantly the ionic acetyl tetrafluoroborate, $CH_3CO^+BF_4^-$, with only a very minor amount of the donor–acceptor complex detectable.

Olah and his co-workers (42) developed two independent methods for the preparation of stable acyl salts.

Acyl fluorides, when treated with Lewis acid type metal fluorides such as boron trifluoride, phosphorus pentafluoride, antimony pentafluoride, and arsenic pentafluoride, gave stable, well defined acyl salts

$$RCOF + MF_{3,5} \rightleftharpoons RCO^+MF_{4,6}^-$$

$$R = CH_3, C_2H_5, i\text{-}C_3H_7, t\text{-}C_4H_9, C_6H_5; M = B, P, As, Sb$$

Equimolar quantities of acetyl, propionyl, and benzoyl fluoride and the appropriate Lewis acid fluoride (BF_3, PF_5, SbF_5, AsF_5) were allowed to react in 1,1,2-trifluorotrichloroethane (Freon 113) or difluorodichloromethane (Freon 12) solutions at temperatures ranging between -78 and $0°$.

Acyl salts can also be prepared without the use of the sometimes inconvenient fluorides by the simple metathetic reaction of the corresponding acyl chlorides (bromides) with the appropriate anhydrous complex silver salts.

$$RCOCl + AgMF_{4,6} = RCO^+MF_{4,6}^- + AgCl$$

$$R = CH_3, C_2H_5, C_6H_5; M = B, As, P, Sb$$

The infra-red spectra of the acyl salts studied are characterized by a strong $C=O$ stretching vibration at about 2300 nm (38,40–42). A second infra-red absorption at about 2200 nm in solution spectra of systems containing the acetyl cation was identified as that of the diacetoacetyl cation $[(CH_3CO)_2CHCO^+]$ (43,44) formed by acetlyation of intermediate ketene in the system.

H^1 and C^{13} n.m.r. spectra (42,45), electron spectroscopic studies (46), and X-ray structural determinations (47) all substantiate the linear structure of the acetyl cation involving resonance forms,

$$CH_3-C^+=O \longleftrightarrow CH_3-C\equiv \overset{+}{O}$$

with the carbonyl carbon displaying substantial electron deficiency. Whereas the complex fluorides (such as fluoroantimonate or tetra-fluoroborate) studied under stable ion conditions are predominantly the ionic acyl salts, under usual Friedel-Crafts conditions in acyl halide–Lewis acid systems the acyl cations are generally in equilibrium with the corresponding donor–acceptor complexes

$$CH_3-C\overset{O \rightarrow AlCl_3}{\underset{Cl}{}} \; \rightleftharpoons \; CH_3CO^+AlCl_4^- \; \rightleftharpoons \; CH_3-C\overset{O}{\underset{Cl...AlCl_3}{}}$$

3. Sulfonylating Agents

The sulfonylation reaction can be regarded as an analog of the acylation reaction, in which a sulfonyl group is substituted for a carbonyl group and the product is a sulfone instead of a ketone. Olivier (48) was the first to study the mechanism of the sulfonylation reaction and the structure of the addition complex between aluminum chloride and benzenesulfonyl chloride, in sulfonyl chloride as solvent. The existence of an 1:1 addition complex was demonstrated. There has been no determination of the structure of the complex but, according to Jensen and Goldman (49), the following structures a, b, and c are possible.

$$\overset{+}{O}:\overset{-}{AlCl_3} \qquad\qquad O^- \qquad\qquad O$$
$$| \qquad\qquad\qquad | \qquad\qquad\qquad \|$$
$$R-\overset{}{S}-Cl \; \rightleftharpoons \; R-\overset{+}{S}{}^+-\overset{-}{Cl}:\overset{-}{AlCl_3} \; \rightleftharpoons \; R-\overset{}{S}{}^+AlCl^-$$
$$\| \qquad\qquad\qquad \| \qquad\qquad\qquad \|$$
$$O \qquad\qquad\qquad O \qquad\qquad\qquad O$$

$$a \qquad\qquad\qquad b \qquad\qquad\qquad c$$

Jenson and Goldman (49) studied in detail the mechanism of the sulfonylation reaction and, based primarily on kinetic evidence, considered a and c to be present in equilibrium, c being the effective sulfonylating agent.

Burton and Hopkins (50) claimed the use of sulfonylium perchlorates as reagents in sulfonylation reactions. It should be pointed out, however, that in the metathetic reaction of sulfonyl chlorides with silver perchlorate the formation of silver chloride does not necessarily prove the formation of an ionic complex, as covalent perchlorates can also be formed.

$$RSO_2Cl + AgClO_4 \xrightarrow{-AgCl} \; + RSO_2^+ClO_4^- \; \text{ or } \; RSO_2OClO_3$$

Klages and Malecki (51) subsequently studied "tosyl perchlorate" in the interaction of p-toluenesulfonyl chloride (bromide) with silver perchlorate in nitromethane solution. They concluded that sulfonyl cations, in contrast to acyl cations, are very electrophilic and react even with weakly nucleophilic anions to give sulfonyl halides. Thus p-toluenesulfonyl bromide with anhydrous silver tetrafluoroborate in benzene gave only the sulfonyl fluoride and no aryl sulfone was formed. Toluene with p-toluenesulfonyl chloride and silver perchlorate, however, gave p,p'-tolyl sulfone. As no direct physical observation of tosyl perchlorate was made, no conclusion can be reached as to whether it should be considered the covalent ester p-$CH_3C_3H_4SO_2OClO_3$ or the ion salt p-$CH_3C_3H_4SO_2^+ ClO_4^-$.

Lindner and Weber report (52) formation of the p-N,N-dimethylaminobenzenesulfonylium ion by the reaction of p-N,N-dimethylaminobenzenesulfonyl chloride with silver hexafluoroantimonate in sulfur dioxide solution.

$$p\text{-}(CH_3)_2NC_6H_4SO_2Cl + Ag[MF_6]$$

$$\xrightarrow[SO_2]{-25°} [p\text{-}(CH_3)_2NC_6H_4SO_2^+][MF_6^-] + AgCl$$

Olah, Ku, and Olah (53) studied a series of sulfonyl halide–antimony pentafluoride complexes and found them to be oxygen-coordinated donor–acceptor complexes. Only in the case of exceptionally stabilizing groups, such as p-N,N-dimethylaminophenylsulfonyl (52) and p-methoxyphenylsulfonyl (54) systems, were long-lived sulfonyl cations formed.

4. Nitrating Agents

Nitric acid is ionized in strong proton acids such as H_2SO_4, $HClO_4$, HF–BF_3, HF–SbF_5, etc. (55,56) forming, in certain cases even isolatable, nitronium salts.

$$HNO_3 + 2HClO_4 \rightleftharpoons NO_2^+ClO_4^- + H_3O^+ClO_4^-$$

$$HNO_3 + HF + 2BF_3 \rightleftharpoons NO_2^+BF_4^- + BF_3 \cdot H_2O$$

Sulfur trioxide forms with nitric acid in nitromethane solution nitronium hydrogen disulfate.

$$HNO_3 + 2SO_3 \rightleftharpoons NO_2^+HS_2O_7^-$$

Nitronium tetrafluoroborate and related complex fluoro nitronium salts are very reactive nitrating agents (56,57) which are presently commercially available.

Nitryl fluoride forms stable nitronium complexes with Lewis acid halides such as BF_3, SbF_5, AsF_5, NbF_5, TaF_5, etc. (58).

$$NO_2F + MF_{3,5} \rightleftharpoons NO_2^+ MF_{4,6}^-$$

Nitryl chloride also forms ionic nitronium salts with certain metal chlorides, such as $SbCl_5$. With anhydrous silver salts it gives the expected metathesis products

$$NO_2Cl + AgMF_{4,6} \longrightarrow NO_2^+ MF_{4,6}^- + AgCl$$

Alkyl and metal nitrates were reported to form various complexes with Lewis acid halides (59). A reinvestigation of the complex formation with boron trifluoride showed that nitronium tetrafluoroborate is the stable nitronium salt formed (60).

$$3AgNO_3 + 8BF_3 \longrightarrow 3NO_2^+ BF_4^- + 3AgBF_4 + B_2O_3$$

White, stable but moisture-sensitive complexes are obtained when nitrogen oxides react with boron trifluoride under anhydrous conditions (61,62). Various investigators have proposed the existence of the complexes

$$N_2O_3 \cdot BF_3 \qquad N_2O_4 \cdot BF_3 \qquad N_2O_5 \cdot BF_3$$
$$N_2O_3 \cdot 2BF_3 \qquad N_2O_4 \cdot 2BF_3$$

with the structures shown (63–65).

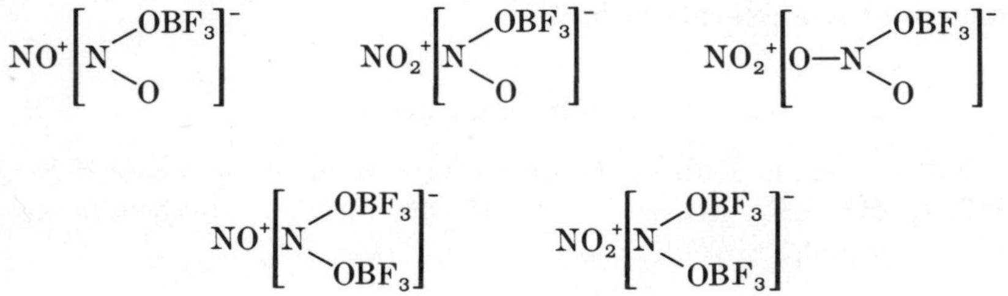

$$NO^+\left[N\begin{matrix} \diagup OBF_3 \\ \diagdown O \end{matrix}\right]^- \qquad NO_2^+\left[N\begin{matrix} \diagup OBF_3 \\ \diagdown O \end{matrix}\right]^- \qquad NO_2^+\left[O—N\begin{matrix} \diagup OBF_3 \\ \diagdown O \end{matrix}\right]^-$$

$$NO^+\left[N\begin{matrix} \diagup OBF_3 \\ \diagdown OBF_3 \end{matrix}\right]^- \qquad NO_2^+\left[N\begin{matrix} \diagup OBF_3 \\ \diagdown OBF_3 \end{matrix}\right]^-$$

Raman and infra-red spectra and X-ray powder diffraction patterns of the reaction products of BF_3 with N_2O_3, N_2O_5, and N_2O_4 in nitromethane or sulfur dioxide solution have shown, however, (65) that the predominant products are, $NO^+ BF_4^-$, $NO_2^+ BF_4^-$, and a mixture of these two ionic compounds. The stoichiometry of the reactions is

$$3N_2O_3 + 8BF_3 = 6NO^+ BF_4^- + B_2O_3$$
$$3N_2O_4 + 8BF_3 = 3NO_2^+ BF_4^- + 3NO^+ BF_4^- + B_2O_3$$
$$3N_2O_5 + 8BF_3 = 6NO_2^+ BF_4^- + B_2O_3$$

III. The General Concept of Carbocations. Differentiation and the Role of Carbenium and Carbonium Ions in Friedel-Crafts Reactions

Before we discuss the reactions of the reactive electrophiles formed by reactant–catalyst interactions with substrates, we must consider in some detail the general concept of carbocations essential to an understanding of the mechanism of Friedel-Crafts reactions. After direct observation of stable, long-lived carbocations, generally in highly acidic (superacid) systems became possible (66), extensive studies of these ions led to the recognition of the general concept of hydrocarbon cations.

The general definition of carbocations (2) was developed *based on the realization that two distinct classes of carbocations* (it is the logical name for all cations of carbon compounds since the negative ions are called *carbonions*) *exist.*

Trivalent ("classical") carbenium ions contain an sp^2-hybridized *electron-deficient* carbon center which tends to be planar in the absence of constraining skeletal rigidity or steric interference.   (It should be noted that sp-hybridized, linear acyl cations and vinyl cations also show substantial electron deficiency of carbocation centers).

Penta- or tetracoordinated ("nonclassical") carbonium ions contain five- or four-coordinated carbon atoms bound by three single bonds, and a two-electron, three center bond (either to two additional bonding atoms or involving a carbon atom to which they are also bound by a single bond).

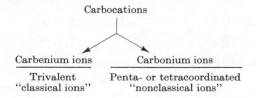

Concerning the carbocation concept, it is regrettable that in general usage the trivalent, planar ions of the CH_3^+ type were for long called *carbonium ions*. If the name is considered analogous to the names of other *onium ions* (ammonium, sulfonium, phosphonium, etc.), it should relate to the higher valency state carbocation. The higher valency state carbocations, however, clearly are not the trivalent but the *pentacoordinated cations* of the CH_5^+ type. The German and French literature indeed frequently used the carbenium ion naming for the trivalent cations. If we consider these latter ions as protonated carbenes, the naming is indeed correct (67), and we can simply *differentiate carbenium (trivalent) and carbonium (penta- or tetracoordinated) ions.* It should be pointed out, however, that carbenium ion should be used only for trivalent ions and not as a generic name for *all* carbocations.

Experimental evidence for observation and differentiation of trivalent alkylcarbenium ions from pentacoordinated carbonium ions comes from n.m.r. (H^1 and C^{13}), infra-red, Raman and photoelectron spectroscopic (68) study of their superacid (SbF_5, SbF_5–FSO_3H, HF–SbF_5) solutions (or matrixes) (69).

Table I summarizes the H^1 and C^{13} n.m.r. parameters for the carbenium ion center in a series of carbenium ions, and Fig. 1 shows the p.m.r. spectra of some representative ions

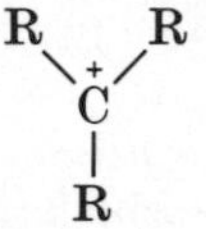

$$\begin{array}{c} R \quad\ \ R \\ \diagdown \overset{+}{\ }\diagup \\ C \\ | \\ R \end{array}$$

trivalent carbenium ion

Data are characterized by substantially deshielded chemical shifts with coupling constants (J_{CH}) that indicate sp^2 hybridization.

In contrast to the highly electron-deficient trivalent carbenium centers, carbonium ions contain penta- or tetracoordinated centers. The parent carbonium ion is CH_5^+ (the methonium ion, carbonium ion).

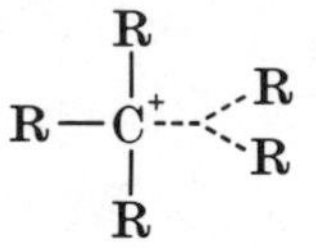

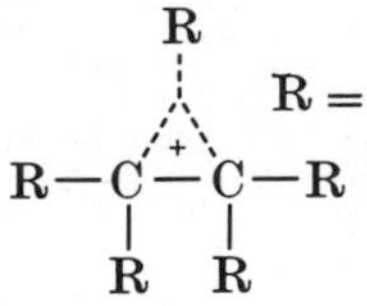

pentacoordinated   tetracoordinated
carbonium ions

R = H or allyl

TABLE I.   Characteristic n.m.r. parameters of alkylcarbenium ions in SbF_5–SO_2ClF soluti⟨
at $-70°$

Ion	δ_{H^1}						$\delta_{C^{13}}$			
	CH^+	J_{+CH}	J_{+CCH}	$-CH_2$	$(-CH_3)_{2,3}$	$-CH_3$	C^+	$(-CH_3)_{2,3}$	$-CH_2$	$-C$⟩
$(CH_3)_2CH^+$	13	169	3.3		4.5		-125.0	132.8		
$(CH_3)_3C^+$			3.6		4.15		-135.4	145.3		
$(CH_3)_2C^+CH_2CH_3$				4.5	4.1	1.94	-142.6	148.6	135.7	184⟩

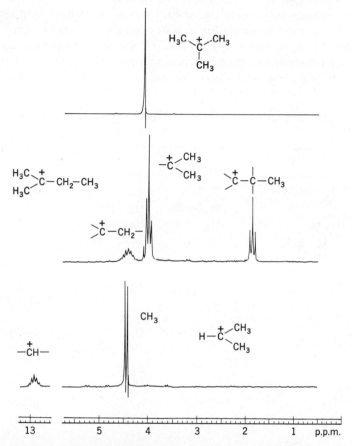

Fig. 1.   Proton magnetic resonance spectra in SbF₅–SO₂C2F solution at −60°
of the trimethylcarbenium ion (top), dimethylethylcarbenium ion
(middle), and dimethylcarbenium ion (bottom).

as CH_3^+ (methenium ion, methyl cation, carbenium ion) is the parent
for trivalent carbenium ions.

Aliphatic carbonium ions of the CH_5^+ type are indicated by the
superacid chemistry of the corresponding alkanes (70) (5-donors) and
by mass spectrometry (71). They are also known in certain ion–
molecule reactions in the gas phase, but no direct observation in
solution has so far been achieved.

More rigid cycloaliphatic, particularly bicyclic, systems provide
examples of directly observable, stable carbonium ions.

The most disputed of nonclassical carbonium ions, the 2-*norbornyl cation* can be generated from 2-*exo*-halonorbornanes in SbF_5–SO_2ClF solution under long-lived ion conditions and at low temperature (72). At $-156°$ it exists in a non-exchanging, static form. H^1 and C^{13} n.m.r. spectroscopy (Table II) indicate that the bridging penta-coordinated methylene carbon atom is tetrahedral in nature and carries little charge. The methine carbons, to which bridging takes

TABLE II.   Characteristic H^1 and ^{13}C n.m.r. parameters of non-classical carbonium ion centers

	δ_{H^1}		J_{CH}		$\delta_{C^{13}}$	
	H_A 3.05	H_B 6.59	H_A 145.8	H_B 184.5	C_A 171.4	C_B 68.5
	H_A 3.25	H_B 7.04	H_A 218.9	H_B 193.8	C_A 159.8	C_B 67.9
	H_A 3.24	H_B 7.48	H_A 216.4	H_B 192.3	C_A 157.6	C_B 78.9

place, are tetracoordinated, the charge delocalized mostly into the methine bonds. This n.m.r. spectrum (Fig. 2) was the first experimental observation of a nonclassical carbonium ion, formed by C–C σ-bond delocalization, *i.e.*, the σ-route to the symmetrically delocalized ion. In other words, the process can be visualized as an intramolecular Friedel-Crafts alkylation process, *i.e.*, of the C-1–C-6 σ-bond by the developing electron-deficient center at C-2. It should also be mentioned that the same ion is observed when generated by

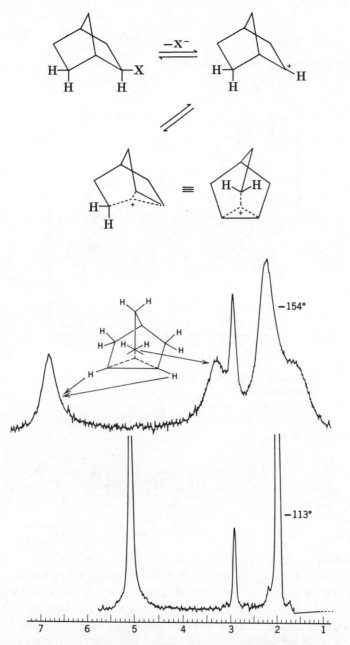

Fig. 2.   Proton magnetic resonance nmr spectra (100 MHz.) of the norbornyl cation in SbF_5–SO_2ClF–SO_2F_2 solution at temperatures between -113 and $-154°$.

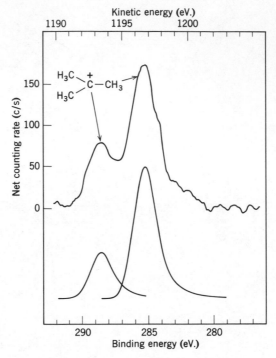

Fig. 3. Carbon $1s$ photoelectron spectrum of t-butyl cation (lower curve computer-resolved).

the alternate π-route from β-Δ^3-cyclopentenylethyl halides (*i.e.*, by π-bond alkylation).

Recently, Olah and Mateescu also applied core electron spectroscopy (ESCA) to the study of carbocations either in frozen superacid solutions or as isolated salts (73). This method allows direct measurement of carbon $1s$ electron binding energies. As the charge distribution within carbocations causes increasing binding energies with increasing positive charge localization, highly electron-deficient classical alkyl and cycloalkyl carbenium centers (such as in t-butyl and adamantyl cations) show about 4-eV. $1s$ binding energy differences from the remaining less electron-positive carbon atoms. (Fig. 3). In nonclassical carbonium ions, such as the norbornonyl ion,

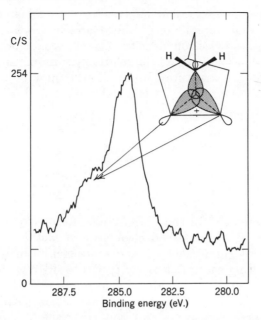

Fig. 4.   Carbon 1s photoelectron spectrum of the norbornyl cation.

there is no such highly electron-deficient carbon center and the photoelectron spectrum indicates only two modestly electron-positive carbon atoms (the tetracoordinated methine atoms, to which bridging takes place), separated by about 1.5 eV. from the remaining carbon atoms, with the bridging pentacoordinated carbon showing no detectable electron-deficient "shift" (Fig. 4).

The bonding of the carbonium center is considered to involve three two-electron covalent bonds with the fourth bond being a two-electron three-center bond (2). This type of bond is involved in pentacoordinated carbonium ions such as CH_5^+ and the norbornyl

$$
\begin{array}{c}
\text{H} \\
| \\
\text{H} \overset{\text{C}}{{-}} \text{H} \\
| \\
\text{H} + \text{H}^+
\end{array}
\rightleftharpoons
\left[
\begin{array}{c}
\text{H} \quad \text{H} \\
\text{H} \overset{\text{C}}{{-}} \\
\text{H} \quad \text{H}
\end{array}
\right]^+
$$

For simplicity
triangular dotted lines are used
to depict the three-center bonds,
as full straight lines are used to
symbolize two-center, two-electron bonds

$$
\left[\text{H}_3\text{C} \cdots \begin{array}{c} \text{H} \\ \text{H} \end{array} \right]^+
$$

or

$$
\left[\text{H}_3\text{C} \begin{array}{c} \text{H} \\ \text{H} \end{array} \right]^+
$$

cation.   Thus the interaction involves the main lobes of the covalent bonds (front-side interaction).   Since an electrophile attacks the points of highest electron density, attack occurs on the covalent bonds themselves and not on the relatively unimportant back lobes.

Of the possible structures for the methonium ion with $D_{3h'}$, C_{4v}, C_s, D_{2h} or C_{3v} symmetry, Olah, Klopman, and Schlosberg (74)

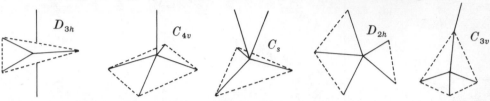

suggested preference for the C_s front-side protonated form, based on consideration of the observed chemistry of methane in superacids (hydrogen–deuterium exchange and more significantly polycondensation, indicating ease of cleavage to CH_3^+ and H_2) and also on the basis of self-consistent field calculations (18).   More extensive calculations (75), including *ab initio* calculations utilizing an "all-geometry" parameter search (76), confirmed the C_s symmetry structure to be favored.   This structure is about 2 kcal./mole below the energy level of the C_{4v} symmetry structure, which in turn is about 8 kcal./mole favored over the trigonal bipyramidal D_{3h} symmetry structure.   At the same time it should also be recognized that ready interconversion of stereoisomeric forms of CH_5^+ is possible by a pseudo-rotational type process or, as Muetterties (77) recently suggested in discussing stereoisomerization processes of this type, by "polytopal rearrangements" (also called "polyhedral rearrangements").   We prefer to call intramolecular carbonium ion rearrangements "bond-to-bond rearrangements" (see subsequent discussion).   Hydrogen–deuterium scrambling observed in superacid solutions of deuterated methane and homolog alkanes strongly indicates such processes.

It is of interest to note that isoelectronic boron compounds can be used as model compounds for both carbenium and carbonium ions. Trimethylboron can be used for comparison with the trimethylcarbenium ion, and similarly BH_5 is a suitable model for CH_5^+ (2).

BH_5 is indicated in the acid hydrolysis of borohydrides (78a). When the hydrolysis is carried out with deuterated acid, not only HD, but also H_2, is formed. This indicates that the attack of D^+ is on the B–H bond, followed by polytopal (bond-to-bond) rearrangement before cleavage takes place. The ease with which diborane exchanges hydrogen for deuterium when treated with deuterium gas, also indicates the formation of the three-center-bonded analog, BH_3D_2

The protonation of methane to CH_5^+ can serve as a prototype for electrophilic reactions of alkanes. It should be emphasized that if steric interference (as is the case in interaction of tertiary C–H bonds of isoalkanes with tertiary carbenium ions) is substantial, the formation of the triangular three-center bond must be highly unsymmetrical, although still not to be considered to be linear, i.e. to lie along an extension of the interacting bond (78b).

The preferred direction of attack by the electrophile should thus not be considered to be always the same and may well vary from compound to compound (depending on the reaction system

involved). The various possible configurations of pentacoordinated carbonium ions do not differ very much in stability (as in CH_5^+).
The back-side-substituted trigonal bipyramidal form is considered,
however, the least favorable. It is unlikely that there is any
"inherently" preferred configuration in the front side attack; it
depends on individual reaction conditions.

IV. The Reaction of the Electrophile with the Substrate

So far we have discussed the interaction of the catalyst with the
reactant, forming a carbocationic (or related electrophilic) reagent.
This reagent, because of its substantial electron deficiency, reacts
with the electron donor (nucleophilic) substrate which can be a
π-, σ-, or n-donor. Friedel-Crafts reactions are of particular interest
in hydrocarbon chemistry, thus we discuss first the more reactive
π-donor hydrocarbons (aromatics, alkenes, and alkynes) and subsequently the less reactive but increasingly important σ-donor
alkanes. As heteroatom n-donor substrates generally react wth
ease, frequently even without catalysts, they are not considered
separately in this discussion.

π-Donor Systems

A. Aromatic Substitution (79)

Electrophilic aromatic substitution is one of the most studied of
organic reactions (80). Excellent reviews are available, and it is not
the purpose to give here a general summary of the field. Instead an
attempt is made to concentrate on mechanistic aspects of current
interest, emphasizing conclusions that can be reached from recent
work.

A single-step, direct substitution mechanism would be expected
to involve a kinetic hydrogen isotope effect in deuterated (or tritiated) aromatics, but in many systems studied such an isotope effect
was not observed. This observation, together with substantial evidence for the formation of ionic intermediates, leads to the generally
accepted two-step mechanism. Attack of an electrophile on the
aromatic compound forms an intermediate cyclohexadienyl cation
[Pfeiffer-Wizinger complex (81a), Wheland intermediate (81b), Brown
σ-complex (81c) or, as suggested by Doering (82), benzenonium ion,

although it should be now recognized that the systematic name for trivalent cyclohexadienyl cations is benzenium ion (2)]. The product is then formed *via* proton elimination forms the intermediate

These benzenium complexes have been well reviewed (85) and their characteristics well established.

It is generally assumed, as stated, for example, by Ingold (83), that cationic cyclohexadienyl intermediates represent satisfactory comparison models for the transition states of electrophilic aromatic substitutions.

1. Substrate and Positional Selectivity

As discussed by Brown in his extensive studies (84), there exists a simple linear relationship between relative rates of numerous electrophilic aromatic substitutions and the relative stabilities of the related σ-complexes. He therefore concluded that the transition state for these reactions generally resembles a σ-complex and the reactions comply with the Brown selectivity relationships (see Chapter V). Considerable direct experimental evidence has since been obtained to prove that σ-complexes do exist as stable intermediates (85). However, their existence does not necessarily prove that the transition states are always closely related to the σ-complexes.

It was Dewar (86) who originally proposed that π-complexes may play an important role in electrophilic aromatic substitutions. He suggested the mechanism

Consequently, the rate of substitutions should parallel the stability of π-complexes.

Olah and co-workers (87) reported a series of electrophilic aromatic substitutions involving strongly electrophilic reagents which showed low substrate selectivity expressed as low $k_{toluene}/k_{benzene}$ rate ratios, but at the same time high positional selectivity (generally high *ortho/para* isomer ratios with only a small amount of the *meta*-isomer formed). We expressed the view that in these systems the

TABLE III.   Relative rates of substitution of ben‍z

Methylbenzene	σ-complex stability HF–BF$_3$[85g]	π-complex stability Ag^{+}[a]	bromination		chlorinat‍
			Br$_2$ in 85% CH$_3$CO$_2$H	Br$_2$ in CF$_3$CO$_2$H[b]	Cl$_2$ in CH$_3$CO$_2$
Benzene	1	1.0	1	1	
Me‍thyl	790	1.5	605	2,580	34
1,2-Dimethyl	7,900	1.8	5,300	50,800	2,03
1,3-Dimethyl	1,000,000	2.0	514,000		180,00
1,4-Dimethyl	3,200	1.6	2,500	9,080	2,00
1,2,3-Trimethyl	2,000,000	2.4	1,670,000		
1,2,4-Trimethyl	2,000,000	2.2	1,520,000		
1,3,5-Trimethyl	630,000,000	2.6	189,000,000		30,000,00
1,2,3,4-Tetramethyl	20,000,000	2.6	11,000,000		
1,2,3,5-Tetramethyl	2,000,000,000	2.7	420,000,000		
1,2,4,5-Tetramethyl	10,000,000	2.8	2,830,000		1,580,00
Pentamethyl	2,000,000,000		810,000,000		134,000,00

transition state of highest energy (which determines substrate selectivity) is of a π-complex nature followed by σ-complex formation determining positional selectivity.

This concept attracted interest (80a) but was also criticized. In criticism (88) it was suggested that the above results could be a consequence of incomplete mixing before fast reaction with the very reactive electrophilic reagents. More recent work however, discussed subsequently, clearly indicates that we are not dealing with an experimental artifact but are encountering a quite general type of electrophilic aromatic substitution displaying low substrate but at the same time high positional selectivity.

2.   Correlation of Reaction Rates with Complex Stabilities

Brown and Stock found excellent correlation between relative rates of halogenation (and other electrophilic substitutions) and relative stabilities of σ-complexes (80c).   There was, however confusion as to what are the proper values for the relative stabilities of the σ-complexes, particularly for benzene and toluene for which these values are not easily determined.   For example, Breslow (80a) and March (80i) list the relative σ-complex stabilities of toluene and benzene as 7:1 (based on data of ref. 85e), whereas the value according to Mackor, Hofstra, and van der Waals (85g) is 790:1.   The basicities of methylbenzenes were first measured by McCaulay and Lien (85d) by means of competitive extraction experiments in which methylbenzenes and their comparison with σ- and π-basicities

chlorination		Halogenation		Nitration		Alkylation	
in CN[d]	Br$_2$ in CF$_3$CO$_2$H[b]	Br$_2$–FeCl$_3$ CH$_3$NO$_2$[e]	Cl$_2$–FeCl$_3$– CH$_3$NO$_2$[f]	NO$_2$$^+BF_4$$^+$– sulfolane[g]	HNO$_3$ in sulfolane[h]	C$_6$H$_5$CH$_2$Cl– AlCl$_3$– CH$_3$NO$_2$[i]	i-C$_3$H$_7$Br– AlCl$_3$– CH$_3$NO$_2$[j]
1	1	1.0	1.0	1.0	1	1.0	1.0
30	2,445	3.6	13.5	1.6	20	3.2	2.0
		3.9	38.0	1.7	61	4.2	2.2
	247,000	5.6	110.0	1.6	100	4.6	2.8
	14,200	4.3	43.9	1.9	114	4.3	2.7
			15.9	2.7	400	5.2	0.3

Andrews, L. J., and R. M. Keefer, "Molecular Complexes in Organic Chemistry," Holden-Day, Francisco, 1964.
Olah, G. A., S. J. Kuhn, S. H. Flood, and B. A. Hardie, J. Am. Chem. Soc., **86**, 1039 (1964).
Baciocchi, E., and G. Illuminati, Gazz. Chim. Ital., **92**, 89 (1962).
Andrews, L. J., and R. M. Keefer, J. Am. Chem. Soc., **81**, 1063 (1959).
Olah, G. A., S. J. Kuhn, S. H. Flood, and B. A. Hardie, J. Am. Chem. Soc., **86**, 1044 (1964).
De la Mare, P. B. D., J. T. Harvey, M. Hasson, and S. Varma, J. Chem. Soc., 2756 (1958).
Olah, G. A., S. J. Kuhn, and S. Flood, J. Am. Chem. Soc., **83**, 4571 (1961).
Hoggett, J. G., R. B. Moodie and K. Schofield, J. Chem. Soc. B, 1 (1969); Chem. Commun., (1969).
Olah, G. A., S. J. Kuhn, and S. Flood, J. Am. Chem. Soc., **84**, 1688 (1962); G. A. Olah and N. rchuk, ibid, **87**, 5786 (1965).
Olah, G. A., S. H. Flood, S. J. Kuhn, M. E. Moffatt, and N. A. Overchuk, J. Am. Chem. , **86**, 1046 (1964).

mixtures of methylbenzenes competed for a limited amount of acid (HF–BF$_3$). Kilpatrick and Luborsky (85e) calculated the basicity constants from conductance experiments (in HF). Mackor, Hofstra, and van der Waals (85g) critically redetermined the basicity constants by the distribution technique, but under much more dilute conditions than used by McCaulay and Lien, allowing determination of activity coefficients for the solutions. The data of Mackor, Hofstra and van der Waals (85g) can be considered the most reliable set of relative σ-basicity data for benzene and alkylbenzenes.

Table III summarizes the relative σ- and π-basicities of benzene and the methylbenzenes and the relative rates of some electrophilic aromatic substitution reactions. In Table III the correlation between relative rates of certain halogenations and σ-basicities is excellent. These data seem to strengthen the conclusion that for these reactions the transition state resembles the Wheland intermediate (σ-complex). However, not all the rate data correlate with relative σ-basicities. In the early 1960's, Olah suggested that the rates of several observed electrophilic aromatic substitutions are related to π-basicities and not σ-basicities of the substrates.

Substitutions that showed relative rates related to π-basicities of the aromatic substrates included nitronium salt nitrations, Friedel-Crafts type alkylations, bromination, and chlorination using strongly electrophilic halogenating agents ($FeCl_3$ or $AlCl_3$ catalysts). Many additional electrophilic aromatic substitutions reported in the literature (such as the classic Ingold nitrations (89), Brown's alkylations (90), and other reactions) also cannot be considered to follow closely σ-basicities, but show intermediate character or much closer relationship to π-basicities of the aromatic substrates.

In view of the discussed relationship of substitution rates and complex stabilities (which are significantly different for σ- and π-complexes, as shown in Table III), the wide acceptance of the view that transition states of electrophilic aromatic substitutions generally resemble σ-complexes is not founded, particularly in case of reactive Friedel-Crafts systems.

3. Effect of Electrophilicity of Reagents on Selectivity

Recent work by Olah's group (92,93) has proved that the transition states of electrophilic aromatic substitutions are not rigidly fixed, always resembling the σ-intermediates, but frequently represent a much earlier state on the reaction coordinate resembling starting aromatics (i.e., π-complexes).

a. Benzylation

It was found that it is possible to vary in a systematic way the electrophilicity of reagents, such as alkylating agents, by introducing suitable substituents. This causes a regular change in the nature of the alkylation transition state from a σ-complex type to one resembling a π-complex. A typical system studied was the titanium

$$CH_3-C_6H_5 \; + \; ClCH_2-C_6H_4-X \xrightarrow{\ TiCl_4\ } CH_3-C_6H_4-CH_2-C_6H_4-X$$

tetrachloride-catalyzed benzylation of toluene and benzene with substituted benzyl chlorides (92).

The data in Table IV show the effect of substituents both on substrate selectivity (as reflected in the $k_{toluene}/k_{benzene}$ rate ratios) and

TABLE IV. TiCl$_4$-catalyzed benzylation of benzene and toluene
with substituted benzyl chlorides XC$_6$H$_4$CH$_2$Cl at 30°

| XC$_6$H$_4$CH$_2$Cl | k_t/k_b | methyldiphenylmethane isomer (%) | | | |
		Ortho	Meta	Para	1/2 Ortho/ Para
p-NO$_2$	2.5	59.6	6.2	34.2	0.87
o-F	4.8	39.3	7.8	52.9	0.37
m-F	4.6	37.7	7.8	54.5	0.35
p-F	8.7	43.0	6.5	50.5	0.43
o-Cl	4.6	38.8	7.2	54.0	0.36
m-Cl	6.4	43.1	6.5	50.4	0.43
p-Cl	6.2	40.1	5.0	54.9	0.37
H	6.3	40.5	4.3	55.2	0.37
o-CH$_3$	19.1	24.6	3.4	72.0	0.17
m-CH$_3$	7.8	41.3	2.5	56.2	0.37
p-CH$_3$	29.0	31.4	2.1	66.5	0.24
2,4,6-(CH$_3$)$_3$	39.1	15.9	3.0	81.1	0.10
o-CH$_3$O	60.3	23.2	1.3	75.5	0.15
m-CH$_3$O	13.2	47.5	2.3	50.2	0.47
p-CH$_3$O	97.0	28.6	1.5	69.9	0.20
2,4,6-(CH$_3$O)$_3$- 3,5-(CH$_3$)$_2$-	136.0	18.3	1.1	80.6	0.11

on positional selectivity (as shown by isomer distributions, primarily the *ortho/para* ratios).   Electron-donating substituents *ortho* and *para* to the benzylic center increase the $k_{toluene}/k_{benzene}$ rate ratio and at the same time decrease the *ortho/para* isomer ratio of the methyldiphenylmethanes formed, *i.e.*, *para*-substitution becomes predominant. In contrast, electron-withdrawing substituents decrease $k_{toluene}/k_{benzene}$ and increase the *ortho/para* isomer ratio.

b. Acylation

The effect of substituents on the electrophilicity of the reagent, causing a change in the nature of the transition state from a σ- to a π-complex nature, was also shown in other typical Friedel-Crafts substitutions, such as acylations (93).

Friedel-Crafts acylation of aromatic hydrocarbons, such as toluene and benzene, are characterized by generally high selectivity of the reactions reflected in high rate ratios (generally $k_{toluene}/k_{benzene} > 100$) and related predominant *para*-substitution.   The latter was always considered a consequence of the steric requirements of the acylating agents, causing large steric *ortho*-hindrance (80,84).

When acetylation and benzoylation of toluene and benzene with acyl and substituted acyl halides were studied, the results clearly proved the importance of substituents in the electrophilic substituting agent in influencing its electrophilicity and through it both substrate and positional selectivity. Tables V and VI summarize the data.

TABLE V. Relative rates and isomer distributions in the boron trifluoride- and aluminum chloride-catalyzed acetylation of toluene and benzene in carbon disulfide solution at 25° (16)

Acyl halide	Catalyst	k_t/k_b	% Ortho	% Meta	% Para
HCOF	$BF_3{}^a$	34.6	43.3	3.5	53.2
CH_3COF	$BF_3{}^a$	130.0	1.2	1.1	97.7
CH_3COCl	$AlCl_3$	141.0	2.5	2.0	95.5
CH_3CH_2COCl	$AlCl_3$	89.2	2.6	3.1	94.3
$(CH_3)_2CHCOCl$	$AlCl_3$	84.4	3.2	2.4	94.4
$CH_2ClCOCl$	$AlCl_3$	78.2	11.1	2.3	86.6
$CHCl_2COCl$	$AlCl_3$	51.4	17.3	3.2	79.5

^{a} Reaction carried out without solvent.

TABLE VI. Relative rates and isomer distributions in the aluminum chloride-catalyzed benzoylation of toluene and benzene with substituted benzoyl chlorides in nitromethane solution at 25° (16)

Benzoyl chloride	k_t/k_b	% Ortho	% Meta	% Para
C_6F_5COCl	16.1	35.4	7.8	56.8
$2,4\text{-}(NO_2)_2C_6H_3COCl$	29.0	42.4	3.0	54.6
$3,5\text{-}(NO_2)_2C_6H_3COCl$	38.9	9.4	6.8	83.8
$p\text{-}NO_2C_6H_4COCl$	52	9.5	1.2	89.3
$2,5\text{-}F_2C_6H_3COCl$	96.2	12.3	1.6	86.1
C_6H_5COCl	153.5	8.1	1.2	90.7
$p\text{-}CH_3C_6H_4COCl$	164.4	7.8	1.1	91.1
$p\text{-}FC_6H_4COCl$	170	8.0	1.1	90.9
$2,4,6\text{-}(CH_3)_3C_6H_2COCl$	196	2.3	0.6	97.1
$p\text{-}CH_3OC_6H_4COCl$	233	14.9	1.5	83.6

Changing not only the electrophilicity of the reagent but also the nucleophilicity of the aromatic substrate can cause the same effect. This is reflected in the extensive literature of Friedel-Crafts acylations (80d), showing that the high selectivity observed in the reactions of toluene compared to benzene does not increase rapidly with further methyl substitution of the aromatic ring, as would be expected on the basis of known σ-basicities (85). Thus with increasingly more basic aromatics even a relatively weak electrophile results in the transition state becoming early on the reaction coordinate, resembling more starting aromatics than intermediates.

Steric effects, as suggested, can play an important role in affecting the isomer distributions in Friedel-Crafts acylations (particularly *ortho*-substitution), but they cannot be considered the only major reason for the observed isomer distributions.

c. Sulfonylation

Many additional examples similar to those discussed can be found. Some illustrative results of sulfonylation (93) are summarized in Table VII.

TABLE VII. AlCl$_3$-catalyzed sulfonylation of toluene and benzene with substituted benzenesulfonyl chorides (93)

	k_t/k_b	% Ortho	% Meta	% Para
p-NO$_2$C$_6$H$_4$SO$_2$Cl	2.8	51.0	($\sim$8)	41
p-ClC$_6$H$_4$SO$_2$Cl	7.5	37.0	($\sim$8)	55
C$_6$H$_5$SO$_2$Cl	9.0	28.4	8.7	62.9
p-CH$_3$C$_6$H$_4$SO$_2$Cl	17.0	14.5	<1.0	85.5
p-CH$_3$OC$_6$H$_4$SO$_2$Cl	83.0	5.6	<1.0	94.4

The effect of substituents in the benzenesulfonyl chlorides again affects both substrate and positional selectivity in a manner similar to that discussed previously in case of benzylation with substituted benzyl chlorides and benzoylations with substituted benzoyl chlorides.

d. Sulfonation

Cerfontain's (91) sulfonation data (Table VIII) show increasing electrophilicity of the reagent with increasing acid strength. It is again reflected in a regular change in both substrate and positional selectivity, attributed by Cerfontain to two different sulfonating agents (H$_3$SO$_4^+$ at low and H$_2$S$_2$O$_7$ at high sulfuric acid concentrations).

TABLE VIII.   Sulfonation of benzene and toluene with sulfuric acid of varying strength (91)

% H_2SO_4	k_t/k_b	% Ortho	% Meta	% Para
77.8	106	21.2	2.1	76.7
81.5	57			
84.3	47	38.8	2.6	58.6
89.1	25			
99–100 in $C_6H_5NO_2$	5.1[a]	50.2	4.9	44.9

[a] Stubbes, F. J., C. D. Williams, and C. N. Hinshelwood, *J. Chem. Soc.*, 1065 (1948).

e.   *Thiolcarboxylation*

Olah and Schilling studied rates (competitive as well as noncompetitive) of the $AlCl_3$-catalyzed arylthiolcarboxylation of toluene and benzene with arylthiolchloroformates in methylene chloride solution at 25°.

$$ArH + ClCSAr' \xrightarrow{AlCl_3} ArCSAr' + HCl$$
$$\quad\;\; \| \qquad\qquad\qquad \|$$
$$\quad\;\; O \qquad\qquad\qquad O$$

Table IX summarizes the data obtained, showing that electron-withdrawing substituents, which increase the electron deficiency of

TABLE IX. $AlCl_3$-catalyzed arylthiolcarboxylation of toluene and benzene with substituted pheylthiolchloroformates

	k_t/k_b	% Ortho	% Meta	% Para
C_6F_5SCOCl	3.3	46.6	5.6	47.8
$2,4$-$(NO_2)_2C_6H_3SCOCl$	7.3	57.9	1.5	40.6
4-$NO_2C_6H_4SCOCl$	28.5	31.6	1.2	67.2
4-BrC_6H_4SCOCl	74.1	10.7	1.0	88.3
4-ClC_6H_4SCOCl	76.2	9.2	1.1	89.7
4-FC_6H_4SCOCl	98	6.3	0.9	92.8
C_6H_5SCOCl	95	5.9	0.8	93.3
4-$CH_3C_6H_4SCOCl$	140	5.8	0.9	93.3
$1,3,5$-$(CH_3)_3C_6H_2SCOCl$	209	5.9	0.7	93.4
$(CH_3)_5C_6SCOCl$	243	5.7	0.8	93.5
4-$CH_3OC_6H_4SCOCl$	310	4.4	0.7	94.9

the incipient carbenium ion center, cause increasingly low substrate selectivities and increased *ortho/para* isomer ratios, whereas electron-donating substituents have an opposite effect.

f. Halogenation

Electrophilic halogenation of toluene and benzene particularly emphasizes the great variation that can be observed depending on the nature of the halogenating agent. The data in Table X indicate the great diversity of substrate and positional selectivity that can be obtained in various systems. No single set of numerical data can thus be considered "characteristic" of electrophilic halogenations (88f), which cover numerous reactions of differing nature.

g. Nitration

Electrophilic nitration of reactive aromatics, such as toluene and benzene, must be considered a substitution reaction with the transition state resembling more starting hydrocarbons than intermediates. This is in keeping in accordance with Ingold's fundamental work (80h,89) proving that the nitrating agent in usual electrophilic nitrations is the highly reactive nitronium ion (NO_2^+).

Ingold, Hughes, and their co-workers (89), using reaction media composed of 40% nitric acid and 60% acetic anhydride or nitromethane, found in nitrations of toluene–benzene mixtures a rate ratio ($k_{toluene}/k_{benzene}$) of 23–27. The isomer distribution in the nitration of toluene was found to be 57% *ortho*, 4% *meta*, and 29% *para*-nitrotoluene.

As nitrations of benzene and toluene were too fast to measure by absolute rate studies, Ingold used the competitive method of rate determination to establish their relative rates, $k_{toluene}/k_{benzene}$.

Olah and Kuhn (94) subsequently developed a new efficient nitration method by using stable nitronium salts (such as the tetrafluoroborate) as nitrating agents. Nitronium salt nitrations are also too fast to measure their non-competitive rates, but the use of the competition method showed low substrate selectivity, e.g., $k_{toluene}/k_{benzene}$ of 1.6. On the basis of the Brown selectivity rules, if the fast reactions followed a σ-complex pattern they would also have a predictably low positional selectivity (with high *meta*-isomer content). However, the observed low substrate selectivities were accompanied by high discrimination between available positions (typical isomer

TABLE X. Electrophilic halogenation of toluene and benzene with reagents of varying nature

Halogenating agent	Catalyst	Solvent	k_t/k_b	% Ortho	% Meta	% Para	Ref.
Cl_2	$FeCl_3$	CH_3NO_2	13.5	67.8	2.3	29.9	a
HOCl	H^+	H_2O	60.0	74.6	2.2	23.2	b
Cl_2		CH_3COOH	344.0	59.8	0.5	39.7	c
		CH_3CN	1650	37.6	0.5	62.4	d
Cl_2	$FeCl_3$	CH_3NO_2	2445.0	33.6	0.15	66.4	e
Br_2		CH_3NO_2	7.1	71.8	1.6	27.3	f
HOBr		H_2O–dioxane	36.2	70.3	2.3	27.4	g
Br_2	$ZnCl_2$	CH_3CO_2H	148.0				h
Br_2		CH_3COOH	605.0	32.9	0.3	66.8	i
Br_2		CF_3CO_2H	2580	17.6	0.2	82.4	i
I^+	(electrochemical)	CH_3CN	2.0	47.0	6.0	47.0	j
ICl	$ZnCl_2$	CH_3CO_2H	140.0				h

[a] Olah, G. A., S. J. Kuhn, and B. A. Hardie, J. Am. Chem. Soc., **86**, 1055 (1964).
[b] See Table I, footnote f.
[c] Stock, L. M., and H. C. Brown, J. Am. Chem. Soc., **81**, 5615 (1959).
[d] See Table I, footnote d.
[e] Stock, L. M., and A. Himoe, J. Am. Chem. Soc., **83**, 1937, 4605 (1961).
[f] See Table I, footnote b.
[g] De la Mare, P. D. B., and J. T. Harvey, J. Chem. Soc., 36 (1956).
[h] Keefer, R. M., and L. J. Andrews, J. Am. Chem. Soc., **78**, 5623 (1956).
[i] Brown, H. C., and Stock, L. M., J. Am. Soc. Chem., **79**, 1421, 5175 (1957); Brown, H. C., and R. A. Wirkkala, J. Am. Chem. Soc., **88**, 1447 (1966).
[j] Miller, L., E. P. Kayawa, and C. B. Campbell, J. Am. Chem. Soc., **92**, 2821 (1970).

distributions of nitrotoluenes were *ortho/meta/para* = 66%:3%: 31%). Consequently, a *meta*-position seemed to be sevenfold deactivated compared to a benzene position, giving a partial rate factor of $m_f = 0.14$. These observations are inconsistent with any mechanism in which the individual nuclear positions compete for the reagent (in the σ-complex formation step).

In explanation, Olah suggested the formation of a π-complex in the first step of the reaction followed by conversion to σ-complexes (which are of course separate for the individual *ortho*-, *para*-, and *meta*-positions), allowing discrimination in orientation of products. The use of the competition method of rate determination was criticized (88b), primarily on the assumption that the nitronium salt reagents used were too reactive to allow real differentiation under competitive reaction conditions and low substrate selectivity was the consequence of fast, indiscriminate reaction before uniform mixing of the reagents could be achieved. It was claimed that if very low concentrations of nitronium salts and very efficient mixing were used reactions would give results comparable to "normal"σ-route nitrations (88b). Olah and Overchuck (88b), however, showed that, under conditions leading to higher substrate selectivity the highly diluted solutions did not contain any more nitronium salts, because the impurity levels of the solutions exceeded the concentration of the reactive nitronium salts. Furthermore, competitive experiments in highly efficient high-speed flow systems, including stopped-flow techniques, with reaction times as short as 0.002 second and varying the toluene/benzene mole ratio in the experiments from 1 : 10 to 10 : 1, led to no significant change in either positional or substrate selectivities (88b,95). Nitronium tetrafluoroborate nitrations with four competing aromatics (toluene, benzene, fluorobenzene, and chlorobenzene) gave also only very slightly changed $k_{toluene}/k_{benzene}$ ratios (88a).

Coombes, Moodie, and Schofield (95) recently found from kinetic nitration studies in 68.3% H_2SO_4 $k_{toluene}/k_{benzene}$ to be 17.2. By increasing the acid strength, the rate ratio decreased in 75.3% H_2SO_4 to 7.2 and in 77.7% to H_2SO_4 to 5.0. These investigators concluded that in acid more concentrated than 68.3% reactions proceed by what was termed as "encounter rate control." Such limiting encounter rates were reached with more reactive aromatics (starting with xylenes) even in weaker acid media. A similar disappearance of aromatic substrate reactivity differences is indeed observed in the nitronium nitration of methylbenzenes (87). Thus there seems to be no discrepancy between the conclusions of Moodie and Schofield and Olah's results. Aromatic substitutions on encounter may be considered exothermic reactions involving early, π-complex type transition states of highest energy. Differences in substrate selectivity of related aromatics (such as toluene and benzene) become insignificant, whereas positional selectivities (determined in subsequent σ-type transition states) remain high.

Whereas electrophilic nitrations of toluene and benzene show less variation in both substrate and particularly positional selectivity than previously discussed alkylations and acylations, by no means can selectivity be considered constant. Table XI summarizes some

TABLE XI. Nitration of toluene and benzene

Nitrating agent	Solvent	k_t/k_b	% Ortho	% Meta	% Para	Ref.
$NO_2^+PF_6^-$	CH_3NO_2	1.6	68.2	2.0	29.8	10
$NO_2^+BF_4^-$	sulfolane	1.7	65.4	2.8	31.8	10
$NO_2^+BF_4^-$	CH_3CN	2.3	69	2	29	10
HNO_3	80% H_2SO_4	4.8				20
	77% H_2SO_4	5.0				20
	75.3% H_2SO_4	7.2				20
	68.3% H_2SO_4	17.2	60	3	37	20
HNO_3	CH_3NO_2	21	58.5	4.4	37.1	2h,12
	$(CH_3CO)_2O$	23	58.4	4.4	37.2	
CH_3COONO_2	CH_3CN	44	63	2	35	18
HNO_2	sulfolane, H_2SO_4	37	61.6	2.9	35.5	11b

representative values of nitration of toluene and benzene. Thus we must conclude that if a single nitrating agent (i.e., the nitronium ion) is the effective reagent in all nitrations, as argued by Ingold (89), its activity must be dependent on the medium. Substrate selectivity changes reflect the differing reactivity of the nitronium ion in different media.

There exists the possibility that certain nitronium ion precursors (such as protonated acetyl nitrate or nitryl halide–Lewis acid halide complexes) can also be nitrating agents in their own right, without first forming the nitronium ion.

Not only the electrophilicity of the reagent but also the nucleo-

TABLE XII. Competitive nitration of nitrobenzene and nitrotoluenes with $NO_2^+PF_6^-$ (18)

	Relative rate		% dinitrotoluene					
	CH_3NO_2	H_2SO_4	2,3–	2,4–	2,5–	2,6–	3,4–	3,5–
Nitrobenzene	1	1	$o/m/p$ in $CH_3NO_2 = 10 : 88.5 : 1.5$; in 96% $H_2SO_4 = 7.1 : 91.5 : 7.4$					
o-Nitrotoluene	384	545	57.4		1.7	40.9		
			71.2			28.8		
m-Nitrotoluene	91	138	42		18.6		35.8	3.6
			28.4		9.9		60.1	1.6
p-Nitrotoluene	147	217		99.8			0.2	
				99.8			0.2	

philicity of the aromatic substrate can affect the relative position and heights of the involved transition states. Whereas nitrations cannot be widely varied by changing the reactivity of the nitrating agent, selectivity of the reactions can be substantially changed by suitable deactivating substituents in the aromatics, which cause the relative height of the barrier to σ-complex formation to increase significantly and thus cause the "late" σ-type transition states to become of highest energy. This is well demonstrated in comparing nitration of nitrobenzene and nitrotoluenes (96,97,98). High substrate and positional selectivity (with preference for para-nitration relative to the methyl group) indicate the σ-pattern of the reactions. There is also an interesting difference in selectivities when nitronium salt nitrations are carried out using aprotic (nitromethane) or protic (96% H_2SO_4) solvent media (as shown in Table XII). In the latter system the nitroaromatic substrates must be at least partially protonated, thus decreasing the n-donor interaction with the reagent NO_2^+ and subsequent O→C nitro group migration. In superacid media (such as FSO_3H–SbF_5, Magic acid) this effect is even better shown. Nitration of nitrobenzene gives 4.4% o-, 90% m-, and 5.6% p-dinitrobenzene.

Electrophilic nitrations resemble other aromatic substitutions in all aspects. Because it is more difficult to vary the reactivity of the nitrating agents, the change in selectivities is best achieved by varying the nucleophilicity of the aromatics.

4. The Role of Intermediate Complexes

The question must be raised to what degree the well-investigated arenium ions (σ-complexes) are involved in the pathway of electrophilic aromatic substitutions. In 1956, Olah, Kuhn, and Pavlath (85f) reported the preparation of a series of σ-complexes formed by the reaction of HF–BF_3, DF–BF_3, alkyl fluoride–BF_3, and acyl fluoride–BF_3 systems, as well as of $NO_2F + BF_3$, with aromatics. One such example was the formation of the C_2H_5F–BF_3 complexes of akylbenzenes, such as mesitylene.

In a detailed low-temperature, infra-red, ultraviolet, and n.m.r. spectroscopic and boron isotope exchange study, Nakane, Natsubori, and Kurihara (99) recently reinvestigated the toluene–ethyl fluoride–boron trifluoride complex. They concluded that the primary complex formed at $-80°$ is not a σ-complex but a termolecular oriented π-complex (with the C–F bond in ethyl fluoride not being broken). Only when the temperature is raised does the ionization and σ-complex formation take place. Zollinger et al. (100) were indeed first able to show the existence of a related oriented π-complex (as a separate entity from a σ-complex) in the iodination of the sterically hindered 2-naphthol-6,8-disulfonic acid and a σ-complex in the bromination of the same substrate. Zollinger's and Nakane's findings, as interesting as they are, relate only to reversible, weak outer complex formations. These do not lead directly to products. It should be emphasized that no inner π-complexes (aronium ions, see later discussion) have so far been directly observed, whereas σ-complex intermediates have been isolated and studied by spectroscopic methods.

5. The Nature of the Transition States

In order to explain experimental data on substrate and positional selectivity in Friedel-Crafts type aromatic substitutions, it must be suggested that the transition state of highest energy can resemble either the intermediate arenium ion (σ-complex) or be more of aronium ion (π-complex) nature.

In the interaction of an electrophile with an aromatic substrate, a weak reagent–substrate complex [outer complex (101)] is formed first. The formation of such complexes is reversible and does not lead to substituted products. Aromaticity is not lost in such complexes, as indicated by spectroscopic studies (as in the work of Zollinger and Nakane mentioned) showing only slight reversible changes in the aromatic substrates.

As the reagent moves closer to bonding distance, the highest-lying occupied aromatic π-orbital containing an electron pair overlaps with the empty orbital of the electrophile, forming a two-electron three-center bond (π-complex). The formed complex is indeed a bridged tetracoordinated carbonium ion (aronium ion)[102]. Opening of the three-center bond of the aronium ion leads to the trivalent arenium ion[102] (σ-complex) intermediate and thus bonding of the entering electrophilic reagent to an individual ring position, accounting for positional selectivity (directing effect).

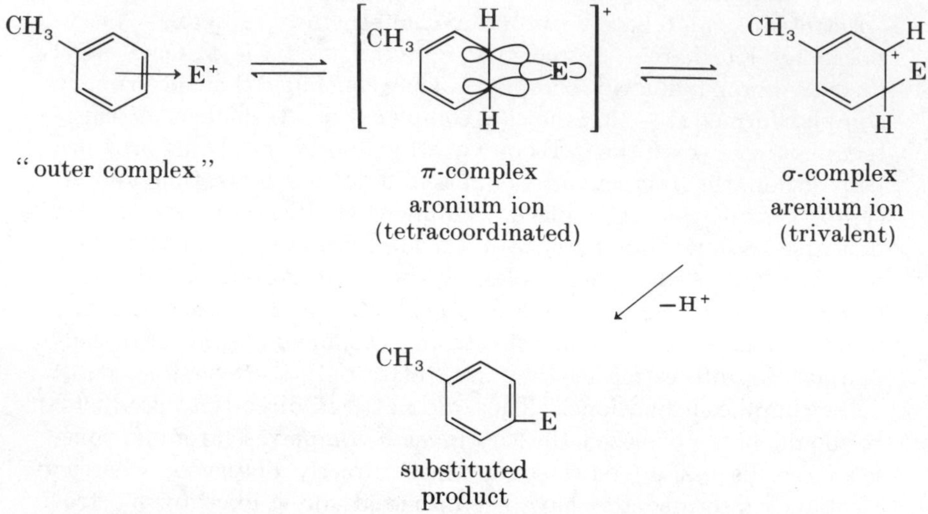

"outer complex"

π-complex
aronium ion
(tetracoordinated)

σ-complex
arenium ion
(trivalent)

$-H^+$

substituted
product

In reactions with relatively weak electrophiles, or in reactions with weakly basic aromatics, the transition state lies late on the reaction coordinate, resembling intermediate σ-complexes. In reactions with strongly electrophilic reagents, or with strongly basic aromatics, it lies *early*, resembling starting aromatics, thus being more of a π-complex in nature.

Figure 5 depicts the potential energy diagrams for reactions involving early and late transition states, respectively. The proton elimination side is not shown in these diagrams for simplicity. The left side of Fig. 5 shows the energy profile in cases in which low substrate but high postitional selectivities are operative. For such reactions the transition state lies early on the reaction path and resembles an aronium ion (oriented π-complex). Whether there is a separate π-complex minimum (corresponding to an intermediate) following the transition state of highest energy, thus separated by a second but lower energy maximum from the benzenium ion (σ-complex) intermediate, cannot be ascertained at this time, due to lack of direct observation of aronium ion complex intermediates. Data showing low substrate, but high positional selectivity substitutions, however, indirectly indicate such separate minima.

The right side of Fig. 5 shows the energy profile in cases in which the *late* transition state is predominantly of an arenium ion (σ-complex) nature.

The outlined mechanism for aromatic substitutions is in accordance with the "Hammond postulate" (103), which states: The transition state of a single-stage reaction is generally more similar to either the

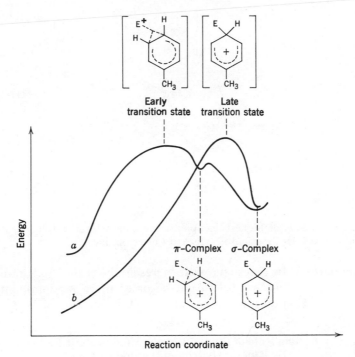

Fig. 5.   (a) Potential energy curve (left side) of reaction with early transition
state resembling starting aromatics (π-complex); (b) potential energy
curve (left side) of reaction with late transition states resembling σ-com-
plex.   For simplicity the proton elimination side of the reaction is not
shown.

reactant or the product, whichever is at the highest energy level.
If the postulate is applied to electrophilic aromatic substitution,
the transition state of highest energy should resemble more the
intermediate the higher the potential energy of the latter, and the
starting aromatics, the stronger the electrophile with high potential
energy.

Concerning the directing effect, as in the case of the methyl group
in toluene, consideration of the symmetry of the highest occupied
molecular orbital involved in the interaction with the electrophile
gives a good indication (Fig. 6).   As interaction can take place only
on bonds formed by orbitals bearing the same sign, i.e., 1-2, 1-6, 3-4,
and 4-5, but not 2-3 and 5-6; attack on the former two can lead only
to ortho-substitution, whereas the latter give para- (and less meta-)
substituted products.   As later the transition state, as more it will
resemble the arenium ion.   Para-substituents stabilize most arenium
ions, thus in reaction with late transition states para-substitution

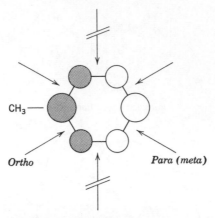

Fig. 6. Symmetry of highest occupied molecular orbital of toluene. Arrows show possible direction of attack by electrophile.

predominates.   In reactions with increasingly earlier nature of the transition state, *ortho*-substitution becomes more predominant and *meta*-substitution also increases (to a much lesser degree).

6. The Role of Proton Elimination as Reflected by Kinetic Hydrogen Isotope Effects

The question of kinetic hydrogen isotope effects in electrophilic aromatic substitutions has been well reviewed (104), and therefore we comment on only one specific aspect.   In substitutions in which the transition state is of a π-complex nature, no primary kinetic hydrogen isotope effects are expected; experimental evidence so far bears this out (104f).   As replacement of hydrogen by deuterium in aromatics enhances the π-basicity, it is of interest to note that nitration (also some other substitutions) of perdeuteriobenzene gave a small but significant inverse secondary isotope effect ($k_H/k_D = 0.89 \pm 0.03$), the heavy compound reacting faster (87).   As pointed out by Halevi and Ravid (105), this observation strongly strengthens the suggested π-complex type transition state concept of the reaction.

B. Reactions of Alkenes and Alkynes

The general reactivity of π-systems (olefins, acetylenes, and aromatic hydrocarbons) toward electrophilies in addition and substitution reactions is based on the π-electron donor ability of the unsaturated C=C or C≡C bonds and π-aromatic systems (106, 107).

In the previously discussed case of aromatic substitution, we emphasized the importance of the initial formation of two-electron, three-center bonded carbonium ions utilizing a π-electron pair from the aromatic ring.   In other words, electrophilic reactions were considered to initiate on the involved bonds (*i.e.*, between atoms) where the highest electron density occurs, and not at the atom themselves. In line with the same concept, electrophilic reactions of alkenes (additions or substitutions) again can be considered to involve initial interaction of the electrophile with the electron donor π-electron pair of the double bond.

Two-electron, three-center bonded carbonium ion formation is thus again considered to be involved in the initial step of electrophilic reactions of π-olefinic and acetylenic systems.

Protonation of olefins, the essential first step in acid-catalyzed isomerization, alkylation, hydration, and related addition and substitution reactions, is considered to involve the initial overlap of the two carbon atom p-orbitals (of the π-bond) with the empty hydrogen s-orbital, forming the two-electron, three-center bond of the alkeneprotonium ion, a hydrogen-bridged carbonium ion (108).

$$
CH_2=CH_2 \;\rightleftharpoons\; \left[\begin{array}{c} CH_2\text{---}CH_2 \\ | \\ H \\ \\ \Updownarrow \\ \\ CH_3CH_2^+ \end{array} \right]^+ \;\rightleftharpoons\; {}^+CH_2CH_3
$$
$$
H^+
$$

In unsymmetrically substituted olefins the three-center bond formation should be usymmetrical (oriented), caused by electronic and steric effects.   Furthermore, theoretical calculations (109) and lack of experimental observation indicate that hydrogen bridged alkeneprotonium ions (as well as alkyl bridged alkenalkonium ions) should be generally considered transition states and not intermediates. Subsequent opening of the three-center bond gives carbenium ion intermediates of highest stability.   The empirical Markownikoff rule can thus be easily deduced from such interaction when considering that, for example, electron-donating methyl substituents distort the three-center bond, increasingly directing the electrophilic reagent to occupy a position closer to the carbon carrying higher charge density.

$$(CH_3)_2C{=}CH_2 \; + \; H^+ \;\rightleftharpoons\; \left[(CH_3)_2C\overset{\overset{\displaystyle H}{|}}{\underset{}{\cdots\cdots}}CH_2 \right]^+ \;\rightleftharpoons\; (CH_3)_2C^+{-}CH_3$$

Alkylation, nitration, etc., of π-bonds is also considered to involve similar interaction with formation of three-center-bonded tetra-coordinated carbonium ion transition states.

$$\left[\begin{array}{c} CH_2\cdots CH_2 \\ | \\ R \end{array} \right]^+ \qquad \left[\begin{array}{c} CH_2\cdots CH_2 \\ | \\ NO_2 \end{array} \right]^+$$

ethenenal-
konium ion   ethenenit-
ronium ion

The long-lived ethenebromonium ion and other alkenehalonium ions in contrast are not three-center-bonded ions, but three-membered ring halonium ions with covalent C–halogen bonds (110). These ions are clearly intermediates and the three-center-bonded halonium ions, similar to previously discussed aronium ions could be the tran-

σ-complex π-complex

sition states leading to them, although existence of three-center-bonded halonium ion intermediates also cannot be ruled out. The distinction between the two different types of ions is not always clear, although it could be made, for example, in case of observable intermediates. Nuclear magnetic resonance data, for example, of the observed three-membered ring halonium ions and the three-center-bonded ethenemercurinium ion (111) (as model compound) are quite different. Even if in the latter case as a result of additional d-orbital participation the bonding is somewhat different from what would be the case in a three-center-bonded halonium ion, increased p-character of the carbon atomic orbitals in a tetracoordinated carbonium ion should be reflected by substantial deshielding of the chemical shifts (caused also by higher charge density on these atoms).

$$H_2C\overset{\displaystyle}{\underset{\underset{\displaystyle Br}{+}}{\underline{\qquad}}}CH_2 \qquad \begin{array}{l}\delta_H \, 5.53\\ \delta_{13C} \, 120.8\end{array} \qquad H_2C\overset{}{\underset{\underset{\displaystyle Hg^{2+}}{|}}{\cdots\cdots}}CH_2 \qquad \begin{array}{l}\delta_H \, 7.68\\ \delta_{13C} \, 57\end{array}$$

In electrophilic halogen additions to olefins, the electrophilic halogen is considered to interact through its vacant atomic orbital with the π-bond, forming a two-electron three-center-bonded ion (transition state) which then by utilizing a non-bonded electron pair of halogen gives the three-membered halonium ion intermediate (directly observed in spectroscopic studies). Subsequent displacement by halide ion from the back side results in the generally observed *trans*-stereospecificity of the addition.

$$CH_2{=}CH_2 \;+\; \text{`` Br}^+\text{''} \;\rightleftharpoons\; \left[H_2C{\cdots}CH_2 \;\; Br \right]^+ \;\rightleftharpoons\; H_2C\overset{+}{\underset{Br}{\diagdown\!\diagup}}CH_2 \;\xrightarrow{\;Br^-\;}\; \overset{}{\underset{Br}{|}}CH_2{-}\overset{Br}{\underset{}{|}}CH_2$$

three-center- bonded   halonium ion
halonium ion (tran   intermediate
sition state)

Strating, Wierenga, and Wynberg (112) recently isolated a bromonium ion salt of adamantylidencadamantane, a highly hindered olefin, by reacting the olefin with 2 moles of bromine. Subsequent studies by Olah, Schilling, and Westerman (113) by H^1 and C^{13} n.m.r. spectroscopy showed that the Br_2 and Cl_2 complexes of adamantylideneadamantane, similar to Ag^+ complexes, are indeed three-center-bonded alkenonium ions (π-complexes), whereas the H^+ complex (in superacids) is the rapidly exchanging alkenium ion (σ-complex). Thus in these cases a direct differentiation of alkenium (σ-complex) from alkonium ions (π-complex) was achieved in electrophilic reactions of alkenes.

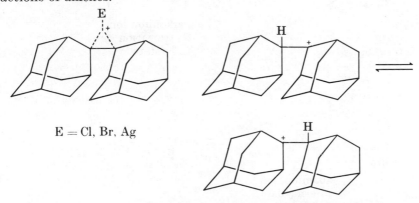

E = Cl, Br, Ag

Olah and Lin (114) however prepared three-membered-ring, σ-bonded halonium ions in direct electrophilic halogenation of olefins by reacting tetramethylethylene and ethylene, respectively, with cyanogen iodide (bromide)–antimony pentafluoride in SO_2 or SO_2ClF solution at $-78°$.

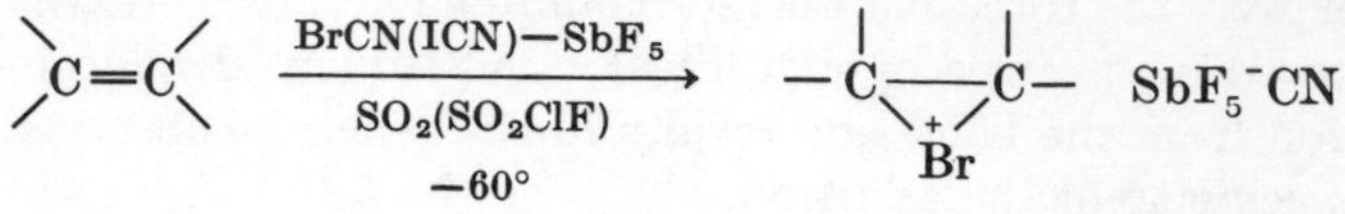

$$\underset{/}{\overset{\backslash}{C}}=\underset{\backslash}{\overset{/}{C}} \quad \xrightarrow[\substack{SO_2(SO_2ClF) \\ -60°}]{BrCN(ICN)-SbF_5} \quad -\underset{\underset{Br}{\overset{\backslash \overset{+}{} /}{}}}{\overset{|}{C}} \underline{\qquad} \overset{|}{C}- \quad SbF_5^-CN$$

In recent studies by Olah and Hockswender (115) on the rates of bromination of alkenes, it was shown that by varying the nature (*i.e.*, electrophilicity) of the brominating system the selectivity of the reacting alkenes was varied, reflecting resemblance of the transition state of the reactions either to the intermediate bromonium ions (σ-complex) or to starting alkenes (π-complex).

Electrophilic reactions of π-aromatic systems and alkenes, as discussed, both involve as the initial electron donation step by a π-electron pair to the vacant orbital of the electrophile. Addition to a carbon–carbon double bond of an alkene initially involves formation of a two-electron three-center-bonded carbonium ion (alkonium ion). Opening of the three-center bond leads to formation of a σ-bonded carbenium ion intermediate (alkenium ion).

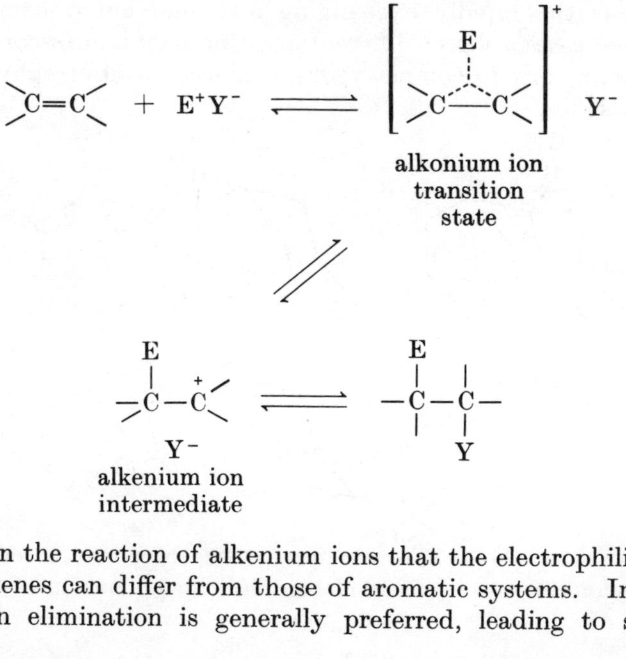

alkonium ion
transition
state

alkenium ion
intermediate

It is in the reaction of alkenium ions that the electrophilic reactions of alkenes can differ from those of aromatic systems. In the latter proton elimination is generally preferred, leading to substituted products, although it should be mentioned that recent studies [Vaughn *et al.*, (116) and Myhre (117)] have shown that benzenium ions can also react with nucleophiles, leading to addition products. Alkenium ions formed *via* electrophilic attack on an olefinic π-bond more frequently undergo addition, as for example in the well-studied electrophilic addition reactions of alkenes (hydrogen halides, halogens, acid-catalyzed hydration, etc.).

However, alkenium ions can also undergo proton elimination, thus giving substituted alkene products. The Nenitzescu type Friedel-Crafts acylation of alkenes (118), as well related alkylations (aryl-alkylations) (119) are examples of this pathway.

$$RCH=CH_2 + CH_3COCl\cdot AlCl_3 \quad [R\overset{+}{C}H-CH_2COCH_3]AlCl_4{}^-$$

$$-H^+ \diagup \qquad \diagdown Cl^-$$

$$RCH=CHCOCH_3 \underset{-HCl}{\rightleftharpoons} \quad \underset{\underset{Cl}{|}}{RCH}-CH_2COCH_3$$

$$R_2C=CH_2 + C_6H_5CHClCH_3 \xrightarrow{ZnCl_2} \left[\underset{\underset{C_6H_5}{|}}{R_2\overset{+}{C}}-CH_2CHCH_3\right] ZnCl_3{}^-$$

$$-H^+ \diagup \qquad \diagdown Cl^-$$

$$\underset{\underset{C_6H_5}{|}}{R_2C}=CHCHCH_3 \underset{-HCl}{\rightleftharpoons} \quad \underset{\underset{Cl}{|}}{R_2C}\underset{\underset{C_6H_5}{|}}{CH_2CH}CH_3$$

Whereas relatively few data are available on the study of Friedel-Crafts reactions of *alkynes* (120), it can be reasonably assumed that the mechanistic considerations of the reactions are similar to those of alkenes (and π-aromatic systems).

As the initial step of the reaction again involves π-electron donation by the alkyne to the electrophile, formation of a three-center-bonded alkynonium ion is involved, which then forms a vinyl cation. The considerably increased strain in forming an alkynonium ion, as compared to an alkenonium ion, in electrophilic reactions of alkynes can

$$-C{\equiv}C- \quad \rightleftharpoons \quad \left[-C{=}C- \atop E \right]^{+} \quad \rightleftharpoons \quad {}_{E}{>}C{=}\overset{+}{C}{<}$$
$$\underset{E^{+}}{}$$

well explain why alkynes generally are considerably less reactive than alkenes.

σ-Donor Systems

A. Reactions of Alkanes

The reactivity of olefins, acetylenes, and aromatic hydrocarbons toward electrophiles is based on their π-electron donor ability. Shared (bonded) σ-electron pair donation by single bonds represents the other major type in electrophilic reactions of hydrocarbons. Contrary to frequent textbook references to electrophilic aliphatic substitution, authenticated examples are restricted to reactions involving organometallic compounds, such as organomercurials. The only "pure" electrophilic substitutions of alkanes reported have been recently observed hydrogen–deuterium exchange and protolytic cleavage reactions in superacid media (121).

Recent studies relating to the reactivity of hydrocarbons in superacids and by Olah and co-workers (122,123) and on reactions with strongly electrophilic reagents resulted in the discovery of the *general electrophilic reactivity of covalent C–H and C–C single bonds of alkanes (cycloalkanes). This reactivity is due to what is considered the general σ-donor ability (σ-basicity) of shared (bonded) electron pairs (single bonds) via two-electron, three-center-bond formation.* It was observed that C–C and C–H single bonds of all types (*i.e.*, tertiary, secondary, and primary) show substantial general reactivity in electrophilic reactions, not only in protolytic processes (isomerization, hydrogen–deuterium exchange, protolysis) but also in typical Friedel-Crafts reactions such as alkylation, nitration, and halogenation. These observations open up a new area of chemistry in which alkanes and cycloalkanes can be used as substrates in a wide variety of Friedel-Crafts reactions. Also, saturated single bonds in general can undergo electrophilic reactions.

1. Protolysis and Hydrogen-Deuterium Exchange

With superacids alkanes readily undergo protolytic reactions involving tertiary, secondary, and primary C–H, as well as C–C, bonds 121–122). In isoalkanes tertiary C–H bond reactivity exceeds that

of secondary (and primary) C–H and of C–C bonds. In n-alkanes the C–C bond reactivity is generally found to exceed that of the C–H bonds. It is also apparent that steric factors affect σ-basicity as they do π- or n-basicity.

Bartlett, Condon, and Schneider (124) as well Nenitzescu, Avram, and Sliam (125) have observed the aluminum halide-catalyzed intermolecular hydride abstractions in which the electrophile (carbenium ion or proton) can remove a tertiary hydrogen atom together with its bonding electron pair.

$$R_3C - H + R^+ \text{ (or } H^+) \; \rightleftharpoons \; R_3C^+ + H:R \text{ (or } H_2)$$

The assumption has been generally made that the hydrogen atom is removed with its electron pair through what amounts to a linear transistion state $\overbrace{R\text{---}H\text{---}H}^{+}$.   Lewis, Hawthorne, and Symmons (126) must be credited with having first made the suggestion that the transition state in hydride abstraction could be considered triangular instead of linear: ". . . [The] electrophilic reagent attacks the C–H bond and the reasonable mode of attack is on the electrons of the bond.   A triangular transition state is therefore proposed."   This suggestion, however, went relatively unnoticed, probably because there was no experimental data available to substantiate it and because it was considered that steric hindrance, particularly in the interaction of a tertiary carbenium ion with an isoalkane, would allow only linear interaction.

The question of the mechanism of the hydride abstraction was not further considered until 1967 when Olah and Lukas (121a) reported the protolytic ionization and hydrogen–deuterium exchange of alkanes and cycloalkanes in superacids such as $FSO_3H–SbF_5(SO_2ClF)$. In independent work Hogeveen $et\ al.$ (121b) made similar observations in $HF–SbF_5$ solution.   These studies demonstrate that in superacid systems not only tertiary and secondary, but also primary C–H bonds and C–C bonds undergo facile protolytic cleavage.   To account for the observed protolytic reactions of alkanes, Olah, Klopman, and Schlosberg (127) suggested that protolyses preferentially involve front-side attack through formation of triangular transition states similar to the case of protonation of methane to CH_5^+.

In contrast, Hogeveen, Gaasbeek, and Bickel (128a) initially rejected triangular transition states in hydride abstractions by arguing

$$CH_4 + H^+ \; \rightleftharpoons \; \left[CH_3\text{---}\!\!\underset{H}{\overset{H}{\diagup}} \right]^+ \; \rightleftharpoons \; CH_3^+ + H_2\uparrow$$

that such transition states would suffer considerable strain. (However subsequently Brouwer and Hogeveen also accepted them (128b). They also stated that they found no evidence for exchange of molecular hydrogen (deuterium) with the acid systems they used (generally SbF_5 in excess HF), which they considered further evidence against triangular transition states. At the same time they showed the reaction of carbenium ions, under stable conditions, with molecular hydrogen (128,129) but considered the reaction to involve a linear transition state.

$$R^+ + H\!\!-\!\!H \rightleftharpoons (R\text{-}\text{-}\text{-}H\text{-}\text{-}\text{-}H)^+ \rightleftharpoons R\!\!-\!\!H + H^+$$

Subsequently however Olah, Shen, and Schlosberg (130) showed that in stronger superacids molecular hydrogen (deuterium) readily undergoes hydrogen–deuterium exchange, even at room temperature, considered to involve a triangular transition state in the protolytic process.

$$\begin{matrix} H \\ | \\ H \end{matrix} + D^+ \rightleftharpoons \left[\begin{matrix} H \\ \\ H \end{matrix}\!\!\cdots\!\!D\right]^+ \rightleftharpoons \begin{matrix} H \\ | \\ D \end{matrix} + + H^+$$

In the studies of Olah and co-workers, it was then demonstrated that protolysis of single bonds, including C–H as well as C–C bonds, is a general reaction as shown in the case of neopentane (122).

$$\left[(CH_3)_3C\text{-}\!\!\cdot\!\!\cdot\!\!\overset{H}{\underset{CH_3}{\cdots}}\right]^+ \longrightarrow (CH_3)_3C^+ + CH_4$$

$$CH_3\!\!-\!\!\overset{\overset{\displaystyle CH_3}{|}}{\underset{\underset{\displaystyle CH_3}{|}}{C}}\!\!-\!\!CH_3 \quad \xrightarrow[\text{HF—SbF}_5 \ \ 2]{\text{FSO}_3\text{H–SbF}_3 \ \ 1}$$

$$\left[(CH_3)_3CCH_2\text{-}\!\!\cdot\!\!\cdot\!\!\overset{H}{\underset{H}{\cdots}}\right]^+ \rightleftharpoons \left[CH_3\!\!-\!\!\overset{\overset{\displaystyle CH_3}{|}}{\underset{\underset{\displaystyle CH_3}{|}}{C}}\!\!-\!\!CH_2^+\right]$$

$$(CH_3)_2\overset{+}{C}CH_2CH_3 + H_2$$

Furthermore, they showed that three-center bond formation is characteristic of all electrophilic reactions at single bonds.

As a result of the oxidizing ability of antimony pentafluoride (which is subsequently reduced to SbF_3), the possibility of redox processes must always be kept in mind when dealing with systems containing SbF_5. In superacids, and particularly at low temperature, the equi-

librium concentration of SbF_5, for example, in a $HF-SbF_5$ system is low. Furthermore, the protolytic reactivity of alkanes is well demonstrated using deuterated superacids. One electron transfer from an alkane single bond is a relatively high-energy process, necessitating 4–5 eV. more activation energy than a comparable electron transfer process from an olefinic or aromatic compound, and consequently is not probable in low-temperature solution chemistry. It is significant to mention that many of the protolytic reactions discussed can also be carried out with acid systems other than those containing SbF_5 · $HF-TaF_5$ and $HF-BF_3$ are, for example, useful superacids in certain reactions. As the redox potentials of the metals in these systems are high, it seems to be further established that alkanes react in superacid solutions *via* single-bond protolytic processes. This is also indicated by results of experiments using deuterated superacids (or alkanes).

2. Alkylation

Despite frequent literature references to electrophilic alkylation of alkanes by olefins, from a mechanistic point of view, these reactions must be considered alkylations of the olefin by the carbenium ion derived from the isoalkane by intermolecular hydride transfer. The suggested reaction mechanism [by Schmerling (131)] can be demonstrated by the products formed, for example, in the reaction of propylene and isobutylene with isobutane.

$$R_2'C{=}CH_2 \xrightarrow{\;H^+\;} R_2'\overset{+}{C}CH_3 + R_3CH \rightarrow R_2'CHCH_3 + R_3C^+$$

$$R_3C^+ + R_2'C{=}CH_2 \rightarrow R_2'C^+{-}CH_2CR_3 \xrightarrow{\;H^-\;} R_2'CHCH_2CR_3$$

Products do not contain 2,2,3-trimethylbutane or 2,2,3,3-tetramethylbutane, which would be the expected direct alkylation products (124,125).

In the Bartlett-Nenitzescu-Schmerling intermolecular hydride abstraction from a tertiary isoalkane by a carbenium ion, the transition state either could be considered linear or triangular (*i.e.*, frontside attack on the C–H bond). The latter would not be symmetrical because of a steric effect between the carbenium ion and the tertiary isoalkane. With an even strongly distorted triangular transition state (*i.e.*, the reaction taking place on the C–H bond and not at the hydrogen atom), it becomes obvious that cleavage can result not only in intermolecular hydrogen transfer (a more proper name than hydride transfer, as no free hydride ion is involved) but also *via* proton elimination, in direct alkylation.

It should be emphasized that there is no necessity to suggest a common transition state for hydrogen abstraction and alkylation, only that triangular transition states of a related nature are involved. Products depend on the nature of reactants, reaction conditions, and stability of products, in addition to other factors such as strain in the formation of transition states affecting the ratios of reactions 1 and 2.

$$R'_3CH + {}^+CR_3 \rightleftharpoons \left[R'_3C\text{---}\overset{H}{\underset{CR_3}{\diagup\diagdown}} \right]^+ \begin{smallmatrix}1\\\longrightarrow\\ \\2\end{smallmatrix} \begin{smallmatrix}R'_3CCR_3 + H^+\\ \\R'_3C^+ + CR_3H\end{smallmatrix}$$

R, R' = allyl

In the case of tertiary-tertiary systems (a tertiary isoalkane reacting with a t-alkyl cation), a symmetrical trigonal three-center transition state would be highly strained, and its formation therefore is impossible. However, again it is unnecessary to consider a symmetrical transition state in the reactions. The tertiary carbenium ion could easily approach the tertiary C–H bond in such a way as to form an unsymmetrical transition state, which can account for the great preference for intermolecular hydrogen transfer over alkylation in the reaction. The reaction of stable alkylcarbenium ions with alkanes was advantageously studied in sulfuryl chloride solution at low temperature (generally as low as $-78°$). A small amount ($\sim 2\%$ of the C_8 fraction) of 2,2,3,3-tetramethylbutane was detected in the reaction of t-butyl fluoroantimonate with isobutane (132).

$$(CH_3)_3C^+ + HC(CH_3)_3 \rightleftharpoons (CH_3)_3C\text{---}C(CH_3)_3 + H^+$$

This observation is of substantial importance since no other way than direct alkylation of the C–H bond of isobutane by the t-butyl cation can account for its formation. That direct alkylation of alkanes can indeed be carried out with ease was further shown when isobutane was reacted with the less bulky isopropyl fluoroantimonate, or when propane was reacted with t-butyl fluoroantimonate, giving in the primary alkylation product up to 12% 2,2,3-trimethylbutane (132). As intermolecular hydrogen transfer is faster than alkylation, and the isopropyl cation more reactive than the t-butyl cation, the alkylation reaction in both systems is considered mainly to be the propylation of isobutane.

$$(CH_3)_2CH_2 + \overset{+}{C}(CH_3)_3 \rightleftharpoons (CH_3)_2CH^+ + HC(CH_3)_3$$

$$(CH_3)_3CH + \overset{+}{H}C(CH_3)_2 \rightleftharpoons (CH_3)_3CCH(CH_3)_2$$

The alkylation products obtained are not only those derived from the starting alkanes and carbenium ion, but also those from the alkanes and carbenium ions formed by intermolecular hydrogen transfer, which generally is a faster reaction than alkylation. Since carbocations also undergo isomerization, and since intermolecular hydrogen transfer produces new alkanes from these ions, alkylation products obviously are increasingly complex. Generally, the bulkier tertiary carbenium ions can attack the more shielded tertiary C–H bonds only with great difficulty. The processes mentioned, however, produce with ease secondary and incipient primary ions in the systems, which then alkylate C–H and C–C bonds (*alkylolysis*). The alkylation reactions of alkanes with stable carbenium ion salts in low-nucleophilicity solvents produce a variety of products which, however, can be quantitatively analyzed with modern methods (gas chromatography, mass spectrometry). Since olefins are formed under the reaction conditions only to a minimal degree, if at all, and at low temperatures ($-78°$) and with short reaction times (< 30 seconds), isomerizations are considered of lesser importance; indeed, *direct alkylation of alkanes* was established in these systems.

Alkylation of alkanes was also effected with the recently described CF_3F–SbF_5 and C_2H_5–SbF_5 complexes (133,134) [which can be considered methyl and ethyl fluoroantimonates] and dialkylhalonium ions (135). It is also reasonable to suggest that in conventional Friedel-Crafts systems direct alkylation of alkanes can play a role, although in systems in which olefin formation is possible alkene reactivity vastly exceeds that of alkanes.

Singlet carbenes are also strong electrophiles, and their insertion reactions into C–H (and in some instances C–C) single bonds can also be considered σ-bond alkylation.

3. Nitration

The general concept of electrophilic reactivity of single bonds (σ-donors) was also demonstrated by Olah and Lin (123) in nitration of alkanes (cycloalkanes). To avoid the possibility of any free-radical reaction resulting from the use of nitric acid and to avoid acid cleavage of reaction products, nitrations were carried out with stable nitronium salts, such as $NO_2^+PF_6^-$, in solvents such as methylene chloride–sulfolane (123). In the case of methane (in which the nitration product—nitromethane—is not acid-sensitive). anhydrous HF or FSO_3H was also used as solvent.

$$R\text{---}H \xrightarrow[\substack{CH_2Cl_3\text{--sulfolane,} \\ 25°, \\ \text{dark}}]{NO_2^+PF_6^-} R\text{---}NO_2 + H^+$$

The electrophilic nitration of alkanes by nitronium ion again involves attack at both C–H bonds, resulting in substitution (*i.e.*, nitration) and at C–C bonds, causing *nitrolysis*.

Since tertiary and secondary nitroalkanes are cleaved with ease by strong acids, aliphatic electrophilic nitration is effected by protolytic

$$CH_4 \xrightarrow{\ NO_2^+\ } \left[CH_3 \cdots \overset{H}{\underset{NO_2}{<}} \right]^+ \longrightarrow CH_3NO_2 + H^+$$

$$\left[(CH_3)_3C \cdots \overset{H}{\underset{NO_2}{<}} \right]^+ \longrightarrow \begin{matrix} (CH_3)_3CNO_2 \\ + \\ H^+ \end{matrix}$$

$$CH_3-\overset{\overset{\displaystyle CH_3}{|}}{\underset{\underset{\displaystyle CH_3}{|}}{C}}-H \xrightarrow{\ NO_2^+\ }$$

$$\left[(CH_3)_2\overset{\overset{\displaystyle H}{|}}{C} \cdots \overset{CH_3}{\underset{NO_2}{<}} \right]^+ \xrightarrow{\ F^-\ } \begin{matrix} CH_3NO_2 \\ + \\ CH_3CH(F)CH_3 \end{matrix}$$

cleavage of products and therefore generally cannot be carried out in nitrating acid media.

4. Chlorination

Chlorination of alkanes can also be carried out, similarly to other discussed Friedel-Crafts type acid-catalyzed reactions.

In contrast to other electrophiles, the nature of positive halogens has not yet been sufficiently defined. Olah *et al.* (136) were successful in carrying out chlorination of alkanes, including methane, at low temperatures (-78 to $0°$) by using catalysts such as antimony penta-fluoride and low-nucleophilicity solvents such as SO_2ClF. The conditions used do not favor radical formation, although it must be pointed out that electrophilic chlorine (Cl^+ or more probably Cl_2^+) can be considered a radical cation. At the same time the chlorine atom ($Cl\cdot$) is a strong electrophile, thus the differentiation between ionic and radical chlorine is not as clear as in the case of other substituting agents.

In the Friedel-Crafts type chlorination of alkanes Lewis acid catalysts (SbF_5, $AlCl_3$, $FeCl_3$, etc.) cause increased dehydrochlorination of secondary and tertiary alkyl chlorides with subsequent chlorine addition, leading to vicinal dichlorides.

VII. Conclusions

Friedel-Crafts reactions can be characterized in a general sense as acid-catalyzed electrophilic reactions. The mechanistic concepts governing reactions of electrophilic reagents (carbenium ions or polarized donor–acceptor complexes, as well as related sulfonyl, nitryl, etc. systems), with both π- and σ-donors can be summarized by the assumption that the electrophile (being an electron-deficient, *i.e.*, electron-seeking, reagent) interacts with positions of highest electron density in the substrates. Electron densities (both π- and σ-) are at a maximum in the major lobes of the overlapping orbitals, *i.e.*, between atoms. The back lobes of the involved orbitals in σ-bonds represent only relatively minor electron density (less than 10%) and therefore it is considered that reactions of both π- and σ-donor systems start at the *involved bonds via* two-electron three-center-bonded carbonium ion formation (tetra-coordinated in the case of alkenes, pentacoordinated when alkanes are the donor molecules). The three-center bonds of carbonium ion transition states (or intermediates) subsequently open, forming in the case of reactions with π-systems trivalent carbenium ions which subsequently form substitution (*via* proton elimination) or addition products.

Alkonium ions, formed in reactions of alkanes with electrophiles, undergo cleavage of the three-centered bond, leading to substitution or cleavage products.

$$R_3CH + E^+Y^- \rightleftharpoons \left[R_3C\!-\!\!\!<^{\overset{H}{\underset{E}{}}} \right]^+ Y^- \longrightarrow \begin{array}{c} R_3CY + HE \\[2ex] R_3CE + HY \end{array}$$

1497470700

$$R_3C\!-\!CR_3 + E^+Y^- \rightleftharpoons \left[R_3C\underset{\overset{|}{H}}{\diagup\!\!\!\diagdown}CR_3 \right]^+ Y^- \longrightarrow R_3CH + CR_3Y$$

Thus there is no difference in principle between reactions of π- and σ-donor systems, only the more available π-bonds generally react more readily and therefore milder reaction conditions are sufficient to bring about their reactions. σ-Bonds become electron donors only under more forcing conditions, necessitating stronger electrophilic reagents or catalysts.

Friedel-Crafts reactions are generally considered to be typical electrophilic reactions. The acidic catalyst enhances the electrophilic nature of the reagent which then reacts with the nucleophilic substrates (arenes, alkenes, alkanes). On closer examination, however, the question arises whether these reactions do indeed always involve electrophilic attack by an electrondeficient reagent, rather than being more displacement reactions, with the nucleophilic substrate displacing the polarized reagent complex.

The former path can be, for example, considered to involve the formation of trivalent carbenium ion in the interaction of an alkyl halide (preferably a tertiary or reactive secondary halide) with the acid catalyst. The carbenium ion then reacts as a strongly electron-deficient reagent with the π-donor substrate (aromatic or olefin), utilizing a π-electron pair of the latter:

$$(CH_3)_3CF + SbF_5 \qquad\qquad (CH_3)_3CCH\!=\!C(CH_3)$$

$$\Updownarrow \qquad\qquad\qquad \uparrow -H^+$$

$$(CH_3)_3C^+ \; SbF_6 \xrightarrow{\;CH_2=C(CH_3)_2\;} (CH_3)_3C\!-\!CH_2\!-\!\overset{+}{C}(CH_3)_2$$

$$\Big\downarrow C_6H_6$$

$$\left[\underset{}{\overset{C(CH_3)_3}{\diagup\!\!\diagup}} \right]^+ \rightleftharpoons \underset{}{\overset{H\quad C(CH_3)_3}{\bigoplus}} \xrightarrow{\;-H^+\;} \underset{}{\overset{C(CH_3)_3}{\bigcirc}}$$

The transition state of a reaction involving electrophilic attack by an electron-deficient carbenium ion (a typical E_2 reaction) thus involves two electrons from the electron donor substrate and an initial three-center bonded (carbonium ion forming) interaction. In this regard there is no difference between reactions of π- or σ-donor substrates. However, because of the much greater availability of π-electron pairs, π-donors react more readily than do σ-donors. Furthermore, alkenes react more readily than arenes, since in alkenes the aromatic resonance energy need not be overcome.

Most Friedel-Crafts reactions probably do not proceed by this pathway, however, because they do not involve complete formation of a highly electron-deficient, free carbenium ion type reagent to interact with the substrate.

Primary alkyl halides, for example, generally form only polarized donor–acceptor complexes with acidic catalysts of the Friedel-Crafts type, but not free alkylcarbenium ions. When reacted with aromatic substrates, such as benzene and toluene, these complexes undergo displacement reactions. A good example of this is the reaction of benzene with a 1-halopropane. The alkylation products contain a significant amount of π-propylbenzene. Absence of isomerization of the propylating agent strongly indicates an S_N2 type displacement reaction, since prior ionization of the halide inevitably would lead to isomerization to the 2-propyl cation, and subsequent alkylation to isopropylated products:

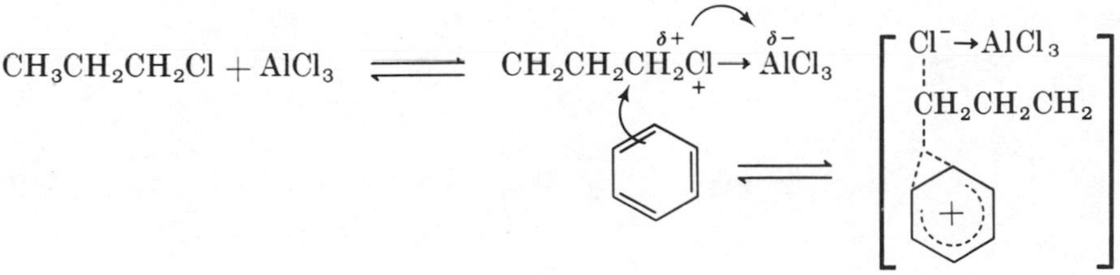

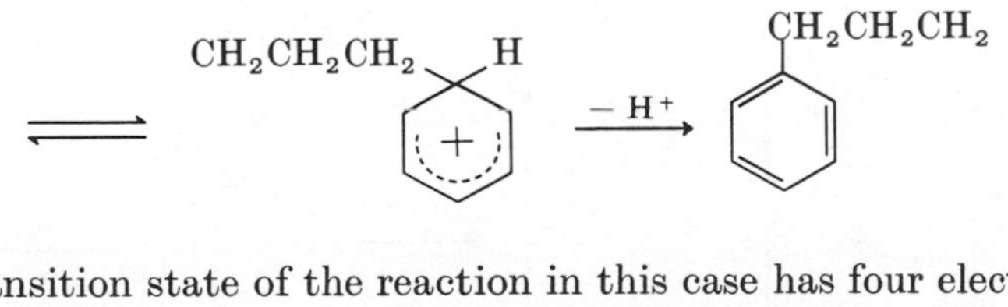

The transition state of the reaction in this case has four electrons at three centers, hence is closely related to the S_N2 type. The nucleophile is a π-nucleophile, and the displaced group can vary widely.

In *acylation* reactions a similar situation exists. Furthermore, even if an acylium ion, such as the acetyl cation, would be completely formed before reaction with the nucleophile (aromatic hydrocarbon, olefin, or acetylene), it would need to undergo a displacement reaction itself.   Acyl cations are linear resonance stabilized ions

$$R C \equiv \overset{+}{O} \longleftrightarrow R \overset{+}{C} = O \longleftrightarrow \overset{+}{R} = C = O$$

in which the oxonium forms predominate. Reaction of an acyl cation, therefore, can be considered as a displacement reaction, in which a π-electron pair is displaced to the oxygen atom.

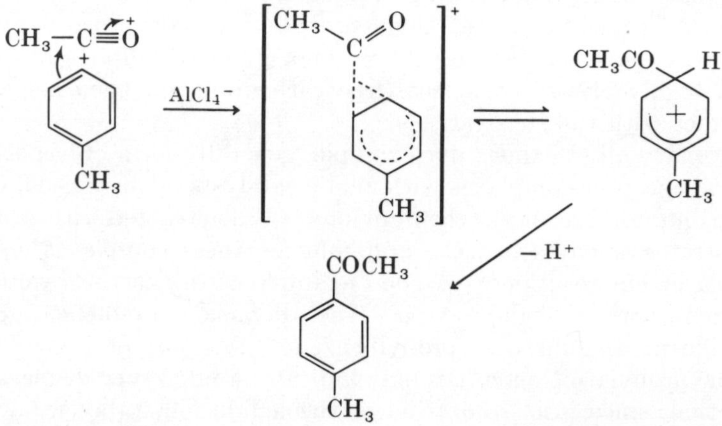

In Friedel-Crafts acylations with acyl halide-Lewis acid halide complexes, displacement can involve the halide ion:

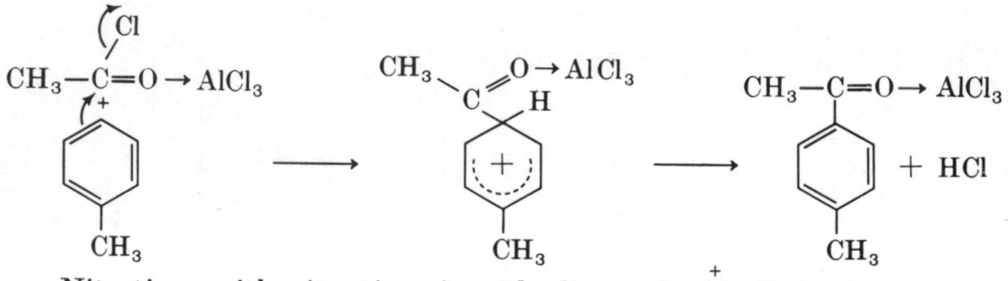

Nitrations with nitronium ion (the linear $O = \overset{+}{N} = O$ ion is not an electron-deficient cation) similarly involve π-electron displacement from the π-bond to oxygen:

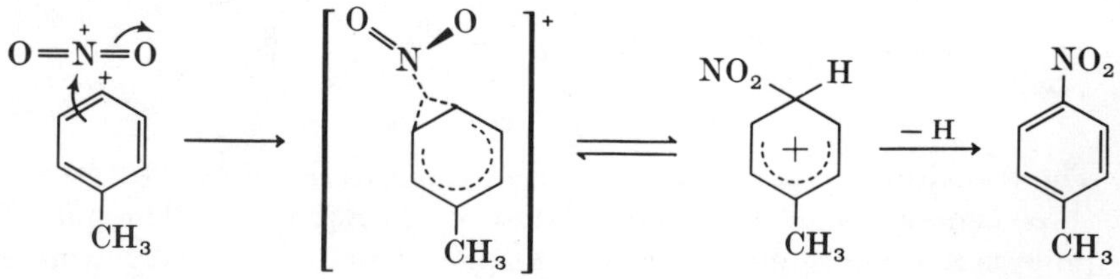

σ-Donor systems (alkanes and other saturated hydrocarbons) are much weaker electron donors than π-systems; consequently in their Friedel-Crafts reactions only electrophilic attack by a powerful electron-seeking (deficient) reagent is possible. Consequently, it is only in these reactions that "pure" S_E2 type electrophilic substitutions can be substantiated. In reactions of π-donor hydrocarbons the displacing ability of the nucleophile on a polarized complex or π-electrophile must always be considered, and it may be of more general importance than so far recognized.

References

1. For a discussion of intermediate complexes in reactions of the Friedel-Crafts type see: Olah, G. A., and M. W. Meyer, in "Friedel-Crafts and Related Reactions," G. A. Olah, ed., Vol. I; Wiley-Interscience, New York, 1963, pp. 623–765; G. A. Olah, in "Reaction Mechanism," Special Publ. No. 19, The Chemical Society, London, 1965.
2. For the general concept of carbocations, the logical name of all hydrocarbon cations, as the related anions are called carbanions, and their naming system, see: Olah, G. A., *J. Amer. Chem. Soc.*, **94**, 808 (1972).
3. For a comprehensive treatment see the monograph series, "Reactive Intermediates in Organic Chemistry," G. A. Olah, ed., Wiley-Interscience, New York, 1968–1972.
4. For a comprehensive review see "Carbonium Ions," G. A. Olah and P. v. R. Schleyer, eds., Vols. I–IV, Wiley-Interscience, New York, 1968–1972.
5. Bethell, D. and V. Gold, "Carbonium Ions, An Introduction," Academic Press, New York, 1967.
6. Brown, H. C., H. Pearsall, and L. P. Eddy, *J. Am. Chem. Soc.*, **72**, 5347 (1950).
7. Wertyporoch, E., and T. Firla, *Ann.* **500**, 287 (1933).
8. Olah, G. A., S. Kuhn, and J. Olah, *J. Chem. Soc.*, 2174 (1957).
9. Fairbrother, F., *J. Chem. Soc.*, 503, (1945).
10. Byrne, J. J., Ph.D. Thesis, Purdue University, 1957; *Dissertation Abstr.*, **18**, 1976 (1958).
11. Nelson, H. M., *J. Phys. Chem.*, **66**, 1380 (1962).
12a. Olah, G. A., Conference Lecture at 9th Reaction Mechanism Conference, Brookhaven, N.Y., August 1962.
12b. Olah, G. A., Abstracts 142nd National Meeting of the American Chemical Society, Atlantic City, N.J., September 1962, pp. 45–46.
12c. Olah, G. A., W. S. Tolgyesi, J. S. McIntyre, I. J. Bastien, M. W. Meyer, and E. B. Baker, Abstracts A, XIX International Congress of Pure and Applied Chemistry, London, June 1963, pp. 121–122.
12d. Olah, G. A., *Angew. Chem.*, **75**, 800 (1963).
12e. Olah, G. A., C. D. Nenitzescu's 60th Birthday Issue, *Rev. Chim.*, **7**, 1139 (1962).

12f. Olah, G. A., W. S. Tolgyesi, S. J. Kuhn, M. E. Moffatt, I. J. Bastien and E. B. Baker, *J. Amer. Chem. Soc.*, **85**, 1328 (1963).

12g. Olah, G. A., E. B. Baker, J. C. Evans, W. S. Tolgyesi, J. S. McIntyre and I. J. Bastien, *J. Am. Chem. Soc.*, **86**, 1360 (1964).

12h. Olah, G. A., American Chemical Society 1964 Petroleum Award Lecture Reprints, Division of Petroleum Chemistry, American Chemical Society, Vol. 9, No. 7, C31–51 (1964).

12j. Olah, G. A., "Intermediate Complexes and Their Role in Electrophilic Aromatic Substitutions," Conference Lecture at Organic Reaction Mechanism Conference, Cork, Ireland, June 1964, Special Publ. No. 19, The Chemical Society, London, 1965.

12k. Olah, G. A., and C. U. Pittman, Jr., *Advan. Phys. Org. Chem.*, **4**, 305 (1966).

13. Nakane, R., O. Kurihara, and A. Natsubori, *J. Phys. Chem.*, **68**, 2876 (1964); R. Nakane, A. Natsubori, and O. Kurihara, *J. Am. Chem. Soc.*, **87**, 3597 (1965).

14. Olah, G. A., J. DeMember, R. H. Schlosberg, and Y. Halpern, *J. Amer. Chem. Soc.*, **94**, 156 (1972).

15. Olah, G. A., and P. Schilling, unpublished results.

16. DeHaan, F. P., and H. C. Brown, *J. Am. Chem. Soc.*, **91**, 4844, 4854 (1969).

17. Olah, G. A., and J. R. DeMember, *J. Am. Chem. Soc.*, **91**, 2113 (1969); **92**, 718 (1970).

18. Rosenbaum, J., and M. C. R. Symons, *Proc. Chem. Soc.*, 92 (1959); *Mol. Phys.*, **3**, 205 (1960).

19. Finch, A. C. M., and M. C. R. Symons, *J. Chem. Soc.*, 378 (1965).

20. Olah, G. A., C. W. Pittman, Jr., R. Waack, and M. A. Doran, *J. Am. Chem. Soc.*, **88**, 1488 (1966).

21. Olah, G. A., C. A. Cupas, M. B. Comisarow, and C. U., Pittman, Jr., *J. Am. Chem. Soc.*, **87**, 2997 (1965).

22. Olah, G. A., C. U. Pittman, Jr., and M. C. R. Symons, in "Carbonium Ions," G. A. Olah, and P. v. R. Schleyer, eds., Vol. I, Wiley-Interscience, New York, 1968, p. 153.

23. Deno, N. C., in "Progress in Physical Organic Chemistry," Vol. II, p. 129, S. G. Cohen, A. Streitwieser, Jr., and R. W. Taft, Jr., eds., Wiley-Interscience, New York, 1964, and reference given therein.

24. Vernon, C. A., *J. Chem. Soc.*, 423 (1954).

25. Taft, R. W., Jr., and P. Riesz, *J. Am. Chem. Soc.*, **77**, 902 (1955).

26. Roberts, W. J., and A. R. Day, *J. Am. Chem. Soc.*, **72**, 1226 (1950).

27. Gillespie, R. J., *Accounts Chem. Res.*, **1**, 202 (1968), and references therein.

28. Woolf, A. A., *J. Chem. Soc.*, 433 (1955).

29. Thompson, R. C., J. Barr, R. J. Gillespie, J. B. Milne, and R. A. Rothenbury, *Inorg. Chem.*, **4**, 1641 (1965).

30. Brouwer, D. M., and E. L. Mackor, *Proc. Chem. Soc.*, 147 (1964).

31. Hogeveen, H., *Rec. Trav. Chim.*, **86**, 1061 (1967).

32. Olah, G. A., and E. Namanworth, *J. Am. Chem. Soc.*, **88**, 5327 (1966).

33. Olah, G. A., J. Sommer, and E. Namanworth, *J. Am. Chem. Soc.*, **89**, 3576 (1967).

34. Olah, G. A., *Chem. Eng. News*, **45**, 76, March 27, 1967; G. A. Olah, and Y. Halpern, *J. Org. Chem.*, **36**, 2354 (1971).

35. Olah, G. A., and D. H. O'Brien, *J. Am. Chem. Soc.*, **89**, 1725 (1967).
36. Olah, G. A., D. H. O'Brien, and C. U. Pittman, Jr., *J. Am. Chem. Soc.*, **89**, 2996 (1967).
37. Seel, F., *Z. anorg. allgem. Chem.*, **250**, 331 (1943).
38. Susz, B. P., and J. J. Wuhrmann, *Helv. Chim. Acta*, **40**, 722 (1957).
39. Pfeiffer, P., "Organische Molekulverbindungen," 2nd Ed., Enke Verlag, Leipzig, 1927, p. 104.
40. Susz, B. P., and J. J. Wuhrmann, *Helv. Chim. Acta*, **40**, 971 (1957).
41. Cook, D., *Can. J. Chem.*, **37**, 48 (1959).
42a. Olah, G. A., and S. Kuhn, *Ber.*, **89**, 866 (1956).
42b. Olah, G. A., S. J. Kuhn, W. S. Tolgyesi, E. B. Baker, *J. Am. Chem. Soc.*, **84**, 2733 (1962).
42c. Olah, G. A., W. S. Tolgyesi, S. J., Kuhn, M. E. Moffatt, I. J. Bastien, and E. B. Baker, *J. Am. Chem. Soc.*, **85**, 1328 (1963).
42d. Olah, G. A., *Rev. Chim.*, **7**, 1139 (1962).
43. Germain A., A. Commeyras, and A. Casadevall, *Chem. Commun.*, 633 (1971).
44. Olah, G. A., K. Dunne, and T. Hockwender, to be published.
45. Olah, G. A., and A. M. White, *J. Am. Chem. Soc.*, **89**, 3591 (1967).
46. Mateescu, Gh. D., L. Riemenschneider, and G. A. Olah, *J. Am. Chem. Soc.*, **94**, 2529 (1972).
47. Boer, F. P., *J. Am. Chem. Soc.*, **88**, 1572 (1966); 90, 6706 (1968).
48. Olivier, M. S. C. J., *Rec. Trav. Chim.*, **35**, 166 (1915).
49. Jensen, F. R., and G. Goldman, in "Friedel-Crafts, and Related Reactions," Vol. III, G. A. Olah, ed., Wiley-Interscience, New York, 1963–1964, p. 1319.
50. Burton, H., and H. B. Hopkins, *J. Chem. Soc.*, 4457 (1952).
51. Klages, F., and F. E. Malecki, *Ann.*, **691**, 15 (1966); *Ber.*, **96**, 2057 (1963).
52. Lindner, E., and H. Weber, *Ber.*, **101**, 2832 (1968).
53. Olah, G. A., A. T. Ku, and J. A. Olah, *J. Org. Chem.*, **35**, 3925 (1970).
54. Olah, G. A., S. Kobayashi, and J. Nishimura, *J. Am. Chem. Soc.*, **95**, 564 (1973).
55. Goddard, D. R., E. D. Hughes, and C. K. Ingold, *J. Chem. Soc.*, 2559 (1950).
56. Kuhn, S. J., and G. A. Olah, *J. Am. Chem. Soc.*, **83**, 4564 (1961).
57. Olah, G. A., S. Kuhn, and A. Mlinko, *J. Chem. Soc.*, 4257 (1956); G. A. Olah, and S. Kuhn, *Chem. Ind.*, **98**, (1956); G. A. Olah, S. J. Kuhn, and S. H. Flood, *J. Am. Chem. Soc.*, **83**, 4571 (1961).
58. Cook, D., S. J. Kuhn, and G. A. Olah, *J. Chem. Phys.*, **33**, 1669 (1960).
59. For a summary see: Topchiev, A. V., "Nitration of Hydrocarbons," Pergamon, Oxford, 1959.
60. Olah, G. A., and J. A. Olah, unpublished results.
61. Kuhlman, F., *Ann.*, **39**, 319 (1891).
62. Sprague, R. W., A. B. Garrett, and H. H. Sisler, *J. Am. Chem. Soc.*, **82**, 1059 (1960); earlier studies are summarized.
63. Bachman, G. B., and T. Hokama, *J. Am. Chem. Soc.*, **79**, 4370 (1957).
64. Bachman, G. B., and J. L. Dever, *J. Am. Chem. Soc.*, **80**, 5871 (1958).
65. Evans, J. C., H. W. Rinn, S. J. Kuhn, and G. A. Olah, *Inorg. Chem.*, **3**, 857 (1964).
66. For reviews see, Olah, G. A. *Chem. Eng. News*, **45**, 76 (March 27, 1967; *Science*, **168**, 1298 (1970).

67. For a review and discussion see: Jennen, J. J., *Chimia*, **20**, 309 (1966).

68. For summaries see: Olah, G. A., and C. U. Pitman, Jr., *Advan. Phys. Org. Chem.*, **4**, 305 (1966); G. A. Olah, and J. A. Olah, in "Carbonium Ions," Vol. II, G. A. Olah and P. v. E. Schleyer, eds., Wiley-Interscience, New York, 1970, p. 175.

69. Magic Acid is a registered trade mark of Cationics, Inc., Cleveland, Ohio.

70a. Olah, G. A., and J. Lukas, *J. Am. Chem. Soc.*, **89**, 2227, 4739 (1967); **90**, 933 (1968).

70b. Olah, G. A., and R. H. Schlosberg, *J. Am. Chem. Soc.*, **90**, 2126 (1968).

70c. Hogeveen, H., and A. F. Bickel, *Chem. Commun.*, 635 (1967); H. Hogeveen, C. J. Gaasbeek, and A. F. Bickel, *Rec. Trav. Chim.*, **88**, 703 (1969); H. Hogeveen and C. J. Gaasbeek, *ibid.*, **87**, 319 (1968).

71a. Talroze, V. L., and A. L. Lyubimova, *Dokl. Akad. Nauk. SSSR*, **86**, 909 (1952).

71b. Field, F. H., and M. S. B. Munson, *J. Am. Chem. Soc.*, **87**, 3289 (1965), and references therein.

72. Olah, G. A., A. M. White, J. R. DeMember, A. Commeyras, and Ch. Y. Lui, *J. Am. Chem. Soc.*, **92**, 4627 (1970), and references therein.

73a. Olah, G. A., Gh. D. Mateescu, L. A. Wilson, and M. H. Gross, *J. Am. Chem. Soc.*, **92**, 7231 (1970).

73b. Olah, G. A., Gh. D. Mateescu, and L. Riemenschneider, *J. Am. Chem. Soc.*, **94**, 2529 (1972).

74. Olah, G. A., G. Klopman, and R. H. Schlosberg, *J. Am. Chem. Soc.*, **91**, 3261 (1969).

75a. Gamba, A., G. Korosi, and M. Simonetta, *Chem. Phys. Letters*, **3**, 20 (1969).

75b. Van Der Lugt, W. Th. A. M., and P. Ros, *Chem. Phys. Letters*, **4**, 389 (1969).

75c. Mulder, J. J. C., and J. S. Wright, *Chem. Phys. Letters*, **5**, 445 (1970).

75d. Kollmar, H., and H. O. Smith, *Chem. Phys. Letters*, **5**, 7 (1970).

76a. Dyczmons, V., V. Staemmler, and W. Kutzelnig, *Chem. Phys. Letters*, **5**, 361 (1970).

76b. Dedieu, A., and A. Veillard, presented at the 21st Annual Meeting of the French Physical Chemical Society, Paris, September 1970; A. Veillard, personal communication.

76c. Lathan W. A., W. J. Heher, and J. A. Pople, *Tetrahedron Letters*, 2699 (1970).

76d. Lathan, W. A., W. J. Heher, and J. A. Pople, *J. Am. Chem. Soc.*, **93**, 808 (1971).

77. Muetterties, E. L. *J. Am. Chem. Soc.*, **91**, 1636 (1969).

78a. Olah, G. A., P. Westerman, Y. K. Mo, and G. Klopman, *J. Am. Chem. Soc.* **94**, 7859 (1972).

78b. Olah, G. A., and J. A. Olah, *J. Am. Chem. Soc.*, **93**, 1256 (1971).

79. Based on the review by Olah, G. A., *Accounts Chem. Res.*, **4**, 240 (1971).

80a. Breslow, R., "Organic Reaction Mechanisms," W. A. Benjamin, New York, 1969.

80b. Stock, L. "Aromatic Substitution Reactions," Prentice-Hall, Englewood Cliffs, N.J., 1968.

80c. Stock, L. M., and H. C. Brown, *Advan. Phys. Org. Chem.*, **1**, 35 (1963).

80d. Olah, G. A., "Friedel-Crafts and Related Reactions," Vols. I–IV, Wiley-Interscience, New York, 1963–1964.

80e. Norman, R. O. C., and R. Taylor, "Electrophilic Substitution in Benzenoid Compounds," American Elsevier, New York, 1965.

80f. De La Mare, P. B. D., and J. Ridd, "Aromatic Substitutions—Nitration and Halogenation," Academic Press, New York, 1959.

80g. Berliner, E., *Prog. Phys. Org. Chem.*, **2**, 253 (1964).

80h. Ingold, C. K., "Structure and Mechanism in Organic Chemistry," 2nd Ed., Cornell University Press, Ithaca, N.Y., 1969, pp. 264–417.

80i. March, J., "Advanced Organic Chemistry, Reaction, Mechanisms, and Structure," McGraw-Hill, New York, 1968, pp. 376–441.

81a. Pfeiffer, P., and R. Wizinger, *Ann.*, **461**, 132 (1928).

81b. Wheland, G. W., *J. Am. Chem. Soc.*, **64**, 900 (1942); "The Theory of Resonance," Wiley, New York, 1944.

81c. Brown, H. C., and J. D. Brady, *J. Am. Chem. Soc.*, **74**, 3570 (1952).

82. Doering W. v. E., M. Saunders, H. G. Boynton, H. W. Earhart, E. F. Wadley, W. R. Edwards, and G. Laber, *Tetrahedron*, **4**, 178 (1958).

83. Ingold, C. K., "The Transition State," Special Publ. No. 16, The Chemical Society, London, 1962, p. 119.

84a. For a summary see ref. 80c.

84b. Brown, H. C., and K. L. Nelson, in "Chemistry of Petroleum Hydrocarbons," B. T. Brooks, S. S. Kurtz, C. B. Board, and L. Schmerling, eds., Vol. III, Van Nostrand, Reinhold, New York, 1955, pp. 465–578.

85a. Brown, H. C., and H. W. Pearsall, *J. Am. Chem. Soc.*, **74**, 191 (1952).

85b. Brown, H. C., and W. J. Wallace, *J. Am. Chem. Soc.*, **75**, 6267, 6268 (1953).

85c. Klit, A., and A. Langseth, *Z. Phys. Chem.*, **176**, 65 (1936).

85d. McCaulay, D. A., and A. P. Lien, *J. Am. Chem. Soc.*, **73**, 2013 (1951); *Tetrahedron*, **5**, 186 (1959).

85e. Kilpatrick, M., and F. E. Luborsky, *J. Am. Chem. Soc.*, **75**, 577 (1953).

85f. Olah, G. A., S. J. Kuhn, and A. Pavlath, *J. Am. Chem. Soc.*, **80**, 6535, 6541 (1958); *Nature*, **178**, 693 (1956).

85g. Mackor, E. L., A. Hofstra, and J. H. van der Waals, *Trans. Faraday Soc.*, **54**, 66, 186 (1958).

85h. Olah, G. A., *J. Am. Chem. Soc.*, **87**, 1103 (1965).

85i. Gold, V., and F. L. Tye, *J. Chem. Soc.*, 2172 (1952).

85j. Reid, C., *J. Am. Chem. Soc.*, **76**, 3264 (1954).

85k. Dallinga, G., E. L. Mackor, A. Hofstra, and A. A. Verrijn Stuart, *J. Mol. Phys.*, **1**, 123 (1958).

85l. MacLean, C., and E. L. Mackor, *J. Mol. Phys.*, **4**, 241 (1961); *Discussions Faraday Soc.*, **34**, 165 (1962).

85m. Birchall, T., and R. J. Gillespie, *Can. J. Chem.*, **42**, 502 (1964).

85n. Perkampus H. H., and E. Baumbgarten, *Angew. Chem.*, **76**, 965 (1964); H. H. Perkampus, *Advan. Phys. Org. Chem.*, **4**, 195 (1966).

85o. Ref. 85.

85p. Brouwer, D. M., E. L. Mackor, and C. L. MacLean, in "Carbonium Ions," G. A. Olah, and P. v. R. Schleyer, eds., Vol. II, Wiley-Interscience, New York, 1970, Chapter 20.

86. Dewar, M. J. S., *J. Chem. Soc.*, 406, 777 (1946); *Nature*, **176**, 784 (1954); M. J. S. Dewar, "The Electronic Theory of Organic Chemistry," Oxford University Press, London, 1949.

87. See Table III footnote *g*; G. A. Olah, S. J. Kuhn. S. Flood, and J. C. Evans, *J. Am. Chem. Soc.*, **84**, 3687 (1962); see Table I, footnote *i*.

88a. Ritchie, C. D., and H. Win, *J. Org. Chem.*, **29**, 3093 (1964).

88b. Tolgyesi, W. S., *Can. J. Chem.*, **43**, 343 (1965) (for a rebuttal see G. A. Olah, and N. Overchuk), *ibid.*, **43**, 3279 (1969).

88c. Ridd, J. D., "Studies on Chemical Structure and Reactivity," Methuen, London, 1966, p. 133.

88d. Brown, H. C., and R. Wirkkala, *J. Am. Chem. Soc.*, **88**, 1453 (1966).

88e. Cerfontain, and H. A. Telder, *Rec. Trav. Pays-Bas.* **86**, 377 (1967).

88f. S. W. Caill and F. P. Corriu, *Chem. Comm.* 1251 (1967); *Tetrabedron.* **25**, 2005 (1969).

88g. Ingold, C. K., "Structure and Mechanism in Organic Chemistry," 2nd Ed., Cornell University Press, Ithaca, N.Y., 1969, p. 290.

89. Ingold C. K. *et al.*, *Nature*, **158**, 448 (1946); *J. Chem. Soc.*, 2400 (1950), and subsequent papers.

90. Brown, H. C., and M. Grayson, *J. Am. Chem. Soc.*, **75**, 6285 (1953); H. C., Brown, and B. A. Bolto, *ibid.*, **81**, 3320 (1959); H. C. Brown and G. Marino, **81**, 3308 (1959); for a review see ref. 80c.

91. For a summary see: Cerfontain, H., "Mechanistic Aspects in Aromatic Sulfonation," Wiley-Interscience, New York, 1968.

92. Olah, G. A., M. Tashiro, and S. Kobayashi, *J. Am. Chem. Soc.*, **92**, 6369 (1970).

93. Olah, G. A., and S. Kobayashi, *J. Am. Chem. Soc.*, in press.

94. Olah, G. A., and S. J. Kuhn, *Chem. Ind.*, **98** (1956); *J. Chem. Soc.*, 4257 (1956); *J. Am. Chem. Soc.*, **83**, 4564 (1961).

95. Coombes, R. G., R. B. Moodie, and K. Schofield, *J. Chem. Soc.*, *B*, 800 (1969).

96. Olah, G. A., and H. C. Lin, unpublished results.

97. Rate studies of nitration of nitrotoluenes were reported by: Westheimer, F. H., and M. S. Kharasch, *J. Am Chem. Soc.*, **68**, 1871 (1946).

98. J. C. D. Brand and R. P. Paton, *J. Chem. Soc.*, 281 (1952); J. G. Tillett, 5142 (1962).

99. Nakane, R., A. Natsubori, and O. Kurihara, *J. Am. Chem. Soc.*, **87**, 3597 (1965); **88**, 3011 (1966); R. Nakane, T. Oyama, and A. Natsubori, *J. Org. Chem.*, **33**, 275 (1968); **34**, 949 (1969); R. Nakane and T. Oyama, *J. Phys. Chem.*, **70**, 1146 (1966).

100. Christen, M., W. Koch, W. Simon, and H. Zollinger, *Helv. Chim. Acta.*, **45**, 2077 (1962).

101. Using Mulliken's definition: Mulliken, R. S., *J. Amer. Chem., Soc.*, **72** 600 (1950); **74**, 811 (1952); *J. Phys. Chem.*, **56**, 801 (1952).

102. For a definition of carbocations, including differentiation of benzonium from benzenium ions, see ref. 2.

103. Hammond, G. S., *J. Am. Chem. Soc.*, **77**, 334 (1955).

104a. Melander, L. C. S., "The Use of Nuclides in the Determination of Organic Reaction Mechanism," Reilly Lectures, Vol. XI, University of Notre Dame Press, Notre Dame, Ind., 1955.

104b. Melander, L. C. S., "Isotope Effects on Reaction Rates," Ronald Press, New York, 1960.

104c. Zollinger, H., *Experientia*, **12**, 165 (1956); *Ann. Rev. Phys. Chem.*, **13**, 400 (1962).

104d. Zollinger, H., *Advan. Phys. Org. Chem.*, **2**, 163 (1964).

104e. Berliner, E., *Progr. Phys. Chem.* **2**, 253 (1964).

104f. Cerfontain, H., H. J. Hofman, and A. Telder, *Rec. Trav. Chim.*, **83**, 493 (1964); G. A. Olah, *J. Tenn. Acad. Sci.*, **40**, 77 (1965).

105. Halevi, E. A., and B. Ravid, *Pure Appl. Chem.*, **8**, 339 (1964).

106. For reviews see: Ingold, C. K., "Structure and Mechanism in Organic Chemistry," 2nd Ed., Cornell University Press, Ithaca, N.Y., 1969.

107. March, J., "Advanced Organic Chemistry: Reactions, Mechanism and Structure," McGraw-Hill, New York, 1968.

108. Taft, R. W., Jr., *J. Am. Chem. Soc.*, **74**, 5372 (1952) seems to be the first to have suggested the initial formation in the acid-catalyzed hydration of isobutylene of a proton π-complex of the type described originally by M. J. S. Dewar [*J. Chem. Soc.*, 406 (1946) and *Bull. Soc. Chim. France*, C75 (1951)] in which ". . . the proton is embedded in the π-orbitals which extend above (or below) the plane of the C—C double bond."

109. For example, see: Radom, L., J. A. Pople, V. Buss, and P. v. R. Schleyer, *J. Am. Chem. Soc.*, **94**, 5935 (1972) and references given herein.

110. Olah, G. A., J. M. Bollinger, and J. Brinich, *J. Am. Chem. Soc.*, **90**, 2587 (1968); G. A. Olah and J. M. Bollinger, **89**, 4744 (1967); **90**, 947 (1968).

111. Olah, G. A., and P. R. Clifford *J. Am. Chem. Soc.*, **93**, 1261, 2320 (1971).

112. Strating, J., J. H. Wieringa, and H. Wynberg, *Chem. Commun.*, 907 (1969).

113. Olah, G. A., P. Schilling, and P. Westerman, unpublished results.

114. Olah, G. A., and H. Lin, unpublished results.

115. Olah, G. A., and T. Hockswender, unpublished results.

116. Blackstock, D. J., M. P. Hartshorn, A. J. Lewis, K. E., Richards, J. Vaughn, and G. J. Wright, *J. Chem. Soc.*, *B*, 1212 (1971).

117. Myhre, P. C. personal communication.

118. For a review see: Nenitzescu, C. D., and A. T. Balaban, in "Friedel-Crafts and Related Reactions," G. A. Olah, Ed., Vol. III, Wiley-Interscience, 1969, p. 1033.

119. Olah, G. A., S. J. Kuhn, and D. G. Barnes, *J. Org. Chem.*, **29**, 2685 (1964).

120. For a review see: Franzen, V., in "Friedel-Crafts and Related Relations," Vol. II, G. A. Olah, Ed., Wiley-Interscience, New York, 1964, p. 413.

121a. Olah, G. A., and J. Lukas, *J. Am. Chem. Soc.*, **89**, 2227, 4739 (1967); **90**, 933 (1968); G. A. Olah, and R. H. Schlosberg, **90**, 2726 (1969).

121b. Hogeveen, H., and A. F. Bickel, *Chem. Commun.*, 635 (1967); H. Hogeveen, C. J. Gaasbeek, and C. J. Bickel, *Rec. Trav. Chim.*, **88**, 703 (1969); H. Hogeveen, and J. Gaasbeek, *ibid.*, **87**, 319 (1968).

122. Olah, G. A., Y. Halpern, J. Shen, and Y. K. Mo, *J. Am. Chem. Soc.*, **93**, 1251 (1971).

123. Olah, G. A., and H. C. Lin, *J. Am. Chem. Soc.*, **93**, 1259 (1971).

124. Bartlett, P. D., F. E. Condon, and A. Schneider, *J. Am. Chem. Soc.*, **66**, 1531 (1944).

125. Nenitzescu, C. D., M. Avram, and E. Sliam, *Bull. Soc. Chim. France*, 1266 (1955), and earlier references therein.

126a. Lewis, E. S., and M. C. R. Symmons, *Quart. Rev.*, **12**, 230 (1959).

126b. Hawthorne, M. F., and E. S. Lewis, *J. Am. Chem. Soc.*, **80**, 4296 (1958).

127. Olah, G. A., G. Klopman, and R. H. Schlosberg, *J. Am. Chem. Soc.*, **90**, 2726 (1968); **91**, 3261 (1969).

128a. Hogeveen, H., C. J. Gaasbeek, and A. F. Bickel, *Rec. Trav. Chim.*, **88**, 703 (1969).

128b. Brouwer, D. M., and H. Hogeveen, Adv. Phys. Org. Chem. **9**, 179 (1972).
129. Bickel, A. F., C. J. Gaasbeek, H. Hogeveen, J. N. Oelderik, and J. C. Platteeuw, *Chem. Commun.*, 634 (1967).
130. Olah, G. A., J. Shen, and R. H. Schlosberg, *J. Am. Chem. Soc.*, **92**, 3831 (1970).
131. Schmerling, L. *J. Am. Chem. Soc.*, **66**, 1422 (1944); **67**, 1778 (1945); **68**, 153 (1946).
132. Olah, G. A., and J. A. Olah, *J. Am. Chem. Soc.*, **93**, 1256 (1971).
133. Olah, G. A., J. R., DeMember, and R. H. Schlosberg, *J. Am. Chem. Soc.*, **91**, 2112 (1969).
134. Olah, G. A., J. R. DeMember, R. H. Schlosberg, and Y. Halpern, *J. Am. Chem. Soc.*, **94**, 156 (1972).
135a. Olah, G. A., and J. R. DeMember, *J. Am. Chem. Soc.*, **91**, 2113 (1969); **92**, 718 (1970); **92**, 2562 (1970).
136. Olah, G. A., Y. K. Mo., R. Renner, and P. Schilling, to be published.

CHAPTER VII

Practical Applications and Future Outlook

I. Introduction

Since their discovery the Friedel-Crafts reactions have been valuable tools in the laboratory as well as in the industrial syntheses of a large variety of organic compounds. Their versatility is possibly matched only by the Grignard reactions. Such divergent methods as the Friedel-Crafts reactions would be expected to find a great deal of commercial use but, for one reason or another, many of these reactions remained for a long time solely as laboratory processes, serving only to build up the catalog of organic compounds which were later referred to by chemists seeking methods of synthesis for compounds needed in commerce. The uses of the Friedel-Crafts reaction in the chemical industry are many, but having regard to the academic importance and prominence given to these reactions, large-scale use is still comparatively limited.

Although practically all types of compounds can be produced by the Friedel-Crafts type reactions—directly or by variations thereof—industrial applications primarily concern the preparation of hydrocarbons and ketones. There is also demand for intermediates for further processing, but these can often be obtained by more direct methods. Probably the most important hydrocarbons produced by Friedel-Crafts alkylation are ethylbenzene (from which styrene is obtained) and cumene (of considerable importance as the hydroperoxide in one of the manufacturing methods for phenol). Dodecylbenzene and related alkylates have acquired an increasing role, giving widely used detergents on sulfonation. Simple aryl ketones have always been made by Friedel-Crafts synthesis or by its variations. The simplest of these ketones, acetophenone, however, offers an illustration of changes that have taken place in the use of this reaction in chemical industry. The original process employed the ketone synthesis from benzene, aluminum chloride, and acetyl chloride or acetic anhydride. When ethylbenzene became available in quantities, the oxidation of this compound by air gave a much cheaper product although purification was more difficult. The oxidation of cumene to produce phenol *via* cumene hydroperoxide always gives acetophenone as a by-product, together with some dimethylphenyl carbinol and acetone. Lately, acetone by-product disposal has made this process less attractive and considerable interest has arisen in the oxidation of toluene *via* benzoic acid to phenol, and in direct oxidation of benzene. Another factor which affects the use of the Friedel-Crafts reaction for the preparation of ketones of this type is the relatively high cost per mole of aluminum chloride due to molecular complex formations. Probably the best commercial use of the ketone synthesis is the production of *o*-benzoylbenzoic acid for anthraquinones. Very good yields are obtained and owing to the high molecular weight of the product the cost of the aluminum chloride is not so important.

II. Early Development

1. First Applications

Like so many other chemical reactions, the Friedel-Crafts reactions found little industrial use until a commercial demand arose for products resulting from them and a cheap source of the catalyst (aluminum chloride) was available. Much could be said concerning the factors contributing to this demand, such as the First World War, the oil, synthetic dye, and automobile industries, etc., but the greater emphasis prior to the Second World War should be placed on the accomplishment of producing cheap aluminum chloride.

Although the Friedel-Crafts syntheses may be applied with equal facility both to the aliphatic and aromatic series of compounds, most of the early work following the discovery of the reaction on an aliphatic system was concentrated on aromatic materials. This possibly accounted for the long-held and erroneous belief that the Friedel-Crafts methods of synthesis could be applied only to aromatics. The reasons why the aliphatic compounds received little attention at the end of the nineteenth century are not clear, but probably the main interest at that time was in such fields as dye manufacturing and it was not until after the first decade of the twentieth century had been marked by the rise in popularity of the gasoline engine that there was any need for petroleum refining and cracking. Thus the wide field of possible applications lay dormant for thirty years after the first French and British patents were issued to Friedel and Crafts.

A contributing factor to this state of affairs was the lack of aluminum chloride on an industrial scale. The matter has been reported by Willemart (40) in these terms: "... in 1855 Deville, at the bidding of Napoleon III, had made a temporary demonstration on a pilot plant scale to the 'Société Générale de Javel'; the method of preparation was the primitive one used by Oerstedt (32) and conceived by Gai-Lussac and Thénard, namely the action of chlorine on a mixture of alumina and carbon at red heat. The several hundred kilogrammes of aluminum chloride thus produced served for Deville to prepare, by Wöhler's reaction, the first ingots of aluminum, which were exhibited at the World Fair in the following year."

2. Production of Aluminum Chloride

The commercial production of anhydrous aluminum chloride started in Germany in the years prior to the First World War.

The production of $AlCl_3$ was first achieved in the United States on a commercial scale by the Gulf Refining Company and Sable-Sayre in 1916. The Dow Chemical Company made aluminum chloride during the First World War emergency, but did not continue in its manufacture thereafter.

Two methods generally used were those of McAfee (27) and Wurster (41). The former utilizes a carbon–bauxite mixture which is allowed to react with chlorine:

$$Al_2O_3 + 3C + 3Cl_2 \rightarrow 2AlCl_3 + 3CO$$

the latter employs the reaction of phosgene (or carbon monoxide and chlorine) on bauxite:

$$Al_2O_3 + 3COCl_2 \rightarrow 2AlCl_3 + 3CO_2$$

There are today several commercial methods of preparing anhydrous aluminum chloride. The purity of the finished product may depend upon the starting material, which may be aluminum scrap, alumina, or bauxite. Both chlorine and hydrogen chloride are used industrially. Three processes are of particular importance: (1) the action of chlorine on a mixture of bauxite with carbon, which acts as a reducing agent, basically the McAfee-Wurster process (the bauxite is calcined, mixed with coal or coke, briquetted, and chlorinated at 1600° F); (2) the action of chlorine on aluminum scrap; and (3) the action of hydrogen chloride, chlorine, or phosgene on alumina. The continuous chlorination of γ-alumina in a fluid-bed reactor by a mixture of Cl_2 + CO (thus containing phosgene) in the presence of sodium–aluminum chloride as catalyst was developed by Hille and Dürrwächter (17,18) and is used commercially on a large scale in

Germany.   The product made from aluminum metal is of the highest
purity, but the use of low-iron bauxite or alumina gives a product of
suitable purity for many uses.   For highest purity the product is
resublimed.   There are innumerable patents and processes for the
preparation of anhydrous aluminum chloride, but the majority of
these are not of commercial importance.

III. Development Prior to World War I and between the two World Wars

The first large-scale use of aluminum chloride was the McAfee
process of cracking petroleum hydrocarbons for the production of
gasoline (28).   This process, however, has been almost entirely
replaced by thermocracking, as well as catalytic cracking, wherein
the catalyst is more stable and longer lasting.   In Europe prior to
1914 the Friedel-Crafts reactions were used to some extent, especially
in the preparation of high-priced medicinals and perfume bases, but
no really large-scale use for aluminum chloride is to be found prior
to World War I.   In fact Ulmann's *Encyclopaedia of Technical
Chemistry* of 1914 gives less than a page to the aluminum chloride
catalyzed reactions and lists only acetophenone, methylanthra-
quinone, and synthetic musk as products prepared by the Friedel-
Crafts reactions.   It lists the Scholl reaction as a Friedel-Crafts type
reaction wherein aromatic nuclei are coupled together.

The development of the reactions of the Friedel-Crafts type
concerning industrial application during the two World Wars was
reviewed by Britton (6).   He gave a partial list of chemicals made
commercially at this period using aluminum chloride as catalyst.

Acetophenone	Chloroacetophenone	Phenothiazine
p-Amylphenol	Chlorobenzoylbenzoic acid	Phenothioxine
Anisaldehyde	Diethylbenzene	β-Phenylethyl alcohol
Benzaldehyde	Diphenylmethane	Propiophenone
Benzophenone	Ethylbenzene	Tricresyl phosphate
Benzoylbenzoic acid	Isopropylbenzene	Triphenylchloromethane
p-t-Butylphenol	Methylacetophenone	Triphenyl phosphate

However, the Friedel-Crafts processes are not necessarily the only
methods available for the preparation of these products.   An
example of a Friedel-Crafts process replaced by a more economic air
oxidation is the preparation of acetophenone from ethylbenzene.

Early uses of the Friedel-Crafts reactions are to be found in the dye
industry, the pharmaceutical industry, and in perfume manufacture.

In many cases the same reactions are in use today, more than 80 years after the discovery of the method. More recent uses, besides the preparation of numerous intermediates, include such products as:

Anthraquinone and derivatives	Dodecylbenzene (detergents)	Keto acids
Aryl ketones	Ethylbenzene (for styrene)	Pharmaceuticals and perfumes
Butyl rubber	Ethyl chloride (for TEL)	Polybutylenes
Dyestuffs	Gasoline (Isomerization)	
	Insecticides	

IV. Fields of Present-day Applications

1. Petroleum and Petrochemical Industry

In spite of the variety of uses already mentioned, most of the aluminum chloride produced is used in the petroleum and petrochemical industry. The main fields of applications involve cracking, alkylation, isomerization, and polymerization.

A. Cracking, Refining, and Additives

It was largely due to the work of McAfee (25,26) that development of the cracking ability of aluminum chloride occurred. At the time of the First World War there was a need to increase the efficiency of gasoline production and to avoid the accumulation of relatively large amounts of high-boiling hydrocarbon by-products. The first catalytic cracking practised on a commercial scale was that in which heavy oils were distilled, with agitation, over anhydrous aluminum chloride. The process was applied most successfully to naphthenic oils. With about 3–5% of aluminum chloride, the mixture was heated slowly at atmospheric pressure until liquid temperatures of approximately 225° were attained. Low-boiling hydrocarbon products of the reaction, largely in the gasoline and kerosene range, distilled over; they were saturated and highly refined. The aluminum chloride cracked gasolines were characterized by moderately high octane numbers, absence of olefins, and high response to tetraethyllead. Catalytic cracking in the presence of aluminum chloride had the advantage over thermal cracking of operating at lower temperatures, preventing the formation of olefins and simultaneously desulfurizing the crude petroleum by decomposition of organic compounds of sulfur. This latter property of aluminum chloride was strictly one of refining since desulfurization of fuel was essential to avoid fouling of engines. By its polymerizing action aluminum chloride effects further refining of fuel by eliminating unsaturated constituents which again may lead to inefficient engine operation. Isomerization due to aluminum chloride causes normal hydro-carbons to be transformed into branched molecules, which are used in aliphatic alkylations to produce high-grade gasoline.

The cracking of gas oils to gasoline by mild heating (225–300°, atmospheric pressure) over a few per cent of aluminum chloride was practised commercially until it was superseded by more efficient catalytic crackings. It is still used for drastic refining of lubricating oil stocks to remove the more aromatic type molecules and those containing oxygen, nitrogen, and sulfur. It was again employed during the Second World War for isomerization of normal butane to isobutane.

Catalytic cracking using more advanced solid acidic catalysts was first operated commercially in 1936, and in many respects is superior to thermal cracking, partly because it does not require high pressure. In the earlier processes the vaporized feed, at a temperature of 470–490° and under a pressure of 1.5–3.5 atm., was passed through a fixed bed of catalyst which required frequent treatment with hot air to remove coke. Continuous processes were subsequently developed in which the catalyst, either in the form of small pellets ("moving-bed" process) or as finely divided particles ("fluid" process), moves continuously between the reactor and a vessel where the catalyst is regenerated.

The cracking of hydrocarbon oils occurs over many types of catalytic materials, but high yields of desirable products are obtained with hydrated aluminum silicates. These may either be activated (acid-treated) natural clays of the bentonite type or synthesized silica–alumina or silica–magnesia preparations (30). Their activity to yield essentially the same products may be enhanced to some extent by the incorporation of small amounts of other materials, such as the oxides of zirconium, boron (which has a tendency to volatilize away on use), and thorium. Both the natural and the synthetic catalysts can be used as pellets or beads, and also in the form of powder.

Neither silica nor alumina alone is effective in promoting catalytic cracking reactions. In fact, they (and also activated carbon) promote hydrocarbon decompositions of the thermal type. A mixture of anhydrous silica and alumina, or anhydrous silica with hydrated alumina, is also essentially non-effective. A catalyst having appreciable cracking activity is obtained only when prepared from hydrous oxides, followed by partial dehydration (calcining). The small amount of water remaining is necessary for proper functioning. Studies made with deuterium oxide have shown that this water is involved in hydrogen transfer reactions between the catalyst and hydrocarbon molecules; these begin considerably below cracking temperatures (15,16).

The chemical changes taking place in the hydrocarbons during the cracking reaction are quite similar to those occurring at lower temperatures in polymerization, isomerization, and alkylation over acid catalysts, such as H_2SO_4, H_3PO_4, $ZnCl_2$, $AlCl_3$, BF_3, etc. The maximum acidity of a silica–alumina complex is obtained when the ratio of aluminum to silicon is 1, while the ratio for maximum catalytic activity is 2 (39). This has been attributed to a certain degree of catalytic effectiveness of the free Al_2O_3 present, presumably not dependent upon an acidic condition. The active catalyst is assumed to be a polymer of a hypothetical aluminosilicic acid $(HAlSiO_4)_n$. This acid is presumably very stable and is distributed in such a manner as to provide active centers on the surface (3,30,31), their number and intrinsic activity determining the effectiveness. Commercial synthetic catalysts are amorphous and contain more silica than is called for by the above formula; they are generally composed of 10–15% Al_2O_3 and 85–90% SiO_2. The corresponding natural materials, such as montmorillonite, a non-swelling bentonite, are hydrosolicates of aluminum, with a well-defined crystal structure and an approximate composition of $Al_2O_3 \cdot 4SiO_2 \cdot H_2O$. Some of the newer catalysts contain up to 25% alumina; longer active life is reported.

The feedstock for catalytic cracking plants is usually a fraction obtained by vacuum distillation of a primary distillation residue (or sometimes a heavy gas oil). Catalytic cracking processes give a greater yield of gasoline, and a smaller quantity of residual oil and gaseous hydrocarbons than thermal cracking processes. Furthermore, the gasoline produced has better anti-knock characteristics. Feedstocks with a high sulfur content are permissible, since some desulfurization, with production of hydrogen sulfide, takes place.

Some type of chemical treatment (refining) to remove or alter the constitution of impurities in petroleum products is usually necessary to produce a usable material. Depending upon the particular treatment used, one or more of the following purposes are achieved: (1) improvement of color; (2) improvement of odor; (3) removal of sulfur; (4) removal of gums; (5) improvement of stability to light and air. Of these, removal of sulfur and improvement of stability are usually the factors governing the treatment employed.

In the early days of refining, aluminum chloride treatment did achieve some of these objectives.

10*

Strong sulfuric acid is now widely used to produce water-white products, but it has the great disadvantage of also dissolving or reacting with valuable olefins. Hydrogen fluoride is also gaining increasing importance in refining and desulfurization.

Lubricating oils may be refined in the same manner as the lower hydrocarbons. Various types of acid catalysts have been suggested for this purpose on account of their polymerizing and desulfurizing effects; in addition, the condensing action of such catalysts allows the preparation of synthetic lubricating oils.

In the field of additives the preparation of "Paraflow" has acquired substantial importance. "Paraflow", developed by the Standard Oil Company, is a condensation product of a chlorinated paraffin wax and a polynuclear hydrocarbon such as naphthalene, anthracene, or phenanthrene (24). The addition of small quantities of these products (0.5–2.0%) to lubricating oil prevents the growth of seed crystals and renders the oil remarkably fluid at low temperatures. "Santopour", developed by the Socony-Vacuum Oil Company, is a product with similar use prepared by the condensation of phenol with chlorinated paraffin wax followed by further condensation with phthalyl chloride (34).

B. Polymerization

Rapid progress has been made in the development of polymerization processes for converting by-product unsaturated hydrocarbon gases produced in cracking into liquid hydrocarbons suitable for use as high-octane motor and aviation fuels and for petrochemicals, e.g., C_{12} tetramer for detergents. Propylene, normal butylene, and isobutylene are the olefins usually polymerized.

$$2CH_3-\underset{\underset{CH_3}{|}}{\overset{\overset{CH_3}{|}}{C}}=CH_2 \longrightarrow CH_3-\underset{\underset{CH_3}{|}}{\overset{\overset{CH_3}{|}}{C}}-CH_2-\overset{\overset{CH_3}{|}}{C}=CH_2 \xrightarrow{C_4H_8} C_{12}H_{24}$$

Diisobutylene   Triisobutylene

$$2CH_3-CH=CH_2 \longrightarrow CH_3-\overset{\overset{CH_3}{|}}{CH}-CH=\overset{\overset{CH_3}{|}}{CH} \xrightarrow{2C_3H_6} C_{12}H_{24}$$

Dipropylene   Tetrapropylene

Polyisobutylenes are produced by the catalyzed polymerization of liquid isobutylene in a solvent at very low temperature. In Germany, during the Second World War, polymerization was effected in liquid ethylene at $-80°$ to $-100°$ in the presence of boron trifluoride (7).

$$n \begin{pmatrix} CH_3 \\ | \\ C=CH_2 \\ | \\ CH_3 \end{pmatrix} \xrightarrow{BF_3} \begin{pmatrix} CH_3 \\ | \\ -C-CH_2- \\ | \\ CH_3 \end{pmatrix}_n$$

Polyisobutylenes can be obtained as viscous oils, sticky gums, or rubberlike materials. They find use as plasticizers, for improving the viscosity index of lubricating oils, and for coating fabrics.

C. Alkylation and Isomerization

Alkylation processes are exothermic and are fundamentally similar to polymerization, differing in that only part of the charging stock need be unsaturated. It is based on the reactivity of the tertiary carbon of isobutane with olefins such as propylene, butylenes, and amylenes. The product "alkylate" is a mixture of saturated, stable isoparaffins distilling in the gasoline range, which becomes the principal component of many high-octane gasolines.

In the alkylation of olefins with paraffins and isoparaffins, sulfuric acid, anhydrous hydrogen fluoride, and aluminum chloride–hydrocarbon complexes (sludge catalysts) are used preferentially as catalysts.

During the Second World War the need for high-octane aviation gasoline increased tremendously and as a consequence alkylation and isomerization methods of aliphatic hydrocarbons underwent rapid development. At the same time aromatic alkylation was also further developed and large-scale industrial processes replacing aluminum chloride with other more versatile and easily handled catalysts, such as anhydrous HF, were developed.

One of the outstanding commercial processes employing HF as a catalyst is the alkylation of isoparaffins with olefins to produce branched-chain paraffin hydrocarbons boiling in the gasoline range. The resulting alkylate has excellent anti-knock properties and became a major component (25–60%) of the fuel for aircraft engines.

Under war pressure, the HF alkylation process was carried from the semi-pilot plant scale directly to an operation responsible for a vital part of the aviation gasoline supply. Twenty-seven plants with a designed capacity of 54,000 barrels of alkylate per day have been built in the U.S.A. Their operating record has been such as to establish HF alkylation firmly as a major unit in the petroleum industry.

During HF alkylation of isoparaffins with olefins (23) the principal changes which occur in the reaction mixture are apparently the following:

1. Direct interaction of an olefin and an isoparaffin, either of which may be charged as such or may be the result of previous reactions;
2. Dealkylation of paraffins of high molecular weight;
3. Polymerization of olefins to form olefins of higher molecular weight;
4. Depolymerization reactions;
5. Hydrogen transfer reactions between olefins to form saturated hydrocarbons and tars of low hydrogen content;
6. Hydrogen transfer reactions between paraffins and olefins to form a paraffin corresponding to the olefin and an olefin corresponding to the paraffin;
7. Probably by the above reactions, net isomerization, particularly of the alkylate product, may occur.

As HF is also an effective desulfurizing agent, it is quite understandable that a number of different processes take place simultaneously to produce the high-octane aviation gasolines.

Isomerization has become of the utmost importance in furnishing the isobutane needed for making "alkylate" as a basis for high-octane gasoline.

$$CH_3-CH_2-CH_2-CH_3 \rightarrow CH_3-\overset{\overset{\displaystyle CH_3}{|}}{C}H-CH_3$$

There are two basic processes for the isomerization of butanes, namely, the vapor- and liquid-phase processes. Shell's vapor-phase process uses a granular catalyst containing aluminum chloride. Liquid-phase isomerization includes Standard Oil of Indiana's "Isomate" process, Shell's liquid-phase process, and Universal Oil Products' liquid-phase process. The "Isomate" process uses an aluminum chloride–hydrocarbon complex; Shell's liquid-phase process employs a combination catalyst comprising aluminum chloride, antimony trichloride, and hydrochloric acid; while the U.O.P. process uses supported aluminum chloride with hydrochloric acid.

Isomerization processes using a solid catalyst are carried out adding hydrogen chloride at a rate of about 3–5% of the charge to activate the catalyst.

When a liquid catalyst complex is used, any one of a number of hydrocarbons may be used as a complexing agent. It is said that to be active the complex should contain at least 60% aluminum chloride. This process is usually carried out at a lower temperature than is the case with solid catalysts.

D. Catalytic Reforming and Aromatization

Reforming is a treatment given a low-octane and usually straight-chain gasoline to improve its octane number.   It involves both cracking and isomerization.   Straight-chain gasoline and light naphthas generally have very low octane numbers.   By sending these fractions to a reforming unit and giving them a light "crack" their octane number may be increased.   This upgrading may be accomplished in part by an increase in the volatility (reduction of molecular size) or chiefly by the conversion of n-paraffins to isoparaffins, olefins, and aromatics, and naphthenes to aromatics.   The nature of the final product is, of course, influenced by the source (and composition) of the feedstock.

Straight separation of aromatics from petroleum is not usually attractive because of the low concentrations and the difficulty of separation from naphthenes of similar boiling range.   Most of the processes for the production of aromatics from petroleum, therefore, include methods of converting naphthenes to aromatics, e.g., by dehydrogenation; these processes are known by the general name of catalytic reforming.   In contrast to this type of process, deep thermal cracking of mainly paraffinic fractions is also carried out to produce aromatics; this is the basis of the catalytic aromatization (Catarole) process.   Under the conditions used in this process, paraffins are cracked to form intermediates—probably mainly butadiene—which combine to give aromatics.   Aromatization of naphthenes and cyclization of paraffins also occur.

Earlier types of catalytic reformer, the so-called *hydroformers*, employed a molybdenum oxide catalyst on an alumina base and coking of the catalyst required frequent reactivation by controlled burning of the carbonaceous deposits.   More recent catalytic reformers employ a platinum-containing catalyst; there are two types of process: non-regenerative processes in which the catalyst is not normally regenerated, e.g., *platforming*, and regenerative processes, e.g., *ultraforming*.

Typical of the many reactions which take place are the following:

(a) Dehydrogenation

(b) Isomerization and dehydrogenation

CH$_3$

Isomerization $\longrightarrow$ Dehydrogenation $\longrightarrow$ + 3H$_2$

Benzene

(c) Cyclization (ring closure) and dehydrogenation

$$CH_3(CH_2)_5CH_3 \longrightarrow \quad + 4H_2$$

CH$_3$

E. Hydrodealkylation

It is difficult to obtain aromatics with a high proportion of benzene by catalytic reforming owing to the shortage of suitable feedstocks, but toluene and xylene can be produced in abundance. A new processing technique called hydrodealkylation has been introduced to redress this imbalance by converting toluene to benzene and methyl-naphthalenes to naphthalene. For this purpose, toluene must be available at a substantially lower price than benzene. Such a state of affairs is unlikely to occur—except in North America—in the immediate future owing to the rising demands for toluene for the manufacture of isocyanates and for new processes such as a direct oxidation of toluene to phenol. Taking a longer term view, however, the potential toluene production by catalytic reforming is enormous, probably exceeding ten million tons per annum. Xylenes can also be converted to benzene by hydrodealkylation if it is economically attractive to do so.

There are many versions of the hydrodealkylation process, but all employ the same principle of catalytic or thermal reaction of toluene with hydrogen, producing benzene and methane.

$$C_6H_5CH_3 + H_2 \rightarrow C_6H_6 + CH_4$$

Side reactions involve hydrogenation of the benzene nucleus and complete destruction of the aromatic nucleus to give methane.

$$C_6H_5CH_3 + 3H_2 \rightarrow C_6H_{11}CH_3$$
$$C_6H_5CH_3 + 10H_2 \rightarrow 7CH_4$$

Small quantities of polyphenyls are also formed. High pressures and temperatures increase the reaction rate and the conversion per path, but if the conditions are too severe the overall yield may suffer. Typical conditions for making benzene from toluene are a pressure of 50–70 atm., a temperature of 550°, and hydrogen and toluene in a molar ratio of approximately 2 : 1. A conversion of 30–50% of toluene per path is generally obtained. Catalysts which have been mentioned for this process include cobalt molybdate and chromia–alumina. Thermal dehydroalkylations are also widely used. The hydrogen supply is almost universally derived from catalytic reformers. Although few details of this new process are known, the catalyst applied combines a catalytic hydrogenation catalyst with a dealkylation catalyst of the acidic oxide type.

The rapidly expanding hydrodealkylation plants in the United States have practically eliminated the previously large coke benzene, and naphthalene requirements.

F. Anti-knock Compounds

Ethyl chloride is used primarily in the manufacture of tetraethyllead. The predominant route to ethyl chloride is through addition of hydrogen chloride to ethylene. This reaction is carried out preferentially in the liquid phase. The reactants are fed in a mixed gas stream to a reactor operating from $-5°$ to $+55°$ and at pressures from 1 to 9 atm. Aluminum chloride catalyst is added in a solution of ethyl chloride. The ethyl chloride produced is removed as a liquid, flash-evaporated to separate it from impurities, and further purified by distillation.

2. Surface-active Agents (Detergents)

In the class of *anionic* surface-active agents, the alkylarene sulfonates (commonly called alkyl arylsulfonates) were the first surface-active agents to be developed. They still comprise the largest volume of detergents produced. The alkylated aromatics (detergent alkylates) needed for the manufacture of the sulfonates are generally prepared by Friedel-Crafts alkylations of benzene, naphthalene, or other aromatics (35).

Dodecylbenzene is typical of this class of surface-active compounds. It is clear from the nature of the raw materials used to prepare them, however, that the alkylbenzene sulfonates are not prepared from chemically pure compounds. The alkylation of benzene to dodecylbenzene was previously carried out with a kerosene fraction, and the product frequently was called kerylbenzene. In recent time, tetrapropylene has been used as an alkylating agent, improving the purity and properties of the alkylates. Aluminum chloride or hydrogen fluoride are the most frequently used catalysts.

$$C_6H_6 + (C_3H_6)_4 \xrightarrow[\text{HF}]{\text{AlCl}_3} C_6H_5C_{12}H_{25}$$

In recent years, the alkyl arylsulfonates have accounted for more than half of the total tonnage of surfactants produced annually in the United States.

The alkylbenzenes used in the manufacture of synthetic detergents are mostly of the polypropylated type. The alkyl portion of the molecule representing polypropylene is essentially $C_{11}-C_{13}$, or the tetramer of propylene. In the alkylation with $AlCl_3$ an excess of benzene is used to minimize the formation of the dialkyl derivative. Hydrogen chloride is added to function as a promoter, and water serves as a proton donor. Hydrogen fluoride is an exceedingly suitable catalyst for the alkylation. Although the alkylation can be carried out using sulfuric acid as catalyst, the products are contaminated with undesirable olefins which are difficult to remove.

The dodecylbenzene produced is then sulfonated with oleum and the product is neutralized with caustic soda to give sodium dodecylbenzene sulfonate.

Among the simplest and longest known alkylaromatic surface-active agents are the propylated naphthalene sulfonates. These were invented in Germany during the First World War in an attempt to develop soap substitutes from non-fatty raw materials. Isopropyl alcohol, naphthalene, and sulfuric acid were heated together, according to the original process, forming a mixture of mono-, di-, and polypropylnaphthalenesulfonic acids (1). In the alkylation stage readily available isopropylating agents, such as isopropanol, n-propanol, n-propyl, or isopropyl halides, diisopropyl ether, or propylene itself, may be used. As catalysts the usual strongly electrophilic alkylation catalysts have all been described. These include sulfuric acid of various concentrations, $AlCl_3$, $ZnCl_2$, BF_3, HF, etc.

Alkylnaphthalene sulfonates are fairly good "wetting-out" agents for fabrics and are used largely for general wetting-out purposes in textile mills. They do not exhibit any appreciable detergent power for soiled fabrics until relatively high concentrations are reached (1–2% or more). Their strong wetting power for glass, porcelain, and other solid surfaces makes them a useful ingredient in dishwashing powders and household scouring compositions. They are also used in making dispersions and emulsions, having been used to a considerable extent in the emulsion polymerization of synthetic rubbers in Germany during the Second World War. Other important uses are in metal cleaning, preparatory to plating or finishing, and as a spreading ingredient in insecticides.

Recently, butylated naphthalenesulfonates have become increasingly important.

Fatty-acid phenones are also intermediates in the preparation of wetting agents, such as alkylaromatic sulfonates, and are formed by the condensation of fatty-acid chlorides with benzene in the presence of aluminum chloride (33). Thus stearyl chloride and benzene yield stearophenone.

$$C_{17}H_{35}COCl + \bigcirc \xrightarrow{AlCl_3} C_{17}H_{35}CO-\bigcirc$$

Besides benzenoid type alkylated aromatics (alkylbenzenes, alkylnaphthalenes, etc.), certain alkylated heterocyclic aromatics are also used as intermediates in the manufacture of detergents.

Thiophene is alkylated by alkyl halides to the α- or β-alkylation products, depending on reaction conditions, in the presence of aluminum chloride or boron trifluoride (22).

Sulfonation of the alkylate yields surface-active agents.

$$\begin{array}{c} HC \!-\!-\!-\! CH \\ HC \quad\quad CH \\ \diagdown S \diagup \end{array} + RX \xrightarrow[\text{BF}_3]{AlCl_3 \text{ or}} \begin{array}{c} HC \!-\!-\! CH \\ HC \quad\quad C\!-\!R \\ \diagdown S \diagup \end{array} \xrightarrow{H_2SO_4} \text{sulfonate}$$

The lower polymers of isobutylene are used in alkylating phenol for making non-ionic detergents.

Several widely used surface-active agents are derived from p-(1,1,3,3-tetramethylbutyl)phenol, a compound made by alkylating phenol with diisobutylene. Sulfuric acid is a frequently used catalyst.

$$(CH_3)_3CCH_2C(CH_3)\!=\!CH_2 + C_6H_5OH \rightarrow p\text{-}(CH_3)_3CCH_2C(CH_3)_2C_6H_4OH$$

Because of its chain structure and the reactivity of the OH, octyl phenol is an excellent starting material for the preparation of anionic, cationic, or non-ionic surface-active agents.

Polyethylenoxy surfactant intermediates are formed in the reaction of phenols with alcohols or olefins in the presence of Friedel-Crafts catalysts (42). More specifically, phenols for conversion to these non-ionic surfactants are those alkylated with C_{18} olefins (19).

Many aralkylcarboxylic acids are good surface-active agents prepared by the oxidation of long-chain ketones, the products of Friedel-Crafts syntheses. Examples may be the acylation of dodecyl- or dibutylbenzene with acetyl chloride followed by hypochlorite oxidation of the ensuing alkyl ketone.

Sulfation of C_{10}–C_{20} straight-chain olefins

anionic surfactants. The requisite polyolefins are the products of polymerizing ethylene or propylene with various Friedel-Crafts catalysts (37).

3. Agricultural Chemicals and Insecticides

In the field of agricultural chemicals so many new compounds are synthesized that it is difficult to keep up with their development. Therefore only a few representative compounds can be mentioned.

Still the largest scale insecticide prepared by a Friedel-Crafts type condensation is p,p-dichlorodiphenyltrichloroethane (DDT).

$$2C_6H_5Cl + CCl_3CHO \xrightarrow{H_2SO_4} ClC_6H_4{-}CH{-}C_6H_4Cl$$
$$\underset{CCl_3}{|}$$

Many analogous compounds of DDT have been prepared and some of them, such as the p,p'-difluoro derivative, have achieved some importance (29).

Chlorinated p-chloroethylbenzene (Lucex), a DDT substitute produced in Germany during the Second World War, was made by the chlorination of p-chloroethylbenzene.

The condensation of p-chlorophenol with 2,4,5-trichlorobenzyl chloride yields 3,4,6,3'-tetrachloro-6'-hydroxyphenylmethane, a compound used to protect fabrics against moths and to prevent the development of microorganisms.

Phenoxathiin is produced when sulfur reacts with diphenyl oxide in the presence of aluminum chloride (Ferrario reaction) (11,38).

Phenoxathiin

It has been used as a stomach poison to replace lead arsenate in combating harmful insects, but there is some evidence that it also functions as a contact insecticide.

Phenothiazine and its oxidation products possess considerable fungicidal action.

Pentachlorophenol, prepared by the $FeCl_3$ or $AlCl_3$ catalyzed chlorination of phenol, is used as an ingredient in wood preservatives and as a mothproofing agent.

4. Elastomers, Plastics, and Plasticizers

The rapidly expanding plastics industry has made extensive use of the polymerizing and condensing effects of aluminum chloride, boron trifluoride, and other Friedel-Crafts catalysts. The polymerization and copolymerization of olefins, such as ethylene, propylene, amylene, and octylene, obtained from petroleum cracking with such dienes as butadiene and isoprene, have yielded products with varying properties from lubricants to elastomers (synthetic rubberlike materials) to compounds used as coating or insulating materials, constituents of glues, paints, lacquers, films, etc.

Friedel-Crafts copolymerization of butylene and isoprene at low temperatures gives a synthetic rubber of excellent properties (butyl rubber).

Butyl rubber is the copolymer of isobutylene and isoprene taken in a ratio in the range of 10 to 1, respectively. The reaction is catalyzed with aluminum chloride and is rapid and exothermic. The process is carried out at low temperature, about $-80°$, because under these conditions higher molecular weight polymers are produced. The reactants are diluted with methyl chloride and cooled to $-80°$ and with intense agitation a dilute solution of anhydrous aluminum chloride in methyl chloride is added. The process is continuous, the product being separated from the equilibrium feed–polymer mixture in a flash tank. Unreacted feed materials thus recovered are recycled.

Besides their use as polymerization catalysts, Friedel-Crafts methods also find widespread application in the preparation of monomers both for polymerizations and polycondensation.

The largest scale individual plastic material is polystyrene. Friedel-Crafts alkylation plays an important role in its preparation. Benzene is ethylated in the presence of aluminum chloride with ethylene to ethylbenzene.

$$C_6H_6 + CH_2{=}CH_2 \xrightarrow[\substack{C_2H_5Cl \text{ or} \\ HCl}]{AlCl_3} C_6H_5CH_2CH_3 \xrightarrow{-H_2} C_6H_5CH{=}CH_2$$

Ethylbenzene is subsequently converted to styrene, either by dehydrogenation (Dow process) or catalytic air oxidation to acetophenone, followed by reduction and dehydration (Union Carbide process).

Dialkylated aromatics, like p-xylene and, more recently, p-diiso-propylbenzene, are the starting materials for oxidation to terephthalic acid.

$$\text{C}_6\text{H}_6 + 2\text{CH}_3\text{CH}{=}\text{CH}_2 \xrightarrow{\text{H}_3\text{PO}_4} \underset{\text{CH}_3\text{CHCH}_3}{\overset{\text{CH}_3\text{CHCH}_3}{\bigcirc}} \xrightarrow{\text{oxid.}} \underset{\text{CO}_2\text{H}}{\overset{\text{CO}_2\text{H}}{\bigcirc}}$$

Ethylene dichloride condenses with benzene in the presence of aluminum chloride to form long-chain polybenzyls (14).

The lower members of this series find use as synthetic lubricant oils, whereas the higher members are pliable compounds which can be subjected to the normal fabricating techniques employed in the plastics industry.

Aluminum chloride has been found valuable in the condensation of phosphorus oxychloride and phenols to make triaryl phosphates, which are used as plasticizers (4).

Phosphorus oxychloride reacts with o-chlorophenol (or its sodium salt) to yield tri-o-chlorophenyl phosphate.

$$O{=}P \begin{cases} O\text{—C}_6\text{H}_4\text{Cl} \\ O\text{—C}_6\text{H}_4\text{Cl} \\ O\text{—C}_6\text{H}_4\text{Cl} \end{cases}$$

Compounds of this type are good plasticizers for nitrocellulose and are sometimes used to reduce the flammability of wood and fabrics by impregnation. In a similar manner ethylene oxide yields tris-(β-chloroethyl) phosphate (8), used as a plasticizer for many cellulose derivatives.

$$\text{POCl}_3 + 3\ \underset{O}{\overset{\text{CH}_2\text{—CH}_2}{\triangle}} \xrightarrow{\text{AlCl}_3} O{=}P \begin{cases} \text{OCH}_2\text{CH}_2\text{Cl} \\ \text{OCH}_2\text{CH}_2\text{Cl} \\ \text{OCH}_2\text{CH}_2\text{Cl} \end{cases}$$

5. Dye Intermediates

The Friedel-Crafts syntheses find one of their most diversified applications in the dye industry. Practically all anthraquinone and substituted anthraquinones are manufactured by the acylation of

benzene and its derivatives by means of phthalic anhydride.   Ben-
zoylbenzoic acids are formed and then converted to the anthraquinone
derivatives by dehydration with sulfuric acid.

Alkyl- and halogen-substituted anthraquinones are generally prepared
by using substituted benzenes, such as toluene, xylenes, and mono-
and   dichlorobenzenes.   Less   frequently,   substituted   phthalic
anhydrides are also used.

Many other dyestuffs and dyestuff intermediates are products of
Friedel-Crafts condensations.   In fact the first known industrial
application of the reactions was developed in 1908 by the Badische
Anilin- & Sodafabrik in Germany for the preparation of indanthrene
orange (in the pyranthrone series) and anthraflavone.

Michler's ketone, an important intermediate in the manufacture
of many triphenylmethane dyes, is made (13) by first condensing
dimethylaniline with phosgene:

Zinc chloride is then added, without isolating the intermediate, and
the second stage of the reaction proceeds:

The excess dimethylaniline required (for binding the hydrogen
chloride) is later recovered.

An important type of dyestuff is that containing an indole nucleus.
An intermediate required for one member of this type is 1-methyl-2-
phenylindole, obtained by a Friedel-Crafts condensation of N-
methylaniline and phenacyl chloride in the presence of zinc chloride.

Isomerization from the 3-phenyl to the 2-phenyl derivative is readily accomplished by heating with zinc chloride.

Several quinonoid coloring materials are manufactured through Friedel-Crafts syntheses. A yellow vat dye is produced when naphthalene is condensed with benzoyl chloride in the presence of aluminum chloride, followed by ring closure brought about by the introduction of oxygen, again in the presence of aluminum chloride (12).

1,5-Dibenzoyl-
naphthalene

3,4,6,9-Dibenzo-
pyrene-5,10-quinone

Other hydrocarbons, such as pyrene, may be used instead of naphthalene (10). Brown and yellow vat dyes are formed when 1,4- and 1,5-di-(α-anthraquinonylamino)anthraquinones, respectively, are heated with aluminum chloride (5).

Fluorescein is made by the condensation of resorcinol (2 moles) and phthalic anhydride in the presence of a dehydrating agent, such as zinc chloride. The most widely sold of the xanthene dyes is eosin (tetrabromofluorescein), which is an acid dye for wool and silk, and also for cotton with a tin or alum mordant. However, its chief application is for the preparation of ordinary red writing and stamping inks.

In vat dye production aluminum chloride is also used as a catalyst to produce carbazole type compounds made by the condensation of an aminoanthraquinone with a haloanthraquinone.

Condensation reactions promoted by anhydrous aluminum chloride are of particular industrial importance in the dyestuff field. Some of these syntheses depend upon the fact that aluminum chloride will promote dehydrogenation of labile hydrogen with resultant ring closure. The pioneering work of Scholl contributed extensively to this development. Thus Friedel-Crafts methods are used in the dye industry both for the preparation of intermediates and of direct dyes.

6. Aromatic Intermediates

Many aromatic intermediates are produced by various Friedel-Crafts type methods using acidic catalysts.

The production of ethylbenzene was discussed in connection with its major use: the manufacture of styrene.

There was a large demand for cumene during the Second World War, when it was used as an aviation gasoline component. Cumene is produced by the alkylation of benzene with propylene, which reacts more readily than ethylene. The reaction can be carried out in the liquid phase using sulfuric acid as catalyst or in the vapor phase using supported phosphoric acid catalyst. The major outlet for cumene now is as a base material for phenol production in the hydroperoxide process, although it is also used in the manufacture of α-methylstyrene.

Long-chain alkylbenzenes, as discussed, are produced as intermediates in the production of alkyl arylsulfonates.

Aromatic hydrocarbons are chlorinated in substitution reactions. Nuclear substitution occurs when the chlorination is carried out in the presence of halogen carriers, such as iron, phosphorus, or iodine.

Chlorobenzene is used on a large scale as an intermediate in the production of phenol by the Dow process and also in the production of DDT. Dichlorobenzenes found application as insecticides and moth repellents, in the production of 2,5-dichloroaniline, and as dye intermediates. Chlorotoluenes are also used as intermediates in dye manufacture. 2,4-Dichlorotoluene is used in the synthesis of the anti-malarial drug, mepacrine.

A variety of nitroaromatic intermediates, including nitrohalobenzenes, are used as dye intermediates and find application in other fields too.

Many other dye intermediates are manufactured by the use of Friedel-Crafts syntheses and include acetophenone, benzaldehyde, alkylphenols, aromatic acids, benzophenone, and their derivatives.

Benzaldehyde can be made by a modified Gattermann-Koch synthesis (21). During the Second World War, Germany, wishing to economize on toluene (another starting material for benzaldehyde preparation), used this method extensively. Addition of hydrogen chloride was found to be unnecessary since adequate amounts were formed in the reaction. Carbon monoxide at 90 atm. pressure was simply condensed into benzene (at 45–50°) containing a suspension of aluminum chloride. Many other aromatic aldehydes have been produced in a similar manner.

7. Pharmaceuticals and Cosmetics

Acetophenone, obtained from benzene by acetylation (or oxidation of ethylbenzene) was previously used as a hypnotic (Hypnone), but now finds more use as an intermediate in the manufacture of various drugs. As an example, the preparation of the antihistamine known as Doxylamine may be quoted. This is synthesized from acetophenone, pyridine, and dimethylaminoethyl chloride (36).

Hexylresorcinol, used as a germicide, is prepared by acylation of resorcinol by caproic acid in the presence of zinc chloride, followed by reduction.

$$\text{resorcinol} + HOOC(CH_2)_4CH_3 \xrightarrow{ZnCl_2} \text{acyl product} \xrightarrow{Zn/Hg} \text{hexylresorcinol}$$

Phenolphthalein, widely used as a laxative, is also made by a Friedel-Crafts type acylation.

$$\text{2 phenol} + \text{phthalic anhydride} \xrightarrow{H_2SO_4} \text{Phenolphthalein}$$

Another medicinal preparation used was 4′-aminobenzosulfamino-4-N,N-dimethylsulfamide (Uleron) prepared from 4-acetylaminobenzenesulfanilide and N-dimethylaminosulfuryl chloride followed by deacetylation.

At one time this drug was used extensively for the treatment of gonorrhea, but has since been superseded.

The antiseptic, thymol, is obtained through the alkylation of m-cresol with isopropyl choride at $-10°$. At $30°$ an isomer, carvacrol, is produced.

Numerous intermediates are used in the manufacture of the ever-increasing number and variety of pharmaceuticals which utilize methods of the Friedel-Crafts type, but detailed discussion of these is outside the scope of the present review.

In the field of cosmetics a variety of synthetic detergents is rapidly replacing many applications of conventional soaps. Intermediates used in the manufacture of perfumes include such aromatic ketones as benzophenone, p-methylacetophenone, and β-acetylnaphthalene, all of which are made by Friedel-Crafts condensations. In addition

β-phenylethyl alcohol, an important ingredient of artificial rose-oil, is made by the condensation of benzene with ethylene oxide in the presence of aluminum chloride (9):

$$CH_2\text{—}CH_2 \diagdown O \diagup \quad + \quad \bighexagon \quad \xrightarrow{AlCl_3} \quad \bighexagon CH_2CH_2OH$$

The use of low operating temperatures in this procedure avoids the formation of undesirable by-products, such as dibenzyl.

Hydrocarbons used either as end products or intermediates in perfumery include t-butyl-m-xylene (from isobutylene and m-xylene) and 5-α,α-diethylpropyl-m-xylene (from methyldiethylethylene and m-xylene).

8. Halogenated Solvents

One of the manufacturing methods of carbon tetrachloride is that from carbon disulfide and chlorine, catalyzed by ferric chloride.

$$CS_2 + 3Cl_2 \xrightarrow[30°]{FeCl_3} CCl_4 + S_2Cl_2$$

$$CS_2 + 2S_2Cl_2 \xrightarrow[60°]{FeCl_3} CCl_4 + 6S$$

Recently, interest in the manufacture of carbon tetrachloride from phosgene has been expressed (14a). According to one patent (20a), carbon tetrachloride is formed from phosgene in yields of up to 86% over Friedel-Crafts catalysts, such as $AlCl_3$ and $FeCl_3$, at pressures up to 186 atm.

$$2 COCl_2 \rightleftharpoons CCl_4 + CO_2$$

Using carbonyl chloride fluoride or carbonyl fluoride as starting material chlorofluoromethanes (Freons) can also be obtained.

Carbon tetrachloride serves as the starting material for the preparation of Freon type compounds, such as dichlorodifluoromethane (Freon 12), trichlorofluoromethane (Freon 11), and trifluorochloromethane (Freon 13). Chlorinated paraffins, such as methyl chloride, ethyl chloride, ethylene chloride, ethylidene chloride, 1,1,2-trichloroethane, acetylene tetrachloride, pentachloroethane and hexachloroethane are manufactured both in addition chlorinations (of olefins) and substitution chlorinations (of alcohols) using acidic catalysts $(AlCl_3)$.

Chlorobromomethane (a fire extinguisher) is made by bromination of methylene chloride in the presence of aluminum or aluminum chloride.

$$6CH_2Cl_2 + 2Al + 3Br_2 \rightarrow 6CH_2ClBr + 2AlCl_3$$

The antimony pentahalide catalyzed fluorination of haloaliphatic compounds by hydrogen fluoride or antimony trifluoride (Swarts reaction) is used on a large scale in the manufacture of fluorochloroalkanes, used under the "Freon" trade name as refrigerants and aerosols.

$$CCl_4 + 2HF \xrightarrow{SbCl_5} CCl_2F_2 \text{ dichlorodifluoromethane (Freon 12)}$$

$$CHCl_3 + 2HF \xrightarrow{SbCl_5} CHClF_2 \text{ chlorodifluoromethane (Freon 22)}$$

V. Conclusions and Future Outlook

In the reviewer's opinion the outlook for the practical applications of Friedel-Crafts type reactions is bright. Although the reactions have developed into some of the most versatile and widely used laboratory methods, until recently their industrial application has been more limited. The largest scale application is in olefin alkylation (for high-octane gasoline), for the preparation of ethylbenzene (for styrene monomer), and in polymerization (butyl rubber). Most of the other applications, as, for example, the preparation of ketones and other intermediates, are still seriously hindered by the fact that large amounts of catalyst per weight of compound are needed and catalyst recovery, concerning aluminum chloride, is not feasible. However, with the introduction of catalysts which can be easily recovered (HF, BF_3, etc.) and catalysts of the acidic oxide (silica–alumina) type, basically Friedel-Crafts type solid acid catalysts, the situation is beginning to change. Another potential field of future development is to find and apply more selective catalysts. Aluminum chloride, still the most widely used catalyst in the Friedel-Crafts field, is generally unable to show any selectivity in systems containing more than one functional group. Thus many potential alkylations, such as chloroalkylation with dichloroethane, alkylation with vinyl chloride, etc., are not feasible with aluminum chloride, which affects both functional groups. Catalysts which could effect selectively a reaction of one of the functional groups without affecting the others may extend considerably the scope and application of Friedel-Crafts type reactions (for a more detailed discussion of the topic see Chapter V). The fact that catalysts can be recovered in simple and easy steps may allow the application of Friedel-Crafts reactions in fields previously not possible because of catalyst cost. Solvents providing homogeneous reactions and thereby eliminating secondary effects due to selective solution of products in the catalyst layer (e.g., further alkylations of the monoalkylate due to their better solubility in the acid catalyst layer, isomerizations, etc.), coupled with a better understanding of the reaction mechanism to allow the preparation of uniform isomers or to direct the reaction to the necessary mono- or disubstituted products without having to deal with rather complex mixtures, are further like fields of development of the reaction.

The possibility of easy catalyst recovery and the added advantages in selective reactions of avoiding undesirable by-products and difficult separation problems, together with a more advantageous product: catalyst ratio, may easily allow some promising catalysts to take over more and more of the field still dominated by aluminum chloride. The realization that Friedel-Crafts reactions are indeed general acid-catalyzed processes offers increasing possibilities in the application of all types of acid catalysts in Friedel-Crafts type reactions. Beside sulfuric acid and anhydrous HF, other acid catalysts systems, such as polyphosphoric acid, supported phosphoric acid, phosphoric acid–BF_3, and many other related acid systems, are gaining in importance and will undoubtedly receive large-scale industrial application in the future. Specific attention should be given to catalysts of the acidic cation exchanger, silica–alumina, and related acidic oxide type, which can be considered as solid acid catalysts providing possibilities of easy recovery. These systems will clearly receive more and more attention.

VI. Addendum

In the eight years since I wrote the preceding remarks about the future of Friedel-Crafts Chemistry, I have not changed my view concerning the bright future I foresee for the field. *Solid acid catalysts* made rapid progress in their importance, particularly in the petrochemical industry. *Bifunctional catalysts*, combining the dehydrogenating-hydrogenating ability of metal (noble metal) catalysts with that of high acidity sites, found particular importance in isomerization processes. *Isomerization* and *alkylation* achieved new importance in obtaining high octane rating gasoline fuels with decreased amount or elimination of lead additives (due to environmental pollution problems).

Superacid catalysts appeared in the last years and promise to acquire substantial importance. Superacid catalysts of the HF–SbF_5 and FSO_3H–SbF_5 type, because of their very high reactivity, can bring about isomerization of saturated hydrocarbons generally at temperatures considerably *lower* than conventional protic acid or Friedel-Crafts catalysts, thus allowing, for example, more highly branched (and therefore more advantageous as high octane gasoline component) isomeric hexane and heptane equilibrium mixtures to be obtained.

References

1. Badische Anilin- and Sodafabrik, Germ. Pat. 336,558 (1917).
2. Balle, G., H. Wagner, and E. Nold, U.S. Pat. 2,195,198 (1940); *C.A.*, **34**, 5093 (1940).
3. Ballod, A. P., I. V. Patsevich, A. S. Feldman, and A. V. Frost, *Proc. Acad. Sci. U.S.S.R.*, *Chem. Sect.*, **78**, 509 (1951).
4. Bass, S. L., U.S. Pat. 2,033,916 (1936), *C.A.*, **30**, 2991 (1936).
5. Bradley, W., and C. R. Thitchener, *J. Chem. Soc.*, (1953), 1085.
6. Britton, E. C., *Am. Chem. Soc.*, *News Ed.*, **19**, 251 (1941).
7. C.I.O.S. Report XXVII, p. 85.
8. Daly, A. J., and W. G. Lowe, Brit. Pat. 475,523 (1937); *C.A.*, **32**, 3512 (1938).
9. Farbenindustrie, I. G., Germ. Pat. 523,436, (1931).
10. Farbenindustrie, I. G., Brit. Pat. 382,877 (1932), *C.A.*, **27**, 4411 (1933).
11. Ferrario, E., *Bull. Soc. Chim. France*, (4) **9**, 536 (1911).
12. F.I.A.T. Final Report 1313, Vol. II, 121.
13. Fierz-David, H. E., and L. Blangey, *Fundamental Processes of Dye Chemistry*, Interscience, New York, p. 139, 1949.
14. Fulton, S. C., and L. A. Mikesba, U.S. Pat. 2,072,107 (1937), *C.A.*, **31**, 2811 (1937).
14a. Gleniser, O., *Angew. Chem.*, **15**, 823 (1963).
15. Hansford, R. C., *Ind. Eng. Chem.*, **39**, 849 (1947).
16. Hansford, R. C., P. G. Waldo, L. C. Drake, and R. E. Honig, *Ind. Eng. Chem.*, **44**, 1108 (1952).
17. Hille, J., Germ. Pat. 817,457 (1940).
18. Hille, J., and W. Dürrwächter, Germ. Pat. 1,061,757 (1958); *Angew. Chem.*, **72**, 850 (1960).
19. Kooijman, E. C., U.S. Pat. 2,567,848 (1951); *C.A.*, **46**, 2577 (1952).
20. Koster, W. R., *Chem. Ind. News*, Sept. 5, 129 (1960).
20a. Kung, F. E., U.S. Pat. 2,892,875 (1956).
21. Larson, A. T., U.S. Pat. 1,989,700 (1935); *C.A.*, **29**, 1834 (1935).
22. Linn, C. B., U.S. Pat. 2,624,742 (1953); *C.A.*, **47**, 3588 (1953).
23. Linn, C. B., and A. V. Grosse, *Ind. Eng. Chem.*, **37**, 924 (1945).
24. MacLaren, F. H., and T. E. Stockdale, U.S. Pat. 2,057,104 (1936); *C.A.*, **30**, 8600 (1936).
25. McAfee, A. M., U.S. Pat. 1,127,465 (1914).
26. McAfee, A. M., *Ind. Eng. Chem.*, **7**, 737 (1915); *Chem. Met. Eng.*, **13**, 542 (1915).
27. McAfee, A. M., U.S. Pat. 1,217,471 (1917).
28. McAfee, A. M., U.S. Pat. 1,478,444 (1913); *Chem. Met. Eng.*, **36**, 422 (1929).
29. Metcalf, R. L., *Organic Insecticides*, Interscience, New York, 1955.
30. Oblad, A. G., T. H. Milliken, Jr., and G. A. Mills, *Advances in Catalysis*, Vol. III, Academic Press, New York, p. 199, 1951.
31. Oblad, A. G., S. H. H ndin, and G. A. Mills, *J. Am. Chem. Soc.*, **75**, 4096 (1953).
32. Oerstedt, H. C., *Pogg. Ann.*, **5**, 132 (1825).
33. Ralston, A. W., U.S. Pat. 2,089,154 (1937); *C.A.*, **31**, 6769 (1937).
34. Reiff, O. M., and D. E. Badertscher, U.S. Pat. 2,048,465-6 (1936); *C.A.*, **30**, 6185 (1936).

Index

571